# PHYSIOLOGIA

# PHYSIOLOGIA

*Natural Philosophy in
Late Aristotelian and
Cartesian Thought*

Dennis Des Chene

*Cornell University Press*

Ithaca and London

PUBLICATION OF THIS BOOK WAS ASSISTED BY A GRANT
FROM THE PUBLICATIONS PROGRAM OF THE NATIONAL ENDOWMENT
FOR THE HUMANITIES, AN INDEPENDENT FEDERAL AGENCY.

First published 1996 by Cornell University Press.

Library of Congress Cataloging-in-Publication Data
Des Chene, Dennis.
Physiologia : natural philosophy in late Aristotelian and
Cartesian thought / Dennis Des Chene.
p.  cm.
Includes bibliographical references and index.
ISBN 0-8014-3072-0 (alk. paper)
1. Physics—Philosophy. 2. Aristotle—Influence. 3. Descartes,
René, 1596–1650—Influence. 4. Philosophy, Medieval. 5. Philosophy
of nature. I. Title.
QC6.D43 1995
113—dc20                                                        95-32631

Printed in the United States of America

# Contents

# Contents

# Preface

It sometimes happens that the book one sets out to write is not the book one eventually writes. So it was with this book: what began as an introductory pair of chapters in a book on Aristotelian and Cartesian psychology augmented itself into a book devoted to Aristotelian and Cartesian philosophy of nature. There was, it turned out, no compendious study of the sixteenth- and early seventeenth-century Aristotelian philosophers who, at least until the time of Locke, dominated the teaching of the universities of Europe. Despite the efforts of Gilson and some of his successors, their commentaries and *cursus,* immense and forbidding, have remained largely unknown territory. Especially among analytic philosophers, the Latin world of the early modern period has until recently been all but forgotten. This work is an initial effort to make that world more familiar.

The disadvantages of ignorance are several. It is impossible to discern what is new and what is not in Descartes's work without a thorough knowledge of those earlier works from which he first learned his metaphysics and natural philosophy. Or rather, since a distinction of new and old is too crude, it is impossible to assess his use of the resources available to him; nor can one securely divine from the Cartesian corpus alone the significance to be attached to terms and propositions whose intended audience often consisted of Aristotelians. It is, moreover, on any but the most naive view of the timelessness of philosophical questions and solutions, fruitless to evaluate his arguments without such knowledge. I would add, finally, that the Aristotelians, contrary to the caricatures often painted of them by their opponents, were careful and serious thinkers; some, like Suárez, are worthy of study independent of their importance to later philosophers.

One aim of this work is to make ignorance of Aristotelianism disreputable. There are indeed obstacles in the way of anyone now who undertakes to study the Aristotelians: the language, the disputational format of their writings, the dependence, explicit or not, of later texts upon earlier texts, and the consequent requirement that one should have also studied the dominant Medieval figures. But those obstacles are effectively no more difficult to overcome than those that present themselves, say, to the student of Kant or Hegel, and overcoming them opens the way to five centuries of rich and profound philosophical argument.

A work whose gestation is lengthy acquires many debts, personal and intellectual. One of the pleasures of finishing is to acknowledge those debts. Every student of Renaissance philosophy will, first of all, build on the work of his or her predecessors. Historians like Étienne Gilson, Anneliese Maier, Charles Schmitt, Tullio Gregory, and Edward Grant have brought to light and interpreted for modern readers the enormous and complex world of Medieval and Renaissance philosophy. Closer to home, Gilson, Dan Garber, Roger Ariew, Alan Gabbey, John Schuster, Geneviève Rodis-Lewis, Jean-Luc Marion, Pierre Costabel, Jean-Louis Armogathe, and many others have enlarged the horizons of the study of Cartesian philosophy of nature. This work—and for that matter the idea of this work—would have been impossible without theirs.

Among those who have read drafts of the manuscript, or who have provided comments and supports along the way, I thank first my *lectrix prima* Mary Des Chene, whose sharp eye no infelicitous phrase evades. She now has more Aristotelian philosophy than any anthropologist needs. I have benefited from comments from and discussions with Mark Rigstad, Thanos Raftopoulos, Mauro Dorato, and Natalie Brender. Natalie has become an able answerer of obscure queries while assisting me in my research. Ed Minar's conversation and questions helped me as I was thinking out the larger project of which this book is the first segment. Ira Singer read and commented on Part I. Peter Achinstein and Rob Rynasiewicz will, I hope, see echoes of bygone lunch conversations in Part II. Jerry Schneewind's belief in the value of a history of philosophy not subservient to present-day conceptions has reinforced my own. The late David Sachs presented to all who knew him a day-to-day exemplum of philosophical curiosity and rigor; I regret that he was not able to see his encouragements come to fruition. Intermittent conversations with Steve Menn have proved fruitful, as has the reading of some of his work in manuscript. Roger Ariew and Emily Grosholz offered knowledgeable advice and criticism. Roger Haydon, the editor for Cornell University Press, has supported the work since its earliest stages: sine qua non.

The project from which this book has grown began with work at Stanford University under Arnold Davidson, Peter Galison, and Stuart Hampshire.

During that time I received support from the philosophy department; a fellowship from the Mrs. Giles S. Whiting Foundation in 1986–1987 funded my fifth year of graduate study. In 1984–1986 a graduate fellowship at the Stanford Humanities Center yielded a fine office and the intellectual stimulation of meeting scholars from diverse disciplines. An Eli S. Lilly Foundation teaching fellowship funded one semester's leave-replacement in the spring of 1993, a crucial time in the writing of this book. Jacqueline Mitchell coordinated the program at Johns Hopkins University; she was, moreover, a source of encouragement through the year of the fellowship. Research materials without which this book would have been impossible were acquired for me through the alchemical skills of Alan Braddock and others at the interlibrary loan office at the Milton S. Eisenhower Library. I thank them for their efficiency and patience.

Earlier versions of material from §6 were given as talks at the University of British Columbia and at Yale University. I thank Paul Russell and Paolo Mancosu for their invitations and hospitality; I have since benefited also from reading Paolo's work on Descartes's geometry. A version of §8.1 was presented, at the invitation of Sharon Kingsland, at the History and Philosophy of Science Colloquium at Hopkins in the fall of 1993.

DENNIS DES CHENE

*Baltimore, Maryland*

# Abbreviations and
# Orthographical Conventions

## Citations

References to primary sources are by author, short title, and editor or edition (where necessary). Secondary sources are referred to by author and year. I have tried where possible to note original dates of publication or composition. Texts from primary sources are cited first by the smallest numbered or named subdivision, and then by volume (if there is more than one) and page or column number. The edition will be indicated only if I have referred to more than one. In page and column references, the following symbols are used:

| | |
|---|---|
| p | page or *folium* |
| r, v | recto, verso |
| a, b | first, second column |
| A, B, C . . . | top-to-bottom divisions of the page, when marked |
| bis | Coimbra *In Phys.* |
| f, ff | as usual, but superscripted to avoid confusion |

I have converted Roman numerals, when they indicate the *primary* pagination of a text (as in Buridan's works), into Arabic numerals. Misprinted page numbers, chapter numbers, and so forth are silently corrected. Line numbers are separated from page numbers by a period.

Medieval and Renaissance texts, by reason of their manuscript origins and their mode of presentation, are often articulated into a hierarchy four or five layers deep. Commentaries are divided by book, sometimes by chapter, and then into *lectiones* or *textus,* which are short passages to which a portion of commentary is devoted. Commentaries on the *Sentences* of Peter Lombard, a mainstay in the curriculum of the thirteenth and fourteenth centuries, are divided, like the *Sentences* themselves, into books, *distinctiones, quæstiones,* and *articula,* with further subdivisions as required. Many commentaries also include *quæstiones* on standard topics associated with some part of the work commented on. References to them are by book, chapter, and question. Such references, I should note, are largely independent of edition; readers should be able to locate them in whatever text is available. Names for the parts of texts are abbreviated as follows:

| | |
|---|---|
| a | article, *articulum* |
| A, B, C . . . | subsection of question (in Ockham) |
| ad1, ad2 . . . | subsection of article |
| c | chapter, *caput* |
| com | commentary, *commentarium* |
| disp | *disputatio* |
| d, dist | distinction, *distinctio* (in *Sentences* commentaries) |
| ex | *exercitatio* |
| lect | *lectio* |
| lib | book, *liber* |
| pr | proposition |
| q | question, *quæstio* |
| resp | *responsum* |
| text | *textus* |
| tr | *tractatus* |

Thus the passage in Fonseca *In meta.* that begins at 1:710C of the 1615 edition will be referred to as '4c2q1§1'.

Since the titles actually given to commentaries on Aristotle (and, in the case of Calcidius, on Plato) vary from edition to edition (and from manuscript to manuscript) and are typically bestowed by their editors, not by their authors, such works are cited under uniform titles of the form '*In X*', where *X* is the abbreviated title of the work commented on: e.g., *In meta.* for a *Metaphysics* commentary, *In de An.* for a *De anima* commentary. Commentaries on the *Sentences* of Peter Lombard are cited under the uniform title *In Sent.*

## Orthography and Translations

In reproducing printed Latin texts, I have observed the following conventions:

1. The letters 'u', 'v', 'i', 'j' have been changed to accord with modern practice. The first letters of sentences and proper names have been capitalized; other capitalization (e.g., of words like 'Philosophy') has been

preserved. I have also kept the accentuation occasionally used to distin-guish, e.g., *cum* meaning 'with' from *cùm* meaning 'when'.

2. Manuscript abbreviations, which are often carried over into printed editions, have been expanded. Other abbreviations—for example, those of the names or authors or works—have been left as printed, or expanded with brackets, as 'Alb[ertus]. Mag[nus].' *Philosophus,* as a proper name, always denotes Aristotle; *Commentator* always denotes Averroes.

3. Obvious misprints have been silently corrected. Where a variant might be significant, it has been noted.

4. When modern editions are quoted—the Adam-Tannery Descartes, for example—the orthography and accidentals of the edition are preserved.

All translations are mine unless otherwise indicated. The desiderata have been to preserve sentence structure wherever possible and to be consistent in terminology. Certain words, however, notably *ratio* and *ordo* in certain constructions, require fairly wide departures if the result is to make sense in English. Brackets are used to indicate insertions. All emphases are in the original. Since sentences, marked with a period, can extend for half a page or more, I have sometimes silently broken them up, replacing colons or semicolons with periods.

The generic pronoun in the sixteenth and seventeenth centuries was 'he'. To use 'she' in paraphrase or exegesis of authors from that period would be anachronistic. So too the use of 'she' to denote God. I have used 'she' as a generic pronoun only when writing *in propria persona.*

# PHYSIOLOGIA

# Introduction

Si les phénomènes ne sont pas enchaînés les uns aux autres, il n'y a point de philosophie.

—Diderot, *De l'interprétation de la nature,* ¶58

## 1. Interpretations of Nature

The physicist who in an idle moment opens Aristotle's *Physics* is likely to be disconcerted by the content of that work. Only in the seventh book will she feel entirely at home, when Aristotle at last gets down to the business of stating quantitative rules for the comparison of motions. But the *Physics* deals with such things only glancingly; its primary and essential subject is what Aristotle calls the "principles" of nature: matter and form, the four causes, natural change, and nature itself. To those principles the Aristotelian commentaries and cursus I study in the first part of this work devoted much of their effort. For them *physica* covered a little of what we call physics and quite a bit of what we call metaphysics. The Aristotelians had their own principled distinction between the two: metaphysics was the study of being qua being, physics the study of corporeal or natural being. But in practice the topics just referred to were treated by works in both disciplines. The rather fluid boundary region between them, which is populated not by particular facts or generalizations but by the vocabulary and schemes in which those are cast, is what I call the philosophy of nature.

It is a somewhat neglected region in the history of science and philosophy, especially in the period I examine.[1] Historians have tended to magnify the importance of the parts of Aristotelian natural philosophy which correspond to the concerns of modern physical science. Large efforts have been devoted to the medieval theory of impetus and the quantitative rules of the

---

1. A brief list of general works on Aristotelian natural philosophy in the Middle Ages and Renaissance is given in §1 below. For the relation of Descartes to the Aristotelians of the Schools, see especially Gilson 1913, Gilson 1984, and Ariew 1992.

[1]

Oxford Calculators, and to the cosmologies of Kepler and Copernicus. The consequence, perhaps unintended, has been to displace the center of gravity of Aristotelianism. Although questions that exhibit some continuity with what are now regarded as genuinely physical questions were certainly not unimportant to them, the Aristotelians spent as much, and sometimes more, labor and ink on the definition of natural change, the intension and remission of qualities, the existence of substantial form.[2] Aristotelian philosophy of nature, moreover, contains the principles common to *all* natural philosophy, and not just to the part that became our physics. Though the Aristotelians believed, as many philosophers do now, that physics provides the foundation of all the sciences, other branches, including the study of living things, were as well developed. In disputed questions about the philosophy of nature, the generation and corruption of living things and mixtures, and the peculiarities of the human soul, figure as prominently as do properly physical instances.

An interest in quickening rather than in moribund science may well be excused. But it does not merely put Aristotelianism in false perspective. It also inclines the historian to partiality in delineating the task that Descartes and his peers confronted. One aim of this work is to urge that the profoundest and historically most effective part of Descartes's project has to do neither with method (whose relation to Descartes's practice is at times tenuous, and which was in any case not the most significant part of his legacy), nor with the geometrization of nature (a means, not an end), nor yet with experiment (which Descartes did not make central to his strategies of persuasion, as Boyle and the Royal Society did later), but with constructing, from prime matter upward and from God downward, a functional equivalent to the Aristotelian philosophy of nature.

To understand how the project was conceived and executed, one must reinsert it into the dense and turbulent context of the early seventeenth century. For Descartes the textbooks and commentaries of Aristotelianism, his chief resource and competitor, provide much of that context. The comprehension of his vocabulary will be helped by reference to them. That is important enough: even where Descartes innovates he starts from the tangled, dispute-ridden network inhabited by those terms, and is obliged, if he would make himself understood, to take it into account. But beyond that, a second level of significance becomes accessible through the study of the

2. The Coimbrans' commentary on the *Physics* devotes its entire first volume, some 400 pages, to the first three books (out of eight). In their second volume, almost a third of the 374 pages is devoted to questions about the eternity of the world, its creation, and the existence of a creator. Toletus is more evenhanded; nevertheless *Physics* 1–3 receives 106 out of 245 folia, with book 8 (the questions about eternity, and so forth) taking up 34 more. In Arriaga's widely circulated *Cursus* of 1632, material coming from *Physics* 1–3 occupies 172 of 250 pages in the *Disputationes physicæ*.

textbooks: the marking of territory in the intellectual landscape accomplished by asserting a systematic array of positions. Not all of those positions need be explicitly stated. To affirm that matter is *res extensa* rules out, for example, the Thomist thesis that matter is *pura potentia*. It also leads to grave discord with the Catholic dogma of the Eucharist. Not for nothing did the Theological Faculty at Louvain censure in 1662 the assertion that in bodies there is only "motion, rest, place, figure, and size," which "appears to subvert the Holy Sacrament of the Altar."[3] The sense of the intellectual position a philosopher chose to take up or created, like the sense of social position conveyed by one's choice of dress, cannot be grasped except by understanding its implications, explicit or implicit, in relation to those of other positions seriously tenable at the time.

The philosophy of nature, in fact, was a kind of clearinghouse in which physics, metaphysics, and theology could meet and negotiate their claims, much less needed now that those disciplines have gone their own ways. Historians of ethics have described the project of moral philosophy in the seventeenth and eighteenth centuries as one of devising a secular ethics. The civic ideal of tolerance promoted endeavors to devise a theory of moral obligation requiring few presuppositions about God and his works. But even though philosophers eventually succeeded in divorcing ethics from theology, the forms ethics took in the interim bear the impress of the particular religions and religious disputes they were struggling to accommodate themselves to. Similarly the project in natural philosophy begun by Descartes and others eventually resulted in a secular and unmetaphysical physics. Although the result now bears few signs of its gestation, its earlier stages can be understood only in relation to the particular religious context that surrounded them. The content of specific generalizations, like the law of falling bodies or the finite velocity of light, may itself not require that context to be understood; but the implications such generalizations were thought to have for the philosophy of nature do.

In the two parts of this work I will trace a number of themes. The first and most basic is that of *natural change* and *agency*. Aristotelianism understood natural change as the expression of potentialities. That expression could be frustrated, naturally in some instances, only by miracle in others. But it is always directed, always *from* one state *to* another. It is also always directed in the sense of proceeding from an agent to a patient. Sheer becoming has no place in Aristotelian physics; to change is to be changed. The history of an individual, then, is ordered by the relation of its *potentiæ* to their actualizations; among individuals there is a second order determined by relations of

3. Du Plessis d'Argentré *Collectio* 3pt2:303b. The third condemned thesis reads: "Extensio Corporis est attributum ejus naturam essentiamque constituens," with a reference to *PP* 1§53 (ib. 304a). The Rector of Paris issued a similar condemnation in 1691 (3pt1:149b).

agency and patiency. Those relations were definitive of natural kinds. Heat, for example, is the most active of elemental qualities, dryness the most passive. More generally, substantial forms are classified first according to what they do, and only secondarily according to what can be done to them.

The second theme is that of the *structure of material substance*. Matter and form, the two principles of substance, were ramified by the Aristotelians into a rather complicated layering of prime matter, quantity and dispositions, substantial form, and active powers. A host of questions in the commentaries lay out the definitions of and distinctions among these components. In those questions it is clear that the positions taken had not only to save the phenomena but to achieve a balance between what one might call the Manichaean and Platonist extremes. The material world had to be, on the one hand, not so divorced from God that it could not share in his goodness or submit to his rule; it had, on the other hand, to be not so dependent as to be deprived of genuine agency altogether. Questions of structure lead sooner or later to questions about power.

The third theme is that of *finality*. Aristotle, notoriously, included ends among the four causes laid out in the second book of the *Physics*. By the end of the sixteenth century, however, Aristotelians had come to limit the scope of the final cause in a number of ways; in particular, it can act on inanimate things only insofar as they are instruments of rational agents. But they do not, and could not, abandon finality altogether. It is embedded in the very notion of natural change; it is essential to distinguishing *per se* causes, which alone admit of scientific understanding, from *per accidens* causes, or what Aristotle calls 'chance'; it is the ordering principle not only of the histories of individuals but of nature as a whole.

Descartes and the other proponents of the new science regarded the Aristotelian entities, especially prime matter and substantial form, as not only superfluous but incoherent. Yet the textbooks show that, however gratuitous such entities may seem from an empirical standpoint, they could be defended on a priori grounds. Even the egregious doctrine of real accidents founders not in contradiction but in obscurity. But even if a priori arguments are set aside, matter and form were applicable to a vast range of phenomena, celestial, meteorological, biological, and psychological. If, as some philosophers of science have argued, some explanation proceeds by way of unification, then Aristotelianism, by its standards at least, had explained a great deal.

Descartes could have ignored the Aristotelian definition of change, its notions of matter and form, its appeal to ends, had he been content to remain within what was called mixed or middle mathematics, under which his early work in music and optics would have been subsumed. But he wanted to refashion not just those subordinate disciplines, but the master discipline itself. *Le Monde,* his first major attempt at an interpretation of nature, and above all the *Principles* were intended to supplant Aristotelian physics. He had, therefore, explicitly to confront the themes I have just laid out. To define matter as *res extensa* had, no doubt, the epistemological advantage of permitting the admirable methods of geometry to be applied to natural things. But because it forced, in Descartes's view, all other accidents of matter to be mere modes of extension, it also had the effect of radically simplifying the structure of material substance. There had already been a tendency, visible in Suárez, to ascribe to matter alone such accidents as quantity and figure, along with the qualities that "disposed" or "proportioned" matter to receive substantial form. Active powers, on the other hand, remained with form—at least those that could not be explicated in terms of dispositive qualities. Descartes goes much further: powers are reduced to dispositions, and dispositions, which in Aristotelianism are constituted by but not reducible to primary qualities, are reduced to "configurations" of quantity. The avenue to that reduction is *simulation:* if a configuration answers to the phenomena, if the world of *Le Monde* appears to its inhabitants as this world appears to us, then a physics of *res extensa* and its modes has been proven sufficient.

A further consequence of the reduction is to eliminate, at least prima facie, agency from material nature. In Aristotelian physics, figure and quantity are prototypically passive. For that reason the nominalists, though they held that matter and quantity are not distinct, did not propose also that qualities be reduced to figure. That would be, as the Aristotelians reiterated in their criticisms of atomism, to deny agency in nature altogether. In Cartesianism, the identification of agents and patients becomes a matter of convenience, because there are no active powers and because motion is relative. But paring away active powers, however attractive an economy it may have seemed, had its price. The trouble brought by simplifying a traditionally asymmetric notion of causal efficacy, and evidenced in Leibniz and Malebranche, came partly because Cartesian physics yielded only symmetrical interactions. Although Newtonianism did reintroduce attractive and repulsive forces, by then the philosophical damage, to which Hume's doubts testify, was well advanced.

The directedness of natural change, and with it the contrast of potential and actual, are likewise banished in the Cartesian restriction of natural properties to figure, size, and motion. Even in Aristotelianism, figure is an anomaly among qualities, in part because, unlike hot and cold, it provides no ground for attributing tendencies. Figure is changed only as a side effect—*per accidens,* as the Aristotelians would put it—consequent upon other, directed changes. Cartesian matter, moreover, is, from an Aristotelian standpoint, at every instant entirely actual. Hence even at the most basic level, that of particular changes to individuals, the Aristotelian concept of natural change fails to apply.

Directedness also fails to apply, at the highest level, to nature as a whole. The Aristotelian maxim "Nature does nothing in vain," because it attributes to nature an *intrinsic* economy, has no place in Cartesian physics. The appearance of economy must be an accidental consequence of the operation of bodies according to laws, or else referred not to nature but to God. The absence of directedness might even go so far as to rule out providence. Leibniz was quick to criticize the suggestion in the *Principles* that the universe might proceed through all possible states: it would ruin, he said, any notion of progress. Indeed the relation of God to the world in Cartesianism is curiously static, limited to the creation of matter endowed with a fixed quantity of motion and thereafter to conservation, entirely uniform, of its existence and concurrence in its operations.

Descartes does, of course, appeal to notions of perfection. Perfection figures largely in the causal proofs of the existence of God in the *Discourse* and the *Principles*. The perfection of nature as a whole serves in the fourth *Meditation* as a tentative excuse for God's having given him an imperfect soul. But in such uses the notion is divorced from becoming: substances, for example, are more perfect than attributes, but substances are not attributes made perfect. Hence even when Descartes uses a notion that had been imbued with finality, it seems purged of its associations, leaving only a fixed hierarchy of modes of existence.

Desmond Clarke begins his work on Descartes's scientific method by proposing to treat Descartes as a "practising scientist who, somewhat unfortunately, wrote a few short and relatively unimportant philosophical essays" (Clarke 1982:2). That is, of course, an exaggeration. It is closer to the truth to treat Descartes as what his contemporaries would have called a physiologist: in our dialect, a philosopher of nature. His metaphysical forays, even his reluctant theological interventions, are all in keeping with the scope of physics as one finds it in early seventeenth-century textbooks. Far from being unfortunate or inappropriate to a "practising scientist," they are entirely in keeping with the profession of a natural philosopher. More than that, the epistemological concerns that philosophers now often take to be central in Descartes's work are, I think, subordinate to his philosophy of nature. If a revolution in method had been his sole ambition, we would have had a second Ramus, not a Descartes. Ramus hoped to supplant the Aristotelian dialectic and rhetoric with his own. Descartes not only devised a new method, but a new interpretation of nature.

## 2. Aristotelianism

In keeping with a number of recent works on Descartes, then, I emphasize the specifics of the Aristotelian context in which he was educated and against which he defined himself.[4] Renaissance Aristotelianism is a complex phenomenon. A history even of interpretations of the *Physics* or *De anima* would require a volume to itself. To keep this book within reasonable bounds I have restricted my attention to a small group of *central texts*. Written by university professors for the required series of lectures on the Aristotelian corpus, these works include running commentaries on Aristotle's text, or *quæstiones* on more or less standard topics suggested by it, or both.[5]

The *quæstio*, a form inherited from the Scholastics, was originally the record of a *disputatio* or formal debate on a given topic.[6] It began with a question, usually in yes-or-no form, like 'Does prime matter exist?'. In the most traditional version, found in Thomas's *Summa*, for example, what followed would be a statement of the *wrong* answer, marked by phrases like 'videtur quod' ('it seems that'), with one or more arguments for it. Next would come a brief statement of the *right* answer, marked by phrases like 'sed contra' ('but against this') or 'respondeo' ('I answer'), accompanied traditionally by an authority or two to back it up. The heart of the question was next: a laying out of distinctions required for understanding the discussion, and a series of *conclusiones*, with arguments, which together comprise the true view. Sometimes—in the Scholastics almost always—the question would conclude with refutations of the arguments for the wrong view.

The form could be elaborated in various ways, but the basic framework persists. It has the advantage of clarity. Arguments pro and con, conclusions, objections are often numbered; refutations follow the order of the arguments they refute. One could no more mistake one's place in the discussion than mistake the nave for the apse in a Gothic church. The disadvantage of the form for the modern reader is that in large doses it is tiring to read, and

4. Throughout this work the term 'Aristotelian' will denote the Western Christian philosophical tradition that was predominant in the Schools and elsewhere from the thirteenth century until the mid-seventeenth century. More specifically, it will denote the later part of that tradition, from around 1550 until its dissolution. Phrases like 'of Aristotle' or 'in Aristotle' will denote Aristotle's own texts. I should note once and for all that the Aristotle here presented makes no pretension to be the real Aristotle. The views attributed to him are, except where otherwise indicated, those he was *believed* to hold, and not necessarily those he actually held. On the use of the term 'Aristotelian', see Grant 1987. I am inclined to agree that 'Aristotelian' denotes a *population* in something like the sense that word has in evolutionary biology, rather than a species having a fixed essence. But it is not clear what will serve as the analogue to reproductive isolation.

5. On textbooks in Renaissance Aristotelianism, see Brockliss "Aristotle," 1981a, Brockliss "Philosophical teaching," 1981b, Grafton 1981, Brockliss 1987, Schmitt 1988.

6. On the earlier history of the *quæstio* and the *disputatio*, see Marenbon 1987.

an author's systematic views will be parceled out among responses to dozens or hundreds of *quæstiones*. In commentaries, moreover, *quæstiones* were scattered at more or less conventional sites along the way. In *Physics* commentaries, for example, a standard question on natural limits to quantity was typically put in the first book, long before the machinery needed to answer it has been encountered in the *Physics* itself.

Here the systematic rearrangement of *quæstiones* according to topic in Suárez, Eustachius, and John of St. Thomas shows its advantage. Such works, often called *cursus*, came in the seventeenth century increasingly to take the place of commentaries like those of the Coimbrans and Toletus.[7] The new arrangement made it possible for an author, if he so chose, to incorporate extended discussions of the new science. In the central texts, however, there are limited signs even of sixteenth-century developments.[8] The Jesuit teachers, following the declarations of the Council of Trent, were preoccupied with the production of texts that would counter the teachings of schismatics and that would provide a firm philosophical foundation for Catholic theology. Later on, works like Eustachius's *Summa*, because they presented an accessible Aristotelianism, gave to the opponents of the Schools the advantage of a definite target.

The central texts were written to serve the teaching of Aristotle in the universities. The curriculum of the first three years of study consisted in a systematic exposition of the *Logic*, the *Physics*, *De generation et corruptione*, *De anima*, the Metaphysics, the *Nicomachean Ethics*, and selected other works in natural philosophy, including *De cælo* and the biological texts.[9] We may lament the rigidity of the curriculum, but we can only envy its coherence. A *Physics* course could presuppose knowledge of the *Logic*, a *De anima* course that of the *Logic* and the Physics, and so forth. Toletus, for example, can presuppose in his *Physics* commentary the terminology that would have been laid out in a *Logic* course. In reading a commentary, one must keep in mind, then, that terms not explained on the spot are likely to have been explained

---

7. Such *cursus* continued to be produced for use in Catholic seminaries into the twentieth century, receiving new impetus from the encyclical *Æterni Patris* promulgated by Leo XIII in 1879 (see Denzinger 1976, *3135–3140; Hickey 1919, 1:vi; Steenberghen 1974).

8. There is, for example, an exceptional reference to Vesalius in a question on the eye in Toletus's *De anima* commentary; Pomponazzi, who argued in the early 1500s that reason alone could not prove the immortality of the soul, is mentioned by Toletus in a question on that subject. The Coimbrans occasionally refer to such near-contemporaries as Scaliger and Fracastoro. One recent development that *is* reflected in the central texts is the recovery and translation into Latin of the ancient Greek commentators, notably Johannes Philoponus and Alexander of Aphrodisias.

9. On the curriculum of the *collèges de plein exercice* in France, of which Descartes's alma mater, La Flèche, was one, during the seventeenth century, see Brockliss 1981b. On La Flèche in particular, Rochemonteix 1899 is still the standard reference. The philosophy curriculum is outlined at 4:21[ff].

elsewhere and that often a phrase like *per se* or *formaliter* was shorthand for distinctions or doctrines expounded at length elsewhere.

The coherence of the curriculum reflects that of the material. The Aristotle of the Schools was the product of several centuries of refinement, whose guiding principle was that all the genuine works of Aristotle were effectively one, as if written in a day. One finds no trace in Aristotelian interpretation of any notion of development, no Higher Criticism. The *Categories*, now thought to be an early work, are treated on the same footing as the later *Metaphysics* or the still later biological works. If two passages seem to be in disagreement, that cannot be put down to a change of mind. Instead they must be reconciled. Determined exegesis had long noted and rendered harmless the "contradictions" sedulously cataloged by Gassendi in his *Exercitationes*.[10] It is indicative of the amnesia in which Aristotelianism was already beginning to be shrouded that such criticisms were taken seriously and repeated ad nauseam by the lesser lights of the new science.

The rationale of each work, and thus of each branch of natural philosophy, was given by its place in Aristotle's system. Though there were controversies about the division and order of the sciences, it was agreed that each had its allotted place. The structure of natural philosophy, moreover, reflected that of nature it-self. Toletus's description of its division is at once a classification of the sciences and of their objects:

> What is contained in natural Philosophy is either about the principles or about the things which are composed out of them. The *Physics* is about the principles of all natural things and their common properties; the rest of the works are about what is composed out of them. Now what is composed is either a simple body, which is not constituted from others, or composite and mixed. If it is simple, it is either incorruptible, like the heavens, which are treated in the first two books of *De cælo,* or corruptible, like the elements, which are the concern of the last two books. [. . .] As for composites, because generation and corruption, and not only they, but also the simple elements themselves, are common to all, *De generatione* first discusses the one and then the others. Of composites, some are inanimate, some animate. Inanimate composites are treated first, and then animate. Among inanimate things some are sublime, and are called "meteors," and occur above us, like winds, rain, rainbow, haloes, and the like. The books of the *Meteors* are about them. Some are beneath us in intrinsic parts of the earth, like metals, stones, which are treated in the *Mineralia.* As for animate things, because the soul is common to them, it is treated first of all in the three books of *De anima,* and then certain things that proceed from the soul, namely, sleep, waking, youth, age, life, death, and

10. See, for example, the *Solutiones contradictionum* of Marcantonio Zimara, included in Averroes *In Phys., Opera* 4:464v[ff], and *In Meta., Opera* 8:401[ff].

the like are treated in the *Parva naturalia*. After those subjects, animate things themselves: of which some are animals, some plants. Animals and their kinds are extensively discussed in the *Historia animalium*, and in *De partibus animalium*. Finally there is *De plantis*. (Toletus *In Phys.* "Prolegomenon," q2; *Opera* 4:6rb)[11]

I quote at length in part to show how comprehensive was the philosophy whose foundations Descartes hoped to overturn. The passage hints also at the degree to which, from the eleventh century forward, the Aristotelian sciences had been systematically integrated so as to form a nearly seamless whole. The resulting natural philosophy had no gaps; it omitted nothing of importance. To a degree that the Vienna Circle could only dream of, Aristotelianism already had been a unified science, comprehending everything in this world and out of it.

Such was the imposing monument that virtually all seventeenth-century philosophers confronted—perhaps not throughout their careers, but certainly so long as they remained in school. It was more than comprehensive. Through centuries, it had, with the occasional prodding of anathemas, been reshaped until its parts gave at least the appearance of decent conformity to one another, to theology, to common experience and ancient authorities. That much we can say in general about the character that Aristotelianism would have presented to the young Descartes. Specifics are, because of his near-obsession with denying any provenance to his thought but that which traced its principles to God and its phenomena to experience, somewhat difficult to come by. What we know is that at La Flèche the texts of Toletus and Suárez were taught, that he very likely read Suárez in the late 1620s, that he certainly did read Eustachius, and that he had looked at Abra de Raconis. The Coimbrans, along with Toletus, he mentions among the texts he recalls from his school days. Arriaga and Fonseca each are mentioned once in his correspondence.[12]

Though all of my central texts have Aristotle at their core, they are not, as one might think from reading Descartes, an undifferentiated mass. Of these texts, Suárez, Fonseca, and Arriaga have the greatest philosophical interest. The Coimbra commentaries vary in quality, the *Physics* commentary being a more polished work than the *Logic*. The *Physics* exhibits, by contrast with Toletus, greater concern with theological questions and more use of patristic and Neoplatonist sources, including pseudo-Dionysius and Hermes Tri-

11. The *Mineralia* and the *De plantis* are pseudo-Aristotelian. Other works attributed to Aristotle in the Middle Ages, like the *De causis* and the *Problemata mechanica*, are omitted. The *De causis*, an Arabic compilation from Proclus's *Theology*, received very few commentaries in the sixteenth century; the *Problemata* several dozen, mostly by secular authors. On pseudo-Aristotle in the Renaissance, see Schmitt 1986.

12. In addition to the references cited in n.1, see Sirven 1928, c.1.

smegistus. Toletus's *Physics* commentary is a solid piece of work that exhibits originality on certain questions. Abra de Raconis occasionally diverges from most of the other authors: on the relation of action to *motus,* for example, he favors a more Scotist view than the Jesuits. Eustachius's *Summa quadripartita,* which Descartes called "the best book ever written on this matter," is, to put it bluntly, *not.* It is a kind of *Cliff's Notes* condensation, mainly of the Coimbrans, from whom Eustachius sometimes takes whole sentences verbatim. It is extremely sparing in its citation of authorities (though that was becoming standard in *cursus*), it often gives no arguments for its conclusions, and it rarely considers alternatives or objections. I am inclined to think that Descartes, who had no patience for details, little regard for authority, and an aversion to dialectic, liked it because it was unequivocal, comprehensive, and short (he regrets that the Coimbra commentaries are too long for him to use in his project—never completed, perhaps never started—of a blow-by-blow *Auseinandersetzung* with Aristotelianism). Though the *Summa* is a good listing of many of the statements Aristotelians thought were true, it is rather less useful for understanding *why* they thought those statements were true, or what the stakes were in affirming or denying them. For that one must go elsewhere.

The exposition of Aristotelian natural philosophy that occupies the lion's share of this work is synoptic and, despite its length, strictly limited in scope. It does not purport to be more than an exposition in depth of what the various figures listed above thought about the themes I am interested in. In particular I do not attempt to find a line of development over the half-century or so between Toletus and Arriaga, nor do I trace filiations among these thinkers. The exposition covers, moreover, with such detours as are needed to make the arguments comprehensible, only material corresponding to the first three books of Aristotle's *Physics:* matter and form, substance, nature, and *motus.*

That, as the reader will see, is more than enough. Arguments on *motus,* matter and form, finality, and nature were complex and the positions eventually taken often quite intricate. But it is there that one must locate the issues upon which Descartes believed, correctly I think, that he differed from the Aristotelians most radically.

## 3. Leitfaden

This work has two parts. The first, on Aristotelian natural philosophy, is structured around the three themes I have mentioned: *agency, the structure of material substance,* and *finality.* I examine the notions of *potentia, actus,* and *motus* and the contrast of active and passive in §2, completing that discussion in the introduction to matter and form in §3.1. The rest of §3 is devoted to substantial form, one of the two "incomplete" substances that make up individual material substances. In §4 I study the other incomplete substance, matter, and its relations—evidently of great importance to Descartes—with quantity; §4.3 is devoted to the role of figure in Aristotelian physical explanation. The exposition of the structure of material substance is completed in §5, which issues in a blueprint, so to speak, of *ens naturale,* the object of physics. I also look at what was to become a key notion in Descartes's physics, *dispositio.* In §6 I take up the appeal to ends and to final causes in Aristotelian natural philosophy, examining the arguments for their existence and their mode of causality, and then considering some instances of their use. The last section, §7, brings together all three themes in a study of the Aristotelian notions of nature—Nature great, or the cosmos, and nature small, the essence or defining character of individual substances. That section concludes with a treatment of the analogy between artifacts and natural things which Descartes would so heavily promote, an analogy not without peril for traditional views about the causal efficacy of natural agents.

The second part takes up the same three themes in Descartes's natural philosophy. As in Part I, I again start with change. In §8 I examine Descartes's definition of motion, the laws of motion, and the ontology of force. The structure of material substance is taken up in §9, where I study the relations between matter, quantity, and space. In §10 I conclude with a brief look at Descartes's arguments against finality and the final cause.

The Descartes that emerges from these investigations is certainly not the character portrayed at the beginning of the *Meditations,* who divests himself of his rotten old beliefs, to begin afresh with the *I think.* But I doubt that anyone thought he was. The Aristotelians who corresponded with him after the publication of the *Discours* do not exhibit the shock of the new (Garber 1988). They disagree with him, certainly, and occasionally there are misunderstandings, but, as Part I will make clear, in one way or another, they had a place ready for him in their spectrum of possible philosophical positions. Later, after the publication of the *Meditations,* there were harsher confrontations, with Bourdin and Voetius. Descartes's works were eventually placed on the *Index* of prohibited books, with the notation *donec corrigantur*—until corrected. The radical divergence that Descartes thought to obtain between his thought and that of his predecessors only gradually gained recognition and force, though already in 1640 his sometime disciple Regius was brandishing Cartesian ideas against the professors of Utrecht.[13]

Historians have come to realize that fractures in intellectual history are

13. On condemnations of Cartesianism, see Ariew 1994; on Regius, see Verbeek 1992, c.2.

seldom clean, seldom uniform. Entities like Aristotelianism are rarely so homogeneous as their opponents make them out to be. Descartes did believe that the philosophers of the Schools all shared certain foundations—a metaphor more suited to his philosophy than theirs—and he saw clearly enough that he disagreed with those foundations. But the "Aristotelianism" he argued against was, like most such isms, a polemical construction. Even if he had managed not to agree with them on any matter of importance, still his views would have been defined partly by their opposition to a specific set of doctrines.

The attempt to place Descartes, or rather to re-place him, in his intellectual setting, has been in progress for at least a century. Already in 1897 Georg Hertling was writing on "Descartes's relations to Scholasticism"; Étienne Gilson's first ground-breaking work and the indispensable *Index cartesio-scolasticus* are now eighty years old. And for that matter Leibniz had already insinuated that Descartes had not left his predecessors so far behind as he had claimed. What the present work contributes to that project is the specifics of those relations in one relatively limited, though fundamental, area of inquiry: not Descartes and Scholasticism—another invention—, or Descartes and Thomas Aquinas, or Descartes and Aristotle, but Descartes and the texts through which he and his audience came to know Aquinas and Aristotle. The difference, as we will see, matters.

PART I

*Vicaria Dei*

# [1]

## Natural Change

Aristotelianism never doubted that there was a natural order. The phenomena it studies are in large part the regularities of everyday life: fire heats, water left alone cools, animals reproduce according to kind, excess is painful to the senses. There are phenomena that lie outside that sphere—miracles, monsters, the fortuitous. But the scientific study of change relies in the first instance on their regular effects. We know nature first of all not when it is frustrated, but when it succeeds.

The task of natural philosophy is to understand that order, which Aristotelianism founded on the natural necessity of the progression from what is potentially so to what is actually so. That necessity was conceived not as the instantiation of universal law but as that of tendencies to ends. Natural things are divided into kinds according to those tendencies, or, to use an Aristotelian term, their active potentiæ. Behind the set of *potentiæ* belonging to each individual lies a substantial form, that at once individuates it, unites its powers, and determines its destiny. That *vinculum* of identity, efficacy, and ends, which Descartes did his best to unravel, together with the preponderance of powers over laws, is what sets Aristotelian natural philosophy apart from modern science.[1]

I start in §2 with natural change. Aristotelianism distinguishes intransitive change from transitive change. Intransitive change, or becoming, was famously, if opaquely, defined in the third book of the *Physics* in terms of *actus* and *potentia*. Around that definition there arose a controversy whose pri-

---

1. Surveys of Medieval natural philosophy can be found in Lindberg 1978, Wallace 1981, Kretzmann et al. 1982, Schmitt and Skinner 1988, and Copenhaver and Schmitt 1992. Mercer 1993 is a useful survey of key figures, both in the Schools and out. The five volumes of Anneliese Maier's *Studien* are indispensable; abridged English translations of a few of the *Studien* are in Maier 1982.

mary impetus was an attempt to find a place for becoming in a system of substances and accidents that had no obvious place for it. The central texts take note of a reductive view according to which change is nothing other than the successive states of the thing that changes, and a realist response that insists on a real distinction between change and the successive states. Rejecting both, they argue for a compromise: *motus* is not a distinct *thing*, but neither is the difference between the *motus* and the successive states solely a matter of conception. The two are "formally" or "modally" distinct: the *mobile* can exist without moving, but not the moving without the *mobile*.

Transitive change introduces a second kind of directedness. Just as the states of one thing are bound together by the directedness of becoming, so too are distinct things bound together by the relation of agent to patient. That relation is not a matter of how we look at things: the Aristotelian thesis holds that change is in the patient alone. Active powers and passive powers are thus distinct as well, a distinction essential to defining natural kinds. In Aristotelian physics active powers characterize substantial forms, and thus natural kinds; in Cartesian physics there are neither active nor passive powers, and the classification of substances must therefore take another path. I will examine the arguments by which the Aristotelians distinguish active and passive powers, and briefly consider the puzzling "power of resistance," in which we find an Aristotelian counterpart to inertia.

In §3 and §4 I turn to what Aristotle calls the *principles* of natural change: matter, form, and privation. Matter is what persists, form what changes. Privation, the third principle, is the state of the thing at the beginning of change. It is not the mere absence of the form eventually attained at the end of change, but the contrary of that form. The notion of contrary, I will argue, presupposes that properties fall into natural groups, or genera, so that change—with the significant exception of the production of new substances—is between properties in the same group. At the end of §3.1 I combine the principles of change with the earlier notions of *actus* and *potentia* to produce what I call the Aristotelian *scheme* of natural change. A scheme, I should note, is not an explanatory theory in the current sense. It is a template for generating suitably described *explananda*.

The remainder of §3 and the first two parts of §4 examine the two "incomplete" substances that together constitute a material thing: *substantial form* and *prime matter.* In §3.2 I look at arguments for substantial form. In physics it served primarily to provide the unifying ground for the nonarbitrarily conjoined powers of natural substances. We will see later that the phenomena adduced in favor of substantial form were not always handled more satisfactorily in the new science. In §3.3 I show how the Aristotelians could speak of form as substance and suggest one reason for the incomprehensibility of the phrase 'substantial form' to later philosophers.

The next section takes up similar problems about prime matter. In §4.1 I will prepare the way for a new interpretation of the Cartesian thesis that the essence of corporeal substance is extension. That thesis, I will later argue, though it is often thought to rest mainly on methodological or epistemological grounds, also accomplishes two tasks in the philosophy of nature. It lays to rest the Manichaean heresy of a matter capable of acting independently of, and perhaps in opposition to, its creator; it precludes, at least until Spinoza, the later heresy of a matter identical to God. However antiquated such worries may now seem, it is clear from the central texts that a physics that did not allay them, even if only by implication, would not have been acceptable to the Church.

In §4.2 I examine the controversy over the nominalist thesis that prime matter and quantity are not really distinct. Though the central texts deny the thesis, some of them—notably Suárez—allot to quantity a privileged place among physical properties: it is the "first accident" of matter, prior even to substantial form. Much of the argument turns on the interpretation of the Eucharist, an issue that, because it divided Catholics and Protestants, was the object of particular scrutiny by the Church. Here we have explicit evidence that Descartes not only knew the arguments in some detail, but that he was eager to avoid even the appearance of heresy.

I conclude §4 with a survey of Aristotelian doctrine on figure. For the Aristotelians, figure was perhaps the least important of physical qualities. Though earlier philosophers like Ockham had insinuated an identity between form and figure, the central texts all regard it as ineligible to serve as a physical principle. Indeed figure is not only entirely lacking in active power, it is not even the effect per se of any natural change, but always a by-product of change whose end is something else. There are, however, in Nicole Oresme's geometrical representations of intensive quantity, and in the so-called middle sciences of optics and astronomy, hints of a meatier role. Descartes, as will be seen in Part II, promoted the middle sciences and their geometrized ontology from their subordinate place among the sciences of nature.

In §5 I study the causes and effects of form, and in particular the notion of *disposition*. The term *dispositio*, or 'disposition', has a special role in Descartes's reworking of Aristotelian physics. It is what I call a *transfer term*, used in ways that enable Descartes to speak with the vulgar, but ultimately divorced from what in the Aristotelians' use had been essential to it. *Dispositio* indeed signifies for them what Descartes means by it—the arrangement of parts. But it signifies in addition the arrangement of parts *to some end*, and in particular to the reception of a form. Of Descartes's use the most that can be said is that *dispositio*, like 'machine', denotes an arrangement of parts that will, given the laws of nature and a certain action upon it, yield known effects. Its finality, merely apparent, lies not in the thing itself but in the intention of the philosopher who, constructing it hypothetically, wishes to explain those effects, or in the intentions of the divine artificer who made it.

The last three sections take up the notion of *end*. The Aristotelians believed that both individual things and nature as a whole act toward ends, and they defended, though with serious restrictions, Aristotle's inclusion of ends among the four causes (§6). The restrictions amount to the requirement that every attribution of an end must eventually lead back to a rational agent that recognizes the end as good. Inanimate things have ends by virtue of being divine instruments. In contrast with Descartes, the Aristotelians did not take the analogy between art and nature to its limit. Inanimate things, though they are instruments, are not thereby reduced to the condition of artifacts, which the Aristotelians, like Descartes, held to be devoid of activity or *stolidæ*, as the Coimbrans say. Instrumentality is not—so Suárez argues at some length—inconsistent with genuine efficacy. It is therefore within God's power to endow his instruments with active powers; we, on the other hand, being limited to moving bits of matter here and there, cannot.

What is perhaps most striking, once one understands the Aristotelians' arguments, is that they would *agree* that the consequences Descartes draws out of his principles, and in particular from the definition of matter as *res extensa*, are *validly* inferred from them. If, for example, the only accidents of material substance were modes of extension, then material "forms," being nothing more than figures and magnitudes, would not be natures as the Aristotelians understood that term. If substances without rational souls literally *were* artifacts, instead of just being similar to them in figure, then they would indeed lack active powers. Most of the views Descartes adheres to, and many of his arguments *against* substantial form, real accidents, and the other entities he rejected, are to be found in the Aristotelians themselves, but with the crucial difference that an Aristotelian will take the conclusions of such arguments to be *reductiones ad absurdum* of the very same premises that Descartes would have us believe are certain.

# [2]

## *Motus, Potentia, Actus*

For the Aristotelian, common sense, authority, and reason concurred in affirming two compendious maxims about the world around us. According to the first, there are not continuous gradations among things, but distinct kinds, marked off from one another by differences in structure and patterns of behavior. "What could be more known to us," writes Suárez, "than that the sun shines, fire heats, and water cools?" (Suárez *Disp.* 18§1¶6, *Opera* 25:594). Such truths are so obvious that merely to point out that a doctrine would have the effect of denying them was sufficient to prove its absurdity.

The other maxim concerns the changes that help us to discern natural kinds. The Aristotelians, seconded again by common sense and authority, distinguish within the goings-on around us a class of specifically *natural* changes. Natural change has two notable features: once begun, it proceeds spontaneously; and, at a certain stage, it ceases spontaneously. This is most evident in the actions of animals. But even among inanimate things there are many kinds of events that seem to have an inherent *terminus* and that can sometimes be said to be interrupted by other events, and so to be incomplete. A stone, if nothing hinders it, keeps falling until it reaches the ground and then ceases to move. Water cools off to a point and then no more. It is an indication of how thoroughly the concept of nature has been transformed since 1600 that we are not struck by the ubiquity of the pattern: things have powers that, when circumstances are right, are triggered, operate, and cease. To that pattern Aristotelianism was exquisitely attuned.

The regularities we find in nature govern natural change so understood. Such regularities will admit of exceptions. Stones naturally fall to the ground, but sometimes an instance of falling will be interrupted by another natural change. Humans give birth to humans, but once in a while the formation of a human may be frustrated through the action of celestial agents. Still it remains true that the complete or perfected action of human seed issues in a human being and that the existence of monstrous births is derivative upon that of natural births. The same applies generally: the rare, the prodigious, the deviant presuppose an order that is not only regular but normative, the standard against which they are measured and found wanting.

The two maxims support each other.[1] The changes characteristic of natural kinds are just the natural changes they undergo. More precisely: if we take the explanatory principles of particular occurrences to be the natural kinds whose existence is affirmed in everyday experience (and confirmed by more careful study), then the occurrences that are, to use an Aristotelian expression, first of all and per se explicable by reference to them are just those changes that exemplify the order also affirmed by experience.

Distinguish, first of all, between the *inception* of a change and the change itself. The spontaneity of natural change is a spontaneity of the change itself and of its cessation. The Aristotelian does not suppose that the inceptions of a thing's changes can be referred to that thing's nature alone. Self-movement, or what Averroes called the *vis initiativa,* is in fact the delimiting feature of animate things. But however a natural change begins, it seems to proceed and cease in a regular fashion without any further intervention from outside. The entire change, once begun, can be explained in terms of the natural kind to which the subject of that change belongs.

Consider, on the other hand, a change which, though resembling a natural change up to some point, is interrupted, or which misses the mark, as Aristotle says of monsters (*Phys.* 199b4), or which has no regular point of cessation. The entire change, if it can be explained at all, must be explained in terms of the natural kinds both of its subject and of whatever interferes with it. But in general the class of things that can interfere with a certain sort of change seems quite arbitrary. The causes of monstrous births include too much or too little seed, too much heat or too little space in the womb, vivid imaginings by the mother, and unusual configurations of celestial powers (§6.4). Those things would seem to have nothing in common, except of course that they can all interfere with human generation. Although there is some hope for a theory of the normal process of birth, there is no hope for a theory of monsters as such.

Taking directed change to be the primary *explanandum* in natural philosophy thus brings together the two pieces of commonsense wisdom. What the philosopher seeks, in attempting to understand the goings-on of this world, is to redescribe them so that their ends are apparent or, failing that, to analyze them as conjunctures of goings-on that can be redescribed. The

---

1. In the following discussion, I have drawn on Waterlow 1982 and Burnyeat 1981.

argument just made suggests a scheme[2] for that redescription: in any natural change suitably described, there will be a subject, the *mobile,* a point of inception, the *terminus a quo,* and a point of (natural) cessation, the *terminus ad quem.* This section, together with §3.1, has as one of its aims the development of this scheme. In §2.1 I begin with Aristotle's definition of *motus.* In that definition the directedness of natural change is not so much proved as presupposed.[3] Change starts from *potentia,* which can be glossed roughly as capacity or power, and ends in *actus,* the manifestation or exercise of *potentia.* It is to those key terms that I will devote most of §2.1.

But the start and end of change are clearly not all there is to change. What of the change itself? It is, it would seem, neither the *terminus a quo* nor the *terminus ad quem,* but something distinct from both: neither Paris nor Rome, but the way between. The central texts summarize, in questions of some complexity, the traditional standoff between nominalist and realist positions, and offer a kind of compromise between them, which I will develop in §2.2.

The definition of *motus* is silent about the inception of *motus.* It concerns the change itself only. Yet Aristotle, unlike some of his successors, believed that every natural change must be at once an *intransitive* becoming and a *transitive* changing and being changed: no inception without an inceptor. The scheme outlined above must be elaborated: "in Physical change all these are found: an agent, a patient [. . .], and furthermore an acquired form, and a way or medium by which it is acquired (Toletus *In Phys.* 3c4q1; *Opera* 4:84ra). The "way" is the *motus;* the "acquired form," as we will see when the basic scheme is completed in §3.1, is the *terminus ad quem* and *actus* of the change. What is new is the distinction between agent and patient. Added with them are two other notions: *action* and *passion.* To them §2.3 is devoted.

In an Aristotelian setting, the addition of new terms inevitably raises the question of whether the entities denoted by those terms are distinct from entities already acknowledged, and if so how. The central texts, with which Descartes for once entirely agrees, hold that *motus,* action, and passion are not "really" distinct, but only distinct *a ratione,* one and the same entity conceived under various relations.

Since, in the more elaborate scheme, *motus* is referred to two things,

---

2. A precise sense will be given to this term in §3.1. For now it suffices to observe that a scheme is not an explanation or an explanation-type; it is instead a form that through specification yields appropriate descriptions of putative *explananda.*

3. Waterlow, concluding that "Aristotle's preoccupation with nature and purpose appears even in his choice of *definiendum,*" refers the choice to a "conceptual bias" (Waterlow 1982:95). I'm not sure that 'bias' is the right word. As against the world of Empedocles or the world of Plato, in Aristotle's world change is, first of all, explicable and, second, explicable in relation not to something outside the world but to natures within it. That, I take it, is not a bias—if by that one means unreasoned opinion—but a choice founded on the premise that a science of nature qua nature is possible.

rather than one, one may sensibly ask to which it belongs. Aristotle's answer
is that *motus* inheres in the patient alone. There is a genuine asymmetry
between the agent and the patient in any natural change. That asymmetry,
which in Cartesian physics finds no basis, is essential to Aristotelian physics,
above all because natural kinds are distinguished primarily by their active
powers (§7.3).

## 2.1. *Potentia* and *Actus*

Philosophers have recently come to treat notions like *power* and *capacity*
more hospitably than did their immediate predecessors in the heyday of
ontological economy.[4] But there remains a residue of distaste potent
enough to lead one commentator still to speak disparagingly of "dormitive
virtue" when discussing the use of *potentia* in *De generatione et corruptione*.[5] The
notion of *potentia* has no doubt been sullied at times by its use in "pseudo-
explanations and bad theories" (Hussey 1983:xiii). So too, by the end of the
seventeenth century, was the notion of *mechanism,* and by the end of the
eighteenth that of *attractive* or *repulsive force.* Such has been the fate of many
successful notions, especially when their scientific virtues are amplified by
the indulgence of authority.

One reason later philosophers rejected Aristotelian notions is because
those notions were, or were thought to be, not just empirically unfounded
or useless, but unintelligible. In later sections there will be several instances
of such claims. Aristotelian definitions were thought to lead from the obscure to the more obscure. The authors of the Port-Royal *Logique,* arguing
that a definition must serve to give us "a clearer and more distinct idea of the
thing it defines," add that clarity and distinctness is "lacking in a great many
of Aristotle's definitions" (Arnauld *Logique* 166). That such definitions, far
from laying controversy to rest, only abetted it, reinforced the criticism.[6]

4. See Cartwright 1989, Salmon 1984. One litmus test for "Aristotelicity" is that a philosopher should take the laws of nature to depend on the causal properties—capacities, processes,
"influences"—of individuals or types rather than the other way around. Descartes comes out as
spectacularly non-Aristotelian concerning the physical world, decidedly Aristotelian concerning the spiritual.

5. See C. J. F. Williams, in Aristotle *De gen.* (Williams 1982), 138: "As an explanation of *how* a
thing becomes *F,* the statement that previously it was potentially *F* is useless: it is the old tale
about 'dormitive virtue'." I would not argue, of course, that Aristotle is right; but only that just-
so stories about "passages" (Empedocles) or atoms (Democritus) could not have been obviously *better*—except, of course, if the only proper answer to the *how* question is a mechanism.
See §5.3 and §5.4 below.

6. On the role of definitions in argument, see Toletus *In Log., Introductio* 1c5, *Opera* 1:8. If
*potentia* turns out to be no clearer and no better known to us than *motus,* Aristotle's definition
will have sinned against the first law of good definition (see Aristotle *Organon, Topics* 6,
141a26).

The definition of *motus* is a case in point. Arnauld and Nicole single it out, asking "Who has better understood the nature of movement by this definition?" (166). Descartes chides the Philosophers for having explained *motus* in terms so obscure that he is "compelled to leave them in their own language" because he "cannot interpret them."[7] Before that, Gassendi had defied anyone to make intelligible the last phrase *quatenus in potentia* of the celebrated definition (*Exercitationes* 1ex6a5, *Opera* 3:134b). Ramus, for his part, singled out the neologism *entelechia*.[8] Behind those jibes there was, no doubt, severe disagreement about the purpose of definitions. But method is not my concern here. My interest is in understanding the variety of items comprised under the headings *motus, potentia, actus* and the view of existence and change that underlies them, and in an eventual contrast of that view with Descartes's.

The extension of Aristotelian *motus* is suggested in its remarks preceding the definition in the *Physics*. Unlike its cognates in English, *motus* (which translates Aristotle's κίνησις) denotes not just motion in space but several other kinds of change as well. In the *Categories* Aristotle sets forth a classification of the ways in which "being is said."[9] Change, properly speaking, can occur, for reasons I will discuss later, in only four of the ten members of that classification. Change of *substance* is *corruption* or *generation*, change of *quantity* is *augmentation* or *diminution*, change of *quality* is *alteration*, and change of *place* is *local motion*. For the first two, because there is a unique dimension, so to speak, along which change can occur, there are two terms, one positive, one negative. Change of substance, I should note, was for a number of reasons set apart and designated by the term *mutatio*.[10] We thus have the classification in Figure 1.

Aristotle's definition seems to apply to all such changes, whether they are natural in the sense described earlier or not. Part of the difficulty in understanding the definition lies in knowing how broadly the defined notion is to

7. *Le monde* 7, *Œ* (AT) 11:39; cf. *Regulæ* 12 (AT) 10:426. Fonseca, at least, had anticipated such objections, as will be seen shortly.

8. "And, by God, you may swear by all the maggots and bookworms of Apellicon, but you will never come up with a pearl so unfit [ἄλογος] as *entelechia* to show the true genus of physical motion" (Ramus *Schol. phys.* 3c2, p81). Apellicon was a philosopher and bibliophile whose collection of Aristotle manuscripts was removed from Athens to Rome by Sulla in 83 B.C. On the novelty of *entelechia*, see Coimbra *In Phys.* 3c2q1a2, 1:334*.

9. The *Categories* distinguish substances from accidents, whose distinguishing character is that they are said to be "in" or to "inhere" in something else. Accidents are in turn divided into quantity, quality, relation, place, time, position, possession, action, and passion (Aristotle *Cat.* 4, 1b25ff). On substance, see §3.3; on inherence, §5.1.

10. See Aristotle *Phys.* 5c1, 225a. The heart of the argument is that "absolute" generation (generation of substance) is from nonbeing to being: but "non-being cannot be moved, and if so, generation cannot be movement" (225a26). Fonseca examines at some length the reasons why generation and corruption are not *motus*, except improperly with respect to prime matter—"improperly" because prime matter, being pure *potentia* is not, properly speaking, the privation of any form in particular (Fonseca *In meta.* 5c13q9§1, 3:712).

| Category | Motus | | |
|---|---|---|---|
| | Positive | Negative | |
| Substance | Generation | Corruption | *Mutatio* |
| Quantity | Augmentation | Diminution | |
| Quality | Alteration | | *Motus proper* |
| Place | Local motion, *latio* | | |

Fig. 1. Classification of *motus*

be taken. There turns out in fact to be a range of notions of change which can be understood in terms of the definition, based on a range of interpretations of *potentia*, from *logical potentia*, which amounts to nothing more than logical possibility, to *natural potentia*, the primary sense in physics.

Aristotle introduces the definition of *motus* by asserting three "divisions," as the commentators put it. The first and most pertinent is that "there is something that is *in actu* alone, and something that is *in actu* and *in potentia*." That division will be the main topic in what follows. The second division is that of the categories. Since *motus* does not exist outside of things (200b32: *motus absque rebus non est*), it must, presumably, belong to one of the categories. Which category, however, was far from obvious; I will return to that question in §2.2. The third and least important division is of things that are relative: some are relative by virtue of being greater or less in quantity, some by virtue of being agents and patients. Aristotle here hints at an argument made later against those who thought that motus was inequality (see 3c2, 201b18$^{\text{ff}}$).

The stage is thus set for the celebrated definition: *motus* is the *actus* of what is in potentia,[11] insofar as it is *in potentia*. Interpreters customarily

11. See Thomas *In Phys.* 3lect2 (*Physics* 201a9$^{\text{ff}}$); Maier 1958:3, 32. The *translatio nova* of William of Moerbeke, used by Thomas and many after him, reads: "Potentia existentis entelechia secundum quod huiusmodi est, motus est." An Arabo-Latin version, probably due to Michael Scotus, and included with Latin translations of Averroes, reads: "Motus erit perfectio eius, quod est in potentia, secundum quòd est tale" (Averroes *In Phys.* 3c2; 87vb; I should note that the modern division of the books of the *Physics* into chapters differs slightly from that of the Latin translations). The translation included with the Coimbrans' commentary reads: "Eius quod potestate est, quatenus tale est, actus motus est" (Coimbra *In Phys.* 3c2, 1:350).

The Aristotelians would paraphrase the definition in several ways: "Motus est actus entis in potentia, ut in potentia est" (Coimbra *In Phys.* 3c2q1a1, 1:332*) and "Motus est actus entis in potentia, prout in potentia" (John of St. Thomas *Nat. phil.* 1lib3c1 summa, *Cursus* 2:288; Suárez *Disp.* 49§2¶2, *Opera* 26:901; Eustachius *Phys.* 1tr3d4q1, *Summa* 2:161, with *quatenus* for *prout*) were the two most common. Descartes's citations are of the second form, but since it was common property, and might have suggested itself to memory even if he had been taught a different one, his wording cannot be traced with certainty to any source in particular.

added two other passages. One, from the same chapter, reads: "*Motus* is the *actus* of a mobile, insofar as it is *mobile*"; the other, from the next chapter, reads: "*Motus* is the *actus* of that which can act and be acted on, insofar as it is such" (*Phys.* 3c2, 201b5; 3c3, 202b26).[12] The three formulations do not differ much; Aristotle puts them all forward without quite settling on any one of them.

The obscurity of the definitions, and the failure of the surrounding text to elucidate the key term *potentia*, made them an obvious subject for dispute. But the definition itself, as was often the case in ostensibly exegetical debates, was not the real issue. In her study of Medieval explications of the definition, Anneliese Maier concludes that although philosophers from Averroes to Buridan all agreed that the definition was "good," their consensus was illusory. They all find a way to achieve a nominal concord with Aristotle. But in fact each of them "begins the explication of the Aristotelian definition with the analysis of *ens in potentia*, and each tacitly brings to that concept his own interpretation of motion" (Maier 1958:56–57). The same can be said of the central texts. In questions on the definition, the stakes lie not in agreeing or disagreeing with Aristotle, who on this as on most questions is right. They lie in understanding *potentia*, *actus*, and the mysterious compound *actus entis in potentia*.

Start with *potentia*, which is far more problematic than *actus*.[13] If *potentia* were simply logical possibility, and the difference between *potentia* and *actus* merely the difference between what is possibly so and what is actually so, then *motus* would be the becoming-so of that which was not so, but whose being so implies no contradiction. The Aristotelians indeed recognized a kind of *potentia*, which they called *potentia logica* and is just logical possibility.[14] But *potentia logica* is clearly not the notion appealed to in the definition. *Potentia logica* does not admit of degrees, and yet Aristotle speaks of *motus* as an "imperfect" *actus* (201b32). Nor is there any halfway house between being possibly but not actually so and being actually so.

---

12. Coimbra *In Phys.* 3c2q1a1, 332*.

13. For a thorough discussion of various kinds of *potentiæ* and their corresponding *actus*, see Fonseca *In meta.* 9c1q2, 3:514ff. One recent discussion of the issue is Freeland 1986. She concludes that δύναμις in the physical sense of 'capacity' should be distinguished from δύναμις or τὸ δυνατόν (the possible) in some larger sense that would encompass outcomes of violent or accidentally conjoined causes, and perhaps even all that is logically possible (85–86).

14. Suárez *Disp.* 42§3¶9; 26:613; 43 ante §1, *Opera* 26:633, Toletus *Opera* 2:268. "In which is called 'possible', two things seem to be included: one is negative, as it were—namely, non-repugnance [i.e., non-contradictoriness] of being, and this is usually called 'the logically possible', and to it corresponds *potentia logica*, which is so-called because it does not consist in any simple and real faculty, but in the mere non-repugnance of the extremes, and thus is understood in relation to the composition and division by the mind that logic is concerned with" (Suárez 26:613).

A second kind of *potentia* recognized by some Aristotelians, especially the followers of Scotus, and denied by others, was *potentia neutra*.[15] The marble out of which statues are made can be chiseled indifferently into many different shapes. Similarly any natural body in its natural place is indifferent to staying at rest or moving in a circle whose center is the center of the world. In both cases it is moving neither upward nor downward. Some Aristotelians, including Suárez, would say that the marble and the body were *in potentia neutra* to those states, any of which could be called the corresponding *actus*.[16]

*Potentia neutra* was used with respect only to natural causes of change. A block of marble is *in potentia neutra* to having the shape of Michelangelo's *David* provided that some natural, rather than supernatural, cause could give it that shape. The *potentia* so defined is not *potentia logica*. It is, moreover, plausible to speak of being shaped as a change, and even a natural change, since it is brought about by natural causes. Aristotle's stock example of the statue made of bronze and the Aristotelians' of water being heated are both of this sort. One could, then, take the *actus* of *potentia neutra* to be one kind of *motus*.

The corresponding notion with respect to divine action is *potentia obedientialis*.[17] Some authors use the term to denote the readiness of any created substance to receive changes that are not naturally possible, as when the host is converted into the Body of Christ. In that use *potentia obedientialis* is, as Fonseca argues, nothing other than *potentia logica* restricted to supernatural *actus* (*In meta.* 9c1q4§8, 3:537bB). But Suárez, following Thomas, uses the term to denote the potentia that "created things have toward receiving certain *actus* of divine and supernatural power, which by natural causes they cannot receive."[18] To those who object that if the essence of the soul were the subject in which grace was received, then all souls—not just human— could receive grace, Thomas answers that the human soul differs in species from others; being able to receive grace "agrees with the essence of the soul insofar as it is of such a species." The human soul, but not those inferior to

---

15. See Fonseca *In meta.* 9c1q4§1, 3:523; Suárez *Disp.* 43§4¶10[ff], *Opera* 26:647[ff], citing Scotus *In sent.* 2d2q6. Suárez defines *potentia neutra* thus: "When, therefore, a passive *potentia* is not only innate and connatural with its subject, but also has an inclination to the *actus* which it can receive by natural causes, then it is most properly and in every respect a natural *potentia*, and thus matter is a natural *potentia* to form, vision to [visible] species or to the *actus* of seeing, etc. But sometimes there is a *potentia* [i.e., *potentia neutra*] which is indeed innate and intrinsic to its subject, but whose *actus* is not such as to aim at the perfection of that *potentia*, or which lies outside its inclination, even if the *actus* can be induced by created causes within the order of nature."

16. See, for example, *Disp.* 43§4¶14, 17, *Opera* 26:649, 650.

17. Fonseca (*In meta.* 9c1q4, 3:521[ff]) devotes a long question to various aspects of *potentia obedientialis*, some of which I will return to in §7.3.

18. Suárez *Disp.* 43§4¶16, *Opera* 26:649; cf. Thomas *ST* 2pt1q110a4, *Opera* (Parma) 2:439.

it, is in that sense *in potentia* toward grace. Since no natural cause can bring about grace, the soul is not *in potentia neutra* toward grace; but neither is it merely *in potentia logica* toward grace.[19]

In both the heating of water and the reception of grace by the soul something obviously changes. Yet in neither is there an inclination toward the final state. The form of water, as we will see, opposes being heated; the soul untouched by revelation or the light of God knows nothing of grace. When water spontaneously cools, on the other hand, not only does the *potentia* to become cool belong to the water by its nature, but also the *actus* is one to which the water, even in the absence of any external agent, is inclined. We thus arrive at *potentia* in the most properly physical sense: *potentia naturalis*, the *potentia* correlative with natural change, and which is alone acknowledged to be a genuine quality in the *Categories*.[20]

*Actus*, which is sometimes translated as 'actuality', looks far less mysterious than *potentia*. There is a metaphysical sense of the word according to which *actus* is existence, whether in relation to a *potentia* or not. The other sense, which is the only one relevant to the definition of *motus*, is that in which *actus* is existence with respect to some *potentia*. That, I take it, is what Suárez had in mind when he wrote that physical *actus* is so-called "because it actualizes something, as form is the *actus* of matter" (Suárez *Disp.* 13§5¶8, *Opera* 25:416). The basic notion is not *actus* but *actus of.*

The definition says of *motus* not that it is the *actus* of a *potentia* but the *actus* of something that "is *in potentia*." To be an *actus* is, in the physical sense, to be "an accidental or substantial form, in succession or enduringly," while to be *in actu* is to "participate in a form," or to have that form. To be a *potentia* is to be a "principle of acting, or undergoing something," while to be *in potentia* is to have such a principle (Toletus *In Phys.* 3c1text3, *Opera* 4:78va). The central texts emphasize that the phrases '*ens in actu*' and '*ens in potentia*' do not differ "as thing from thing, but as the same thing from itself according to modes of being" (Fonseca *In meta.* 9c1q1§2, 3:513; Toletus 4:78vb).

---

19. Suárez notes that the *potentia* of natural things to be made into artifacts is sometimes also called *potentia obedientialis*. The reason is presumably that just as certain effects are beyond the power of nature as a whole to produce, so some effects—the making of a book, say—are beyond the power of nonhuman nature to produce (¶17, 26:650).

20. *Potentia* in the strict sense of the *Categories* is the "proximate principle of an operation, to which it is by its nature instituted and ordered" (Suárez *Disp.* 43§3¶2; 26:644). The operation in question can be either an action or a passion. The word 'proximate' contrasts *potentiæ* with substantial forms, whose activity is always mediated by accidental forms (see §5.4), and with prime matter, which is pure *potentia* but only in a broader sense. The phrase 'instituted by its nature' serves to distinguish *potentiæ* from other species of quality, which though they are principles of operation are not defined primarily in terms of their operation. Such are the sensible qualities, color, light, and so forth; and the elemental qualities, hot, cold, wet, dry.

The water that is *in potentia* hot and the water that is *in actu* hot are the same water, whether hot or having only the power to become so.

*Motus* is the *actus* of a being *in potentia*. So far so good. But then Aristotle adds the phrase Gassendi didn't like: insofar as it is *in potentia*. The Coimbrans, citing Averroes, note that two entities seem to be touched on by the definition: "the heat, which is acquired *in actu,* and the acquisition or flow [*fluxes*] of the heat."[21] The obvious question is: which is the *motus?* In terms used by the Aristotelians themselves, is *motus* the *forma fluens,* the form itself acquired in passing, or the *fluxus formæ,* the "way" or "tending" of that form toward another? Each view had weighty authorities to cite on its behalf. But the central texts tend to favor the second. That question tended to be taken up in close connection with a second question. Suppose that *motus* is indeed the *fluxus* and not the form. In an Aristotelian context one still must decide whether, and how, the *fluxus* and the form differ. They could be two things (res), in which case they would be said to be "really" distinct. Or they could be one thing in every sense, but conceived in different ways. Or, finally, they could be what was called "formally" or "modally" distinct, in the way that the spherical shape of the earth differs from the earth itself not merely as one conception of a thing from another, but as a modification differs from the thing modified. That question is the topic of the next subsection. Here I examine the question of whether *motus* denotes a *fluxus* rather than a form.[22]

The common opinion, as I said, is that the entity defined in the definition of motus is the *fluxus,* not the form.[23] The Coimbrans, rejecting Averroes'

---

21. Coimbra *In Phys.* 3c2q1a1, 1:332 (part of which is included in Gilson 1912:187–188); Averroes *In Phys.* 5c3, com9, *Opera* 4:215B. Averroes frames the question as one about the category to which *motus* should be assigned (Maier 1958:21ff). Maier concludes that in the understanding of Avicenna and Averroes, "the *ens in potentia* of the original formulation [of the definition] is no longer an *existens in potentia* but an *exiens* [going out, issuing] *de potentia in actum.*" The shift prepared the way toward interpreting *motus* as *fluxus* rather than form. Some recent commentators argue that such a shift makes the definition circular (Kosman 1969:41; Maier 1958:23). But it rests, as the Coimbrans' argument shows, on the reasonable assumption that no form, considered in itself, changes or is a change; in the formulations favored by the central texts, *fluxus* is not a process. See Toletus's remark, cited below, distinguishing the "flow" of oil from the *fluxus formæ.*

22. On the controversy over *forma fluens* and *fluxus formæ,* see Maier 1958, c.2; Wallace 1972; Murdoch and Sylla 1978:213–222; Wallace 1981, c.4; Sarnowsky 1989, §4.2.

23. See Toletus *In Phys.* 3c4q1, *Opera* 4:84ra; Eustachius *Phys.* 1tr3d4q1, *Summa* 2:161. Abra de Raconis, citing Suárez, identifies *motus* as a "successive passion," rather than a *fluxus.* But Suárez's position is not quite that *motus* is a passion (i.e., belongs to the category of passion) *rather than* being a *fluxus.* It is that he cannot see any real difference between them: "illud [i.e., the claim that *motus* does not belong to the category of passion] semper mihi visum est difficile, quia non video in quo constituatur differentia inter passionem et motum, si cum proportione et secundum completas rationes suas sumantur" (Suárez *Disp.* 49§2¶4, *Opera* 26:901). Later Suárez writes that if one considers *motus/passio* in abstraction from subject and source, that can be called a "fluxus," but it is in no way distinct from the *passio* (¶14, 904).

view that motus is the form (which is also supported by Albert and by at least one passage in Thomas), argue:

> Nevertheless the contrary opinion is truer and more common [. . .], namely that *motus* is the acquisition itself, or the flux, of form. [. . .] *Motus* according to its own account [*ratio*] is not a form *per se;* nor is it the form and the way or acquisition together, but the acquisition or tendency to form itself. The proof of this is that form *per se* is not change [*mutatio*], and so it is not *motus.* Again, form and flux together [. . .] is something composed *per accidens* out of flux and form. Such composites are not the object of proper philosophical definitions [. . .] Since nothing else is left over, in which the *ratio* of *motus* would consist, it therefore consists in a flux or tendency (*In Phys.* 3c2q1a1, 1:332)

Toletus likewise concludes that the *motus* can only be the *via ad formam* (*In Phys.* 3c4q1, *Opera* 4:84ra,vb). I should note that *fluxus* does not mean 'flow' in the sense that oil is said to flow. Forms don't flow in that sense. The *fluxus* is, as the next subsection will show, rather a mode of existence that a form can have, as warm-on-the-way-to-hot is the quality *warm* existing, so to speak, *toward* hot, which is different from being warm-on-the-way-to-cold.

Toletus, more thorough than the Coimbrans, relies on a certain kind of divisibility peculiar to *motus* to make his case. There are, he says, four ways in which motus can be called "divisible" (84rb–va):

(i) by virtue of the extension of the *mobile;*
(ii) by virtue of proceeding through spatial parts or degrees of intensity;
(iii) by virtue of being the successive acquisition of places, or of qualitative or quantitative forms;
(iv) by virtue of occurring in successive instants of time.

Of these four, the third belongs to *motus* and to *motus* alone. The first is simply borrowed from the *mobile;* the second, because it applies to quantities or qualities independent of change, also applies to *motus* only at second hand; the last is distinct because the successive acquisition of the same quantity (in growth) or quality (in alteration) can occur over different lengths of time. The successive acquisition of forms, in other words, cannot be identified with the succession of instants through which a *motus* occurs.

Now to finish with the definition. It must exclude what is *in potentia* $X$ but not now becoming $X$, as well as what was *in potentia* $X$ but is now wholly $X$ *in actu*.[24] The qualifying clause 'insofar as it is *in potentia*' is supposed to do

---

24. The phrase *entis in potentia* (being in *potentia*) excludes not only all instantaneous changes, like generation, corruption, and creation, but also "all perfected *actus*, like maximum coldness, maximum heat, which leave no power [*potestatem*] in the subject to acquire any further part [of the perfected accidental form]" (Coimbra *In Phys.* 3c2q1a1, 1:333*). Such *actus* can only be the *termini* of change.

that, as Aristotle's subsequent discussion shows (201a29–201b5). The *actus* in the definition is the "imperfect" *actus* of hotness in a being that was hot entirely *in potentia* and yet is still not yet hot entirely *in actu*. It is not the perfect *actus* of heat in water that has turned to steam, or the *actus* of the water qua water before it began to be heated. In the former, nothing is *in potentia* hot any longer; in the latter, nothing is *in actu* hot yet. Water is indeed the *actus* of a *potentia*, as we will see, namely the *potentia* of matter to form. But as such it is, like the heat of the steam, entirely *in actu*, and not *in potentia* any longer. There seems, in other words, to be something that is the *actus* of something in *potentia*, so described, when and only when there is heating, and not before or after.[25]

The "imperfect" heat in water that is being heated is, I should note, imperfect or "inchoate," as Fonseca puts it, not merely by virtue of being an *intermediate* degree of heat: "For if the heat which is introduced into a piece of wood is considered solely with respect to its definition, from which follows a certain perfection in this instant, it will clearly not satisfy the definition of *motus* or of the acquisition of form [. . .] But since it is taken [in the definition of *motus*] in such a way that not only is it viewed with respect to being introduced but also with respect to being about to be introduced, it will now satisfy in every way the true and genuine definition of *motus*."[26] If, for example, there is an interval of rest in the middle of the heating, only at the moment of its introduction will the intermediate degree of heat reached in that interval fully satisfy the definition of *motus*, since only then will it have been "just introduced."

One can well understand why later philosophers thought that something that should be obvious was being shrouded in mystery. The discussion has an air of the *via negativa* about it. Aristotle at one point remarks that it is "indeed difficult to grasp the nature [of *motus*], for it is necessary that one should put it either with privation, or with *potentia*, or with what is in every way and plainly *actus*, but it seems that *motus* can be none of these" (*Phys.* 3c2text15, Coimbra *In Phys.* 3c2 *explanatio*, 1:352). Fonseca forestalls that objection: "First of all [one might object to] the obvious extreme obscurity of the definition, which is so great, that while a definition should be a thing not just equally known but more known than the thing defined, this [defini-

25. "The only time at which the buildability of the heap *is* essential to what is happening is when it is getting built into a house [. . .] So, if the definition makes any sense at all, it gives the right result in the case of building and by analogy in all other cases" (Hussey 1983:62, *ad* 201b5).

26. "Nam, si calor, qui introducitur in lignum, ea tantum ratione consideretur, qua certam perfectionem in hoc instanti consecutus est, non habebit sane rationem motus, seu acquisitionis formæ, sed formæ certæ perfectionis iam acquisitæ. At, cum ita accipitur, ut non tantum spectetur, qua ex parte introductus est, sed etiam qua est immediate introducendus, iam omnino habebit veram & germanam motus rationem" (Fonseca *In meta.* 5c13q9§2, 2:719).

tion] brings so much darkness that after it is proposed and we want to explain what it is, right away we seem to be ignorant of what we knew well enough before" (Fonseca *In meta.* 5c13q9§3, 3:719). His answer is (i) that *motus* is well enough known in its species (e.g. running, walking, etc.), but not in general; (ii) that because it is a sort of mixture of the actual and the potential (*Phys.* 201b33) but with more of the potential, since its *terminus*, so long as the *motus* exists, is *in potentia* only, and since we know the *actus* of a thing before and better than we know its *potentia*, *motus* is particularly hard to know; (iii) it is, moreover, known only by analogy (see Aristotle's remarks on matter at *Phys.* 1c7, 191a8ff) and is "variable and inconstant" in nature (§4, 3:721).

Grasping what cannot be fixed is no doubt a perennial problem, but for the Aristotelians it took an especially acute form. The categories of being established at the very beginning of Aristotle's logic have no place for change, but only for agents, subjects of change, and the results of change. The difficulty only grew worse when, in the aftermath of nominalism, the furniture of the universe was increasingly limited to "individual substances, qualities, and quantities."[27] In that setting, one could, it would seem, do no more than to say what change is *of,* and not what change *is.* The tortuous exegetical maneuvers of the Aristotelians, the aporetic close to Aristotle's own discussion in *Physics* 3c2, are witnesses to the severity of the problem.

One further point can, however, be gleaned from the Aristotelians' discussions. Actus, as I said, is existence, but existence viewed in relation to a *potentia* from which the *actus* is thought to issue. It is thus also the perfection or completion of what had been imperfect or incomplete. The coincidence of existence and perfection, which is perhaps the most fundamental expression of the teleological orientation of Aristotelian natural philosophy, must confront the fact that some changes, or what seem to be changes, lead to imperfection or even, as in corruption, to nonexistence.

Fonseca and the Coimbrans both take note of this, but Fonseca sees the problem more clearly:

> Diminution, or decrease of quantity, and remission of quality are not inchoate and imperfect actus tending to greater perfection. On the contrary [they are] perfect *actus* progressing toward less perfection. The same can be said of corruption, which does not progress from the imperfect to the perfect, but from the perfect to the imperfect and in all ways toward nonbeing. (Fonseca *In meta.* 5c13q9§3, 3:719)

---

27. Murdoch and Sylla 1978:215. To their list, however, one must add intrinsic place, or *Ubi.* Suárez, for one, argues that *Ubi* is a genuinely intrinsic property (*Disp.* 51§1, *Opera* 26:972ff).

Of the several not entirely satisfactory responses Fonseca offers to the objection, the one that interests me is this:

> [To the objection] one may say that it is quite probable that Aristotle did not want to include in the definition *motus* that are, so to speak, deleterious [*objectivos*], namely decrease, remission, and corruption: [. . .] because they are not *per se* intended by Nature, which imitates the condition of its Author (*He creates*, says *Sapientia* c.1, *so that all things should be*, and not so that they should give up their existence, unless *per accidens* and by a certain consequence [i.e., of other changes that are intended by Nature]). (Fonseca *In meta.* 5c13q9§4, 3:721aC)

The Coimbrans likewise suggest that perhaps such changes "are not true and positive *actus*, but merely corruptive," and are therefore not comprised in the definition (Coimbra *In Phys.* 3c2q2a2, 1:334*).

One response, then, to the observation that there are changes that do not tend to perfection is just to deny that they are, properly speaking, *motus*. Though all changes, natural or not, are imperfect *actus*, natural changes—or the *actus* of natural *potentia*—are, so to speak, the most perfect among changes; unlike violent changes, for example, they can sustain themselves without the constant influx of an external efficient cause. The *terminus* of natural change is, by comparison with that of violent change, good not only with respect to Nature as a whole, but also with respect to the thing that changes. It is not surprising that Christian philosophy, resonant to a theology of providence and salvation, and anxious to "overcome," as Blumenberg puts it, the threat of gnosticism (Blumenberg 1988, pt.2 c.1), should have found the agathism of Aristotle, intensified into optimism, to its liking.

## 2.2. Independent Existence of *Motus*

The *definiendum* in Aristotle's definition, then, is the *fluxus* of the form. But *fluxus* need not signify anything other than the form itself. That question remains to be settled. It had, as Aristotelian questions often do, a lengthy genealogy, rather more complicated than usual. A quick sketch of it will be necessary to understand the terms in which the Aristotelians laid out the alternatives and the arguments brought to bear on them.[28]

28. What follows is a history of the question as it could have been gleaned from the central texts. A full history of the question from Avicenna to Blasius of Parma, which quotes many of the relevant passages, is found in Maier 1958, c.2. Brief expositions are found in Wallace 1981, c.4 (through Soto), Adams 1987, c.19 (a concise and clear account, relying on Maier, of the question up through Ockham and Buridan), Sarnowsky 1989, §4.2 (Albertus Magnus, Buridan, and Albert of Saxony).

That there was such a thing as change, and even natural change as Aristotle understood it, no one doubted.[29] Aristotle himself dismisses from the domain of natural philosophy the arguments of Parmenides and Melissus (*Phys.* 1c2, 185a13ff, 2c1, 193a1ff). 'Is there *motus?*' is not a genuine physical question. The genuine question is '*How* is there *motus?*'

The underlying issue is certainly not peculiar to Aristotelianism. Descartes scholars have argued over a quite similar issue in Cartesian physics.[30] Though it is often put in terms of the relativity of motion, and though even in Scholastic discussions one can find philosophers who notice the perceptual relativity of motion,[31] in the questions I am about to discuss, a less proleptic way of putting the issue is this: when a thing changes, it has one state at the beginning, another at the end.[32] If the change is uninterrupted, the same holds of each segment into which the change can be divided. Hence it is reasonable, if change is indefinitely divisible, to suppose that at each moment the thing has a state distinct from those it had or will have. Change therefore entails a succession of states. But *is* it that succession? And is the change at each moment just the state at that moment? If so, then once we are given things, their states, and the order of those states, we have all we need to describe the world and what happens in it. Change certainly occurs: in that sense it is real enough, but it is derivative upon other things.

Such would be the view, I imagine, of most philosophers now.[33] It was essentially Ockham's, and perhaps Descartes's. But some philosophers have held that change is something in addition to things, their states, and the succession of those states.[34] The questions I am about to examine concern the truth of that claim, both *secundum Aristotelem* and *secundum veritatem.*

29. Abra de Raconis includes a question on "whether there is in fact *motus*," in which he refutes the "sophisms" of Zeno (Abra de Raconis *Phys.* 240). That seems to have been unusual.

30. For a thorough airing of the issues, see Garber 1988, c.6; I return to the question in Chapter 8, below.

31. Buridan, *after* he has demonstrated that local *motus* is distinct from the thing that moves and the places it moves through, argues that local *motus* is not called 'local' because *locus* is necessary to it but because "it cannot be perceived unless there appears to be change of place or orientation [. . .] just as those who are on ships travelling simultaneously and at the same velocity do not perceive that they are moving" (Buridan *In Phys.* 3q7, p51ra). Clearly Buridan does not believe that the perceptual relativity of *motus* argues against its being distinct from the *mobile* and its *termini*.

32. By 'state' I mean the set of properties possessed by a thing at a given time. This is not intended to be a rigorous definition. If so-called Cambridge changes (the change that occurs to the North Pole when I move ten feet south) are to be ruled out, then some relational properties will have to be excluded. Adjustments would also have to be made to accommodate periodic changes. But these would not affect the argument.

33. So W. H. Newton-Smith writes that "a change is [. . .] constituted by an alteration in the properties of a persisting object *or* by an alteration of the properties of a region of space" (1980:14ff). I take it that most philosophers would now, quibbles aside, find that unexceptionable.

34. One notable instance is Bergson: "One would have to abstract completely [from the "cinematographic" view of change as a series of states] in order to dissipate at one blow the theoretical absurdities raised by the question of movement [. . .] There is *more* in a transition

The *Physics* would have change occur in three categories—quantity, quality, and place—or four if substantial change is included. Yet the *Categories* put *motus* in the category of *passio:* "Acting and undergoing also admit contrariety [the implied comparison is with qualities]. Heating and cooling, being heated and being cooled, and being affected with pleasure or pain are contraries: and for that reason they admit contrariety. They also admit of degrees, since they are said to be more or less" (*Cat.* 9, 11b1[ff]; cf. Coimbra *In Log.* 1:514 and Toletus *In Log., Opera* 2:166a for Latin versions).[35] Worse yet, the *Metaphysics* seem to say that it is a continuous quantity (*Meta.* 5c13, 1020a27[ff]). That passage at least can be dismissed, since it says that *motus* is quantity only by way of occurring in space and time, not in itself (see Fonseca *In meta.* 5c13q8, 2:705[ff]). But the other two are not so easily reconciled.

Averroes acknowledges both possibilities. The *Categories* view, that *motus* is a passio, he calls the "more famous." Nevertheless, it is false. The true view is that of the *Physics,* according to which *motus* is the "flowing form itself with respect to its being acquired part by part [*forma ipsa fluente secundum acquisitionem partis post partem*]."[36]

Returning to the question in *Physics* 5, Averroes introduces a rather unhappy distinction between the "matter" and the "form" of *motus,* by which he means the entity that is the *motus,* and its definition. He restates the earlier position thus:

> It seemed that this being [i.e., *motus*], which is between *actus* and *potentia* of that genus [i.e., the category of the *terminus*], should belong to that genus, since what is intermediate between a *potentia* and *actus* which belong to that genus, is necessarily of the genus of the actus which is the completion [of the intermediate], differing however according to more or less. [. . .] And it is true to say, according to this way of speaking, that *motus* is in the genus of that toward which it is *motus,* according to its matter. But accordingly as *motus* is a form in that matter [i.e., the so-called matter of the *motus* itself], one must judge that *motus* is a category by itself [i.e., passion]. (Averroes *In Phys.* 5c1com9, *Opera* 4:214L–215B; cf. Maier 1958:65 and Adams 1987:803)

---

than a series of states—that is, of possible cuts—, *more* in movement than the series of positions—that is, of possible stopping-points" (*L'évolution créatrice,* c.4, Bergson 1959:760).

35. One chapter of the so-called *Postprædicamenta* in the *Categories* is also devoted to *motus.* The chapter shows that (i) the contrary of *motus* generally is *quies* or rest; (ii) alteration, though it cannot occur except if local motion or augmentation also occurs, is nevertheless distinct from them (Coimbra *In Log.* 1:563[f]). It figures in the questions I am discussing only by way of providing one argument to show that *motus* does not belong to the category of *passio* (Suárez *Disp.* 49§2¶2, *Opera* 26:901).

36. Toletus *In Phys.* 3c3q1, *Opera* 4:82vb; cf. Averroes *In Phys.* 3com4, *Opera* 4:87D, which is quoted in Maier 1958:63.

*Motus* according to its "matter" belongs to the category of its *terminus*, although according to its "form" it is a passion. Certain later authors took up that distinction to propose a reconciliation of the two views,[37] but most philosophers were not inclined to let the matter rest there. By the end of the thirteenth century two well-defined positions had been marked out: the reductive view of Ockham and the realist view argued for by, among others, Buridan, the Thomist Capreolus, and Paul of Venice.

Ockham's position was that *motus* is in no way distinct from the *terminus*. In his words, "*Motus* is not something else in addition to enduring things [*rebus permanentibus*]" (Ockham *Summula* 3c5 and 6, *Opera philos.* 6:262,265). In order for a thing to be said to move, "it suffices that the *mobile* should continually and without interruption or rest, acquire successively and part by part one thing after another or lose something." Since the arguments he gives are to some extent taken over by the Aristotelians on behalf of their view, I omit them here.

The realist view is that "*motus* is distinguished *in re* from the *terminus* acquired by it" (Coimbra *In Phys.* 3c2q3a1, 1:340). The strong flavor of that claim can be tasted in the following passage from Buridan.[38] Buridan is answering the objection that if *motus*—in this context, local motion— were really distinct from the thing moved, then "God could, having annihilated them, separate and separately conserve the *motus* without the *mobili* and the terminal place, which seems implausible since then there would be *motus* and nothing moving" (Buridan *In Phys.* 3q7, p50ra). His reply:

> I say that it is no more a problem that there should be *motus* and nothing moved or changed than that there should be whiteness and nothing being white. Neither is naturally possible, both are supernaturally possible. But of [the objection] that for there to be local *motus* without place is a contradiction I say that the *motus* of the outermost sphere or of a ship in a river is called 'local' not because it is necessary that according to [that *motus*] they should be changed in place, but because according to the common course of nature all that is moved by this *motus* in fact varies in local or postural habitation [i.e., in place or in orientation]. Yet every

37. Maier 1958:84ᶠ (on John of Jandun, who agrees however that the first view is truer), 95 (on William of Alnwick); cf. Toletus 83rb, which mentions the view without ascribing it to anyone. The formal/material distinction is sometimes used in a logical sense to distinguish a *definiendum* from the subject in which it inheres or which it informs; what is 'formal' in that sense need not be a form in the physical or metaphysical sense. Cf. Chauvin *Lexicon* s.v. 'Formale'.

38. On Buridan's view, see Maier 1958:117–133; a brief summary of this is found in Adams 1987:824ᶠᶠ; Sarnowsky 1989:137ᶠᶠ. On all *motus* except local *motus*, Buridan agrees with Ockham.

*motus* we call 'local' could fail to be local insofar as no place or orientation would be changed to some other. (51ra)[39]

Although Ockham argued that if *motus* were really distinct from the *terminus* there could be heating without the form of heat, and from that concludes that *motus* is not really distinct, Buridan argues that since everyone agrees that there are real accidents, which can exist apart from their subjects (§4.1), there is nothing untoward in supposing a *motus* without a *mobile*. Nor does all local *motus* require a change of place. The outermost sphere, for example, which contains every corporeal thing and every place, cannot change its place relative to anything else, since everything else moves with it. Yet it can and does move: it rotates with unimaginable speed around the earth's axis, and it could move in a straight line.[40]

The central texts, then, arrive at the end of a rich tradition. The conclusion typically reached in them is a compromise between the realism of Buridan and the reduction practiced by Ockham. I should note that in these conclusions the word *terminus* may denote any intermediate form or place during a change.[41]

> *Motus* is not really, but only formally distinct from the *terminus* it aims at. (Coimbra *In Phys.* 3c2q3a2, 1:341)

> Therefore let this be the first conclusion. That way [i.e., toward form] and *motus,* with respect to three of its species, namely alteration, augmentation, and generation, along with their opposites, is not really distinct from the *terminus* or the form [acquired in passing]. [. . .]
> Second conclusion. Local *motus* is not really distinct from the *mobile* which moves. [. . .]

39. The following question (3q8) argues a similar point at greater length: it is not necessary that a local *motus* should have, in addition to its *fluxus,* a *terminus à quo* and a *terminus ad quem.* Every *rectilinear* movement, however, and every movement caused by heaviness or lightness does have extrinsic *termini* (52ra).

40. See *In Phys.* 3q7, *conclusio prima.* This was Buridan's most telling example, since among the propositions condemned by Étienne Tempier, the Bishop of Paris, in 1277 was the proposition "that God could not move the heavens in a straight line. And the reason is because this would leave a vacuum" (see Hissette 1977:118ff, no. 66; Dénifle and Chatelain 1889, no. 49; Buridan refers to the proposition at 50va). No movement of the heavens or outermost sphere could consist in the successive acquisition of new places, because the heavens contain all the places there are. This is true whether 'place' is defined in terms of circumambient bodies or intrinsic *Ubi.* The heavens might seem to be a unique instance: but for any body God could annihilate all bodies except that body, and thereby make that body as if it were the outermost sphere.

41. "If by the total [movement of] heating eight degrees of heat are acquired, the *terminus* of the entire heating will be those same eight degrees; while to the single parts of the heating there correspond the single parts of the heat, namely those which by the single parts of the *motus* were acquired" (Coimbra *In Phys.* 3c2q3a2, 1:341).

Third conclusion. Nevertheless *motus* and its *terminus,* or local *motus* and the *mobile* are distinct in reason and in definition. (Toletus *In Phys.* 3c3q3, *Opera* 4:87ra–b)

I. Conclusion: *Motus* does not really differ from the *terminus* toward which it tends. [. . .]

II. Conclusion: *Motus* is formally distinct from its *terminus.* (Eustachius *Phys.* 1tr3d4q2, *Summa* 2:163)

The arguments against a real distinction are, by and large, those of Ockham and other nominalists; the arguments in favor of a formal distinction are suitably modified versions of the realists' arguments.[42]

Now for the arguments. There are two cases: the distinction between the *motus* and its (intermediate) *termini;* the distinction between the *motus* and the *mobile.* Against a real distinction between *motus* and *terminus,* the texts offer essentially two arguments. The Coimbrans argue, following Ockham, that "if *motus* differed in re from the *terminus,* it could be divinely conserved without it." But *motus* is just the acquisition of the *terminus,* and it would be contradictory to suppose that it could exist without the *terminus.* The second argument is a regress: if *motus* existed separately from the *terminus,* it would itself be caused by some action. That action would amount to a second *motus* and require yet another action, and so on, ad infinitum. Every *motus* would comprise an actual infinity of *motus,* which is impossible.[43]

So much for the real distinction. As for the formal distinction, many things can be asserted of the *motus* that cannot be asserted of its *termini:* that it is fast or slow, successive, has parts (Coimbra a1, 1:340). The *motus,* moreover, exists before its *termini* do, since it is the way toward them. The Coimbrans, I should note, do not hold that a difference in properties entails that two things should be entirely distinct. They may be discernible even if one is merely a mode of the other, as acquisition or *tendentia* is of form (a2, 342).[44]

42. Suárez is exceptional in holding that *motus* and *passio* are identical except in reason (only Avicenna had taken that position before him; Abra de Raconis is the only seventeenth-century text that agrees with him). But except for that, he too agrees that *motus* is not really but only formally distinct from the *terminus.*

43. Coimbra *In Phys.* 3c2q3a2, 1:341*; condensed by Eustachius *Phys.* 1tr3d4q2, *Summa* 2:163; cf. also Suárez *Disp.* 49§2¶5–6, *Opera* 26:902. For the regress argument, see Toletus *In Phys.* 3c3q3, *Opera* 4:87rb.

44. "Respondemus in eorum sententia, qui motum nihil aliud esse, quàm ipsam formam fluentem opinantur, non distingui motum à termino, nisi penes diversum modum eiusdem essentiæ; est enim motus eademmet forma, quæ acquiritur, eademque essentia, nihilo à se ipsa plus differens, quàm quod aliter ac aliter se habeat [cf. Buridan *In Phys.* 3q7, 50va], prout ab esse imperfecto ad perfectum progreditur [. . .] At iuxta aliorum opinionem, quam statuimus, arbitrantium motum propriè, ac formaliter esse ipsum fluxum, seu progressionem ad formam, dicendum erit motum distingui essentia à termino. Enim verò motus secundum suam naturam est quidpiam successivum [. . .]; forma autem, quæ acquiritur [. . .] non ita se habet, etiam interim dum acquisitionem subit; quia si tunc secundum suam essentiam foret successivus,

It follows readily from the second conclusion that *motus* is really distinct from the *mobile*, contrary to what Ockham had argued. As we will see, the Aristotelians held that accidents are really distinct from the substances they inhere in (§5.1). Since *motus* is a mode of the *terminus*, and the *terminus* is (except in generation, which is excluded from these arguments) an accident of the *mobile*, it is distinct from the *mobile* to whatever degree the *terminus* is distinct.

One case, however, requires separate treatment: change of place, or *latio*. *Latio* is indeed really distinct from the *mobile*, for the reasons I have just mentioned. But because the place of a thing, which is the outermost containing surface of a body, is only formally distinct from quantity, *latio* too is only formally distinct from quantity.[45] If, as Ockham famously argued, quantity does not differ from the thing that has quantity, then *latio* turns out to be only formally distinct from the *mobile*.[46]

The authors of the central texts, though they were not nominalists, were certainly cognizant of the nominalist view. They agreed with Ockham that *motus* is not something in addition to *res permanentes*, if by that is meant something really distinct from them. But they were also well aware of the difficulties, some of which I have mentioned, that had been urged against Ockham by Buridan and others. In the formal distinction, which they inherited from Scotus, they found the middle way they were looking for. *Motus* is not a *res*, but a mode of a *res*, which consists not just in being one stage in a succession, but in being a stage that is *about to become* something else. *Motus*, like any mode, depends on something else for its existence; nevertheless it exists in things, independent of the ways in which we conceive them.

## 2.3. Action and Passion

The definition of *motus* does not appear to presuppose that for every *motus* there must be not only a subject in which that *motus* inheres, but an agent or

---

utique etiam post acquisitionem, finito motu eius natura in successione consisteret, quod longe aberrat à veritate": "We answer the view of those who hold that *motus* is nothing other than the *forma fluens* itself by saying that *motus* is not distinct from the *terminus* except insofar as it is a different mode of the same essence. *Motus* is just the same as the form which is being acquired and the same essence, in no way different except that it is different at different times—namely, it proceeds from imperfect to perfect existence [. . .] Concerning the opinion of the others, already mentioned, according to which *motus* is properly and formally the *fluxus* itself, or the progression to the form, it should be stated that *motus* is essentially distinct from the *terminus*. *Motus* according to its nature is something successive; but the form which is acquired is not such, even while it is being acquired, since if it were successive according to its essence, it would also be successive after it was acquired, and its nature would consist in succession after the *motus* was over, which is very far from the truth" (Coimbra *In Phys.* 3c2q3a2, 1:342).

45. Aristotle *Phys.* 4c4, 212a5ᶠ. Cf. Ockham *Q. in phys.* 72, *Opera philos.* 6:598.
46. Ockham *Q. in phys.* 14, *Opera philos.* 6:430. On quantity and matter, see *Summula* 1c13, *Opera philos.* 6:191ᶠᶠ and §4.2 below.

an inceptor of that *motus*. Yet Aristotle and the Aristotelians do not doubt that there must be.[47] Toletus, echoing Aristotle's final words on *motus* in *Physics* 3, writes that "*motus* is twofold: it is a certain *actus*, but from one it proceeds, as from its principle: in another it is received, as in a subject."[48] It is, or entails, both action and passion.

The distinction between acting and being acted on is already presupposed in the definition of 'nature', and a distinction between agents and patients in the explication of the four causes in *Physics* 2. An efficient cause is "that from which comes the first beginning of change or rest, [. . .] and in general the agent is the cause of what is done, the changer of what is changed" (194b29–32). Out of that, perhaps, one can extract a reason, if not an argument. If there were no agents, there would be no "first beginnings" of change, and no changes. An Aristotelian, then, could no more doubt that there are agents and actions than that there is change in the world, and of *that* there can be no doubt: it is "so evident from the perpetual vicissitudes of things [. . .] that to prove it by argument would be superfluous."[49]

Among Aristotelians, that is perhaps all that needed to be said. They seem not to have worried much about the relation of the intransitive *motus* defined in Physics 3c2 and the transitive *motus* presupposed throughout. Common experience showed, in any case, that in most if not all changes there is in fact an agent and a patient. We thus arrive at a scheme for the description of change more elaborate than that suggested in the definition of *motus*. "These five are to be considered," Toletus writes, "in every action: the agent, the patient, the form which is brought about, the *fluxus* of the form, and the various respects or relations consequent upon them" (*In Phys.* 3c3q2, *Opera* 4:86rb).

One indication of the degree to which the Aristotelians found the scheme of transitive action essential is the effort some of them make to apply it even in unpromising cases. Suárez, for example, accepts the standard distinction between immanent and transeunt action. A *transeunt* action is one "that has

47. Waterlow notes that "Aristotle cannot or will not recognize such a thing as *intransitive change that is neither an acting upon nor a being acted upon*" (1982:170). According to Émile Bréhier, the Stoics did. He writes that "when knife cuts flesh, the one body produces upon the second not a new property but a new attribute, that of being cut. [. . .] This manner of being exists in a way at the limit, the surface of being, whose nature it cannot change: it is truly neither active nor passive, since passivity would suppose a corporeal nature that underwent some action" (Bréhier 1928:11–12). This notion, new to philosophy, Bréhier identifies with our notion of 'event' or 'fact'.

48. Toletus *In Phys.* 3c3text18, *Opera* 4:81va; cf. *Phys.* 3c3, 202b26, paraphrased thus by the Coimbrans: "motus est actus eius, quod agere, & pati potest, quatenus tale est." This was quoted earlier as one of the "definitions" of *motus*. *Physics* 3c3 does not argue the equivalence of this and the earlier definitions; it presupposes from the start that in every *motus* there is not only a *mobile* but a *motor*.

49. *Disp.* 18§10¶5, 25:681ff; 12§3¶2, 25:388.

an effect outside the agent itself," while an *immanent* action is one that "has no effect outside the agent."[50] The question he sets himself is this. In a transeunt action—the sun heating a rock, say—the agent brings about a *motus* in something else; that *motus* has a *terminus* that is, since it is outside the agent, distinct from any property of the agent. The warmth of the rock is evidently not also in the sun. The *terminus* is said to be the *terminus* of the action as well as of the motus. In immanent action, on the other hand, since there is no effect outside the agent, it is not obvious that there need be a *terminus*, or that if there is, it must be distinct from the action. Suárez therefore asks whether every action requires a terminus.

There was, to judge by his account, some agreement that immanent actions are qualities of the *potentiæ* in which they inhere: desire, for example, is a quality of the will; cognition a quality of the intellect. The question was whether immanent actions just are such qualities, and so not really actions, or whether desire, say, is the *terminus* of an interior *motus* of the will.[51] To make a long—though by Suárezian standards rather brief—story short, the answer is *yes*. After arguing that God could produce the quality of loving without the aid of the will, and that the quality could be produced in several ways, each of which would constitute a different action, he concludes that desire is (also) an action, "because it is a tendency toward a quality, and by its nature distinct from it" (*Disp.* 48§2¶13, *Opera* 26:877).

Contrary to what many Thomists thought, and what some commentators now believe, Suárez holds that in what Aristotle calls ἐνέργεια (*Meta.* 9c6, 1048b23[ff]), as well as in what he calls κίνησις, there is a *terminus* that must be attained if the change is to be called complete (¶20, 879).[52] The difference between desiring a cake and baking it is not that desiring is not an action or without a *terminus*, but that the *terminus*—the quality of soul which consists in that desire—lasts only so long as the action does (¶21, 879). The scheme, but for that one anomaly, suits immanent actions as nicely as it suits transeunt actions.

50. Suárez *Disp.* 48§2¶1, *Opera* 26:874; cf. Toletus *In Phys.* 3c4q2, *Opera* 4:85vb. Toletus, however, defines an immanent action as one that does not exist outside its "effective principle," which is to say the *potentia* of the agent. The standard examples of immanent action are acts of will or acts of the intellect, like contemplation.

51. Suárez's representative for the view he opposes is the Thomist Soncinas (*Q. meta.* 9q21, p254b[ff]).

52. Ἐνέργεια is usually interchangeable with ἐντέλεχεια. The *Metaphysics* passage is the only place where Aristotle makes explicit the distinction between ἐνέργεια and κίνησις, although there are other texts that seem to call for it. One modern interpreter who emphasizes the passage is Waterlow 1982:184. Fonseca notes, and a look at the commentaries of Thomas and Averroes confirms, that this passage is missing in the earlier Latin translations (Fonseca *In meta.* 9c6 *explanatio*, 3:647; see also the translator's note in Averroes *In Meta.* 9 *ante* com11, *Opera* 8:235H). It had no role, therefore, in Medieval philosophers' interpretations.

Suárez's was far from the only view.[53] But his opponents, those who held that immanent actions are not actions at all but instead qualities, typically did so on the grounds that immanent actions have no *termini*. They did not, therefore, depart from the scheme. They only disagreed about its applicability.

From the scheme of transitive action the central texts drew three conclusions. Together those conclusions show that agency and efficient causality differ significantly under the Aristotelian conception from their counterparts in Cartesian or classical physics. The first is that *motus* is in the patient alone. The second, which follows in short order from the first, is that the agent is impassive in action except per accidens. (That consequence is consistent with the claim in *De generatione* that every agent is acted upon in return [1c7, 324b]. In reaction, the agent is *not* moved qua agent, but qua patient of the coincident re-action of its patient.) The third consequence is that *motus,* action, and passion all inhere in the patient and are at most formally distinct from one another.[54]

1. Motus *is in the patient.* Since *motus* is a mode of its *terminus,* and so must inhere in a subject, and since in the scheme of transitive action two things take part, it makes sense to ask to which of them the *motus* belongs. Aristotle answers unequivocally: *motus* inheres in the patient (*Phys.* 3c3, 202a13$^f$). But since *motus* is the *actus* of a *potentia,* the implication is that the *actus* of the agent's *potentia* inheres in the patient.

Naively one might wonder whether it isn't absurd to say that the *actus* of one thing is in another (202b6). The quick answer is that it all depends: an *actus* is "of" whatever its *potentia* inheres in, "not because it always inheres in it, but because it either inheres in it or is from it" (Toletus *In Phys.* 3c3q2, 4:86vb).

But the objection can be pushed further. There do seem to be actions, like walking for one's health, that perfect the agent. More generally, in *De cælo* Aristotle writes that each thing is "on account of" its operations (286a8). Thomas therefore calls operation the "ultimate perfection" of things (*In de Cælo* 2lect4; *Opera* [Parma] 19:87). But a thing must change if it is to be perfected, and so some actions must change their agents.

Among self-perfecting actions, some are immanent. For them agent and patient coincide, or are parts of the same whole (as when one chooses to

---

53. Fonseca agrees with Scotus, Thomas, and various Thomists that intellection, which is an immanent action of the rational soul, has no *terminus* and is therefore a quality, not an action (Fonseca *In meta.* 7c8q3§3, 3:300aD).

54. Every relation between two things must rest on nonrelational accidents of those things, which are called the *fundamenta* of the relation. My being taller than Tom Thumb rests on my being such and such a height and Tom's being such and such a height. See, for example, Coimbra *In Log.* 1:465.

walk for one's health), or are at most formally distinct. Such actions present no problem. But Aristotle's claim is general: even transeunt actions in some way perfect their agents. Yet that cannot follow from the nature of action itself, or else God, some of whose actions are transeunt, would be perfected, which is impossible.

2. *The agent is impassive.* Toletus agrees that "in general the *actus* of natural agents are also perfections [of those agents], or some sort of way toward their perfection" (4:86vb). The only exceptions he makes are for agents already perfect, like the "perfect craftsman" or God. Yet he has earlier concluded that transeunt action "is in the patient *per se*, but in the agent *per accidens*"—either because it is from the agent, or because the patient reacts. So it seems that without some further explanation, the most one could say is that the agent may be perfected by its actions, not that it will be.

Suárez addresses that point. Transeunt actions of created agents bring only a "certain extrinsic perfection," either because the external things they perfect in turn act on them, or because they preserve themselves in acting (which covers the case of nourishment). The perfection of the agent in acting, then, is properly that of something else, and only denominatively that of the agent (*Disp.* 48§4¶18; on 'denominative', see below).[55] It undergoes no change in acting transeuntly.

The impassivity of Aristotelian agents reinforces the original asymmetry in transitive action: *motus* is in the patient alone, the agent moves but is not itself moved. Here we see a distinctly unmodern element in the Aristotelian theory of efficient causation. In Cartesianism, and in classical physics generally, there is no agent that is *not* acted upon in acting.[56] Or rather: because the ground of that distinction—the fact that the *motus* inheres in the patient alone—is effectively denied, the very ideas of acting and being acting upon lose their grip on the world.

3. *Action, passion, and* motus *are not really distinct.* The third conclusion drawn from the scheme of transitive action is that action, passion, and *motus* not only inhere in the patient alone but are at most formally distinct from each other. Opinions differ about the precise kinds of distinction to be

55. The Coimbrans rescue Thomas's claim by shifting the sense of 'perfection': "That transeunt action also is a perfection of created agents is clear from the fact that, because the greatest embellishment of creatures is to imitate the First Cause, they excel when in acting they communicate something of themselves to others; as Dionysius [. . .] says, the most divine thing of all is to become a co-operant with God [*omnium divinissimum esse Dei cooperatorem fieri*]" (Coimbra *In Phys.* 2c7q14, 1:281; cf. Dionysius *Coel. hier.* 3§2, *Opera* 156: "divinius est omnium, ipsius etiam Dei [. . .] cooperator existat, divinamque in semetipso demonstret operationem, quoad potest, elucentem": "The most divine thing of all is that it should exist as a co-operant with God himself [. . .], and in its very self show forth the shining divine operation").

56. For a survey of action and reaction which includes some Renaissance Aristotelian authors, see Russell 1976.

drawn in the three cases. Suárez, for example, holds, unlike most other authors, that motus and passion are only distinct in reason, or even just in name (Suárez *Disp.* 49§2¶4, *Opera* 26:904). But since a detailed census of views would not, I think, yield further insight into the Aristotelian conception of agency, I will not undertake one.

It suffices to note that the central texts (with the exception of Abra de Raconis) all reject the Scotist contention that "transient action, regarded formally, is the emanation of the form from the agent," and is therefore "received not in the patient, but in the agent." The Scotists' argument is that action differs from passion just by virtue of being a relation of the agent to the patient. But that relation is in the patient, since its *fundamentum* is the *potentia* of the agent (Coimbra *In Phys.* 3c3q1a1, 1:347). In plainer terms: action—many philosophers now would agree—is a relation; that relation holds just because the agent's *potentia* is manifested in the *motus* of the patient. So the relation, even though it is between the agent and the patient, resides in the agent.[57]

To that the Coimbrans reply that to distinguish action from passion it is enough to say that the *motus,* with respect to its source, is action; and with respect to its subject, is passion. For the rest, Toletus's quick way suffices: the ground of the relation of agent to patient is just that the *actus* of the patient is from the agent, not in it (1:349; cf. Suárez *Disp.* 48§2¶20, *Opera* 26:873).

The arguments are abstruse, as often happens when Scotus is in the wings, but they bring to light two points. The first is that action and passion are not relations in the modern sense. A logician now might represent 'Socrates teaches Plato' by the formula '*Tsp*'. Asked what the "action" of Socrates' teaching Plato is, she is likely to reply that it consisted in Socrates' occupying the first slot in the (asymmetric) two-place relation *T*, or, more formally, in his satisfying the open formula '*Txp*', where *x* is a variable-name. The Aristotelian tends to think in reverse fashion. There is a *motus*, a teaching, construed nonrelationally; Socrates' action consists in its being with

57. A relation, contrary to what anyone trained in logic now would think, inheres, like any other accident, in one subject: "A relation is a being [*ens*], whose whole being [*esse*] is with respect to another [*aliud respicere*]." The "peculiar and proper nature of relation, which distinguishes it from other [categories], is constituted by its tending to another as to a *terminus*," while the other categories have their essence in themselves (as substance does) or in a subject (as quality, for example, does). My resemblance to a dog is an accident of me, whose *fundamentum* is, let us say, warm-bloodedness, and the *ratio* of that *fundamentum* is the identity of warm-bloodedness in me and in the dog. Scotus's claim, then, is that what we call 'action' is an accident of the agent whose fundamentum is the active *potentia* whose *actus* is the *motus*, and whose *ratio* is the identity of that *actus* with the *actus* of the passive *potentia* in the patient, which is also the *motus*. No more than anyone else in his period would Scotus have dreamt of calling action a relation in the sense in which the word is now used.

respect to him—the subject in whom the *motus* does *not* inhere—that the *motus* comes to exist.

The second point is that *motus* is the *actus* of the patient quite differently than it is the *actus* of the agent. It can be, and in natural change is, not only an accident inhering in the patient but a perfection of the patient. However obscure Aristotle's definition may be, it is reasonably plain what that definition is about, so long as we consider *motus* with respect to the patient. It is about processes like growth, nourishment, the descent of falling bodies— changes in their respective subjects. But it is not at all clear how a process going on in something else can be thought to be the *actus* of a *potentia* in the agent (or of the agent existing *in potentia*). Suárez at one point says that in acting the agent "expresses or reveals [*declarat*] its perfection" (*Disp.* 48§4¶18, *Opera* 26:893). Though he goes on to say that it does so best in immanent actions or in converting something to its use, the root notion is that of a *manifestation of power.* It is true, we may suppose, that in doing so an agent assimilates itself to God, and thus to the ultimate good. But if that can be sensibly thought of as a kind of becoming, of being-perfected, at all, it is, one would imagine, only tenuously related to the genuine becoming that occurs in the patient.

What holds the two manners of becoming-perfect together, I think, is some notion of unfolding, of coming fully into existence. In the patient that unfolding is instigated, and governed by the ends, of another. Patiency is therefore, however much it may be an expression of some part of one's being, a kind of subordination. In the agent, on the other hand, that unfolding is both spontaneous and autonomous. In successful transeunt action, it amounts to mastery over the patient.

## 2.4. Active and Passive *Potentiæ*

Every change, being transitive, is the *actus* of two *potentia,* one in the agent, one in the patient. Except in self-action, the *potentiæ* belong to distinct things, and are therefore numerically distinct, just as your soul and mine, though one in form, are distinct. But are active and passive distinct *classes* of *potentiæ?* Or is every active *potentia* also passive, and conversely? Since the distinction of matter and form, and with it the classification of natural substances, rests on the distinction between active and passive *potentiæ,* the question is far from idle (see §3.2 and §7.1 below).

It was important also to know whether active and passive included between them all *potentiæ.* From the point of view of the new science of Descartes, the most significant case was *potentia resistendi,* sometimes called *potentia resistiva.* Some Scholastics had argued, for example, that change takes time because the passive *potentia* of the patient to receive an action must overcome a contrary potentia to resist it.[58] The central texts were inclined to show that *potentia resistendi* was either passive *potentia* under another name or else not really a *potentia* at all. The basic pair suffice: the distinction of active and passive is, *if not disjoint,* at least exhaustive.

In general the difference between the two classes was uncontroversial. The Metaphysics simply defines active *potentia* as the "principle of change in another, as other," passive *potentia* as the "principle of being changed by another, as other" (9c2,1046a11$^{ff}$). The only difficulty is the phrase 'as other'. Aristotle added it, according to Thomas and many others, to cover the case of self-action, in which the agent and patient, though numerically one, are, under the designations 'agent' and 'patient', as if distinct.[59] It seems to have been obvious that the classes so defined are distinct. We have seen some examples from the *Physics.* In *De generatione et corruptione* the elemental qualities hot and cold are active, while dry and wet are passive (2c2, 329b24). *De anima* argues, among other instances, that sensation is passive, while the faculty of movement is active.

Suárez, nevertheless, considers the question. Since action and passion are not really distinct, and since agent and patient sometimes coincide, one might well wonder if active and passive *potentia* are really distinct classes. Suárez holds that they are. The argument has two parts. In one Suárez considers whether the classes are coextensive; in the other whether each *potentia* taken by itself is both active and passive. I will detail only the first.

There are, he argues, purely active, but no purely passive, *potentiæ.* The class of active powers properly includes the class of passive *potentiæ.* Experience shows that some *potentiæ* always act transeuntly and do not intrinsically require the action of anything else in order to act (*Disp.* 43§2¶1). Among them are the attractive power of magnets and the explosive power of hot air. Since such *potentiæ* do not undergo their own action, they need not be supposed passive for that reason. Nor does their action presuppose that they themselves are acted on by something else. Skills too are purely active, since once learned, they need not be acted upon in order to act. There is no

---

58. Maier cites a formulation of the principle in William of Auvergne: "[. . .] every operation of this sort [i.e., that takes time] occurs through conflict and a victory of the agent over the patient and of the efficient power [*virtus*] over its contrary [. . .]" (Maier 1955:227, n.1).

59. See Thomas *In Meta.* 9lect1, §1776–1777. Averroes reads the passage differently: Aristotle adds 'as other' to the definition of 'active *potentia*' because "it is manifest that nothing acts on itself" (*In Meta.* 9c2, p227ra[B]). Although it would be out of place to argue the point, Thomas's reading is preferable, since Aristotle explicitly says of animate things that they move themselves (e.g., at *Phys.* 8c4, 255a6).

reason to suppose that such a *potentia* is other than active, since it is changed neither by itself nor by another.

So there are purely active *potentiæ*. What of passive *potentiæ*? Suárez agrees that if *potentia* is taken in the broad sense (*Disp.* 43§3¶2), there are purely passive *potentiæ*. The *potentia* of matter to receive form is passive. But among *potentia* in the strict sense, every passive *potentia* is also active. Suárez's strategy, in brief, is to insist on the definition of *potentia*.[60] A passive *potentia* is not merely any accident that enables a thing to receive a form. It must be "ordered" or "instituted" to that form, and Suárez interprets that strictly. Wetness and dryness are said to be passive (*De gen. et corr.* 2c2, 329b24–25). But they are passive *qualities, not passive potentiæ:* "first because [Aristotle] holds them to be [also] active qualities, although not as efficacious as cold and hot, in comparison with which they are said to be "passive." [. . .] Finally because dryness and wetness are not the sort of qualities that in themselves receive other qualities or forms by which they are actualized—which is the character [*ratio*] proper to passive *potentiæ*—but instead dispose matter for the reception of substantial form, or also other accidents, and according to that reason [*sub ea ratione*] are called not passive *potentiæ* but passive qualities, belonging to the third species" (*Disp.* 43§2¶5, 26:639).[61] What matters here is that although any accident by which a thing is enabled to undergo the action of another is, in the broad sense, a passive *potentia*, the proper sense is much more restrictive. The only unequivocal instances admitted by Suárez are psychological: sensation and the passive intellect. But even these are not *purely* passive, since every "vital" power includes some degree of activity. All putative instances are passive, because they are not ordered by their nature to the reception of a *particular* form, referred to other species of quality, or—in the case of quantity—to other categories.

In the *Metaphysics* and (for some interpreters) the *Physics,* a simple picture is suggested: for each active *potentia* there is a corresponding passive *potentia*.[62] René Le Bossu, the would-be conciliator of Aristotle and Descartes,

60. "A *potentia* is the proximate principle of some operation, to which by its nature it is instituted and ordered" (*Disp.* 43§3¶2; 26:644). Cf. n.20.

61. According to Suárez, passive qualities, the third species of quality in the *Categories*, are called "passive" not because they are principles by which a thing undergoes the action of another, or receives its form, but primarily because they are forms of a sort which are received through passions and because they make a thing receptive to various actions (*Disp.* 42§4¶14). Toletus's account is somewhat different: qualities of the third species are called "passive" either because they cause passions (and are therefore active powers) or because they are the result of passions (*In Log., Prædicamenta* 8, *Opera* 2:156b–157a). What matters here is that if they are qualities of the third species, then their passivity is incidental to their nature. The four species of quality are discussed below (§4.3).

62. Thomas, referring to *Meta.* 9c1 (lect2), writes that "an active *potentia* would be in vain if no passive *potentia* corresponded to it" (*De potentia Dei* 1art2, p5). Since no natural active *potentia* could be in vain, each must have a corresponding passive *potentia*.

writes for example that "if a cannonball has the Active Power to overturn a wall, the wall has likewise the Passive Power to be overturned [. . .] So that Movement is equally the Act of a Being that has an Active Power, and of a being that has a Passive Power" (*Parallèle* c16, p214). Late Aristotelianism does not in fact have so simple a view, and indeed would deny that being able to be overturned by violent motion is a passive power. Suárez, arguing that not every active power has a corresponding passive power, concludes that "it is not necessary that, [even] if active *potentia* were a quality instituted *per se* to act, passive potentia would also be a quality instituted *per se* and first of all to receive [form]. Heat, as an instrumental virtue of fire, acts immediately in the substantial *potentia* of prime matter [i.e., its *potentia* in the broad sense to receive form], drawing out a substantial form from its *potentia;* but as the principal virtue of heating, it acts on the body insofar as the body is affected with quantity [i.e., it acts on the body just because the body has length, width, and breadth], and no other peculiar passive *potentia* corresponds to it" (*Disp.* 43§2¶12, 26:641). Quantity, which is not ordered toward any particular quality and which is always present in corporeal substance, suffices for a thing to be *in potentia* hot. More generally, passive powers, ordered by their nature to the reception of a form, are supplanted by equivocal *dispositiones,* which, though "proportioned" to the forms whose reception they enable, are not defined solely in relation to them. To that degree, matter—by which I mean all that is needed for the reception of form—becomes plastic, defined less in terms of ends, more in terms of "material" qualities lacking teleological implications.

'Active' and 'passive', then, are not contrary. Far from it: when *potentia* is taken narrowly, 'active' includes 'passive'. But do the two exhaust the field? Experience shows that bodies do not always change readily. Every medium resists more or less the motion of things through it (*Phys.* 4c8, 215a25[ff]). Materials variously resist change of shape (*De gen. et corr.* 1c10, 328a36–b5; Zabarella *De rebus nat.* 436D). Air resists the action of fire (Soncinas *Q. meta.* 9q6, p234b). Some philosophers inferred that there is, in addition to the *potentia passiva* or the *dispositio* which assists and receives the action of an agent, a *potentia resistendi* opposing it.[63]

One argument for distinguishing *potentia resistendi* from active and passive potentia was that resistance to *motus* was neither action nor passion.[64] It is

63. Neither Soncinas nor Suárez mentions any names. Zabarella cites Pomponazzi (*De reactione* 5, in *De rebus nat.* p436D). The most important case was the resistance of a medium to local motion, since it figured in Aristotle's rules for the comparison of motions (a general discussion is found in Maier 1955, c.4; for the views of Albert of Saxony and others in the mid–fourteenth-century Paris School, see also Sarnowsky 1989:211[ff]; for those of Mutius Vitelleschi, a Jesuit professor at the Collegio Romano in the 1590s, see Wallace 1981:116–120).

64. Soncinas *Q. meta.* 9q6, secunda ratio; p234.

contrary to both.[65] Soncinas's answer is to argue that "being active and being resistive are relative, since they are said in relation to something, [and] nothing is active or resistive in itself'. So coldness in a patient, insofar as it lessens the heat of something acting on it, is active; but insofar as it retards the action of the heat, is resistive. Zabarella rejects that answer. His proposal is that each form not only acts on its contrary (as heat on cold, and cold on heat) but strives to conserve itself. It impedes the action of whatever would change it, where 'impede' is understood simply to denote a diminution of that action rather than a contrary action. Diminution is not a positive thing; it is a "privation," a nonbeing. Privations do not, in general, require distinct causes, and the self-conserving *nisus* of forms is thus not a *potentia* (Zabarella *De rebus nat.* p440E).

Suárez likewise rejects Soncinas's answer. With Soncinas he notes that one thing may react against the action of another, thereby diminishing the active power of the other. Resistance, properly speaking, is not, however, reaction. As Zabarella says, it is simply a diminution or retardation of the agent's action, and thus mere privation. The reason for it is that any form refuses the forms and actions contrary to itself (*formæ et actioni sibi contrariæ formaliter repugnat; Disp.* 43§2¶10, 26:636). Hence even relatively passive qualities like dryness resist being turned into their contraries.

Soncinas makes *potentia resistendi* a respect under which active power can be considered. This is of course one common way to trim unwanted entities. We have seen it at work already in the treatment of action and passion. But it fails, as Suárez quickly notes, to account for the generality of the phenomenon. Whiteness is never active (no color is); yet whiteness too (so Suárez believes) resists the introduction of its contrary. Suárez's own proposal, oddly enough, makes of what would seem the merely logical feature of *repugnantia* (contrariness or contradictoriness) something that, even if only negatively, functions to alter the course of natural change. He himself recognizes that resistance so conceived is not easily seen to admit of degrees. Yet experience confirms that it does. Though he has answers to that objection, a modern reader will undoubtedly feel more comfortable with Zabarella's proposal. That proposal, with due allowance for a new notion of form, could easily be interpreted as Descartes's first law of motion. Zabarella, however, speaks of the form in the thing as "striving" to conserve

65. One might wonder why *potentia resistendi* is not just an active *potentia* in the patient reacting against the agent. The refutation of that view cited by Soncinas rests on the single example of dryness. Air resists the action of fire by virtue of dryness (its only other elemental quality is heat, which clearly will not resist the further heat of the fire). If dryness were active, it would drive out wetness, being contrary to it. Yet a dry piece of iron, if cold, will not make a wet thing dry. Only if heated will it do so. Heat is, therefore, the active power that drives out wetness, and dryness is not active (*Q. meta.* p234). Soncinas's response, considered in the text above, is to make 'active' a relative rather than an absolute feature of qualities.

itself.[66] In Descartes's world there is no such striving. The *tendance à mouvoir* is, as we will see, defined so as to avoid the appearance of finality.[67]

In the basic Aristotelian scheme, every change comprises an agent and a patient, with the patient receiving, upon the completion of the change, a form from the agent. That picture remains unchallenged. But the corresponding distinction of active and passive *potentia* becomes more problematic. It is too simple to suppose that for every active *potentia* there is a corresponding passive *potentia*. If by 'passive *potentia*' one means trivially that the patient is in some way fit to be acted on, then indeed the correspondence holds. But that would be to overlook the difference Suárez insists on: a passive *potentia* is a quality "instituted by its nature" to receive a particular kind of action. That notion is not trivial. We have seen that passive *potentia* so defined tends to be displaced by qualities whose definition does not refer specifically to the actions they make a thing fit to receive; and by the striving after self-conservation, which, though neither *potentia* nor quality, nevertheless has consequences.

The Aristotelian theory of *motus* includes not one but two kinds of directedness. The first points from *potentia* to *actus*. In intransitive change, the actualization of a *potentia* tends toward completion. Even if frustrated, it points, like an undelivered postcard, to a destination, to its end. This kind of directedness is of course well known. The second points from agent to patient. Aristotelianism, like common sense, sees the arrow hit the target and not a collision of bodies that are equal partners to the event. The agent's action, identified with the *motus* in the patient, ceases with the completion of the form that the agent donates to the patient. Active powers, just because in them that form, belonging to another, is implicated, presuppose the other; and since (see §5.3) the patient must be suitably disposed to receive the form, active powers presuppose such patients as well.[68] The

66. In *De Anima* 2c4, Aristotle argues that every animate form, at least, participates in "divine and immortal being" so far as it can. Hence "all desire it, and on account of it all things do whatever they do, according to nature" (415b1$^f$). Thomas explains that "just as each thing [. . .] when it is in *potentia*, is ordered to the *actus*, and naturally desires it, and when it is in a less perfect *actus* desires a more perfect: so each thing that belongs to an inferior rank [*gradus*] desires to be assimilated to the superior ranks, so far as it can be" (*In de An.* 2lect7; Pirotta ¶315, Leonina 45/1:97). Such explanations make it clear that in general the self-conservation of forms cannot be regarded as an anticipation of inertia.

67. In an early manuscript of Galileo, resistance is defined simply as "permanence in a proper state against a contrary action"; he adds that "I do not differentiate resistance from the thing's very existence whereby it endures." On the other hand, he argues that resistance always has a cause, and that just as animals guard themselves "through appropriate action," so too "even qualities and similar things have a certain natural property by which they more or less resist a contrary" (Galileo *Early notebooks* 244; *Opere* 1:171; the translation is Wallace's).

68. Mark Smith writes concerning vision: "The object itself cannot possibly express its visible nature without the physical intervention of a transparent medium, which, lacking

directedness of transitive change, or better, of active powers, is less often recognized than the directedness of intransitive change. Yet it is one tie by which different natural kinds are bound together—luminous bodies and the medium that receives their light, animals and the stuffs that nourish them, humanity and nature.

In Cartesian physics, both connections are refused. The only glue that *could* hold together successive states of a particle is the persistence implicit in the first law of motion. But the ground of that persistence, which is the fixity of the divine will, is extrinsic to Nature, and thus to the successive states. The possibility of distinguishing agents from patients likewise falls away, and with it the bond between them. With a change of reference frame, the arrow becomes stationary and the target a projectile. Who hits whom is thus a matter of convenience. The distinction between agent and patient, and of active from passive powers, was, as I will argue shortly, the basis in Aristotelianism for the classification of natural kinds. Without that distinction, it would seem, the very idea of natural kind is in jeopardy. What replaced active power in classification was the notion of structure.

---

transparency, could not take on the color that causes visibility" (1981:576). If one believes, as the Aristotelians did, that no active power exists in vain—which is to say without acting (n.62)—then the colored thing's having the power it has entails the existence of the means by which to exercise those powers.

# [ 3 ]

# Form, Privation, and Substance

Few notions met with such uniform rejection among major seventeenth-century philosophers as that of substantial form. Descartes was among the more categorical. In *Le Monde* he writes that "all the forms of inanimate bodies" can be explained "without the need of supposing any other thing in their matter than movement, magnitude, figure, and the arrangement of their parts."[1] Merely by supposing, he wrote some years later, that the insensible particles of which all bodies are composed have the very same properties we perceive in larger bodies, one can do without "prime matter, substantial forms, and all the great baggage of qualities that some philosophers have been accustomed to supposing."[2] Descartes was not the first to declare that forms were superfluous. Others had already issued similar challenges to Aristotelianism. Most of the seventeenth-century philosophers who cast their lot with the new science concurred in banishing substantial forms from natural philosophy.[3] Their efforts met, as everyone knows, with great success. The notion of a mechanism now seems obvious and indispensable; that of substantial form obscure and gratuitous.

Two things stand out when the charge and the facts of the case are compared. Those who objected to substantial form seem frequently to have misunderstood it. It was often held that the very idea of substantial form was incoherent. No form could be a substance, because substances are self-subsistent and form is not. Yet the Aristotelians were careful to explicate 'substance' as applied to form in such a way as to avoid that inconsistency. The misreading was abetted by Aristotelian affirmations about the human

1. AT 11:26; Alq.1:338.
2. *PP* 4¶201; AT 9/2:319–320 (added in the French edition).
3. Leibniz is the great exception (*Discours* §10).

[53]

soul, which was indeed a self-subsistent material form. There were, moreover, among writers outside Aristotelianism, uses of the term 'form' that
aligned it with spirit, making it entirely distinct from matter (see §3.2). The
temptation, in the later seventeenth century, to conceive the relation of
substantial form to matter on the analogy of the relation of Cartesian soul to
Cartesian body was strong. Strong but misguided. So little were substantial
forms in nature capable of separate existence that delicate arguments had
to be brought forward to demonstrate that the human soul was, alone
among them, an exception.

   The second noteworthy fact about the charge is that certain of the problems that substantial form was intended to deal with were handled no more
successfully—and sometimes with less success—than they had been by the
Aristotelians. Chief among them were those of the unity of individuals,
especially of animals, and of defining natural kinds.[4] Showing what makes
an organism *one* was, as recent studies have shown, a primary motivation for
Aristotle's theory of form.[5] Descartes saw that the unity of the human body,
at least, needed an account. But it cannot be said that his view is satisfactory.
As for the unity of animals and plants, he seems hardly to have noticed the
problem, as Leibniz soon pointed out. Nor did he have anything like a
taxonomy of plant and animal kinds.

   A related question was that of the unity of active powers, or—more
neutrally—of essential properties, in a natural kind. Boyle, more perspicuously than Descartes, recognized the difficulty. His solution was to
reject the commonsense view of kinds.[6] He was led thereby to reject yet
another distinction that substantial form had been invoked to explain.
Some change, as will be explained shortly, was held to be *substantial,* and its
result a substitution of one kind for another. Change not so radical was
called *accidental.* Boyle, and Descartes at least by implication, denied the
distinction. With it went one of the more commonly adduced arguments
against corpuscular and atomist physics.

   One last question was that of preferred states. Water is by nature cold, the
Aristotelians believed, and thus will return to its natural state when a source

---

4. Emerton notes, citing Boyle, that some mechanists appealed to a "reinterpreted" notion
of form to supply a "cause of the specificity of bodies." Boyle, who was not alone in this,
admitted an internal "plastic" principle "implanted by the most wise Creator in certain parcels
of matter, that does produce in such concretions as well the hard consistence as the determinate figure" (Emerton 1984:72, 73; cf. Boyle *Works* 1:275–276). 'Hard consistence', or impenetrability, was soon recognized as not comprehended under the Cartesian conception of
matter; 'determinate figure' refers in particular to the regular shapes of crystals. Gassendi had
earlier concluded that "gems seem in a special way to come from a seed" (Emerton 1984:43–
44).

5. See the studies collected in Gotthelf 1985 and Gotthelf & Lennox 1987, as well as Furth
1988, esp. §§11, 15.

6. See Boyle *Papers* 61ᶠᶠ.

of heat is withdrawn from it. Similar arguments could be made concerning the natural places of the elements. Again the short way was to deny the phenomena, as Boyle did. But the question was much deeper than they, or indeed the Aristotelians, recognized. The cooling of hot water to room temperature cannot be wholly explained without the second law of thermodynamics. In many such instances the tools of seventeenth-century physics fell short. Nevertheless substantial form, at least in name, was not kept to explain them.

I will first consider the general question of matter and form, and then substantial form in particular. In *Physics* I, Aristotle argues that the principles of change are form, matter, and privation. The argument, a familiar one, I will call the "substrate argument." What distinguishes Aristotle's version is the notion of privation, and two fundamental themes: that the mutable properties of things fall naturally into genera whose species are contrary to one another, and thus fit to change into one another; that privation of form can be distinguished from its mere negation precisely as a condition tending to it. It is thus that the description of the *termini* of change joins with the definition of *motus* already examined.

The introduction of substantial form and prime matter into natural philosophy was not, whatever later philosophers cared to say, gratuitous. I will examine a number of empirical arguments on its behalf. They center on the conditions under which change is reversible or not, and more fundamentally on the *common fate* of certain accidental properties conjoined in individual substances. Supposing, for the moment, that in material things substantial form corresponds to essence or nature, the arguments indicate that to argue for essences it is not enough to argue for essential properties. Only if essential properties have a common ground or *ratio* in a single form do they comprise an essence.

I then turn to the logical objection, according to which 'substantial form', denoting an item at once subsistent in itself and dependent on another, is an incoherent notion. The incomprehension implied in that criticism stems, I suggest, from the suppression of the 'is' of specification or of *informatio* in favor of the 'is' of identity and the 'is' of inherence, a suppression linked in Descartes's thought with the rejection of substantial forms.


## 3.1. Principles of Change

The task of describing change, as I said, can be divided into two parts. One part concerns change in itself. It issues in the definition of *motus* and the explication of related notions. The other concerns the difference between what was before and what now is. Wherever we speak of change—

understood intransitively—there is something that remains the same and can be reidentified after the change, and something that differs. The question, then, is how to conceive that identity and difference. Aristotle's well-known answer was to regard the thing that changes as a composite of matter and form. These, together with privation, are the principles of change.

The argument of *Physics* I begins, like many of Aristotle's arguments, with a survey of the opinions of his predecessors. For all of them, Aristotle concludes, change is between contraries. The atomists have their solid and empty, their angular and nonangular, Empedocles his love and strife, and so for the rest (1c5, 188a9ᶠᶠ). The conclusion can be argued empirically as well. Not just anything acts on, or is acted on by, or comes from anything. The white comes from the nonwhite, not the cultured (*musicus*), save by accident; it comes, moreover, not just from any nonwhite, but from black or some other color.

One step in the argument deserves notice. Few would deny that when a thing becomes white, what had not been white is now white. Now a cultured thing may be either white or not. If it is not, then—supposing it to be colored at all—it must have been *nonwhite,* which is to say, a color which is not white and which cannot coexist with it. To be cultured is not a way of being colored, even if every cultured thing is colored, and even if being cultured is, like being white, a quality. The argument presupposes that accidents can be arranged into natural groups, or genera, such that within each group the members, or species, are contrary to one another. Color is one such group, temperature another.[7]

It is not easy, however, to see how Aristotle can make his point without begging the question. One might suppose that it would be sufficient to have noticed that black and white never occur together in the same individual. But of course there are many collocations of properties we may never come across. Some simply are rare. Others, more significantly, are ruled out because of regularities holding among properties. No ruby is green; yet 'ruby' and 'green' are not contraries.

The treatment of contraries in the *Metaphysics* does not resolve the

---

7. Since Aristotle speaks of white as coming from "black or from intermediate [colors]," 'contrary' is here used in the broad sense in which any two colors may be said to be contrary. In a stricter sense, only those species in the same genus that are "maximally distant" are contraries (*Meta.* 10c4, 1055a5ᶠ). The terms *genus* and *species*, more commonly used of substantial than of accidental forms, are here used of qualities. Species, I should note, are produced out of genera by differentia. Furth argues that "the original, 'logical' concept of differentia is basically that of a 'difference from'," and thus that of a "contrariety or opposition" (Furth 1988:101, citing *Meta.* 4c2, 1004a21, which in Moerbeke's translation reads "Differentia namque quædam contrarietas est, et differentia diversitas"—cf. Thomas *In Meta.* ad loc.). The *differentia* by which species are distinguished within a genus are contraries, as for example 'two-footed' and 'four-footed' (on *differentia* in biology, see Pellegrin 1987:320–322).

difficulty (see 10c4, 1055a7–9). Suárez, referring to this passage, concludes that "between contraries there is a certain order *per se* ordered [*quidam ordo per se in ordine*] toward transmutation; from the nature of the thing there occurs a transition from one to the other and conversely" (*Disp.* 45§2¶5, 26:742). Contraries are, in other words, defined in terms of change, which must be understood to be nonarbitrary in just the way Aristotle is arguing for in the *Physics* passage cited above.

Rather than fault Aristotle for circularity, it would be better to conclude that the notion of 'contrary' and that of natural change are nourished from the same source. That source is the common experience of particular kinds of alteration, whether from hot to cold, from wet to dry, or—less commonly, though the example is often used—from black to white.[8] There is no explanation for the fact that properties can be sifted out into genera the species of which are transmutable into one another.[9] There is only the testimony of cases, and abstraction from them to the notions of genus, species, and contrary. Just as in the definition of motus Aristotle seeks not an explanation but a perspicuous scheme under which to bring together acknowledged phenomena, here too the claim that change is between contraries provides not an explanation or an analysis but a scheme.

By 'scheme' I mean a general form into which the phenomena must be cast so as to be suitable for explanation. The definition can be interpreted as containing schematic letters, to be filled for each genus: *X*-ing is the *actus* of the *X*-able qua *X*-able. The result is a perspicuous definition of *X*-ing. A remark by Aristotle not long after the definition of *motus* exemplifies what I have in mind: "What *motus* is has been told in general and in detail: for it is clear how each of its kinds is to be defined; alteration is the *actus* of the alterable, insofar as it is alterable" (*Phys.* 202b24). Toletus adds, "and augmentation is the *actus* of the augmentable, as such: and so for the rest" (*In Phys.* 3c3text23, *Opera* 4:82va). So Aristotle defines light as the actus of the transparent qua transparent (*De anima* 2c7, 418b9).[10] Since 'light' here

8. For the moment I am leaving generation and corruption aside, since they present special problems.

9. The case of figure, which seems to fit the criteria I have listed, and yet is said not to undergo alteration, will be taken up in §4.3. The general account given here undergoes certain restrictions in *Physics* 7, where Aristotle argues that only the so-called passive qualities of the third species (sensible and elemental qualities) are alterable per se. Others are either inalterable (virtue and vice), or else alterable only as a consequence of alteration in passive qualities (245b3–9, 246b2–20, etc.). In the latter case, it seems to me, the scheme still applies. Aristotle is not denying that there are *motus* of color or illumination; he is, rather, attempting to show that such *motus* have other, more fundamental, *motus* as their principle.

10. The word here translated 'actus' is ἐυέργεια, not ἐυτελέχεια, but the distinction made elsewhere between these terms is not crucial to my point. Hicks translates it as 'actuality' (*De anima*, Hicks 1907:79); Moerbeke's Latin translation has 'actus' (see Thomas *In de An.* 2lect14, Leonina 45/1:123).

means 'illumination', and 'transparent' means 'illuminable', the definition is simply a version of the scheme presented in the *Physics*.[11]

Something similar holds for the proposition that change is between contraries. Setting aside certain complications that I will come to later, the scheme suggested is this: in any natural change, at least, when a subject *S*, having been *A*, *becomes B*, the forms *A* and *B* must be so described as to be species of the same genus. Only then will the change be amenable to explanation. A perspicuous description of the transmutation of earth into fire, as we learn in *De generatione et corruptione*, is that it is the change of the cold and dry into the hot and dry. 'Earth' and 'fire' themselves are in that respect opaque, one might say nonnatural, descriptions of their referents.

So in change one contrary "expels" another. Yet something remains: although "the not-cultured or the uncultured do not subsist as simple or as united to their subject," "the man subsists when he becomes cultured and he is still a man" (*Phys.* 190a11$^f$). There are two cases to consider, corresponding to substance and accident:

> According to the diverse things that come to be, different subjects must be set up: For in accidental [change] we suppose a first substance, because that is the thing in which all accidents inhere, and which itself inheres in nothing.[12]

> But in substantial [change], in which substances themselves change [. . .] some other subject must be established that persists in the change, which cannot be unless we say the substances themselves are composites of matter and form, and the matter is the subject which according to the various forms of substance is changed. (*In Phys.* 1c7text62, 4:31va)

The argument continues with an induction: the generation of plants and animals is from seed, and in general substances come about by transfiguration, by accretion, by subtraction, by composition, and finally by alteration "according to matter" (*Phys.* 190b9). In each of these processes some material is presupposed, and persists through the change. Supposing that the list is complete, it follows that all change presupposes matter.[13]

The correlative to matter, form, is introduced by way of the notion of *definition*. The subject of change is "not distinct numerically from itself, but

11. Hussey notes that some ancient commentators treated the definition of *motus* as a "definition-schema, to be filled out differently for each type of change" (Hussey 1983:60).

12. This is the definition of 'first substance' in the *Categories* (c5, 2a11$^{ff}$ and 3a7$^{ff}$). A first substance is an individual concrete thing, a 'this something' (*hoc aliquid*, τόδε τι); a second substance is a genus or species (2a14$^{ff}$).

13. Clearly "altered according to matter" (*secundum materiæ*, κατὰ ὕλην) is a wild card, especially since Aristotle argues in *De generatione et corruptione* that the elements themselves can be transformed into each other. Matter, as the subject of substantial change, was supposed to remain fixed during change. See §4.1 below.

distinct according to its *ratio* and definition." When the uncultured becomes cultured, "'the uncultured' denotes the subject together with a form, or a privation, and the *ratio* of one [i.e., the subject] remains, while the *ratio* of the other [i.e. the unculturedness of the subject] goes away" (Toletus *In Phys.* 1c7text60; cf. *Phys.* 190a14–16). The *ratio* of the subject is the human, that of the form or privation is the uncultured. Those *rationes* are evidently different (*Phys.* 190a16), even though their subject is one. Hence matter and form are distinct, at least in reason.

The argument—call it the substrate argument—is familiar, its conclusion almost anodyne. Few natural philosophers would have disagreed with it. The alternative, Suárez argues, is to suppose that in each change there is an annihilation and a *creatio ex nihilo*, which is not only extravagant but also leaves no reason to explain what is produced: "Here also [in generation] it is necessary that a common subject persist, since otherwise the entire preceding alteration, or the heating of the oakum [*stupa*], would be irrelevant to the procreation of the fire, because it would in no way contribute to the fire, if it and its whole subject perished" (*Disp.* 13§1¶6, 25:396). Nor would there be any reason why the destruction of the fuel should be necessary to the creation of the fire, and vice versa (ib. ¶9; 397).[14] At most one could suppose that the fuel was destroyed to make room for the fire; but why it should perish instead of moving out of the way, a much less drastic response, would still be a mystery. To conceive of change as annihilation and creation, while it yields no inconsistency, is in effect to make physics impossible, or relegate it to a study of phenomena.

In the sixteenth and seventeenth centuries, at least, the substrate argument was unchallenged. Some sort of contrast between form and matter was standard.[15] Descartes uses the term to denote the figures of his three elements, figures that are contrasted with the underlying extended substance of which they are modes. For Boyle, the 'form' of a material particle is "an essential modification and, as it were, the *stamp* of its matter," specified as size, shape, motion, and "contexture" (*Papers*, "Origin of forms and qualities," 69). Yet despite the coincidence of terms, Descartes and Boyle

---

14. It should be noted that these are physical arguments. God could, according to his absolute power, destroy and recreate the world at each moment, while making it appear as if the regularities of this world still held. It is, in other words, not logically impossible that change could consist in annihilation and creation; but to suppose so is not only extravagant but renders change inexplicable, or at least requires us always to limit our assertions to the phenomena.

15. See Emerton 1984, c.2. 'Form' was used to denote both "the cause of the specificity of bodies," or structure, and "the physical mode in which order is established in matter," or formative cause (72). Though both uses are found among Aristotelians, the first is primary. Form as formative cause, *semen*, or vital spirit, owes more to the various forms of Platonism (59–65, citing Daniel Sennert and Sebastian Basso). But even if we exclude the latter, there are still a great many non-Aristotelian versions of form.

thought they were contradicting the Schoolmen. What, then, is the difference?

It is with the third of Aristotle's three principles—*privation*—that the Aristotelian twist on the substrate argument appears. The unculturedness of what becomes cultured, the blackness of what becomes white, are not the mere negation of being cultured or white, but their contraries.[16] Unculturedness in a person who is fit to become cultured is indeed the absence of culture. But trees and marmosets lack culture too. As Toletus puts it, "privation means not only the negation of a form but [its negation] in a certain subject [. . .] Privation supposes not only a subject, but an aptitude in the subject to the form, of which it is the privation, so that the negation of the form in a subject apt for it is privation. (*In Phys.* 1q21, 4:42va). Suárez explains that privation, unlike form and matter, is a principle "not by a positive influx [of its being] but only according to a necessary *habitus per se* toward the other" (*Disp.* 12§1¶6, 25:374). It is in one respect more general than contrariety. The stuff of which a human is made does not have a form contrary to the human form. Though of course many things are nonhuman, the human form, like all substantial forms, has no contrary. But it is supposed to be the sort of stuff that is fit to become part of a human. In that rather weak sense, it has the "privation" of humanness. Privation thus covers not only contraries, which figure in alteration and other *motus,* but in a certain way the matter-form distinction itself, which figures only in *mutatio* or generation.

Like *potentia* it includes an essential though unspecific reference to something other than itself—its completion or end. In the previous section the result of change was said to be the *actus* of a *potentia.* That *potentia* we may now identify with the privation of the form received in change, and the *actus* with the form itself. In alteration, augmentation, and local motion the *potentia* is a form contrary to the *actus.* In generation it is a disposition or *habitus* of the matter whose actus is a new substantial form. As the slogan has it, form is the *actus* of matter.[17]

Those identifications, however, signal certain complexities in the notion of directedness that were left unexplored in §2.1. Though each *potentia* must somehow be defined in terms of the corresponding *actus,* that *actus*

16. "Since an object's lack can be rectified by means of a change, the privation is not just any state other than the goal but one on a path leading to it" (Gill 1991:247). The notion of a 'path' leading to the *terminus* must be understood quite broadly. It is rather the agent that determines the path than the patient, as will be seen shortly.

17. The slogan applies, of course, only to forms that can exist only as part of a composite substance. To distinguish them from the forms of angels and God, such forms were sometimes called "material forms." Perez-Ramos calls the notion "lexically self-contradictory" (1988:74). It would be self-contradictory if it meant 'form that is also matter'. But it does not. It means 'form of the sort that must be joined with matter to exist', or 'form that is an *actus* of matter'. In that sense matter and form are not contrary but complementary.

cannot simply be called its end, nor can the relation between them be described simply as a tendency of the *potentia* to a particular *actus*.

1. In alteration the hot comes from its contrary, the cold. Yet it cannot be that the cold tends to the hot, since then the cold elements earth and water would tend to become the hot elements fire and air. Hence to describe the cold as the hot *in potentia* means simply that the cold and the hot, being extremes of the same genus, are suited thereby to become one another. That is a much weaker directedness than one sees in the paradigmatic instance of acorn and oak. It is, all the same, not quite trivial, since it presupposes, as we have seen, that qualities fall naturally into genera.

2. In substantial change, the matter that is about to receive a new form does not lack a form. Matter cannot naturally exist without a substantial form. Hence the same matter must be capable of receiving more than one form, since it loses one and takes on another. Prime matter, the substrate of substantial form, must be capable of receiving all forms. Its *potentia*, then, far from being directed to any one form, is directed indifferently to all.

3. Finally, in certain instances of accidental change the *potentia* of the patient exhibits a similar indifference. The same youth is *in potentia* skilled at grammar and at soldiering. Each skill or *habitus* is an *actus* of the rational soul, which is indifferently *in potentia* to them all. There may be differences in aptitude among pupils, but qua humans they are all equally fit to learn.

These remarks all come from the side of the patient. Passive *potentiæ* vary enormously in the specificity of their corresponding *actus*.[18] Active *potentiæ* in general do not. The *actus* of the teacher who exercises her power to teach grammar is precisely the grammatical skill that her pupil acquires. The active power of the semen has precisely the end of introducing the form of the father into the matter provided by the mother. The activity of heat is just to produce more heat. Unlike passive *potentiæ*, active *potentiæ* are, with one important qualification, unambiguously specified in terms of their *actus*. The qualification concerns so-called rational powers, which are precisely those that can act in more than one way. The *vis motiva* of animals can bring about local motions of all sorts in their parts; the will can produce volitions to any action. Since rational powers, as their name implies, are manifested only in animate things, the study of them belongs to psychology. In physics, active powers are univocal. It is their job, so to speak, to disambiguate privations.

18. 'Passive *potentia*' is here used in the broad sense of 'any principle by which a thing is changed', not the narrow sense in which it is a species of quality (§2.3).

A finished intransitive change, then, can be described either as the complete *actus* of a *potentia* or as the fulfillment of a privation by a suitable form. Transitive change is a joint *actus* of agent and patient (§2.2). Although the privation belongs to the patient alone, the form, since it is the *actus* of both agent and patient, must pertain to both. The Aristotelians say that the form is received *in the patient from* the agent. Yet it cannot always be straightforwardly identified with any prior form in the agent. When water is heated by fire, the form of heat—or better a form, more or less intense—is received by the water from the fire. The elemental heat of the fire is the active power by which it acts. That same active power, when brought into contact with wood, not only heats the wood but consumes it, making more fire, which in turn has the same power to produce more. But water, however much heat it receives, never acquires the power to consume wood so long as it remains water. Hence the heat received by the water cannot be quite the same as the heat in the fire. How, then, does the form received by a patient differ from the form identified as that by which the agent gives or imposes that form on the patient?

The answer seems to vary with the agent and the form. In the example just offered, the answer given by Zabarella is in effect that the elemental heat of the fire and the heat of the water differ only in intensity.[19] But in the transmission of colors, the "intentional species" received from colored things by the senses are different in kind from the colors in the things themselves, since they lack even the feeble activity of colors. It is in general not true that the form received from the agent is the same as that by which it acts; nor is it true without qualification that the two are similar.

What can definitely be affirmed is that, despite the language of giving and receiving, nothing is literally transmitted from agent to patient. The form that is said to be "received" is actually "educed" by the agent from the *potentia* or the privation of that form in the patient (see §5.4). 'Educe' means nothing more than that the agent initiates and sustains the *motus* that, as we have seen, inheres in the patient alone. The formal cause of change is always intrinsic to the patient; the role of the efficient cause is not to impose a form on the patient from outside (as Aristotle's own occasional analogy of wax and stamp, or statue and mold, might lead one to think), but to determine just how a certain mode of being[20] in the patient, as yet

---

19. "The natures and forces [*vires*] proper to these qualities [sc., the four elemental qualities] are the same in all things; for every [instance of] heat warms, and disperses heterogeneous things, and gathers together homogeneous things, and melts and thins congealed things, without any difference except that which is according to the more and the less" ("De qualitatibus elementaribus" 2c6, *De rebus nat.* 531C). Toletus takes the same position, his grounds being simplicity and likeness of effects (*In de Gen.* 2q7, *Opera* 5:319rb).

20. 'Mode of being' is designedly vague: I use it as a placeholder for any of the categories in which change can occur. In substantial change the mode of being *in potentia* some-substance-

potential and indeterminate, will become actual. Much better than the analogy of wax and stamp is that of the teacher who cultivates virtue in her student. Virtue is conceived not as itself an active power but as a persistent modulation, so to speak, of the activity of the will. The teacher transmits nothing, but by word and example elicits from the student habits of judging wisely and acting well that were always potentially there.

We have now, finally, the entire scheme of natural change. It will be useful to summarize it in outline:

1. Intransitive change.

   1.1. Every natural change has a subject in which it occurs, a *terminus a quo* from which it begins, and a *terminus ad quem* (i) which it will attain if not prevented by an external agent, and (ii) at which it will cease.

   1.2. The *terminus ad quem* is a form $X^*$, the *terminus a quo* a privation among whose fulfillments is the *terminus ad quem*. The subject, relative to the form and its privation, is matter.

   1.3. The process of $X$-ing is the imperfect *actus* of $X$ in an $X$-able thing.

   1.4. To be $X$-able is to have a privation of, or a *potentia* to, the form $X^*$ elicited in $X$-ing. The perfected *actus* of that *potentia* is $X^*$, while an imperfect actus is a form contrary to $X^*$ (and thus of the same genus) regarded as tending to $X^*$.

   1.5. To have a privation of $X^*$ is either (i) to have a form contrary to $X^*$ or (ii) to lack $X^*$ and yet be suitably disposed to receive it.

2. Transitive change.

   2.1. Every natural change is the joint *actus* of two things: (i) a patient in which it occurs as an intransitive change, and in which it is the *actus* of a passive *potentia;* (ii) an agent, of which it is the *actus* of an active potentia.

   2.2. The (perfected) *actus* of both *potentia* is the *terminus ad quem* $X^*$.

   2.3. It is an *actus* of the passive *potentia* by virtue of being a specified form from among those that the passive *potentia* is contrary to or suited to receive.

   2.4. It is an *actus* of the active *potentia* by virtue of being specified by the agent insofar as the agent has that *potentia*.[21]

   2.5. Since the *actus* inheres in the patient alone, the agent is never changed, and thus never perfected, by its action, except accidentally (e.g., by reaction).

   2.6. Action is nothing other than the change itself, considered as the *actus* of the agent; and similarly for passion.

The summary captures what is common to the conclusions about natural change affirmed in the central texts. They record, of course, a wide variety of dissenting views. But I think it reasonable to suppose that Descartes, when

---

or-other (but as yet none in particular) is that of prime matter; in other categories, it is a contrary or a disposition in an existing substance.

21. The qualification is to forestall examples such as that of a medicine that cures by virtue of one of its active ingredients and not by virtue of another. The *actus* is that of the first ingredient alone.

he contemplated what the Philosophers thought about change, would have attributed to them the propositions just listed. His own views, as we will see, if seen through the lens of the central texts, would have registered as tending to the nominalist side. There is, however, a significant difference between Cartesian physics and Cartesian metaphysics (including the study of the soul apart from the body). In the physics virtually all of this machinery is dismantled, or retained in name only. To answer a question raised earlier: the difference between the "form" recognized by Boyle and Descartes and Aristotelian form is that by 'form' they mean 'figure' or 'configuration'. No figure has any intrinsic tendency to become any other figure. No figure is, in other words, the privation of another.

In the metaphysics, on the other hand, a great deal of the machinery is retained. It may have been inevitable that where the two met—at what we now call the "mind-body problem"—clarity would give way to obscurity, and certainty would founder in doubt. But even within the physics, as we will see in Part II, there are tensions between the ideal of a science purged of activity and finality, and the science actually practiced, in which the repressed elements continually threatened to return.

## 3.2. Substantial Form and Prime Matter

It is one thing to admit a distinction between matter and form, quite another to admit *substantial* forms and *prime* matter. Since substantial form, with certain qualifications to be added below (§5.3), precedes all others in the constitution of material substance, the corresponding matter was a matter bereft of all forms. We have seen that form is the *actus* of matter, and matter form *in potentia*. The matter that receives substantial form has, therefore, no other *actus*. It is, as the slogan has it, pure *potentia* (but see §4.1).

To argue for substantial form was to argue against a range of alternatives specific to the period. Chief among these were the mixture theory of Empedocles, the atomism traditionally associated with Democritus, the Scotist hypothesis of a *forma corporeitatis* common to all material substances and preceding specific forms, and finally various Neoplatonist theories ascribing the individuation of material substances to a "seminal spirit" or "formative soul."[22] From the standpoint of Aristotelian *quæstiones* on substantial form and prime matter, the alternatives can be divided into (i) those that denied the distinction between substantial and accidental form and (ii) those that

22. 'Seminal spirit' is found in Étienne de Claves's *Paradoxes* (1635), where he is recounting the views of earlier philosophers, including, for example, the chemist Joseph Duchesne (Quercetanus). 'Formative soul' (*anima formatrix*) and 'formative faculty' occur in Kepler's work (*De Nive sex.*, *Werke* 4:275ᶠ, 278). See Emerton 1984:41, 61, 169.

while maintaining the distinction, denied the *uniqueness* of substantial form (and thus its role in specification), or denied that matter and form were genuinely one (thus making each a complete substance).[23] Under (i) fall atomism and mixture theories, under (ii) the Scotist and Neoplatonist theories.

The position argued for, then, was that there is a distinction between substantial and accidental form, *and* that substantial form is unique in each individual, specifies that individual, and must be united to a matter without which it cannot subsist. That characterization, it seems to me, suffices to pick out the Aristotelian notion of form in my central texts from the bewildering multitude of sixteenth-century options. That is the notion against which Descartes directs by far the greater part of his polemical force.

Toletus defines substantial form as "a formal simple *actus,* forming with matter [a thing that is] one *per se,* the principle of the proper operations of the thing" (see Fig. 2).[24] Form is the *actus* of matter, matter the *potentia* of form, because form perfects and determines matter.[25] It is called 'simple' to distinguish it from the whole composite substance, which is in a sense also the *actus* of the matter. The term 'formal' designates the manner in which it perfects the matter, which is to say, by informing (or determining) the matter and uniting with it. That serves to distinguish the form from an efficient cause, which does not unite itself with the matter.[26] The phrase 'one per se' is the key to distinguishing substantial from accidental form: accidental forms are all formal simple *actus* of matter, but substantial form

23. Thus Gilson's verdict (1984:163) that the "monstrous" notion of substantial form, which Descartes could conceive only as a "substance immatérielle [. . .] qui s'ajoute à une substance corporelle [. . .] pour composer avec elle une substance purement corporelle," is a monster of his own making, must be softened. Such monsters abounded, if not in textbook Aristotelianism, at least in precincts not far from it. The French alchemist Quercetanus writes that the "three formall beginnings" (the Paracelsian Mercury, Sulphur, and Salt) "are more spirituall than corporall, yet [. . .] they make a materiall body" (Emerton 1984:182, quoting the 1605 English translation of Quercetanus's major work)—the very contradiction that Descartes saw in the notion of substantial form. Descartes did not, I should add, confuse the Philosophers of the schools with the likes of Quercetanus. But there are, even in them, traces of the "spiritualization" of form, especially in the theories of light and intentional species.

24. "Forma est actus simplex formalis, unum per se cum materia componens, principium propriarum operationum rei. Hæc conclusio est velut definitio explicans ipsius formæ naturam [. . .]" (*In Phys.* 1c9q19, *Opera* 4:41ra). The context makes it clear that the form being defined is substantial, not accidental, form. Almost identical definitions are found in Eustachius (*Physica 1§2q5, Summa* 3:123), the Coimbrans (*In Phys.* 1c9q9), and Suárez (*Disp.* 15§5).

25. "Forma dicitur actus, quia dat esse materiæ perfectum, & determinatum in specie certa: Nam actus materiæ est imperfectus, & communis, ac potentialis, forma determinat & perficit ad certam speciem" (*Opera* 4:41ra).

26. "Dicitur formalis, ut intelligas, quomodo forma det esse materiæ: nempè si ipsa informando, & uniendo sibi materiam [. . .] & hoc ad discrimen causæ efficientis, quæ dat esse, puta, ignis facit lignum ignem, sed non dando seipsum, sed aliquid aliud extra se producendo" (*Opera* 4:41rb).

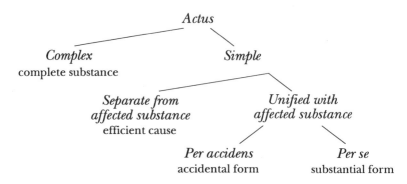

Fig 2. Definition of substantial form

alone enables matter to subsist.[27] It is, finally, the principle (by implication unique) of all the active powers of a thing, and thus of its "operations." That, as we will see, is the focus of the empirical arguments: substantial form is, in a manner yet to be delineated, the cause or origin of all other forms, and uniquely so.

There are, in keeping with that complexity of the concept, a number of standard *quæstiones* devoted to the defense of substantial form. Those that concern the Scotist view I will examine in the next section. Here I take up only those that concern the distinction between substantial and accidental form. Such arguments are directed in part against atomism, in part against skepticism about substantial form within Aristotelianism itself.

The arguments for distinguishing substantial from accidental form fall roughly under four headings: logical differences between substance and accident; the empirical distinction between generation (or corruption) and alteration; the unity of active powers in a natural kind; the existence of preferred states. A detailed examination will yield several fruits. It will show, first, that in the hands of a determined advocate like Suárez, many *experimenta* can be brought forward on behalf of substantial form.[28] Second, it will make somewhat clearer the role of substantial form in natural philosophy, and its relation to more recent notions of essence and natural kind. It will, finally, begin to show the complexity of the Aristotelian theory of material substance. That complexity is sometimes glossed over in brief expositions.

27. "Dicitur, unum cum materia per se faciens, ad excludendas formas accidentales, hæ enim actus sunt simplices & formales, sed cum eo, cuius actus sunt, non faciunt unum per se, sed per accidens: non enim complent accidentia subiectum, at forma complet materiam: est enim materia dimidium ens, ut diximus, ac ob id dicitur facere unum per se cum materia" (*Opera* 4:41rb). On the phrase *unum per se*, see §5.2.

28. Suárez calls them *indicia* and *signa*. "Accidents are called *signa* when they are better known to us than their causes" (Zabarella, *De methodis* 12, *Opera logica* 202B; see also 754E, where the dictum is ascribed to Averroes).

Yet it is entirely pertinent, I believe, to the responses of the earlier opponents of Aristotelianism, who were still in a position to recognize it.

1. *Logical differences.* I have touched on such differences already. Toletus argues:

> There is the greatest difference [*maxime discrimen*] between accidental and substantial forms. Accidental forms have [with each other] not only repugnance [i.e., incompossibility], but a determinate repugnance, as white with black or an intermediate [color], heat with cold, and so forth. This determinate repugnance consists in the fact that for an appropriate subject, either one or the other is found, nor can one leave unless another arrives. Between substantial forms there is a certain repugnance, but it is not determinate, since every substantial form is repugnant to every other equally: nor are they consequent upon a particular matter [*nec circa materiam consequenter se habet*], since matter can exist without the two or three [mutually repugnant] forms. (*In Phys.* 5c1, *Opera* 4:151vb–152ra)

The argument is that the logical space, as it were, of accidents is structured in a way that the logical space of substance-kinds is not. Accidents, as we have seen, are supposed to fall under genera whose character is spelled out reasonably clearly by Toletus. First, what one might call the "generic" accident—'being colored'—is properly predicable only of certain kinds of subject; and a specific accident—'being red'—falling under it is properly predicable of a subject only if the generic accident is. Angels are not colored, hence neither red or blue nor whatever. Second, every thing of which the generic accident is properly predicated has at least one specific accident from within the genus, and no more than one such accident. The apple of Eden, being colorable, was of one and only one shade.

Now even though substance-kinds can be arranged into genera, and then via their *differentia* into species, those genera do not share the character just described. The notion of "proper subject" does not apply, because 'human', though a species of animal, is not predicated of the animal as an accident. More significantly, any matter could, in principle, exist without any of the forms that fall under the genus 'animal': it could be a rock or a flame instead. The only list of substance-kinds of which one may say that a subject *must* be at least one such would be the list of all substance-kinds. An accident is repugnant especially to other accidents within its genus, and that special repugnance is just what is called 'contrariety'. A substance-kind, on the other hand, is indiscriminately repugnant to all other substance-kinds. To put the point linguistically: if I call a thing 'not white', the implication normally is that it has some color, only not white; but if I say a thing is not an apple, there need not be a genus such that my statement implies that it is one of those, only not an apple.

On the face of it, the first part of the argument begs the question. Within the genus 'animal', the *differentia* 'rational', predicated of a generic "animal-stuff," yields the species 'human'. *Differentia* are contraries (see n.7). We therefore have a structure not unlike that of colors: every animal is either rational or irrational, and only animals are rational or irrational. Similarly, 'animate' could be regarded as a *differentia* within the genus 'complex material substance'. Why then is the phrase 'rational animal' not on a par with 'white table'? Why aren't there accidental forms all the way down?

Suárez records a similar point among the arguments against substantial form: "fire, for example, is sufficiently understood in its constituted essence if we conceive a substance having perfect maximal hotness and dryness together, even if the substance which is the subject of those accidents is simple: and this also is enough [to understand] all the actions of fire we experience, and the distinction between fire and water, and the transmutation of each into the other, which seems to consist in the fact that a substance changes from maximal cold to maximal heat, and vice versa. This is therefore sufficient for the constitution, distinction, and action of the elements" (*Disp.* 15§1¶1, *Opera* 25:498). The *differentia* 'hot' and 'dry' suffice to explain what fire is, how it differs from the other elements, and how it acts. Why not, then, say that the substance fire just *is* those *differentia*, conjoined in a material substrate?

Now it could be that whichever substance you consider, and whichever change you consider, there is a kernel that always and everywhere persists, *and* that suffices to individuate that substance from others. The kernel, by hypothesis common to all substances, could be called indifferently matter or form: matter insofar as it is the substrate of change, form insofar as it makes the substance one thing. There would be a "substantial form," but it would have nothing to do with the kind of thing it was. That would be determined entirely by accidental forms.

Or it could be that for no substance is there such a kernel, much less for all. Aristotle's argument for distinguishing form from matter shows only that in each change there is *a* matter and *a* form. Nothing licenses us so far to speak of *the* matter or *the* form of a substance. The example of fire is telling here. Fire is maximal heat combined with maximal dryness. If the heat changes into its contrary cold, one has earth. If the dryness changes into its contrary, wetness, one has air. In each case there is a "matter," namely, that one of the two accidents is not changed. But there is no "matter" that is the common persistent in every change.[29]

----

29. See Furth 1988:221ᶠᶠ for an analysis of elemental mutation that follows out this observation to argue that prime matter is not needed in Aristotle's physics. Zabarella already observes that even in elemental mutation, one of the two elemental qualities (or *symbolæ*) is always preserved (*De rebus nat.* 207F).

I will examine prime matter in §4.1. Here I want to observe that the Aristotelians, and perhaps even Aristotle himself, wanted to preserve two fundamentals. The first is that even the elements can be generated and corrupted. One consequence of that is to preclude any common kernel—at least the Aristotelians thought so. The second fundamental is that for each thing there is a kernel that determines its kind or essence and whose destruction is the destruction of the thing. That kernel is the substantial form. Some changes are indeed annihilations and creations, but of the thing according to its kind—*secundum quid*—and not of its being *simpliciter.*

Aristotelianism thus has to avoid two extremes. It cannot accept a principle of individuation and unification which does not also tally with natural kinds. Nor can it give up the view that there are destructions that are not annihilations, generations that are not creations *ex nihilo.* So the Coimbrans argue that "if form were an accident, there would be no substantial generation if indeed matter cannot come into existence" (*In Phys.* 1c9q9a2). Since matter indeed cannot come into existence (except at the Creation) there would be no substantial generation, which is contrary to experience. But if some destructions are not annihilations, and some changes are not destructions, there must be something that determines both what kind of thing an individual is and when it starts and ceases to exist as that individual. Aristotelianism would like to reserve creation for God and restrict natural agents to generation. But it accepts the doctrine of the transmutation of the elements, which seems to preclude any preservation of structure beyond mere existence. Generation coincides with alteration. Hence the continual puzzles about generation. Hence also the temptation to rase the whole immense structure, and in particular to banish substantial form. Rather than provide an apt descriptive scheme for natural change, it seemed to block the way to a true physics.

Descartes certainly thought so. In the next section we will see that his physics embraces, in a certain way, both the hypothesis of a single substantial form and that of the reduction of generation to alteration, whose implication is that there are only accidental forms. Extended substance indeed has an irremovable form—its quantity, which the Aristotelians numbered among the accidental forms. But even though Descartes, with misgivings, treats extension as an attribute of substance rather than as substance itself, extension, because it is, so to speak, the unique substantial form in nature, admits of neither generation nor corruption. To generate it is to create it, to corrupt it is to annihilate it. To hold that there is one substantial form is tantamount to holding that there are none. What the Aristotelians called "generation" is after all reduced to alteration, and all forms, save the one, are accidental.

But cleansing the stables had its costs. Whatever the difficulties of the doctrine, it saved a great many phenomena. Chief among them, as I have said, were the difference between generation and alteration, the unity of powers, and the existence of preferred states.[30] The empirical arguments, I think, must in retrospect look to yield a stronger case for substantial forms than the logical. The phenomena they adduce cannot be, and in the fullness of time, were not neglected in natural philosophy; and the revisions of mechanism necessary to manage them, although they may not have ushered in a new Aristotelianism, did temper the radical erasures effected by Descartes.

2. *Generation and alteration.* Generation is ordinarily defined as change of substance, alteration as change of quality. To argue that the existence of the two kinds of change so defined implies the existence of substantial form would be question-begging. What one must do is first to argue from experience that there are two kinds of change, show that one kind is best described as change of substance, the other as change of quality, and then make the inference to substantial form.

Suárez, thorough as always, does not overlook the argument. One "index" of "substantial composition" is that generation differs from alteration, as "evident experience" shows. The argument deserves to be stated in full:

> We experience alteration, as, for example, the heating of water or iron, to be sometimes so vehement that the most intense heat is felt in them, and yet if the action of the contrary agent [i.e., an agent impressing upon water or iron a quality contrary to their natural coldness] these things remain whole or almost whole in their substance, and easily revert to their [prior] accidental state; but sometimes alteration proceeds so far that an all-encompassing transmutation of the thing occurs, so that although the agent is removed, the patient can never go back to its pristine state, or recover its prior actions or accidents similar [to those it had]; sometimes also it is changed into viler sensible substances, like ashes, scoria, etc.; and occasionally it is entirely and insensibly consumed, because it is transformed into another more subtle and insensible body; it is therefore an evident sign that alteration is sometimes pure, and remains within the bounds [*latitudo*] of accidental change, but sometimes has conjoined with it a greater change. This cannot be otherwise than because the substantial composite, once the form has withdrawn, itself is destroyed [*dissolvitur*]; therefore substantial forms exist [*dantur ergo substantiales formæ*]. (Suárez *Disp.* 15§1¶12, *Opera* 25:501)

The observation that things naturally cold will revert to that quality after being heated will be the basis for another argument. Here the emphasis is

---

30. See Boyle *Papers* 59[r] and Chauvin s.v. "Forma" for brief, hostile discussions of the arguments.

on changes that preclude reversion: water changed to steam, iron vaporized, bone reduced to ash. Consider the reasons for such changes. The heat itself cannot be the reason. Heat per se does not tend to its contrary: no form tends to its own destruction. So in the original substance something else must have opposed it, and if it is now not opposed, that something must have been removed. That something cannot be any accident of the water. Of its accidents only coldness is contrary to the impressed heat, and by hypothesis that coldness no longer suffices, if indeed it is present at all.

What opposed the heat in the original substance must, therefore, have been its matter or a form distinct from ordinary accidental forms. Matter, however, is indifferent to the accidents that inhere in it. It is, after all, that which "remains under every transmutation." It doesn't play favorites. The thing whose loss results in irreversible change must therefore be a form. But we have just seen that no accidental form will do, or at least no ordinary accidental form. This form has the peculiar feature of opposing heat without being contrary to it. Now it may be that some accidents are inseparable from one another, as rarity from heat, or whiteness from a certain "temperament of primary qualities." One might suppose then that the loss of density, which is inseparable from cold, is the reason for the loss of cold. But that only defers the question, which could just as well be asked about density: why is *it* irreversibly changed? Suárez concludes that the reason must lie in a form prior to accidental forms, "a substantial, and not an accidental, form, since it constitutes the proper essence [of the thing], in which accidental properties connaturally and inseparably inhere" (*Disp.* 15§1¶13, *Opera* 25:502). That form must, in other words, be the unique determining *actus* of matter defined earlier (see n.24 above).

3. *Unity of powers.* One can already see in Suárez's argument that essence, if identified with substantial form, is not a mere list of properties the loss of any one of which must result in the destruction of the individual. Many discussions of essentialism now, especially those that treat the issue in terms of modalities, forego essences in favor of essential properties. The arguments for substantial form which I consider here require, on the contrary, that essences be distinguished from essential properties.

That is, in fact, the heart of the problem. It is not enough to exhibit essential properties. There could be a world in which concrete physical individuals had essential properties but no substantial forms. In such a world it might be true, for example, that a piece of iron cannot persist if its coldness gives way to extreme heat; and that it cannot persist if its dryness gives way to extreme wetness (i.e., if it rusts).[31] What would not be true is

---

31. Iron is of course not an element but a mixture. The explanation of its changes might not be as simple as I am making it here.

that those two had any particular relation with one another. In effect each individual would consist in a bundle of essential properties, and natural kinds would be lists of compossible properties (hot and dry, for example, but not hot and cold).

Such a world is not ours. Two kinds of argument can be made to show that it is not, one concerning the possible causes of accidental forms, the other concerning their nonarbitrary collocation in a single subject.

Suppose, first of all, that in natural individuals certain "proper and peculiar functions" are found, "as in man thought, in horses whinnying, in fire heating" and so forth.[32] Where do they come from? Not from matter, since matter has no efficacy. It will not suffice to suppose that they arise out of whatever accidents matter happens to be endowed with, "either because [matter], on account of its innate inertia and the sterility of bare *potentia* [*insitam inertiam nudæque potentiæ sterilitatem*] brings forth no accidents of itself, or because in sublunary things there is one and the same [matter] and so the same accidents would promiscuously inhere in all" (Coimbra, *In Phys.* 1c9q9, 1:180). The very fact that there are proper functions or active powers requires explanation. Certain capacities are associated with individuals of each natural kind (it is supposed to be manifest that there are kinds). The underlying matter (which is clearly intended to be prime matter) is common to all. So even if there were accidents that inhered in matter as such—even if matter were not *pura potentia*—they too would be common to all. But of course different kinds of thing have different proper functions. Hence something other than matter must be supposed as a reason for their being proper.

The Coimbrans conclude that "the origin of such accidents must be referred to a substantial form, as their source." Why a substantial, and not an accidental, form? We have seen that one part of the definition of substantial form is that it be the cause or origin of all other forms. That is so here, pending a proof of uniqueness. One could, I suppose, argue for uniqueness on grounds of simplicity. But a more forceful argument will be established if we consider that these forms are united in the individuals to which they are proper. The ground of unity, after all, could hardly be other than unique.

Suárez notes, in making the argument, that some accidents may indeed be subordinate to others in the same individual, and originate from them. Colors arise, for instance, from certain combinations of primary qualities.

> But sometimes there is no subordination among them, as with heat and humidity in air, whiteness and sweetness in milk, or the several senses in animals; this multiplicity and variety of properties [. . .] requires one

32. "Motus intrinsecus est via et fluxus, via autem intrinsece respicit terminum" (*Disp.* 49§2¶4, 26:901).

form, in which all are united; otherwise they would be merely accidentally united in the same subject, and if one were removed, the others would not thereby also take their leave. But experience testifies that the opposite holds; which is therefore a sign that such accidents, required in such a subject and being in such number, weight, and measure, do not gain their connection only in respect of the first subject or prime matter, but in respect of some composite, by reason of which a certain order is required among the forms of those accidents. (Suárez *Disp.* 15§1¶14, *Opera* 25:502)

We have, then, properties only accidentally related in themselves that nevertheless share a common fate in certain individuals. Chalk and chocolate suffice to show that whiteness and sweetness are not inseparable. Yet in milk (fresh milk, naturally) they stand or fall together. What is the reason for that? Matter is indifferent. Even if these two were subordinated to a third form, the same question could be raised in relation to that form and yet others inhering in the same individual. Whatever hierarchy there is among accidental forms, the search for a reason for their collocation will wind up insisting that *one* form stand at the top, to account for the whole. Such a form, as I have already argued, will satisfy the definition of 'substantial form'. It will, I should add, not be the universal kernel briefly envisaged earlier, since it is the ground and origin of a *particular* collection of accidents belonging to individuals of a certain kind.

4. *Preferred states.* In water, let us suppose, the independent properties of coldness and wetness are conjoined. Yet some water is warm, even hot, though perhaps never as hot as air (remember that 'water' means *liquid* water, not steam). On what basis, then, does the Aristotelian call it cold? The answer to that question provides one last empirical argument for substantial form.

Some alterations, as we have seen, are irreversible. Others are not only reversible, but tend naturally to be reversed. Hot water "intrinsically reverts to its pristine coldness, as experience attests" (Suárez *Disp.* 15§1¶8, *Opera* 25:500ᶠ). "Some intimate principle" must restore its coldness. No extrinsic principle will do: some are not always present, and of those that are, the air is naturally hot, or if cold will itself endeavor to restore its original heat rather than cooling the water, while the various celestial and universal causes do not have such effects.

The example was in fact quite controversial. Some thought that the coldness remained somehow in some part of the water even as it was heated. Suárez calls that view "frivolous and contrary to experience," but goes on in his usual way to offer reasons against it. Others held that coldness remains in some degree throughout, and reasserts itself once the external cause of heat is removed. Suárez argues that a less intense degree of cold has no innate tendency to become more intense, that the greater degree of heat, say, in water near boiling ought to conquer any remanent coldness, and that in any case there will not be any coldness, since it will have been conquered by the heat.

We now believe that the water, so long as it is warmer than the surrounding air, communicates heat to the air by imparting kinetic energy to the air molecules.[33] Water has no "natural" temperature; its cooling comes about by an extrinsic principle. So the example fails. But the principle it illustrates has many other instances. Many kinds of thing do seem to have preferred states. Excited electrons in an atom "spontaneously" emit photons and return to a "ground" state. Proteins have preferred ways of folding up. Many animals exhibit a tendency within certain limits to regulate their internal states around a certain norm. The distinction between disease and health is as between a "normal" and a "pathological" state. Cure consists partly in removing the agent of disease, partly in the spontaneous restoration of the body to its healthy state. There is, then, a wide variety of natural phenomena in which a thing is taken to return spontaneously from a disturbed or unnatural state to a ground or natural state.

Both the reversibility of some changes and the irreversibility of others are "signs" or "indices" of substantial form. In reversible change a natural property is restored in the continuing presence of others; in irreversible change many or all of them are altered at once and no longer spontaneously revert. The argument, then, does not rest so much on essential properties themselves as on the common fate of the qualities proper to a natural kind. For the Aristotelian, their common fate requires a reason, and that reason is the substantial form.

It may seem uninformative. What more has really been said than that what is united is so for some reason, to which the obscure label 'substantial form' is now attached? One answer is that it is informative not so much through what it affirms as through what it denies. What is denied is that the kinds we encounter in nature are *merely* collocations of accidents;[34] and more specifically that reversion to pristine states can be explained entirely by reference to other accidents. Some accounts are ruled out, and that is

33. Or rather we believe that in the long run the water and air together, considered as a closed system, will approach a condition of maximum entropy, which effectively means that inequalities in temperature will vanish. Toletus, who ascribes the account of cooling argued for by Suárez to Avicenna, agrees instead with Walter Burley that the water is cooled by the surrounding air (Toletus *In Phys.* 2c1q2, *Opera* 4:48). He considers what ought to be a crucial instance: "What if the water were placed in a vacuum, where there was no air or anything surrounding it?" The answer is that the heat would be corrupted by the wetness of the water, but no new cold would be generated, for lack of a generator.

34. Some ancient "bundle" theories of form are discussed in Sorabji 1988, c.4. I do not find that the Aristotelians mention any of them.

progress of a sort. Another answer comes from the theory of generation (§5.3): something other than accidental form is needed to explain the manner in which preexisting stuff is "disposed" so as to become a new kind of thing. Here again it is not single changes but the coordination of many that seems to call for an independent principle.[35]

Even so, such arguments will not likely satisfy philosophers now. One wants to know what the substantial form *is*. That question will not find a straightforward answer. First, because the expected *kind* of answer will not have been sought by the Aristotelians. If one wants an analysis of substantial form into simpler components—what philosophers now call "microstructure"—one wants what Aristotelianism did not in general aim to provide. Even the theory of secondary qualities does not proceed thus. Second, because the only "analysis" Aristotelianism was willing to provide was to describe the active powers associated with a form and the dispositions required for its reception. It is in that way that the soul is studied. Finally, because substantial form is that in which accidental forms inhere. The human soul is that which "lives, feels, and thinks": if one asks what it is apart from what it does, there is no answer. "Almost never can we explicate the essences of things, as they are in the thing [*prout in re sunt*], but only through their being ordered to some property" (Suárez *Disp.* 40§4¶16, *Opera* 26:547).

We find here a familiar paradox. If one asks that an ultimate subject of predication, a hoc aliquid or 'this something' be described apart from that which is truly predicated of it, there is, of course, no answer. Substantial form is, as we have seen, not quite a *hoc aliquid*—only the composite is—but it is that in which all accidental forms inhere. Whatever is truly said of it, save that it is existent and one, is said of an accidental form. Hence the question 'What *is* it?' cannot be answered; to ask it indicates a misunderstanding of its role.[36]

---

35. Mary Louise Gill argues that "active forms," like the soul, "co-ordinate distinct, and often successive, motions toward a goal whose realization depends upon that co-ordination, as in an organism's nutrition and growth" (1991:251). "Passive forms," which would include not only passive potentiæ as the Aristotelians understood them, but also certain active *potentiæ* like heat and *gravitas,* cannot do that. Though Gill's active forms are not necessarily substantial forms in the Aristotelians' sense, the line of argument is similar. What one needs to connect the two are: an argument to the effect that in each thing, or at least each sufficiently complex thing, there is a unique active form that coordinates its activities, and which is "first" in the Aristotelians' sense; and an argument showing that the form so defined can be identified with the thing's essence or nature, or, in other words, that what coordinates the activities of a thing is also that which defines it.

36. "Substance-in-general is, as far as we can know, a mysterious indeterminate stuff which is logically indescribable except that we can say that it supports qualities. However, it was postulated just because it was thought that qualities needed support, so the hypothesis of substance is a paradigm *ad hoc* hypothesis" (Alexander 1985:213, describing the "usual view").

## 3.3. Form as Substance

Earlier I mentioned that some philosophers held that the notion of substantial form was incoherent. The objection, I said, rested on a misunderstanding. I will now examine briefly the response to it. The response turns on two further notions, that of an *incomplete substance,* and the distinction between *inherence* and *formation.* The fate of those two notions, and of the allied distinction between first and second substances, hang together. I will suggest that the misunderstanding arose in part because the 'is' of specification, essential to defining second substances, and to making sense of incomplete substances, is elided in favor of the 'is' of accidental predication.

For the Aristotelians, interestingly enough, calling matter a substance was more problematic than calling form a substance. Questions concerning the substantiality of prime matter are rather more common than questions on form. But the primary difficulty in each case was the same. A substance is self-subsistent, requiring nothing else in order to exist. Yet matter and form—at least corporeal form—each require the other to exist.[37]

Aristotle uses 'substance' in two not obviously compatible ways. In the *Categories,* substance is defined as that which is "neither in a thing, nor of a thing." 'Being in' was standardly taken to denote inherence, the relation between a subject and its properties. In other texts, notably the *Metaphysics,* Aristotle argues that form is substance, and indeed more so than matter (*Meta.* 7c3, 1029a, *De an.* 2c1, 412a6ff).[38] But if the relation of form to matter, or to the composite, is inherence, then form cannot be a substance according to the *Categories.*

To resolve the inconsistency, one could deny that substance is univocal, or that its relation to matter or to the composite is one of inherence. The strategy of Eustachius is effectively to deny that 'substance' has one sense. Substantial form is "a certain substantial *actus,* but incomplete or (so to speak) semisubstantial, and which joined with matter constitutes a single

---

37. Suárez reports the objection thus: "A contradiction [*repugnantia*] seems to be involved when form is said to be both formative and substantial: for either it is a subsistent thing, and requires no sustaining subject, or else it requires one: if the first holds, it cannot be a formative form, because it is contradictory that what is subsistent should be received in another. If the second holds, it is an inherent form, and therefore accidental. Hence there is no substantial form" (*Disp.* 15§1¶2, *Opera* 25:498). The term 'formative' (*informans*) distinguishes material forms from those of angels and God, whose mode of existence does not involve their being the form of something else. To such forms the objection does not apply. *Informans* is also used in contrast with *inherens* to differentiate the way in which substantial form exists in the composite with the way that accidents exist in substantial form or in matter.

38. "He [Aristotle] says first that the subject, which is the first particular substance [i.e., a concrete individual, a "this"] is divided into three: matter, form, and the composite [. . .] The composite as well as the matter and form is called a particular substance" (Thomas *In Meta.* 7lect2, Cathala ¶1276; cf. 1278).

whole substance [*unam integram substantiam constituit*]" (*Physica* 1§2q5, *Summa* 3:123–124). Hence when Aristotle argues that form (or matter) is substance in the *Metaphysics,* he means incomplete substance or "semisubstance." Only the composite is a substance in the sense of the *Categories.*

But one could object that the very idea of "incomplete substance" is contradictory. Suárez insists that it is not: "We can suppose [. . .] that such a genus of being or incomplete substance is given in the nature of things: what contradiction in this could anyone concoct or dream up?" It is not inconsistent, he argues, to suppose that one and the same thing is both the *actus* of something else, as form is of matter, and substantial; indeed the more severe difficulty is to show how matter, which is not even an *actus,* could be substantial. Hence whoever agrees that matter is substance must also agree that form is substance.[39]

The argument, however, seems curiously beside the point. The objection cannot be turned aside merely by arguing ad hominem against those who hold that matter is substance.[40] What we need is a coherent definition of 'incomplete substance' or a reasoned distinction between "informing" and "inhering."

Recall that the key phrase in the standard definition was that in a complete substance substantial form and prime matter are *unum per se* (see n.27). Accidental forms and matter are merely *unum per accidens.* The sense of that contrast can be grasped well enough for now if one considers that substantial form is the origin of accidental forms and that accidental forms are therefore joined with matter only through substantial form.

The composite of substantial form and matter, then, is prior to any combination of matter and accidental form (but see §5.3). Matter, on the other hand, cannot exist without substantial form (nor, of course, can material form exist without matter). The union of matter with substantial form is necessary for there to be an existing thing at all. Union with accidental form, on the other hand, presupposes that there is an existing composite of matter and form. The relation of matter and substantial form, informing, is

39. "But that is clear, whether because nothing can be assigned to either that is repugnant with the other; or because the *ratio* of 'actus' of itself denotes perfection: and if that can without contradiction be conjoined with being accidental, why should it be contradictory to couple it with being substantial? Or finally because substantial being [*entitas*], since it concerns perfection *simpliciter,* seems rather to contradict the *ratio* of 'potentiality' than that of 'actuality': but it is not contradictory with the first, since in prime matter we find [both substance and *potentia*], and so neither is it with the second" (Suárez *Disp.* 15§1¶16, *Opera* 25:503–504).

40. The Coimbrans make a similar argument: "[prime] matter is bare *potentia* [. . .], and yet it is, according to our opponent, a substance; hence form is all the more justly a substance" (*In Phys.* 1c9q9, 1:179). The argument has force only if their earlier argument that matter is bare *potentia* is accepted. Atomists, who hold both that matter is actual and that all forms are accidental, will not be persuaded.

quite different from the relation of matter of accidental form, inherence. To speak of prime matter and substantial form as "incomplete substances," then, would be to mark their special bond with each other.

It may seem at times as if the disagreement of later philosophers is not so much with the notion as with the name. But the disfavor that substantial form fell into, and the misunderstandings it engendered, signal differences between the Aristotelians and their successors that are not merely verbal. Briefly put, those differences were a shift in emphasis in the definition of substance, and a correlative restriction on the 'is' of predication.

Descartes, advising his protégé Regius, writes that by 'substantial form' he understands "a certain substance adjoined to matter, and with it composing a merely corporeal whole, and which is not less, and even more, a true substance, or a thing subsisting by itself [*res per se subsistens*], than matter is, because it is called *Actus,* the other only *Potentia*" (*To Regius* Jan. 1642; AT 3:502). There is nothing here that would surprise an Aristotelian, with one exception. Form is indeed substance, and joined to matter to form a corporeal thing, and it is more of a substance than matter, precisely because it is *actus,* the other merely *potentia.*

What would make the Aristotelian pause is the gloss of 'substance' in this context as 'thing subsisting by itself'. It is indeed not at all hard to find substance so defined. Toletus writes that *singular* substance is called 'substance' "because all accidents inhere in it first, and it itself is what subsists," and later that it is "a nature subsisting by itself [*natura per se subsistens*], and in no way in another" (*In Log., Prædicamentum* 5 [q3], *Opera* 2:100b, 104b). Subsistence per se is the defining feature of individual complete substances like Sappho. But the Aristotelians, building on a distinction made in the *Categories* between "first" and "second" substances, do not make it the defining feature of substance *tout court.*[41]

First substances are singulars, like 'this human' and Sappho. Second substances are genera or species, like 'human' and 'animal'. The distinction between second substances and accidents parallels that between substantial and accidental forms. 'Human' is truly said of Sappho, if the substantial form of humanness informs the individual referred to by that name; 'white' is truly said of Sappho, if the accidental form of white *inheres* in her. Second substances and substantial forms, though they do not subsist per se, share with first substances the feature of not inhering in anything.[42]

We have, then, two defining features of substance: subsistence per se, and the weaker condition of noninherence. Descartes, by making subsistence per se the sole defining feature of substance, effectively denies that second

41. See, e.g., Arriaga *Cursus* 855.
42. Toletus *Opera* 4:78va,b.

substances are in any way substances. That denial is of a piece with his rejection of substantial forms in physics. Substantial forms, like second substances, satisfy only the weaker condition; they do not subsist per se. The proscription has the further consequence of restricting the 'is' of predication to 'being in'. 'Being of', as the relation that holds between individuals and their generic or specific forms, can be done without. Logic is left with the distinction, regarded as exhaustive, between the 'is' of identity and the 'is' of predication.[43]

It would take me too far afield to follow out the suggestion that the denial of substantial forms, or second substances, and the suppression of the 'is' of specification should be taken together as aspects of a single transformation. But I will make one observation. Earlier I noted that one alternative to the Aristotelian theory of substantial form would be a physics in which there was just one kind of form, by which substances were individuated but not specified. In Cartesian physics that form would be extension, which in the *Principles* Descartes calls the "principal attribute" of body. All other physical properties are, or are reducible to, "modes" of that attribute—or, to revert to Aristotelian terms, accidental forms inhering in the one substantial form. The *Categories* distinction between first and second substances, and with it the 'is' of specification, though they might have a place in metaphysics, where there is at least the other attribute of thought, have no work to do in physics. In natural philosophy the 'is' of identity and the 'is' of predication suffice.

The intent of this section has been to expose a body of doctrine much of which is terra incognita even to historians of modern philosophy, and all the more so to philosophers generally. It is a defense, not so much of substantial form itself, as of the inescapability of the functions it performed in natural

---

43. Evidence for such a development is found in the *Logic* of Arnauld and Nicole, first published in 1662. Substance is defined as "that which one conceives as subsisting by itself, and as the subject of all that one conceives in it [*ce que l'on conçoit comme subsistant par soi-même, & comme le sujet de tout ce que l'on y conçoit*]" (*Logique* 1c2, p47). A mode, on the other hand, is "that which, being conceived in the thing, and as not being able to subsist without it, determines it to be in a certain manner [*ce qui étant conçu dans la chose, & comme ne pouvant subsister sans elle, la détermine à être d'une certaine façon*]." A "modified thing" is a substance considered as "determined by a certain manner or mode," as when we speak of a round body. Aristotle's 'being of' has dropped out.

The significant passage comes a bit later: "Our mind, being accustomed to knowing most things as modified [. . .] often divides substance even in its essence into two ideas, of which it regards one as subject, the other as mode." So "one often considers man as the subject of humanity habens humanitatem, and consequently as a modified thing." Every manner of being, in other words, if it is not considered to be the substance itself (as God's infiniteness is God himself) must be conceived as a mode—as "being in" the substance, not "of" it. Thus the copula signifies either identity or inherence, but not specification.

philosophy.[44] Those functions, which include the explanation of preferred states and the common fate of conjoined qualities, were taken up into the new science. We will see how in Cartesian physics the notion later christened 'configuration' by Malebranche took over the last of those functions. Substantial form, moreover, was the source of active powers (see §5.4 for further elaborations on that theme), and thus the ground of natural spontaneity.[45] Descartes refused all spontaneity to nature, preferring to bestow movement on his world with what Pascal called a tweak of God's finger (*Pensées* 77/1001). But it was not, in fact, so easily disposed of, especially when, in the eighteenth century, the divine ghost was banished from the world-machine. The resurgence, in the writings of La Mettrie and Diderot, of a vitalism entirely absent from Cartesian physiology is one testimony to its stubbornness.

44. I should emphasize 'in natural philosophy'. Somewhat artificially, I have neglected arguments for substantial form which appeal to various characteristics of the human soul, and in particular its separability from the body.

45. For a modern Thomist defense of active powers, see Weisheipl 1985:9–10.

# [4]

# Matter, Quantity, and Figure

The substrate argument yields a straightforward distinction between form and matter. With that distinction many philosophers, Aristotelian or not, would have agreed. But the Aristotelians, as we have seen, divided form itself into substantial and accidental. Corresponding to that distinction is a distinction between prime matter and matter broadly understood. Prime matter is the stuff, whatever it may be, which, when joined with substantial form, yields an individual substance. Matter broadly understood is either the composite substance, regarded as the subject of accidental form, or else the already formed and disposed matter necessary to the more perfect among substantial forms: the human soul cannot naturally be joined with anything but a human body.

Matter broadly understood was unproblematic. Prime matter, on the other hand, met with opprobrium even among philosophers friendly to form and matter. Prime matter, they said, is incoherent, incognizable, superfluous. Walter Charleton, an English follower of Gassendi, describes it as "rather *Potential,* than Actual, and absolutely devoid of all *Quantity;* then which we know no more open and inexcusable a *Contradiction"* (*Physiologia* 88). Corpuscularians like Descartes and atomists like Gassendi found the hypothesis of an entirely formless stuff underlying all corporeal substances to be useless, or worse than useless. Far better to suppose, as Descartes did, that matter has the attribute of extension, or, as the atomists did, that the permanent substrate underlying natural change consists of immutable individual substances, each fully specified with respect to a list of fundamental physical properties.

The objections had long been noted within Aristotelianism itself. Standard *quæstiones* are devoted to the existence of prime matter, to its essence, and to our knowledge of it. The complaint about superfluity, though not explicitly addressed, is implicitly answered in refutations of atomism. The cavils of seventeenth-century philosophers were anticipated by Avicenna and Averroes five hundred years earlier, and before that by the Greeks. Aristotelianism had long since accommodated itself to them. It is, therefore, all the more puzzling that, in the first decades of the seventeenth century, they became decisive reasons for throwing off the whole weight of tradition.

I consider first questions about the essence of prime matter. The starting point for all discussions is the definition at the end of *Physics I:* "I call 'matter' the first subject for each thing, from which each thing comes immanently and not accidentally" (*Phys.* 1c9, 192a31$^{ff}$). Aristotle's version of the substrate argument (§3.1) makes it clear that by 'first subject' he means that which persists through substantial change, the subject of generation and corruption and not merely of alteration. Thomas argued that prime matter must be pure *potentia,* not only because it is indifferently receptive to every substantial form, but because it lacks any actuality of its own. All actuality comes from form; and since existence entails actuality, what has no form has no proper existence.

Against that both Scotus and Ockham had argued that matter, as part of complete substance, and as the subject of form, must have an actuality and existence of its own. Scotus (see §3.2) held that all material things share a *forma corporeitatis,* that belongs to matter qua matter and thus actualizes it apart from substantial form. Ockham held not only that matter is an actual entity, but that it is "of itself quantified and endowed with dimensions."[1]

With the exception of the Coimbrans, whose position I find confused, the central texts all reject the Thomist view. Broadly put, their position is that actuality is equivocal. One sort of *actus* consists in existence, the other in determination. Matter is pure *potentia* only in the sense that it is *in potentia* to all forms (whether indifferently so was a further question). It is not pure *potentia* in the sense of lacking existence independent of form. In effect Suárez and the others reject the unrestricted claim that all existence comes from form.

I then turn to quantity. Aristotelian physics is often said to be qualitative, and its successor quantitative. But a better distinction, as Annaliese Maier has argued, is not between a physics of quantity and one of quality, but between a physics of *intensive* and one of *extensive* quantity. Intensive quantity is exemplified by degrees of heat, but the notion was also applied to virtue. It is, unlike Cartesian extension, a mode of quality first of all. From the fourteenth century onward, intensive quantity becomes increasingly the focus of inquiry in Aristotelian natural philosophy. The importance of ques-

1. Ockham *Summula* 1c10, 13; the quotation is from p192. On Scotus's and Ockham's treatment of prime matter, see Adams 1987:639–647. Ockham attributes the view to Averroes, as does Zabarella, who endorses it.

tions concerning intensive quantity is apparent, for example, in the fourteenth-century Paris philosopher Nicole Oresme's "geometric treatment" of alteration.

Nevertheless, quantity plays a subordinate role. The *Calculatores* at Oxford and their successors at Paris excelled in thought experiments. Evidence of their having actually measured anything is hard to come by. Even Oresme made no attempt to measure intensities. The importance of quantity lay not in its being measured, or treated mathematically, but in the fact that quantity stands to matter as quality to form. Each was the primary "assistant" serving to fulfill the ends of the incomplete substance it inhered in. Because matter cannot naturally subsist without quantity, some philosophers came to believe that quantity is essential to matter, or at least that quantity, unlike other accidental forms, precedes substantial form—is presupposed by it rather than the other way around. I will consider in some detail Suárez's rather nuanced view. The argument that quantity is not essential to matter turns, interestingly enough, on the traditional explanation of transubstantiation, one of the few theological controversies in which Descartes permitted himself to intervene.

Quantity could be indeterminate or determinate. Determinate continuous quantity must have what we would now call boundaries. *Figure* is a particular way or mode of determination. It is a real physical quality, but it is, as a mere mode, among the most dependent of qualities. In particular figure is passive *in extremis* (*see* §4.3). Yet since it is found in all matter, and of infinite variety, figure might serve—so the ancient atomists had already surmised—as a *succedaneum* for form. Descartes often emphasized the clarity of our ideas of figure and the certainty with which in geometry we reason about it. But I think it equally important that figure, as a *physical* property, is ubiquitous, infinitely variable, passive, and only modally distinct from extension. To substitute figure for form, as Descartes did, was effectively to eliminate activity from nature and substantiality from form.

## 4.1. The Essence of Matter

Prime matter is "very like darkness, which we perceive when we see nothing, while when we see we do not know it."[2] No doubt part of the difficulty lies in

---

2. *Summa* pt3, *Physica* 1tr1disp2§1q1, 3:119. Allusions to Augustine were commonplace in questions about the cognizability of matter (Suárez *Disp.* 13§6¶4, *Opera* 25:421; Coimbra *In Phys.* 1c9q2a2, 1:154). In the *Confessions* Augustine writes that we "know it in not knowing it or are ignorant of it in knowing it [*nosse ignorando vel ignorare noscendo*]" (*Conf.* 12c5, p218). The association with darkness is brought out by Ægidius Romanus, who writes that "matter has something in it of darkness, since in it all the remarkable splendor of form is lacking" (*Hexam.* 5c3, quoted in Coimbra ib.). Fonseca says that prime matter is called *Tenebra* since, being

attempting to know that from which every mark by which we know individual substances is absent. But the difficulty lies also in the paradoxes that seem to bedevil the concept. Matter was introduced in the first place as the substrate persisting through substantial change. It therefore exists. Whatever exists is actual (since what is *in potentia* only does not exist as such), and whatever is actual is actual by virtue of having form. But prime matter, because it is the correlate of substantial form, can have no substantial form itself. Nor, if all accidental forms presuppose an actual composite of substantial form and matter, can it have any accidental form. Prime matter is *inactual,* it does not exist.[3] But it was supposed in the first place to be the persisting substrate of change, since otherwise generation and creation would coincide. So it would seem that prime matter is both actual and not, both existing and not, both something and not something.

Nevertheless, the existence of prime matter was never seriously in doubt, no more than that of substantial form. Questions therefore concentrated on its essence. Here there were two avenues of escape from the paradoxes. One was to grant form to matter. From that standpoint, atomism amounts to supposing that prime matter consists of a plurality of complete substances. Even if, as the Aristotelians believed, atomism is false, one could still hold that there is a generic form common to all material substances. Such a form, often called the *forma corporeitatis,* was proposed by Avicenna and, in a more restricted way, by Scotus and Henry of Ghent. Matter endowed with it would be an actual existing thing, and the paradoxes would be avoided.

Or one could deny that all accidental forms presuppose substantial form, and that actual existence requires a substantial form. Prime matter could then exist and be the material cause of substance by virtue of an essential accidental form. The most obvious candidate was quantity, since no material substance can exist naturally without it. Prime matter, the permanent substrate of natural change, would be a quasi substance, quantified matter. That was, effectively, the way of Descartes.

---

without form, it cannot be known except by analogy (*In meta.* 1c7q3§8, 1:369E). Seventeenth-century critics of the notion were pleased to repeat these commonplaces.

3. Dupleix states the dilemma succinctly: "Si la matiere premiere est quelque chose elle est substance ou accident. Or elle n'est ny substance ny accident: substance parce qu'il n'y a point de substance (pour le moins materielle et corporelle) sans forme: accident, d'autant qu'estant accident elle ne pourroit pas estre principe ni partie des substances: car la substance est le subject et le fondement des accidens [. . .] Partant il n'y a point de matiere premiere en aucune sorte": "If prime matter is something it is either a substance or an accident. But it is neither a substance nor an accident: [it is not a] substance because there is no substance (or at least no material and corporeal substance) without form; [it is not an] accident, since if it were an accident it could not be either a principle or a part of substances: substance is the subject and ground of accidents [. . .] Hence there is no prime matter of any sort" (Dupleix, *Physique* 2c5a2, p130). For a recent presentation of a similar argument, see Graham 1987.

The view predominant in the central texts, however, is that prime matter has neither substantial nor accidental form. Its essence consists in its being *in potentia* to *all* forms. The paradoxes were avoided not by qualifying the claim that matter is *potentia,* but by showing that it could nevertheless enjoy a sort of actuality. In what follows I will trace the way to the characteristic Thomist thesis that matter is pure *potentia,* and the reasons why the majority of the central texts depart from it. The question, as we will see, concerns not merely the definition of matter, but the genesis of material individuals. The position of the central texts can be summed up thus. Prime matter is

(i)  *Unique:* there is but one substrate of natural change. The atomists and various other ancient philosophers were mistaken in supposing several or even infinitely many kinds of ultimate matter.

(ii)  *Simple:* prime matter is not a composite of substance and form. The Scotists were mistaken in attributing to all material substances a generic "form of corporeality."

(iii)  *Actual:* it is not incoherent to hold both that prime matter is pure *potentia* and that it exists. But it does not exist, as the Scotists again believed, by virtue of a special kind of actuality distinct from that conferred by form. It exists just by virtue of being a component, with substantial form, in complete substances. (Here one finds dissension: the position I am describing is that of Suárez.)

1. *Uniqueness.* Fonseca and Suárez emphasize that every kind of sublunary substance can change into every other kind. Substantial change is in principle ubiquitous. Experience makes it evident that "the elements act mutually among themselves, and one is converted into another, either mediately or immediately, and mixtures also are generated from them, and consequently are resolved into them, so that all sublunary things, according to the virtue [*vis*] of their nature and composition, are mutually transmutable. I say 'according to the virtue of their nature' because it can happen that some parts of the elements are never transmuted, on account of their being in hidden and very remote places, which the actions of contrary agents do not reach" (*Disp.* 13§1¶5, *Opera* 25:396–397).[4] Substances could have been arranged into disjoint genera, within which transmutation was possible but

4. "And if there is given a common subject of all transmutations, it itself cannot be made out of yet another subject, for then it would not underlie every transmutation; therefore if in every transmutation there is given a common subject, and all things (according to us) are transmuted mutually into one another, it follows necessarily that there should be given a first subject not made out of another," and thus a first matter (Fonseca *In meta.* 1c7q2§2; 1:330F– 331A). On the mutual transmutability of the elements, see *De gen.* 2c4, 330a6$^{ff}$; *De cælo* 3c6, 304b23$^{ff}$.

not otherwise.[5] But the elements can be changed each into the others, as Aristotle argues in *De generatione et corruptione*. Provided that all other substances, whether mixtures like wine or animate beings like humans, can be resolved into their elemental components, the mutual transmutability of the elements entails that all other sublunary substances are in principle also mutually transmutable.

Mutual transmutability therefore renders idle any suggestion of specific differences in matter. It is "of itself indifferent to all forms of corruptible things, and to their dispositions; it requires therefore in itself no distinction or multiplication of kinds" (Suárez *Disp.* 13§2¶8, *Opera* 25:401). If it must have a form, that form will have to be universal.

Descartes too holds that all material substances can be changed into one another (*Monde* 5; AT 11:28), unlike atomists like Gassendi, for whom the ultimate constituents of matter are incorruptible.[6] In Descartes's physics, extension occupies the role of prime matter. It occupies all space, underlies all corporeal things, cannot be destroyed or created except by God. Since Aristotelian matter cannot naturally subsist without quantity, only one point, it would seem, separates Descartes from his predecessors: the conceivability, as opposed to the physical possibility, of matter deprived of extension. If, as usual, we put the question in terms of the absolute and ordained power of God, then the difference is this: the Aristotelians believed that God, according to his absolute power, could allow matter to subsist without quantity, while Descartes did not (§4.2). Of course, once Descartes identifies the substrate with extension, or nearly so, he excludes from matter all properties not contained in extension. That cuts deeply enough for the common ground to escape notice. Still he granted the universality of change, and with it the consequence of a single substrate. Instead of declaring prime matter absurd he simulated it.

2. *Simplicity.* The second goal is to show that prime matter, though not an element, is not a unique *substance* underlying all others—that it is not itself composite. There were reasons to believe otherwise.[7] One argument is that in generation what receives substantial form cannot have been entirely devoid of form, since then the form would have been produced from nothing. So even prime matter must have an "inchoative" form out of which

---

5. There are, in fact, two genera. Terrestrial and celestial matter are *not* mutually transmutable (Suárez *Disp.* 13§1¶11). Celestial matter is incorruptible (see Aristotle *Meta.* 12c2, 1069a30ff; *De cælo* 2c1, etc.). It undergoes no substantial change, and a fortiori no change into terrestrial matter. The question therefore does arise whether celestial and terrestrial *materia prima* are of one species (see Suárez *Disp.* 13§11; Zabarella, *De natura coeli* in *De rebus nat.* 270–290).

6. See *Opera* 1:256b, 266b, 273b.

7. I consider only two physical arguments. For other arguments, chiefly concerning the human body, see Soncinas *Q. meta.* 8q8.

individual substantial forms are "educed."[8] Since this argument concerns the production of forms, I defer it until §5.3.

The second argument requires a few preliminaries about form that were not needed earlier. We have seen that in the *Categories* Aristotle divides substances into first and second. First substances are concrete individuals. Second substances are genera and species. The analogy that warrants calling them both 'substance' is that both are subjects of predication: one may equally well say 'Sappho is wise' and 'The human is wise'. The *Physics* and the *Metaphysics* introduce into first substances a structure unmentioned in the *Categories*: each individual is a composite of matter and form (or forms). Instead of being motivated logically, the new distinction is motivated, as we have seen, by an analysis of change.

The question naturally arises how the two accounts are to be reconciled.[9] Genera are divided into species according to differentia like 'winged', 'rational'. In the biological works, the animal and plant worlds are classified in terms of numerous crosscutting criteria like 'terrestrial', 'aquatic', 'two-footed', 'four-footed', and so forth. That system would sit happily with a metaphysics in which the defining form of a thing was a set of accidental forms. But we have seen that substantial forms are not merely collections of accidental forms. So although a human is an animal and rational, the human form is one, not several. We have, then, a problem: how are the obvious relations between substances that are exploited in taxonomy to be reflected in the metaphysics of matter and form?

The second argument in favor of a *forma corporeitatis* rests on one answer to that question. Individuals are grouped under a genus or species by virtue of a common nature, that if the generic term is univocal must be real rather than imposed by reason. But there cannot be such a common nature except if there is a form common to all those individuals. As Averroes says, "all the parts of a definition are forms; the genus is a universal form, while the differentia is a particular form."[10] Taking that claim to be correct for the moment, we then have a straightforward general argument on behalf of a *forma corporeitatis*: "Body, which is the highest genus in the category of sub-

---

8. Although the *forma inchoativa* and the *forma corporeitatis* have different origins and rationales, both terms are used in discussions of Avicenna. On *forma inchoativa* in Albert and Thomas, see Nardi 1960, c.2. There the doctrine is traced back to the *semina* or *rationes seminales* of Augustine (*Gen. ad litt.* 6c11, 9c17) and to Stoic–Neoplatonic doctrines (Nardi 1960:76).

9. Recent commentators usually explain the difference in terms of Aristotle's development (see Furth 1988; on the developmental mode of criticism, see Owen 1978:xv–xvi, and the references there cited).

10. Zabarella, ib. "Definitions are composed of a universal form, which is the genus, and a proper [form], which is the differentia" (Averroes *In Phys.* 2text28, *Opera* 4:59K). "But you may object that Aristotle [. . .] says that the parts of the definition correspond to parts of the thing; but the parts of the definition are the genus and the differentia; hence to those parts there correspond in the defined thing parts that are really diverse, and which are not matter and form alone" (Suárez *Disp.* 15§10¶5, *Opera* 25:537; cf. Aristotle *Meta.* 7c10, 1035b20ff).

stance, is the univocal genus of all perishable bodies, and so there corresponds to it in things a certain form participated in by all such bodies; there exists, therefore, a 'form of body' [*forma corporis*]" (Zabarella *De rebus nat.* 207E; cf. Soncinas *Q. meta.* 8q8, p192a; Coimbra *In Phys.* 1c9q3a2). The "form of body," because it has no contrary form, can never perish; nor can it be generated; and so it will be, as the expression goes, "coeternal" with matter.

What was at stake in the question is best grasped by looking at the arguments against the *forma corporeitatis.* One rests on the by now familiar distinction between substantial and accidental change. If the *forma corporeitatis* suffices to make matter substance, and if there can be but one substantial form in each individual, then all other change—from elemental mutation upward—must be accidental. Indeed, since material substances are never changed except into other material substances, all physical change must be accidental. We have seen, however, that elemental mutation must be distinguished from, say, alteration in degree of heat. There can be, therefore, no *forma corporeitatis.*

But must substantial forms be unique? That is what the second argument denies. The *forma corporeitatis* suffices to make matter a substance, but not to make that substance animate or human. To be animate is to have a second form, the soul. Some philosophers argued that humans have three souls—vegetative, sensible, and rational. The vegetative soul corresponds to the genus *living thing;* the sensible to the genus *animal;* the rational to the species *human.* The same, they believed, must hold of every individual, animate or not: it will have a full range of distinct forms, at once "substantiating" and "specifying," from *genus generalissima* to *infima species.*

To argue against the *forma corporeitatis,* then, one must show that a single individual cannot have several "substantiating" forms. Only the last of those forms, it should be noted, will yield a fully specified individual. The others, as Zabarella puts it, "indeed produce an *actus* [in matter], but an imperfect one, mixed with potentia."[11] One cannot argue, then, simply that a plurality of forms would—absurdly—yield a plurality of fully specified individuals out of one matter. Suárez, recognizing that, rests his argument on the thesis that "there can be no form that produces just generic being."

He offers two arguments, one inductive, one from the *ratio* of substantial form. The inductive argument will likely strike readers now as strange, and is

11. "They [the defenders of Avicenna] would say that substantial form is twofold, one specific and [. . .] ultimate, which is the ultimate form of the species, like man, horse, cow; the other common and general, by which is constituted not the *species infima,* but a genus [. . .] To form of the second sort existence cannot be attributed; only with the specific form can this be done, since the general form indeed produces an *actus,* but an imperfect one, mixed with *potentia* [*cum potestate commistum*]; while the specific form produces a perfected *actus,* requiring no further form in order to exist" (Zabarella *De rebus nat.* 208C–D).

worth noting for that very reason. It is an instance of the ordering of forms (see §6.2). Accidental forms never fail to be ultimately specific. Nothing is just colored; it must be a particular shade of red or blue or whatever. So too for the entirely spiritual forms of angels. But material substantial forms are "a kind of medium [*quasi mediæ*]" between the two, and there is no reason to think them different, so they too must be ultimately specific. The point— angels aside—is that in both instances there is a genus–species distinction, and yet no form of either kind fails to be as specific as it can be (*Disp.* 13§3¶17, *Opera* 25:407).

One could argue that the question is begged. Perhaps there are generic accidental forms too. The argument must rest on an implicit understanding that since the effects of quality can always be explained in terms of specific forms, there is no reason to suppose any generic accidental form. What we see, for example, is always an ultimately specific shade of color, not color in general. Where accidental form is concerned, there is no other reason to suppose generic forms. The *forma corporeitatis*, too, since it occurs in every material substance and can neither be generated nor corrupted, has no physical effects peculiar to it. Experience can yield no evidence on its behalf (*Disp.* 13§3¶14, *Opera* 25:406).[12] We have no a posteriori argument for generic form.

Indeed the very idea of substantial form precludes "substantiating" forms that confer merely generic or incomplete being. "Every substantial form has its proper mode of being [*propriam entitatem distinctam*] in the thing itself, distinct from that of matter or other forms; it is therefore necessary to understand in that mode of being an ultimate and proper difference [*differentiam*], or a reason by which it is distinguished essentially from other forms with diverse essences, with which it agrees with respect to the common reason of being substantial form" (*Disp.* 13§3¶17, *Opera* 25:407). The crucial point is that for each substantial—or "substantiating"—form, its mode of being in its subject (i.e., as informing the subject) must be distinguished from those of other substantiating forms by a difference that amounts to the difference between *species infima*. It will not suffice for two such forms to differ as genus and species. The reason, I take it, is that the genus, being contained in each of its species, is not really distinct from them.[13] The generic form of animality and the specific form of humanity in

12. Suárez notes that one might argue that quantity, which is accessible to the senses, inheres in matter directly, rather than through an individual substantial form, and thus requires that matter have a generic form of its own prior to individual form. But the presence of quantity in matter presupposes that of substantial form (cf., however, §5.3), and so the argument fails.

13. See *Disp.* 6§9¶15ᶠ, *Opera* 25:240ᶠ, which cites Fonseca *In meta.* 5c28q14§3, 2:1078ᶠ. The argument in brief is that a genus is such by virtue of being differentiable into species; so it must first be differentiated specifically before it can serve as a principle of individuation. A

Socrates are not really distinct and share the same mode of being in him. Only another specific form—that of dog, say—would be really distinct from that of humanity. But clearly those two forms cannot coexist in one individual at the same time.

To put the point another way: suppose for a moment that a substantial form were just a list of essential properties, or rather, not just any list but a *complete* (and ordered) list. A genus could not be a complete list, since it can still be added to. Only a *species infima* could be complete, and any *species infima* will contain the genera superior to it. None of them will be distinct from it. *Species infima* of the same genus will be complete lists differing in their last component. But the differentia of species under the same genus are contraries. Hence no one individual could have two *species infima.* In short, of the candidates for substantial form, the genera superior to a *species infima* are all contained in it, while other *species infima* are incompossible with it. Substantial form, therefore, is unique.

The argument might still seem to beg the question. It depends on our agreeing that substantiation and individuation go hand in hand, so that the only substances are (fully specified) individuals. Incompletely specified substances—in the sense in which the *forma corporeitatis*, together with its matter, was earlier considered to be an incomplete substance—are not really substances at all, or if they are, they are only complete substances incompletely described. To go further with this point would take me into the dense and treacherous domain of universals. Suffice it to say that, although Aristotelianism is not celebrated for the austerity of its ontology, the list of things to which its physics accords full-fledged individual existence is in fact rather short. There are concrete individuals like Socrates and Fido, and within each such individual its prime matter and substantial form— though matter and form are, as we have seen, regarded as "incomplete substances," since each requires the other to exist naturally.[14] There are, moreover, certain accidental forms that the Aristotelians held to be really distinct from the substances they naturally inhere in. Everything else depends, either by inherence or by stronger relations of dependence, on them.

The argument against the *forma corporeitatis,* then, exemplifies a more comprehensive refusal to countenance a plurality of substantial forms in

---

*species infima,* on the other hand, includes "all specific, generic, or higher predicates that can be abstracted or conceived in it" (240). To say of something that it is animal is incomplete. One hasn't, so to speak, yet enough information to pick out the thing as 'this so-and-so'. Only the *species infima* will do. But if one has picked it out as 'this human', the predicates 'animal', 'living', 'corporeal', 'substance' are all implied and need not be separately asserted of it.

14. I am setting aside the doctrine of real accidents whose primary role was to explain how transubstantiation was possible. I should also emphasize that the Aristotelians recognized an infinite variety of spiritual substances: human souls, the angels, and God.

one individual. We will see that refusal again in arguments about the unity of the soul. I will close this discussion with two remarks.

Zabarella's argument suggests that individuals might originate in a process that, from faceless prime matter, proceeds through ever more specific genera to the *species infima* and thence to individuals.[15] Aristotelianism acknowledges such an order—an order of specification. But it refuses to translate that order into an order of generation. In the Creation, what God made was creatures like Adam and Eve—fully differentiated concrete individuals. Any suggestion of a successive emanation from an original One or Being, such as one finds in some Neoplatonist and Gnostic cosmologies, is turned aside. The word 'emanation', as we will see, does occur in Aristotelian metaphysics. But it does not denote a succession of ever more specific substantiating forms within individuals, or of one individual from another. It denotes the relation of proper accidental forms to the substantial forms that unite them. Emanation in that sense is not a temporal relation; nor, more important, is it a diversification of substance.

The second remark concerns prime matter. Some philosophers, among them Scotus and Suárez, held that prime matter could exist without form (§5.1). But it could do so only preternaturally, through the absolute power of God, who then takes over the role of form as an extrinsic cause of matter. Matter remains in that circumstance as devoid of proper activity, and as dependent on something else for its existence, as it is in the natural order.

For Gnosticism, and especially for Manichaeism, matter was to some degree independent of God and resistant to his will.[16] By establishing that matter depended entirely on God for its existence and activity, Christian theology, as Hans Blumenberg has argued, attempted to answer the gnostic challenge—in his view unsatisfactorily (Blumenberg 1988, pt.2, esp. 139–149). Thomas and his followers reduced matter to pure *potentia*, lacking even a proper existence. That view leads to paradox; the central texts withdraw to varying degrees from Thomas's position. But they minimize, all the same, the independence of matter, and they remain one with Thomas in according it no active powers.

Descartes, as we will see, preserves that view. Indeed he strengthens it: not only matter but material substance, including animals, is devoid of activity. Yet by virtually identifying matter with extension, he restores to it at least the possibility of independent existence. Though he himself held that created

15. Thus Peter of Tarantasia (Innocent V) writes that the *ratio seminalis* (see n.9) is "like a beginning or seed of the complete form in matter, which by the action of a natural agent is led out from *potentia* to *actus*. It flows or passes from one being to another, until it arrives at the being of the last, and completing, form" (*In Sent.* dist18q1a3, cited in Nardi 1960:76).

16. The Coimbrans write that "the Manichæans, according to St. Augustine [*De natura boni* 18], affirmed that matter was the formative power [*formatrix*] of all bodies" (Coimbra *In Phys.* 1c9q3a2).

substances must at every moment be conserved by God, his definition of matter, by implicitly giving matter a "form of body," opens the way to considering matter again in its own right as self-subsistent. What remained was to restore to it some sort of activity. That was accomplished when Newton ascribed to all matter qua matter a power of attraction.

3. *Actuality*. If prime matter has no form of its own, the paradox remains unsolved. What remains is either to loosen the tie between form and actuality or to show that the actuality enjoyed by matter in composite substance suffices to solve the paradox. I will first lay out the Coimbrans' position, which insists that matter is pure *potentia*, and then summarize the position of the other central texts, which hold that it is not.

The Coimbrans argue that matter "is neither an *actus,* nor something composed of *potentia* and *actus*" (*In Phys.* 1c9q3a1, p156). The argument is by cases. Clearly matter is not a composite, since that would only raise a similar question about the "matter" of matter. If matter is an *actus,* it is either an *actus subsistens* or an *actus informans.* An *actus subsistens* is an *actus* that requires no matter ("à materiæ societate omnino seiunctus"), like God. Clearly matter is not God (though some, "fallen with the Manichæans under the blows of madness" (158), have sunk so low as to affirm this). If it were an *actus subsistens* it would have no need of form. So if it is an *actus* it is an *actus informans.* But then it is the informing *actus* of some subject, which is contrary to its definition as *first* subject. It is, therefore, no *actus* at all.

It is easy to sympathize with those who found the doctrine unintelligible.[17] It is difficult to conceive how something whose essence is *potentia* could so much as exist (just as God, whose essence is *actus,* cannot be conceived *not* to exist). Something of the flavor of the difficulty can be gotten from an argument of Averroes (Toletus *In Phys.* 1c7q14, *Opera* 4:35ra): if the essence of matter were *potentia,* then when form was joined to it, it would be destroyed, since form is *actus* and the opposite of *potentia.*[18] The trouble arises as soon as substantial form is defined as that which "gives being *simpliciter,*" as the expression goes. The contribution of matter to the composite consists solely in being that *to* which form gives being. Like the receptacle or matrix of the *Timæus,* prime matter serves only as the indefi-

---

17. Charleton was by no means the first. Toletus writes that matter "has an *actus* in its own right [*secundum se*], and is not pure *potentia.* This is indeed so certain, that the opposite cannot be understood" (Toletus *In Phys.* 1c7q13, *Opera* 4:34rb). Zabarella, too, writes that "if others are endowed with so much acumen, that they are [. . .] able to imagine this incorporate matter, I for one (confessing my ignorance) cannot do so at all" (Zabarella *De rebus nat.* 218A). By 'incorporate matter' (*materia incorporea*) Zabarella means matter considered in abstraction from its being extended in three dimensions, and in particular matter considered as pure *potentia.*

18. "Si potentia esset in substantia eius [i.e., primi subjecti mutationis], tunc esse eius destruereter ablatione potentiæ" (Averroes *In Phys.* 1text70, *Opera* 4:41F). See also Coimbra *In Phys.* 1c9q3a2, p158 (2d argument).

nite ground of existence, distinguishing those forms which are from those which are not—a sort of reified ∃. But if the office of matter is nothing more than to receive the *actus* of form, and if that is nothing more than coming to be, then it would seem that matter is dispensable in favor of the creative and conserving acts of God. Put bluntly: matter is nothing other than God himself with respect to those acts.[19]

That way lies heresy, or Spinoza. But it is not easy to see why prime matter is not superfluous. Part of the answer is that it does in fact have a proper *actus*. In Toletus and Suárez, the argument for this has, broadly speaking, three stages.[20] The first is to distinguish between an *actus* that consists in reception of form and an *actus* that consists merely in existence. Next they show that the *actus* that consists merely in existence need not be an *actus subsistens,* or the existence of a self-subsistent thing.[21] Then, they distinguish between the perfect or complete *actus* of composite substance and the imperfect *actus* of its components:

> The first is that *actus* which in the genus of being *simpliciter* or of substance is complete, so that it is neither constituted by a physical *actus* distinct from itself [i.e., it is not a material form] nor is actualized by one [i.e., it is not matter], nor does it require such an *actus* to exist. [. . .] [Imperfect *actus*] designates that being which has some actuality, insofar it is actually apart from nothing [*in quantum actu est extra nihil*]. But the actuality it has is incomplete and imperfect, because it is not so sufficient as not to need another *actus,* either to complete it with respect to being *simpliciter* or even in order to exist. (Suárez *Disp.* 13§5¶8, *Opera* 25:416)

Since substantial form too is characterized as an incomplete *actus,* that of matter is said to be "*in potentia* to any perfect *actus* whatsoever" (Toletus *In*

19. Thomas records an argument whose conclusion is that God and prime matter are entirely identical. The argument is just that both, being simple, cannot differ, because difference implies differentia and thus compositeness (*ST* 1q3a8; cf. Coimbra *In Phys.* 1c9q3a2, p158). Thomas's refutation of that argument, which may have come from David of Dinant (see note 26 below), does not touch the issue raised here.

20. Toletus *In Phys.* 1c7q13, *Opera* 4:34rb–35va; Suárez *Disp.* 13§5, *Opera* 25:414ff. Eustachius puts the point in terms of a distinction between the *actus physicus* of form (which is the *actus* that figures in the definition of *motus*) and an *actus metaphysicus* common to both matter and form. The latter seems to consist in nothing more than that matter indeed exists, since it is part of the composite (Eustachius *Summa* pt3, *Physica* 1tr1q3, 3:121). Even the Coimbrans admit as much. What they deny is that the existence of matter in the composite is *its* existence, or in other words, that matter itself can be divided in any way into what is actual and what is merely potential. Since, following Thomas, they make a real distinction between essence and existence, they hold that "even if in the thing [matter] is an existing essence, still that existence is potential, so much so that it is not capable of existence without the mediation of form; and in that sense they can call matter 'pure *potentia*', even in relation to the *actus* of existence [*in ordine ad actum entitativum*]" (ib. ¶7). That, if I understand it correctly, is the view that I have suggested leads toward Spinoza.

21. Suárez *Disp.* 13§5¶5, *Opera* 25:415.

*Phys. 1c7q13, Opera* 4:34va). But that is not the same as being in *pura potentia.*
Even though, as Augustine said, prime matter is "next to nothing" (*prope
nihil: see Conf.* 12c6, p. 219), still it is "created, receives form, and is a
component of the composite, all of which include existence" (¶12, 417).[22]

The crucial point is that matter is, "metaphysically" though not "physically," a composite of *potentia* and *actus.* Physically it is not, because it has no
form. Metaphysically it is, because like form it is a component of substance.
Even so, matter doesn't amount to much: "Matter is an entity of such a sort
that by itself it is not enough to exist without a substantial *actus* perfecting
and actualizing it; so that from the force of its being alone it includes no
formal *actus* either formally or eminently [. . .] Whatever sort of being is in
prime matter, it is entirely in service to the *potentia* for receiving substantial
form; for to that it is primarily and per se instituted."[23] The arguments on
behalf of an *actus* proper to matter show only that the underlying substrate,
the first subject, *whatever it may be,* must be at least metaphysically *in actu.*
There is no reason why God could not play the role. Yet to put God in the
place of matter was deepest heresy, to whose proponents the Coimbrans give
no name.

The passage from Albert to which they refer does name names. Epicurus
and Alexander of Aphrodisias said that "God is matter or does not exist
outside it and everything essentially is God, and Forms are imaginary accidents without true being, and thus they say that all things are the same."[24]
More recently, David of Dinant, whose works were condemned in the Paris
decrees of 1210 and 1215 which forbade the teaching of Aristotle, identified God with the underlying matter of both bodies and souls. In the *Conclusio* to his *Quaternuli,* he writes that "it is therefore manifest that there is
but one substance, not only of all bodies, but also of all souls, and that
[substance] is nothing other than God himself. The substance from which
all bodies are made is called 'Matter' (*Hyle*); the substance from which all

22. "Matter, insofar as it is presupposed by form, and is the subject of generation, is not
entirely nothing, since otherwise generation would arise from nothing; it is therefore a created
entity, and so an actual and existent entity, because creation does not terminate except with
actual being [*entitatem*] and existence" (Suárez *Disp.* 14§4¶13, *Opera* 25:413). Creation has a
natural *terminus ad quem,* which cannot be mere possibility or being *in potentia.* It must therefore terminate in the actual existence of the thing created. The Coimbrans adduce similar
arguments to show that matter can subsist without form, all the while insisting that "although
matter has a proper existence, nevertheless it is pure *potentia*" (Coimbra *In Phys.* 1c9q6a2,
p170).

23. Suárez *Disp.* 13§5¶11, *Opera* 25:417. A "formal" *actus* is what I have been calling an *actus
informans.* To include something "formally" is to have it as part of one's form; to include it
"eminently" is, in this context, to have it within one's power to produce it. "What eminently
contains another is commonly said to require two things: first, that the containing thing be of a
superior nature to the thing contained; second, that what is inferior be found in the superior"
in one of four ways, of which one is "as in its producing cause" (Chauvin *Lexicon* s.v.
'eminenter').

24. Albert *In phys.* 1tr3c13; *Opera* (Inst.) 4pt1:64.

souls are made is called reason or mind (*ratio sive mens*). There it is manifest that God is the reason of all souls and the matter of all bodies" (*Quaternuli* p71; Kurdzialek 1966:410). The writings of David of Dinant were burned; only recently have fragments of the Quaternuli been discovered and published (Kurdzialek 1963; cf. Birkenmajer 1933). But his views are thought to have influenced Catharism in the thirteenth century and John Wyclif in the latter part of the fourteenth. One can infer also from Bayle's article on Spinoza that David of Dinant's views were known in the seventeenth century and were likened to Spinoza's (*Dictionnaire*, s.v. 'Spinoza', Remark A, note 5).[25]

What, then, is distinctive about material form, and why should there be such a thing as matter? It has been asserted that material form is incomplete. Just as matter is instituted to receive it, so too material form is instituted so as to be the actus of matter. It cannot naturally be *in actu* except as the *actus* of matter. But how does 'being the *actus* of matter' differ from 'being actual', or 'existing'?

Suárez's answer is that "nothing is more repugnant to [the nature of] God than the task [*munus*] for which prime matter is posited, which is to receive [form], and to be a passive *potentia*, to be actualized and perfected" (Suárez *Disp.* 13§14¶15, *Opera* 25:413). Matter, unlike God, does not include existence in its essence. But that fails to take the heretic entirely seriously. The heretic may as well deny also that God is pure *actus,* so that the development of the world is the unfolding of God's power. Or deny that matter is in the relevant sense actualized or perfected, as an atomist might. God could still be pure *actus,* as the theologians have it, and yet one with the world. If, like Spinoza, the heretic is willing to assert that whatever is, is necessary, not even the contingency of the world can be argued against him. The Aristotelians, if they foresaw such a defense, would have found it incredible no doubt that anyone should accept the consequences of identifying God and matter. But if Suárez hoped to show that the very idea of matter precludes it, his argument underestimates the cogency of the heretic's view.

There are in Aristotelianism two paths toward answering the heretic. One can show that the first subject of natural change must have properties inconsistent with those of the first cause or *ens perfectissimum*. Chief among those properties is quantity, or the divisibility characteristic of quantified matter. Or one can show that material forms as a matter of fact have properties that entail that they are incomplete and that what completes them cannot be God himself. Here the chief property of interest is corruptibility. Experience shows that the forms of sensible things come and go: what was

25. A brief treatment of David of Dinant can be found in Steenberghen 1991:83–89; fuller treatments include Théry 1925, Grabmann 1941, Kurdzialek 1966; relations with Catharism, Wyclif, and Spinozism are examined in Kalivoda 1966.

wood is now fire, what was a person is now a cadaver. If there are, on the other hand, forms not subject to corruption, like God and the rational soul, there must be a reason for that difference.

The distinction between created incorruptible forms and corruptible forms rests on a distinction between two modes of coming to be and ceasing to be: creation and annihilation on one side, generation and corruption on the other. We have seen that one argument for prime matter was precisely that it would be absurd to suppose that in natural change one thing is destroyed and another numerically distinct thing created. That argument does not rest on the principle that nothing comes from nothing (though the principle is often invoked in arguments on behalf of prime matter). It rests on the observation that the generation of material forms is preceded by changes that prepare the way for it, as heating readies wood to become fire. If generation were creation, those changes would be gratuitous, and anything could come from anything (a point that did not escape the opponents of occasionalism). It is experience that shows that generation is not creation, or corruption annihilation, and experience too that concludes in favor of material forms.

The objections against the definition of matter as pure *potentia* are not allayed by the dissenting opinions of Suárez and the rest. What we have is largely negative. Matter is indeed neither a complete substance, nor quality, nor quantity, nor in any other category.[26] But until we escape the *via negativa*, there would seem to be no response to the heretic. Even aside from that, experience shows that the substrate of natural change cannot be entirely indifferent to all forms. Matter is so partial to quantity that it cannot naturally subsist without it. Celestial and terrestrial matter cannot receive the same substantial forms. Terrestrial matter imposes an order on the reception of forms, so that the human form cannot be received unless those of flesh and blood are already present.[27] It is to such features that we must look in order to explain what is proper to material substance, and thereby answer the heretic.

Descartes, we will see, answered in his usual drastic manner (see §9.2). Supposing that material substance is coextensive with sensible substance, and that we can be certain of sensible substance that it is extended, then matter cannot be pure *potentia*. It is a complete substance. There is no

26. "I say *matter* is what 'of itself' (that is, considered according to its essence) in no way is 'what' (i.e., substance), 'nor quality nor of the other categories by which being is divided or determined'" (Thomas *In Meta.* 7lect2, Cathala ¶1286, paraphrasing *Meta.* 1029a20).

27. "Although prime matter is *in potentia* to all forms, still it takes them on in a certain order. It is first *in potentia* to elemental forms, and by their mediation according to various proportions of mixture is *in potentia* to various [other] forms, and so not just anything can be made out of just anything, unless perhaps through resolution into prime matter" (Thomas *In Meta.* 12lect2, Cathala ¶2438).

paradox in asserting its existence. If by 'form', moreover, you mean substantial form, then to suppose that God fills the office of matter is inconsistent with the idea of a perfect being. Since God is not a deceiver, extension—the unique corporeal form—must underlie sensible accidents. If by 'form' you mean the properties by which natural kinds are differentiated, then matter supplies to such forms the complete substance on whose existence they entirely depend. In Cartesian physics, form is figure, and figure cannot subsist without quantity even with respect to the absolute power of God (see §4.3). Matter supplies, in short, everything to form except particularity. God's conserving power cannot take its place. The heretic is answered, but for one small detail. Descartes made extension an attribute of substance and not itself substance. So one may still ask: of what is it an attribute? With that question the problem of the substrate is revived. Spinoza, for one, did not hesitate to revive it.

## 4.2. Quantity and Prime Matter

Experience creates a presumption in favor of a special relation between matter and quantity. It reveals to us that quantity, like matter, is found in all sensible substances. Only the eye of faith sees that in the Eucharist the body and blood of Christ are present without their quantity. In natural change, accidental or substantial, quantity seems never to be destroyed or created but only augmented or diminished. Quantity seems also to be more fundamental than other accidents. The whiteness and coldness of a snowball have the quantity, extension, and figure of the matter they inhere in.[28] All these phenomena could be accounted for if quantity were essential to matter.

We have seen, moreover, that if matter is regarded as pure *potentia*, it becomes difficult to understand either why it should be posited or why certain forms should require it and not others. If, on the other hand, matter were essentially endowed with quantity, it would bring to form more than bare existence; and one could at least argue from experience that some forms do seem naturally to exist only in extension, while others do not. A final point in favor of a special relation between matter and quantity was the sheer difficulty in conceiving of material substance without it. The Coimbrans, who defend Thomas's view that matter and quantity are distinct, note that "if form could be united immediately and *per se* with matter,

---

28. "Omnia [accidentia] insunt substantiæ, media quantitate, ut omnes philosophi docent; et videtur probari experientia, nam albedo in superficie extenditur, et calor similiter, cum diffunditur per corpus, in quantitate eius extenditur": "All [accidents] inhere in substance by way of quantity, as all the philosophers teaches us; and experience seems to prove this, since white is extended over the surface [of a thing]; similarly heat, when it spreads through a body, is extended in the quantity [of the body]" (Suárez *Disp.* 14§4¶7, *Opera* 25:495).

then it could be united to it, at least by divine virtue, without the mediation of quantity, so that from a rational soul and matter there would result a composite that was one *per se,* and which could be nothing other than human. There would then exist a man without head or heart or other integral parts, and that, it seems, ought not to be conceded" (Coimbra *In Phys.* 1c9q11a1, 1:182). Nevertheless the Coimbrans agree that form can be united immediately with matter and accept the consequence (see §5.2 below). Others, including Descartes, find it absurd.[29]

Physical arguments were overshadowed, however, by a theological doctrine that seemed to preclude any straightforward identification of matter and quantity. The Catholic church held firmly that in Holy Communion the bread and wine of the host, when consecrated, are by God's miraculous act entirely transmuted into the body and blood of Christ. Immense ingenious effort had been expended to reconcile that doctrine with Aristotelian physics. The problem was that sensible accidents are determined by, or "emanate" from, substantial form (§5.4). Yet in the Eucharist the accidents of the bread remain even while the form of bread is succeeded by that of Christ's body.[30] The solution offered by Thomas and others was that the quantity of the bread is miraculously sustained by God in the absence of the matter and form in which it naturally inheres, and the sensible accidents of the bread attach themselves to it. Appearances are saved even while the underlying substance perishes.

Quantity was thus central to the account made dogma by the Council of Trent.[31] Any thesis on the relation between quantity and matter which

29. See, however, the letter to Mesland (9 Feb. 1645) in which Descartes implies that a human soul can be joined with any matter whatsoever, in such a way that the matter can be said to be its body.

30. On Suárez's explication of the doctrine of "real accidents," which was often singled out for ridicule by seventeenth-century philosophers, see §5.1.

31. The relevant canon reads: "Si quis dixerit, in sacrosancto Eucharistiæ sacramento remanere substantiam panis et vini una cum corpore et sanguine Domini nostri Iesu Christi, negaveritque mirabilem illam et singularem conversionem totius substantiæ panis in corpus et totius substantiæ vini in sanguinem, manentibus dumtaxat speciebus panis et vini, quam quidem conversionem catholica Ecclesia aptissime transsubstantionem appellat: anathema sit [If anyone says that in the holy sacrament of the Eucharist the substance of the bread and wine remains with the one body and blood of our Lord Jesus Christ, and denies the miraculous and singular conversion of the whole substance of the bread into the body and of the whole substance of the wine into the blood, even while the *species* of the bread and wine remain (. . .) let him be anathema]" (Conc. Tridentinum, *sessio* 13 (1551), Canon 2 (cf. ib. 4); Denzinger and Schönmetzer 1976, no. 1652, 1642. The doctrine of transubstantiation was reaffirmed as recently as 1950: cf. no. 3891).

It should be noted that the Council did not favor any particular philosophical theory, Thomist or otherwise, that purported to show how the *species*—the sensible qualities—of the host could remain after transubstantiation. Descartes was thus not compelled to accept any existing theory; but he was compelled, if he wished to respect the doctrines of the Church, to show that it was not impossible that the *species* should remain. On the debates surrounding the threatened to undermine the possibility of transubstantiation had for that reason to be rejected or modified. If the quantity of the host, for example, were destroyed along with its matter, then Thomas's solution would no longer serve. It was therefore incumbent on the Nominalists in particular to show that their thesis did not render transubstantiation unintelligible or inconsistent with experience. Descartes, too, had to show that his physics, though it identified matter with extension, could coexist with dogma.

A cursory look at arguments about prime matter and quantity reveals that their grounds were quite various. But the arguments given the most weight cluster around two topics: the relation of quantity to extension, and the role of accidents in generation and corruption. I will defer the second topic until §5. In treating the first I will start with a Nominalist argument purporting to show that matter and quantity are not really distinct, and then examine the responses of the central texts. In them the Eucharist plays the role of an exemplary and inevitable instance.

Quantity was divided first into discrete quantity, or multitude, and continuous quantity, or magnitude. The continuous quantity peculiar to matter, or *extensive* quantity, was then distinguished from the quantity pertaining to qualities alone, or *intensive* quantity.[32] Extensive quantity was generally thought to confer the following characters on material substance:[33]

(i) *Extension:* quantified substance is spatially distributed (*extensum*), or at least capable of being spatially distributed (*extensibile*).

(ii) *Divisibility:* quantified substance can be divided into really distinct parts. A thing is divided, strictly speaking, only "when the parts which were united are conserved separately, once their union is dissolved,"[34] and divisible if it can be divided, naturally or by divine power.[35]

---

Eucharist, especially after the Reformation, see Armogathe 1977, §1; Andresen 1982, esp. 1:548[ff], 2:483; Redondi 1987, c.7.

32. *Intensive* quantity is the mode by which degrees of the same quality differ. Qualities thus exhibit two kinds of *motus:* increase or decrease of intensive quantity, increase or decrease of extensive quantity (diffusion). Whether increase of intensity consists in the addition of parts of the quality or in the destruction of the old and the production of a new quality was much disputed. See Maier 1951, Murdoch and Sylla 1978:231[ff].

33. Another character mentioned by Aristotle—being measurable—plays no role in debates about prime matter. See Fonseca *In meta.* 5c13q1§1, 2:634[f]; Suárez *Disp.* 40§3, *Opera* 26:538.

34. "Dividitur enim proprie res, quando, ablata unione, partes quæ erant unitæ, separatæ conservantur" (Suárez *Disp.* 40§1¶11, *Opera* 26:531). That character applies to anything that has really distinct parts, including composite substance, whose matter and form are really distinct. Substance is excluded by another part of the full definition, which entails that quantity is an accident, not a substance (cf. ¶2, ¶9; Aristotle *Meta.* 5c13, 1020a8).

35. Celestial matter, "although it cannot be divided by any natural agent, can be divided by divine power. The same is to be said of natural minima." In any portion of matter, celestial or not, one can at least by "mental designation" pick out disjoint parts (Suárez *Disp.* 40§1¶12, *Opera* 26:532).

(iii) *Impenetrability:* if two substances have quantity, it is contradictory to
suppose that they occupy the same place. By contrast, spiritual substances, which cannot have quantity, may do so.

The position of the Nominalists[36] was that extensive quantity, or "bulk
quantity" (*quantitas molis*), was "not a thing distinct from substance and
material qualities. Rather the being [*entitas*] of each of them by itself has
that bulk and extension of parts, which is in bodies; that being is called
'matter' insofar as it is a substantial subject, and quantity insofar as it has
extension and distinction of parts" (Suárez *Disp.* 40§2¶2, *Opera* 26:533; cf.
Coimbra *In Phys.* 1c2q2, 1:94). Quantity is nothing other than matter itself,
designated with respect to the distinction among its parts and their consequent extension in space.[37]

The argument is that quantity has no effects beyond those that already
arise from matter itself.[38] If there were such effects, then among them there
would surely be a "real distinction or situation of the parts of substance,
since by the fact alone that a thing is understood to have one part outside
another [*unam partem extra aliam*], both in being and in place, quantity too is
understood" (Suárez ib. ¶3). It remains to be shown that matter is indeed
understood to have one part outside another. But clearly its parts are distinct of themselves. They require no further cause to be outside one another
in being. Moreover, whatever entities are distinguished in being "can also be
constituted in diverse places." They require no further cause to be outside
one another in place. But those are the effects quantity would have if it had
any proper effects. So any supposed distinction between matter and quantity is idle.[39]

Though Descartes treats extension as if it were entirely unproblematic,
the Aristotelians regarded it as comprising two features. A thing is extended
only if it has really distinct parts. Some Aristotelians held that distinctness in
being sufficed for quantity,[40] but the more common view was that distinct-

36. The authorities listed by Fonseca are Petrus Aureolus, Ockham, Gabriel Biel, and Adam
of Wodeham; to them Suárez adds John Major and Albert of Saxony.

37. Ockham himself affirms that "quantity is nothing but a thing's having part outside part
and one part distant in place from another" (*De sacramento altaris;* cf. Maier 1955:183).

38. See, for example, Ockham *Summula* 1c13, *Opera philos.* 6:192–193.

39. Chauvin, who takes the Nominalists' side, writes that matter is per se extended, because
extension is "nothing other than the several parts of matter set outside one another [*nihil aliud
est quàm plures materiæ partes extra se invicem positæ*]" (Chauvin *Lexicon* s.v. 'quantitas').

40. Suárez cites Capreolus (*In Sent.* 2d3a1,3; 2d18q1) and Soncinas (*Q. meta.* 5q19, p75b).
The argument was that quantity is by definition divisibility (an opinion disputed by both Suárez
and Fonseca, although it finds textual support in Aristotle: cf. *Meta.* 5c13, 1020a8), and that
distinctness of parts suffices for divisibility ("divisibilitatem [. . .] solum esse quod una pars non
sit alia"). Since matter cannot but have parts, it follows that matter cannot be conserved
without quantity even by God's absolute power (Soncinas *Q. meta.* 5q22, p80b), a view Suárez
calls "very common among Thomists" (*Disp.* 40§4¶5, *Opera* 26:544), and which he denies.

ness in place, and thus quantity, is not entailed by distinctness in being.[41] The parts must not only be "distinct in being"—which is to say, separable at least by the absolute power of God—but also "distinct in place." The notion of place, as Descartes well knew, was itself not simple.[42] For the moment it will suffice to think of place as what was usually called "extrinsic place." The extrinsic place of a body is, roughly, the surface defined by the bodies circumscribing it, or what one might call its (outer) bounding surface.[43] Hence to be distinct in place is to have a different bounding surface; to be extended is to have distinct parts that in turn have different bounding surfaces.

If the Nominalist argument works, there must be some special reason why distinctness of being among the parts of matter should entail distinctness of place. Ockham's reason is that "the parts of matter can never be in the same place" (*Summula* 1c13, *Opera philos.* 6:191). That comes close to begging the question, at least from the point of view of those who hold that neither matter nor quantity per se is impenetrable. It also runs counter to the standard account of the Eucharist, according to which the *matter* of Christ's body is present in the host, while the *quantity* of the bread or wine is in the place of the host.

But perhaps there is another reason. Take a globe of water. It is homogeneous, yet it has distinct parts. How are those parts different? The only criterion that suggests itself is spatial separation: this part and that part are two if the place occupied by one does not coincide with the place occupied by the other. Here the very idea of part seems to presuppose distance in place. Prime matter is, of course, as homogeneous as can be. Its parts cannot be distinguished by substantial form. Nor can they be distinguished by quality. Only place seems to distinguish them. Matter's distinct parts are distanced parts, and matter is extended per se.

I know of no Aristotelian who mentions such an argument. But something like it may be behind the unease exhibited at the thought that substance deprived of quantity would collapse into a point. How would one identify its parts? In any case, I think the argument can be answered. We have seen that the criterion of real distinction is that two things are really distinct just in case God could, according to his absolute power, preserve one while annihilating the other. Since God's power was limited only by noncontradiction, two things are really distinct just in case it is not contradictory to suppose that the first exists and the second does not, and

41. In addition to Suárez, see Fonseca *In meta.* 5c13q2§3, 2:650E. Suárez and Fonseca credit Paul of Venice (*In meta.* 5c12) with the distinction.

42. Descartes embarks upon what for him are unusually lengthy and careful discussions of place and space in the *Rules* (AT 10:426, 443$^{\text{ff}}$) and in the *Principles* (2§10–15, AT 8$^1$:45–49). See §9.2 below.

43. See Suárez *Disp.* 51§2¶3, *Opera* 25:980.

conversely. If Socrates' body were deprived of its quantity, his head and his feet would not occupy distinct places (though they could still be *present* in different places, as Suárez argues). But they would remain distinct. The presence of a real distinction does not depend on our being able to recognize it. Nor does it require that we can represent clearly the two things as distinct. Distinction, in short, is not subordinate to the discernment of the intellect.

What quantity brings to matter, then, is distinction in place. Earlier I characterized "extrinsic" place as the outer bounding surface of a body. What is sometimes called "intrinsic" place, or *Ubi* ('Where'), is quite different (Suárez *Disp.* 51§2¶4, *Opera* 26:980). The intrinsic place of a thing is "a certain real mode intrinsic to a thing which is said to be somewhere, and from which such a thing has [the property of] being here or there" (§1¶13, 975), or "locally present" (¶14, 976). Although it cannot be explained except through relations of distance like nearness and farness or above and below, it is not itself a relation. Nor is it identical to or caused by extrinsic place (¶18–20, 977). Nor, most important, is it a region in space, where 'space' is understood as an empty receptacle waiting to be filled with bodies (¶9–12, 974–975; cf. Toletus *In Phys.* 4c5q3, *Opera* 4:115ff).

There is, accordingly, a difference between having intrinsic place, or being present at a place, and having extrinsic place, or *occupying* a place. Having extrinsic place implies having quantity. Spiritual things cannot have extrinsic place. But they can have intrinsic place. God, for example, "is present to the corporeal universe, not only by presence [*per præsentiam*], that is, by cognition, and by *potentia* or action, but also by essence or by his substance" (§3¶8, 984). So too for angels, which, because their powers are finite, are present only at finite regions in space (§3¶10, 985). It is not necessary here to understand what being present by *potentia* or by substance might consist in. All that matters is that a distinction between being present at a place and being in or occupying a place be provisionally granted.

Suárez uses the distinction to dissolve the Nominalist argument. He concedes that the parts of matter are distinct in being and that they can be separated in place, even when matter is bereft of quantity, just as two incorporeal beings can be present at different places.[44] Nevertheless he denies that matter must be quantified of itself merely because its parts are separated. Remember that Ockham's argument was that quantity has no effects not already given in matter itself. Suárez's reply is that

---

44. "If, therefore, quantity is taken away, while the substance is conserved, and no local motion occurs in the substance, the substance will remain with the same substantial presence, and with the same distance or nearness to the center and the poles of the world; and so the whole will remain present in the same space, and its parts in the same parts of space" (*Disp.* 40§4¶25, *Opera* 26:550).

(i) it is characteristic of *quantified* matter that its parts *naturally* exist at diverse places;

(ii) it is not sufficient, if one is to attribute that characteristic to matter alone, to point out that the parts of matter are, or can be, present at different places, since incorporeal things too can be, but need not be, present at different places;

(iii) moreover, matter itself if deprived of quantity need not be present at different places;

(iv) so the characteristic in (i) does not arise from matter alone, but from quantity.[45]

Occupying several places, in other words, is not proper to matter, but only to quantified matter. Quantity does have an effect not already given in matter itself.

It remains to be shown that quantity is really distinct from matter. Here the doctrine of the Eucharist comes into play. First a few fundamentals. When the host is consecrated, its *entire* substance, both form and matter, is destroyed, and the entire substance of Christ's body replaces it. Nevertheless the accidents of the original substance remain, as our senses show. The host retains its color, smell, and, more significantly, its figure and quantity. In the standard account, the quantity of the host, after consecration, serves as a quasi matter for its other accidents in the absence of the original matter. The quantity of Christ's body, on the other hand, does not accompany its substance: it is present at, but does not occupy, the place of the host.

From that doctrine one may infer first of all that quantity can subsist without matter or form: "in the holy Eucharist the substance of the bread does not remain, yet its quantity does, sustaining and uniting qualities that are really distinct, like brightness, taste, and so on. Quantity therefore is really distinct not only from the substance of the bread, but also from its qualities" (Fonseca *In meta.* 5c13q2§2, 2:648A). If quantity can persist without matter, it is at least modally distinct from matter. But since no one thinks that matter is a mode of quantity, it must be really distinct (*Disp.* 40§2¶8, *Opera* 26:535).[46]

45. "Aliud est enim esse posse in diversis spatiis, quod duabus rebus etiam incorporeis convenit; aliud vero est naturaliter esse non posse nisi in diversis spatiis, quod duobus Angelis non inest; igitur illud prius non requirit quantitatem, et ideo convenire posset partibus materiæ, etiamsi quantitate privarentur; hoc vero posterius omnino requirit quantitatem. Unde si partes materiæ sine quantitate essent, indifferenter esse possent, vel in eodem Ubi, vel in diversis. Quod ergo sint ita dispositæ, ut necessario requirant ex natura rei situs diversos, id provenit ex quantitate" (Suárez *Disp.* 40§2¶20, *Opera* 26:538). Suárez adds that the reasons why quantity must be really distinct from matter are those he has adduced earlier in discussing the Eucharist.

46. From the definitions of modal and real distinctness, it follows that if $X$ can exist (at least by God's absolute power) without $Y$, then $Y$ is either a mode of $X$ or really distinct from $X$, depending on whether $Y$ cannot or can exist apart from $X$.

Quantity, moreover, is not actual extension in place, but only potential extension. The body of Christ, like all bodies, has quantity. But when present at the place of the host it cannot actually be extended, since that would exclude the quantity of the now departed substance of the host. Still "it is actually so extended and ordered within itself that if it were not supernaturally impeded, it would also have actual extension at that place" (§4¶8, 547). Hence quantity is "a form giving things corporeal bulk [*molem*], or extension," not actually but potentially, at least with respect to the absolute power of God. It is even possible, contrary to what the Nominalists argued, for a body to be deprived of quantity without any local motion of its parts. It will then still be present at every place it was present at before, but neither it nor any of its parts will occupy any place.

It is time to summarize the results so far. Remembering that God's absolute power is limited only by what is logically contradictory, we have the following:

(i)   It is not contradictory to suppose that matter can exist without quantity, or quantity without matter. But matter does not naturally exist with quantity, nor quantity without matter.

(ii)  It is not contradictory to suppose that matter having quantity is not actually extended, or present at a place without occupying it. But matter does naturally occupy every place at which it is present.

(iii) Quantity is a *potentia* of matter, whose *actus* can be impeded supernaturally but not naturally, and whose primary effect is to prevent other quantified things from occupying or being present at the same place.[47]

Matter is naturally extended in place, spatially divisible, and impenetrable, just as experience tells us. But with respect to the absolute power of God, it can be deprived of all of those features.

It must be admitted that unextended, penetrable matter is hard to imagine. There are two difficulties. One is to conceive of matter apart from impenetrability; the other is to conceive of matter apart from extension— apart from being present at all the points of a certain *region* in space. The matter of Socrates is naturally present at all the points of a region that is larger than the region occupied by the matter of his head or finger. How, then, might it be shrunk to a single point, without depriving its parts of their separate identities within the whole?

---

47. The essential definition [*ratio*] of quantity is to be "a form giving to things corporeal bulk [*molem*], or extension [. . .] What 'having corporeal bulk' is, we cannot state, except in relation to [*in ordinem ad*] that effect, which is to expel similar bulk from the same space, not indeed actually, since the effect can be impeded by God's absolute power, while its formal effect is conserved, but in aptitude [*aptitudine*]" (Suárez *Disp.* 40§4¶16, *Opera* 26:547).

Matter in itself is not impenetrable. It is, as we have seen, the substrate of all forms, the "first subject." That definition does not preclude the presence of distinct matters at one place. Quantity, on the other hand, does confer impenetrability. Matter without quantity would continue to be a subject informed by substantial form. Socrates sans quantity remains a rational animal; but his body no longer has the power to prevent others from occupying any region in space. Matter with quantity can, as we have seen, be prevented by God from exercising that power; but power frustrated remains power: it would if it could. Matter without quantity no longer has the power: it would require a miracle to bring it about that other bodies should *not* occupy the place at which it is present.[48] In both instances the result is the same: we have a bit of matter that no longer hinders other bits from occupying any region of space. Whether one would say, in a given instance, that one or the other was true depends on whether there is reason otherwise to believe that it still has quantity. In the theology of the Eucharist, there was general agreement that Christ's body in heaven has all the properties of matter, including quantity. That it does not occupy space in the host, then, must be because God has prevented it from doing so.[49]

Suppose, anachronistically, that every point of material substance were endowed with a repulsive force. So long as a material point has that force it will prevent other material points from being where it is. There seems to be no contradiction in supposing that Socrates' body continues to serve as the substrate of his soul even while the repulsive force is removed from the matter of his body. Like a ghost, Socrates' body could then penetrate or be penetrated by other bodies. Yet it would not have become a spiritual substance: unlike any such substance, it would, if God ceased to act on it miraculously, immediately occupy space.

Ghostly matter is difficult enough to conceive. To conceive of unextended matter was, for many philosophers, not merely difficult but impossible. Imagine Socrates—or for that matter, the world—compressed into a single point. Would not its parts become confused "and all immediately joined together in substantial union," as some Thomists thought?[50] One might suppose that the parts at least of elemental substances, which are entirely homogeneous, are distinguished *only* by their relative positions in

48. "Substantia autem quantitate carens nec posset impellere aliud corpus, nec ab eo impelli, nec etiam posset a suo loco excludi ab alio corpore, quia non esset in loco, quantitativo modo, nec esset impenetrabilis loco cum quolibet corpore" (Suárez *Disp.* 40§1¶10, *Opera* 26:534).

49. The other alternative is seen in the doctrine rather obscurely alluded to by Fonseca: the bodies of saints can "have several parts in the same place, for instance, two hands or two feet, and indeed however many [parts] as they wish, which can be at once in the same place as other bodies" (Fonseca *In meta.* 5c13q1, 2:643D).

50. Suárez *Disp.* 40§4¶28, *Opera* 26:550; cf. ¶19–20.

space. If such a substance were deprived of quantity, and therefore of extension, those parts would coalesce into one.

For more complex substances, at least, the outline of an answer can be
found. In a passage quoted earlier, Suárez holds that a material substance is
"actually so extended and ordered within itself" that if not hindered by
divine intervention it will be present at, and occupy, a more or less extensive
region of space. It has, in other words, an *intrinsic* order prior to the ordering imposed on it by its relations to other material substances. The nature of
that order is suggested by some remarks in Fonseca. He distinguishes the
"order of parts in the whole" from the "order of parts in place." Socrates'
head and feet, for example, are extremities of his body, while his chest is in
the middle. That holds wherever Socrates may be, and no matter what
relative position in space the parts of his body may occupy (Fonseca *In meta.*
5q1§4; 2:643). It is prior to, and determines, the ways his body may actually
be extended in space.

Imagine a crude picture of Socrates' body, in which his hands, feet, head,
and so forth, are designated by dots, and the relations of contiguity among
those parts by lines connecting the dots. Such an object would now be called
a *graph*. Certain properties of graphs hold independent of any particular
embedding in space. A graph is connected, for example, if any two points or
nodes in the graph can be connected by a path along the edges of the graph.
A connected graph is a tree if there is exactly one such path between any two
points. The graph of Socrates is a tree. There is, for example, just one path
from the head to the left foot, and no matter how the graph is drawn, that
path must encounter the torso and the left leg. An extremity of the graph is
a node that meets exactly one edge. Nodes are extremities or not independent of any particular embedding. With due allowance for anachronism,
Fonseca's remark can be taken to express that independence. Socrates'
head and feet are extremities, and the torso lies between them, no matter
what region of space he happens to occupy, and no matter how he is
oriented.

One cannot credit the Aristotelians with twentieth-century mathematical
ideas. But Aristotelianism encourages a view of substances and their parts
which is, in its way, less concrete than that of many seventeenth-century
philosophers. Medieval diagrams of the eye, for example, unlike those
found in Descartes's *Dioptrics,* make little attempt to convey the precise
spatial dimensions of its parts. The diagrams depict what one might call the
*structural* characters of the parts of the eye, defined, as we will see, by their
operations and their "topological" relations.[51] To argue, as Suárez does, that

---

51. See the examples in Lindberg 1970; Reisch *Marg. phil.* 10tr2c9, p421. The drawing in
Reisch (1517), unlike the medieval drawings reproduced by Lindberg, does not represent the
various "tunics" merely by concentric and intersecting circles. But it comes hardly any closer to

a material substance that is not actually extended still has a *potentia* to occupy space, both as a whole and part by part, is, I take it, to recognize that there are determinate relations among the parts independent of its actual existence in space. Such relations cannot be predicated of spiritual substances. There is no graph of the powers of the rational soul, even if these are regarded as parts.

The predominant doctrine concerning condensation and rarefaction also militated against identifying parts through actual spatial relations, or quantity with actual extension. The same "quantity of mass," it was thought, could occupy volumes of quite different size. A passage from the *Sentences* commentary of Franciscus de Marchia expresses the common view: "Quantity of mass in one thing, and the variable extension of that quantity, even while it itself does not vary, is another, as is apparent in the *motus* of condensation and rarefaction. A rarefied body, though no part of quantity is added to it by rarefaction [. . .] occupies a greater place than before, which is not on account of its having a greater quantity now than before, but only on account of extension."[52] Since a vacuum was thought to be impossible, rarefaction could not consist in the dispersion of particles into a larger, otherwise empty space, or in the enlargement of vacuous pores in a spongelike connected mass. It consisted instead in what one might call a "thinning out" of the *same* continuous quantity of matter.[53] In watercolor painting, a brushful of paint can be dabbed densely onto a small spot or spread over half a sheet. The bulk of paint is the same; but its place varies in size, and thus its extension in the sense that more places are occupied by that it, and other bodies excluded. Where we would see, in the expansion of gases when heated, a fixed number of particles, each itself unchanged, dispersed into ever larger spaces, the Aristotelian sees a fluid being smoothly and continuously redistributed. The parts of an expanding body

---

depicting actual spatial relations (cf. the modern drawing in Barlow and Mollon 1982:36 [fig. 2.1]).

52. *In sent.* 4d13q1, quoted in Maier 1955:142, 200. Cf. Toletus *In Phys.* 4c9q11, *Opera* 4:132rb, which resembles closely the definition of Ægidius (see n.56). Rarefaction and condensation are distinct from augmentation and diminution. In augmentation and diminution, quantity is added to or taken from a substance. Animals and plants do not grow by rarefying, but by converting nutrients into their own matter. Air that expands upon being heated, on the other hand, does not increase in quantity; experience shows that it expands even when heated in a closed jar (Toletus *In Phys.* 4c9q11, *Opera* 4:132va).

53. Needless to say, the interpretation of condensation and rarefaction was controversial. The view I summarize, which seems to me to make the most sense, follows that of Ægidius Romanus (see his *In phys.* 4lect17, p96va; quoted in Maier 1949:29–30). Ægidius argued that there were "two kinds of quantity: one by which matter is so and so much, and which is great or little, and one by which matter occupies so-and-so-much space as [being] bigger or smaller." Call the first kind of quantity 'bulk', the second 'volume'. A substance is rare if its bulk occupies a large volume, dense if its bulk occupies a small volume. A given bulk of water, for example, will occupy ten times the volume if transmuted into air. Only its volume changes, however; its bulk remains the same throughout.

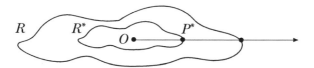

Fig. 3. Condensation

will of course expand with it; their shape need not be preserved, but their graph-theoretic and topological relations will be.

Ockham's picture seems to have been that condensation and rarefaction were what would now be called 'dilations' of quantified matter (see Figure 3). Let $\gamma$ be a constant greater than 0. Take a fixed point $O$ in space, and let $P$ be an arbitrary point distinct from $O$. Let $l$ be the line from $O$ through $P$. On $l$ lay off a segment $OP^*$ whose length is equal to $\gamma$ times the length of the segment $OP$ from $O$ to $P$. The point $P^*$ is the dilation of $P$ with respect to $O$ and $\times$. If the same transformation is applied to all points of a region $R$ the result will be a region $R^*$ whose points are in one-to-one correspondence with those of $R$. Ockham writes that "from the very fact that the parts of a substance or quality are more distant from each other and from the center of the whole, without any other thing arriving [i.e. without the reception of any new form], a body is rarefied; and if [the parts] were less distant before it was then dense" (Ockham *Q. in phys.* 97, *Opera philos.* 6:658). So every part into which the solid matter could be divided must be more distant from its neighbors. The transformation of the region must be continuous and one-to-one. If it is, there will be no interpenetration or destruction of parts; shape will even be preserved if the dilation is uniform (i.e., if $\gamma$ is constant).

Ockham concludes that quantity remains numerically identical through condensation or rarefaction, and that there is no need to suppose, as the atomists did (and as Descartes later does), that condensation and rarefaction are caused by the contraction or expansion of invisible pores. Toletus agrees with the second but not the first: "I judge [. . .] that there is no change according to rarity or density unless something of quantity is lost or newly acquired" (*In Phys.* 4c9q11, *Opera* 4:132vb). Toletus treats rarefaction and condensation on a par with other alterations like heating and cooling. Alteration in general consists in the addition or removal of degrees of quality; the same holds for rarefaction and condensation. Quantity no more remains identical in such changes than do qualities in alteration.

Whichever view one considers, the identity of parts within a body cannot be tied too closely to their actual spatial relations. Neither view precludes the possibility that a body could be shrunk to a single point. In that case there would be no spatial relations by which to distinguish parts (which is why philosophers like Capreolus believed that there would be no parts). But there can still be relations of order—connectedness, extremity, and so forth. Such relations, I take it, may be what Suárez and Fonseca have in mind when they insist that even under such conditions a material substance may remain what it was. In that sense body can be conceived apart from quantity, and without being confused with spiritual substances. As Suárez notes, even when deprived of quantity altogether, a material substance, unlike any spiritual substance, remains the sort of thing that *could* occupy space, and indeed would if its natural state were restored to it.

## 4.3. Figure and Other Qualities

Many interpreters of Descartes have noted that the definition of matter as extended substance and the restriction of its forms to the modes of extension amount to a rejection of active powers in nature. But few to my knowledge have looked at the Aristotelian treatment of figure. Figure is indeed not central to Aristotelian physics. Its role in physical explanation is, by comparison with that of quantity or motion, more restricted and less controversial. The central texts agree that it is entirely passive and never an essential property of any natural thing. Only in artifacts does it come into its own; but artifacts have no natures, no internal principles of change, and are therefore not among the objects of physics. Nevertheless figure recurs in physical arguments as a kind of stand-in or likeness of form, especially substantial form. We find Toletus writing that "the figures of things are strongly analogous [*magnam habent proportionem*] to substantial forms. For as proper substantial forms are consequent upon singular things and species, so too are their exterior figures; so that indeed many have mistakenly judged that substantial form is nothing other than this exterior form" (Toletus *In Phys.* 7c3q3, *Opera* 4:198va). Toletus adds that just as matter and substantial form are composed into a thing that is *unum per se,* so too figure and a "subject proportioned" to it are composed into something that is "somehow" *unum per se* also. One indication of that unity is the use of the same word to denote both people and images of people, as if the form of the image conferred determinate existence on its "matter" as the human form confers existence on prime matter. Figure, moreover, is not alterable. Unlike other qualities, it is not altered per se, but produced anew when some other quality changes.[54]

Figure thus enjoys a kind of double life within Aristotelian physics. It is, on the one hand, a relatively unimportant part of the furniture of the world, never initiating change, or even being changed independently of other more fundamental properties. It is, on the other hand, the character by which artifacts are defined, and through the recurrent analogy of artifacts

54. Aristotle *Phys.* 7c3, 245b9[ff].

with natural objects it becomes a simulacrum, an "index," a representer of substantial forms which, because their activity is always mediated by accidental forms, cannot be immediately known to us. In this section I consider the definition, mode of existence, and physical properties of form, deferring the analogy of art and nature to §7.1.

The Categories, as I have mentioned, put figure among the species of quality. It is, in the words of the Coimbrans, a "quality resulting from the termination of visible quantity in a natural thing, like the external aspect of man or lion" (Coimbra *In Log.* 1:489; Suárez *Disp.* 42§4¶15, *Opera* 26:615). It is the external termination of quantity, "a certain agreement and harmony of the parts of quantified substance [*quanti*]," the internal termination being what we might call the amount of a finite quantity—two feet, say.[55] With figure is included, significantly, form, not the form which is a principle of change, but what was variously interpreted as a special case of figure— whether that of animate as opposed to inanimate things, or of beautiful things, or of corporeal as opposed to mathematical things.[56]

It may be surprising to find figure regarded as a quality. Certainly Descartes, when he inveighs against qualities, did not intend to include figure. Indeed the genus of quality seems a mixed bag, and it is not evident at first sight what holds it together. The four species defined in the *Categories* are:

> (i)  *habitus* or *dispositio,* a property "which is *per se* and first of all ordered to an operation" (unlike substantial form), not however as a primary power of acting (unlike *potentia*), but as "assisting and facilitating" the actualization of such a power (Suárez *Disp.* 44§1¶6; on dispositions, see §5.3 below);
>
> (ii)  *potentia* (§2.3) and *impotentia* (the privation of a *potentia,* like blindness or lameness);
>
> (iii)  "passive" qualities, like heat and cold, rarity and density, colors, and so forth;
>
> (iv)  figure and form.[57]

There were numerous attempts to explain why the four species were all called 'qualities'. Aristotle has it that 'quality' is "that by which a thing is said

---

55. "Duplices autem terminos forma præstat, externos, & internos; externi sunt illi, qui non sunt de genere quantitatis, ut figura, quæ est in quarta specie qualitatis, & nil aliud est, nisi quidam partium quanti inter se concentus, & harmonia; interni vero illi sunt, qui sunt de genere quanti, & certam quanti speciem constituunt, ut bicubitum, tricubitum, & huiusmodi" (Zabarella *De prima rerum materia* 2c9, *De rebus nat.* 196C; cited in Goclenius *Lexicon* s.v. 'figura').

56. The first is suggested by Zabarella (*Opera logica, Tabulæ* 124); the second is attributed to Simplicius and Boethius, and the third, which Toletus favors, to Iamblichus (Toletus *In Log.*, *Prædicamenta* 8, *Opera* 2:158).

57. On the classification of qualities, see Fonseca *In meta.* 5c14q1 and 2; Suárez *Disp.* 42§4; Toletus *In Log.*, *Prædicamenta* 8, *Opera* 2:153ff; and the useful *Tabulæ* of Zabarella, included in his *Opera logica.*

to be quale" (8b25). That is uninformative, to say the least. Later he notes that qualities have contraries, are capable of degrees and provide the ground of similarity and dissimilarity. But the first two do not apply to all four species, and in particular not to figure. The last, though shared by all four species, is not shared by them alone. Aristotle's text leaves the problem unsolved.

The most common answer was that a quality is an accident by which a substance is perfected in being or in acting. Unlike substantial form, quality presupposes a substance and cannot itself constitute a substance from prime matter. It "perfects" substance only in the sense that any actual existing thing must be fully specified in every way that things of its kind can be. In Suárez's careful formulation: "Quality, therefore, seems to be a certain absolute accident, adjoined to created substance in order to complement its perfection, both in existing and in acting" (Suárez *Disp.* 42§1¶5, *Opera* 26:607). How, then, is figure a quality? Every material substance naturally has quantity. Its quantity must be of some definite amount, and in that sense specified or "terminated." The same quantity may be terminated in many different ways, may have many different shapes or figures. Any actual substance must at each moment have one particular figure. Figure completes finite quantity, makes it fully determinate and thus actually existent. It satisfies, therefore, the definition of 'quality'.

Among qualities, however, it has several peculiarities. Suárez admits that figure "would seem to follow upon [*consequi*] quantity" rather than form. If, moreover, quantity exists in matter prior to its union with substantial form (§5.3), then figure too, as a mode of quantity, might precede substantial form rather than complete it. Nevertheless it "follows upon form, insofar as it pertains to the finishing [*ornamentum*] of substance, and can, in a certain way, be of service in action or natural change" (*Opera* 26:607). As Aristotle says in *De anima*, each thing, especially each animate thing, has a figure proper to it (*De an.* 407a23$^{\text{ff}}$). Matter has no power to determine such a figure: it must therefore come from form.

Indeed, figure is interchangeable with form in certain contexts. Ockham, who came closer than most to identifying form and figure, agrees that "in artificial things there is neither matter nor form" when those words are taken in the strict sense. But if by 'matter' one means merely 'that by whose transmutation some proposition now is true that was false before', and by 'form' 'something verifiable afterwards (i.e., after the change)', then figure can be the form of an artifact. So, for example, the figure of a bronze statue can be called the "form" of the bronze, since the bronze was not a statue before and now is one.[58] The example recurs, as well it might, in the attempt of Le Bossu to align Cartesian form with Aristotelian form (*Parallèle*

---

58. Ockham *Brev. sum.* 1c3, *Summula* 2c2; *Opera philos.* 6:23, 216–217.

12, p133). Le Bossu simply substitutes figure for form. For him, as for Descartes, the figure of a statue and the form of its bronze are forms in precisely the same sense. With that none of the central texts could have agreed. Even Ockham denies that 'form' can be applied unequivocally to artificial and natural things. But the possibility was there to be exploited.

The second peculiarity of figure follows from the first. Unlike other qualities, it is peculiarly dependent, not merely on substance, but on another accident—quantity—for its existence. In Aristotle's list, it is last because least, "both in perfection and in its mode of being, so that to some it seems scarcely to merit the univocal designation of 'quality' " (*Disp.* 42§5¶9, *Opera* 26:625). Other qualities can, at least with respect to the absolute power of God, exist apart from quantity (7§1¶19, 25:257). But figure can no more exist apart from quantity than *motus* apart from the *mobile*. Neither has "by way of its own concept sufficient being to be conserved; only by way of a certain identity with that in which they inhere [do they have enough being to be conserved]" (7§2¶10, 25:265). Yet figure is not identical to quantity: quantity can exist without any one of the particular figures it is capable of, though not perhaps without any.

The concept of figure, as we have seen, is of a certain mode by which finite quantity is terminated. Figure is, so to speak, an accident of an accident, an adverb of *quantum*. The quality of heat, on the other hand, is not thus related to any other accident of the substance it inheres in. Though a lump of coal may be hot and square, its hotness is not a way of being *quantum* or of being black. What stands to heat as figure to quantity is intensity or degree: just as a thing cannot be just *quanta*, but must have a definite figure, so too a thing cannot be just hot, but must have a definite degree of heat.

Neither figure, then, nor intensity inheres in bodies independently of other accidents. Since, as we have seen several times, independent existence is the ground of distinction, figure, unlike other qualities, cannot be really distinct from the quantity of which it is a mode. It might be objected that by the same reasoning other qualities too must be modes of quantity. The crucial instance would be one in which matter and qualities remained, while quantity was ablated. But we have seen that the Aristotelians generally agreed that a corporeal substance can exist without its quantity. Its figure will, of course, not continue to exist. But that is because of its peculiar relation to quantity. There is no contradiction—so it was thought—in holding, say, for heat that it exists in a body even while that body lacks quantity. The heat will not occupy space, to be sure, but again there seems to be no contradiction in that consequence.[59]

59. Color—to take an instance discussed in the seventeenth century—might well not be

Descartes, like Boyle, was inclined to ridicule the doctrine of real qualities or real accidents. But it seems to me that the difference discerned by the Aristotelians between "adverbial" accidents and "primary" accidents is genuine. That they expressed the difference in holding that primary accidents are real and adverbial accidents only modes was unfortunate, although the discomfort of later philosophers rests largely on misinterpreting reality as substantial existence. Still the intuition was sound. A real accident is a feature of the world that, though neither a complete individual in its own right nor capable of conferring existence of itself (like substantial form), serves as the basis for further specification, as quantity for figure, or heat in general for intensities of heat. An adverbial accident is such a specification. That is not to say that there is heat in general, or indeterminate quantity, and then also particular intensities or figures. Every actual instance of quantity inheres in a completely specified individual and is thus itself completely specified. But a thing must have quantity if it is to have a completely specified quantity. Though the converse also holds, the same instance of quantity can, as I have already said, be specified in different ways. So although each instance of quantity can exist without the figure it now has, its figure, being the specification of *that* quantity, cannot exist without that instance of quantity. Another instance of quantity could, of course, have an exactly similar figure; but that figure would be numerically distinct from the first. It is in that sense that figure is dependent on quantity, or intensity on heat.

The final peculiarity of figure is passivity. The Coimbrans, arguing that figure has no active power whatsoever, begin with the commonplace assertion that "every natural composite comprises matter and form, of which [matter] has the task of undergoing action, and [form] that of acting." Quantity serves as an "assistant" [*administrum*] to matter, as quality to form, in carrying out those tasks—quantity by way of supplying extension to matter, quality by way of supplying active powers to substantial form. Of the qualities we find in matter, some, like the elemental qualities, generate their like in other substances. Quantity, on the other hand, does not: it "is consequent upon matter, and thus imitates its nature, and exists to receive, not to act" (Suárez *Disp.* 18§4¶3, *Opera* 25:624). Although it is true that an animal in reproducing will generate a quantified substance, its quantity serves only as the point d'appui for the active powers that do the real work of reproduction. Hence there is some reason to hold that quantity is not itself active. But if quantity is not active, then a fortiori figure, which is merely the specifica-

---

capable of existing without quantity. Color was thought to be a quality inhering in the surfaces of bodies. A body without quantity, since it does not occupy any place, has no surface, and so no color either. Of sensible qualities, perhaps only hot and cold would persist in the absence of quantity. Even wetness and dryness, because they are defined in terms of the limits of bodies— dry bodies having definite limits, wet bodies indefinite limits—depend on quantity.

tion of quantity, is not active either (*Disp.* 18§4¶8, *Opera* 25:625; Coimbra *In Phys.* 2c7q19, 1:302).

Figure does affect the way in which active powers operate. A knife must be sharp in order to cut. For that reason some philosophers had argued that figure is a "principle of local motion."[60] Albert, for example, argued that "the acute figure [of an axe, say] is part of the active principle by which it cuts," along with its heaviness and hardness. Similarly Durandus argued that "the figure of a seal is the principle *per se* of the similar figure impressed in wax" (Suárez *Disp.* 18§4¶9, *Opera* 25:626). But Suárez and the Coimbrans both argue that figure contributes to the result only as a "disposition"—a term I will return to (in §5.3): "But one may answer that such figures are only dispositions on the part of, the instrument or body, so that it may move or be moved more easily in such manner, either because it resists the movement less [. . .] or because it is less resisted, as in the motion of cutting; for where an instrument is sharper, it has fewer parts, and thus less resistance occurs" (Suárez *Disp.* 18§4¶9, *Opera* 25:627; cf. Coimbra *In Phys.* 2c7q19, 1:303). The pointedness of a knife consists in its having a small cross section relative to its length and the gradual increase of the cross section as one moves from the point inward. Since resistance varies directly with cross section (see, e.g., Coimbra *In Phys.* 4c9q5, 2:75), a pointed object encounters less resistance than a blunt one. But the mere arrangement of parts (so the argument assumes) cannot of itself initiate the movement. Only the heaviness can do that. Figure contributes not to the existence of the movement, but to its manner.

Figures change when other qualities are altered, and when substances are generated or corrupted, but always as a by-product. In any natural change, as we have seen (§3.1), there is a *terminus ad quem,* a state aimed at, which consists in the reception of a form. That form is the proper effect of the causes of the change: the material cause is its subject, the formal cause the form itself, the efficient cause the agent initiating the change, and the final cause again (with some qualifications) the form. Figure, as we have seen, is never the *terminus* of any natural change. It comes about, to use Toletus's word, *with* change, but not *because* of change (Toletus *In Phys.* 2c2q3, *Opera* 4:52va).

Natural change is paradigmatically change between contrary species under the same genus, as from cold to hot. Toletus divides such changes into "immediate" and "mediate" (Toletus *In Phys.* 7c3q3, *Opera* 4:198rb). Immediate changes are between qualities that "are contraries, and by themselves act and undergo action." Only the elementary qualities satisfy that condi-

---

60. Suárez and the Coimbrans both mention Durandus (*In sent.* 1d45q2n8) and Albert (*In de cælo* 4 ad finem); in support of the opposing view the Coimbrans cite Scotus (*In sent.* 4d1q5), Thomas (*Contra gent.* 3c105, *ST* 2pt2q96a2), and the Thomist commentator Ferrarius.

tion. Mediate changes are between qualities that "do not act of themselves, but by way of the first [qualities], from the mixture of which they arise." All sensible qualities satisfy that condition; they are contrary to one another derivatively, by virtue of the primitive contrariness of the elementary qualities they are composed of. Figures, on the other hand, are neither elementary qualities nor composed from elementary qualities. They are contrary to one another neither primitively nor derivatively, but only in the very weak sense that no body can have distinct figures at the same time. Change of figure cannot be subsumed under the schema that was introduced in §3.1.[61]

The natures of things depend primarily on their active powers (§7.1); the reception of the substantial forms from which those powers emanate depends ultimately on the elemental qualities of the underlying matter. Quantity is indeed a necessary condition for the reception of form (§5.3). But of itself, it is entirely indifferent to the particular forms that inhere with it in prime matter. There are indeed certain limits to the size of complex material substances, especially animals (§6.4). In that sense the determination of quantity, and thus to some small degree figure, has a role in the reception of form. But in general, figure, however important to our knowledge of substances and to the practical arts, has virtually no role in explaining natural change.

That is, nevertheless, not the whole story. Figure has a role in arguments against Atomism, since their identification of figure and form must be refuted. It is also prominent in questions about the motions of the heavens, since Aristotle held that the proper motion of the simple substances of which they are composed was circular. There are a few questions even in terrestrial physics which refer to figure, notably those that concern the shape of the earth. But the figures in question occur as effects of motions whose end is other than the production of figure.

The central texts argue that figure is at most a "disposition" contributing to the effects of active powers. The fourteenth-century Parisian philosopher Nicole Oresme argues, however, that configurations of qualities can have

61. The point is not just that changes in figure are derivative upon other changes. Changes in color or taste are also derivative. But in those cases the quality attained by the change can still be regarded as the *terminus ad quem* of the change, and the state of the thing with respect to that quality before the change as a privation. But (so I read Toletus) the figure attained in change is never that which is aimed at in the change; it comes about incidentally as something else is attained. When bits of earth coalesce at the center of the universe, the figure of the resulting mass is spherical; but sphericity is not part of the *terminus* of the local motion of any of the bits. The case of animal parts is more difficult. It would seem, for example, that when carnivores grow teeth, sharpness as well as hardness would be the *terminus*, since otherwise the teeth would be ineffectual. Nevertheless, it may be that the sharpness of the teeth can be regarded as derivative, this time not upon the actions of the elements, but upon that of the soul, whose end is not the sharpness of the teeth per se but their effectiveness in eating or fighting (cf. the discussion of the shedding of horns by deer in Gotthelf 1987:220, n.32).

proper effects. Contrary to the Coimbrans, who are very much against attributing efficacy to the figures and characters used in divination (*In Phys.* 2c1q7a2, 1:219), Oresme, though skeptical, does not reject the possibility out of hand (*De config.* 1c31, p231). His doubt concerns not the possibility of efficacy, but whether it can be achieved by art, since "it is much more probable that bodies have an efficacy or power [*virtutem*] arising from a natural figuration of active quality than from an artificial figuration of quality." Natural figurations probably do have proper effects. Different kinds of animal, for example, may well have different characteristic configurations of heat. The natural friendship between certain species might result from the "fitting accord between the configurations of the primary or other natural qualities" in each (Oresme *De config.* 1c24,27; p233, 241ᶠ). Differences in the expression of the active powers of things—the sympathies and antipathies, for example, of natural magic—can be correlated with differences in their qualitative figures. In the soul, the seat of memory and imagination can be "figured qualitatively in a variety of ways depending on the diversity of the forms or species which it receives [in sensation]" (ib. 1c31, p249). For Descartes the suitability of spatial figure, in its infinite variety, to provide the terms for a "coding" of physical properties into shapes is the key to his theory of the senses (see *Dioptrique* 4, *Regulæ* 12). And when he enthusiastically affirms in the *Regulæ* that *all* physical "dimensions," qualitative or quantitative, can be represented by lines and plane figures, he repeats one of Oresme's contentions.

Nevertheless in all such instances figure remains a means of representation only. More significant to Descartes was the use of geometric reasoning in music and optics. These were among the "middle mathematics," neither wholly physical nor wholly mathematical; they were also, and not coincidentally, the domains in which Cartesian science came closest to fulfilling the promises made for it. I want here to mention one aspect of the relation between physics and mathematics: the application of the matter-form distinction to mathematical objects.

Toletus, following St. Thomas, distinguishes sensible from intelligible matter:

> By sensible matter [St. Thomas] understands a substance endowed with elementary qualities—heat, cold, wetness, dryness—and composed from them; while by intelligible matter, he understands substance conceived solely by way of quantity, and which is considered in mathematics; so it is commonly said. But as for intelligible matter, it seems to me that one should say that quantity alone is intelligible matter. Note, however, that the Mathematician has as his object not quantity, but figures and forms, which occur in quantity. For as the ironworker is concerned not with iron

but with the form that is made in iron, so too the Geometer [is con-
cerned] not with quantity but with the figures made in it. (Toletus *In Phys.*
1c1q3, *Opera* 4:8vb; cf. 2c2q3, 53va)

The quantity that serves as the matter of mathematics, Toletus adds, is not
quite that which is found in things. Points, lines, and surfaces must be
abstracted from the quantity we encounter in experience before they can be
treated in mathematics (9ra).

The intelligible form corresponding to intelligible matter is figure. The
mathematician "proceeds by form [i.e., external form or figure], that is, in
proving passions by their forms or subjects; for he considers primarily form,
and makes proofs by them" (ib. 9ra, 53va). Why not, one might ask, also
colors or heat? After all, these too have extension. The answer has been
anticipated earlier. Figure alone, unlike other qualities (which also inhere
in quantified matter), is a mode of quantity, and like quantity, it is neither an
effect of natural change nor a cause (*In Phys.* 2c2q3, *Opera* 4:52va). Quantity, moreover, when abstracted from matter, retains its modes (its "passions," Toletus says), including, of course, figure. Other qualities, even if
they inhere in matter, are not modes of quantity, and will be cast aside.

The analogy is this: figure is to quantity as form to matter. Quantity is
therefore said to be the intelligible matter of mathematics. But in middle
mathematics quantity cannot be abstracted altogether from sensible matter.
Music is the arithmetic of sounds, optics the geometry of light rays. The
Coimbrans, defending the thesis that middle mathematics is "more mathematical than physical," note first that optics, music, and astronomy are
commonly included among the mathematical arts, and that "those who call
themselves Mathematicians, customarily treat them." More importantly,
"the affections that they demonstrate concerning their subjects are usually
shown by means of Geometry and Arithmetic." Since the medium of
demonstration is a "kind of form" in the sciences, middle mathematics, as a
branch of knowledge, differs from physics in form if not in matter (*In Phys.*
2c2q1, 1:227). But the argument, it should be noted, ill suits the practice
even of the Schools themselves. Local motion, which is definitely part of the
subject matter of physics, was treated geometrically, and, as we have seen,
there were efforts to extend that treatment to alterations as well.

A more promising distinction is found in the Coimbrans' reply to one of
the arguments for the opposing view. They grant that neither astronomy nor
optics treats quantity in general. To that extent the two sciences differ from
arithmetic and geometry, and can be said to be concerned with substance as
well as quantity. But "they do not investigate [*scrutentur*] their substance."
Instead they touch on substance—light or the ether—only incidentally,

because they consider the quantity only of certain kinds of substance. Their content is tailored to suit the nature of the substances they treat.

The mathematics of music illustrates well what the Coimbrans may have had in mind.[62] Beeckman writes: "I suppose that the nature of the human voice, of pipes, lyres, and any sort of musical instrument is the same as the nature of strings, since experience shows that all voices can be consonant with the voices of strings" (*Journal* 1:54). Beeckman no doubt means that "voice" or sound is the same sort of thing no matter what the source. But the succeeding investigation, after starting with an idealized divided string, consists simply in the comparison of ratios. Not only do the sources drop out, but sound itself, except incidentally as that which motivates the mathematics. When, on the other hand, Beeckman examines properly physical questions about sound—why, for example, two strings tuned to the same pitch resonate with one another, or why a single voice may be pleasant or unpleasant (ib. 166, 177)—mathematics is conspicuously absent.

Mathematically, music consists in the computation and comparison of ratios of whole numbers (see Figure 4). Neither geometry nor algebra is used. Nor are just any ratios of interest: only those generated from small whole numbers are studied, since only such ratios are heard as consonant. The mathematical content of music is thus dictated by the long-known relation of string lengths to perceived pitch and by the felt pleasantness or unpleasantness of sounds. Though in principle it could treat ratios with arbitrarily large prime denominators, or even irrational ratios, it does not, because such ratios are regarded as dissonant in every context, and are thus never used.[63] To that extent the content of music depends on the substances it treats. But since it never concerns itself with either the causes of sound or the physiology of hearing, it can reasonably be called a mathematical branch of knowledge.

Similarly in optics, the nature and causes of light are disregarded; quantity, in the form of "rays" or line segments, surfaces (of reflective objects), and volumes (of refractive media), is again all that remains. Certain physical features of light—that it travels in straight lines, that the sun radiates light along all the radii extending outward from its center—guide the choice of quantities. But once the choice is made, optics no longer concerns itself with physical change as such, or with its causes and principles. When Descartes declares, near the beginning of the *Dioptrics*, that he will not concern himself with the nature of light, he is doing no more than making

---

62. On music theory in the period of Descartes, see Cohen 1984 and Dear 1988, c.6.

63. Beeckman, like Descartes, is somewhat conservative in this respect. Simon Stevin had already constructed the "tempered" scale, in which the ratio between every pair of notes separated by a semitone is $\sqrt[12]{2}$. Beeckman rejected that innovation (To Mersenne 30 Apr. 1630, *Journal* 1:180).

| $\frac{1}{2}$ | Octava. | | | | | | | | |
|---|---|---|---|---|---|---|---|---|---|
| $\frac{1}{3}$ | Duodecima | $\frac{2}{3}$ | quinta. | | | | | | |
| $\frac{1}{4}$ | Decima 5ª. | $\frac{2}{4}$ | Octava. | $\frac{3}{4}$ | quarta. | | | | |
| $\frac{1}{5}$ | Decima7.ª | $\frac{2}{5}$ | 10ª maj. | $\frac{3}{5}$ | 6ª maj. | $\frac{4}{5}$ | Ditonus. | | |
| $\frac{1}{6}$ | Decima 9ª. | $\frac{2}{6}$ | 12ª. | $\frac{3}{6}$ | Octava. | $\frac{4}{6}$ | Quinta. | $\frac{5}{6}$ | Tertia min |

Fig. 4. Intervals (from Descartes, *Compendium musicæ;* AT 10:98; note that the largest prime denominator is 5; intervals with prime denominators of 7 or more were regarded as disonant when played both together and in succession)

explicit what anyone reading his work would have taken for granted. It is not that declaration but the subsequent violation of it that evoked the criticisms of Morin.

What distinguishes middle mathematics from pure mathematics is not its objects but the principle by which those objects, and the propositions to be proved about them, are selected. That principle comes not from the objects themselves (if that is possible), or from simple curiosity, but from certain physical or phenomenal properties of the particular kinds of bodies to which a branch of middle mathematics is to be applied. Music deals in ratios because it is the science of consonant sounds, optics in rays and surfaces because it is the science of light. Middle mathematics, because it concerns itself with quantity and its modes, is, from the Aristotelian point of view, inept to determine the natures of things. Quantity and its modes are neither the causes nor the effects of change. Though essential to every physical thing, they are among the least physically significant of its properties. To study them alone, or to postulate an object for physics that had only such properties, would be, for an Aristotelian, to give up the project of natural philosophy.

Indeed, a comparison of Descartes with Oresme and his imitators raises an issue that I will return to in the next part. It is clear that for Oresme the configurations of qualities, though real enough, and *representable* by geometric figures, are not—and could not be—themselves such figures.

Qualities, though by virtue of their intensities analogous to quantities, are in no way reduced to quantities. The issue, then, is not so much why Descartes thought that physics had to employ magnitude and figure in representation. It is why he thought it had to conceive of the *things represented* as nothing other than magnitudes and figures. The methodological virtues of a geometric treatment do not entail that one regard the objects treated as exclusively geometric, or even—as the example of Cusa shows—as quantities. So even though Descartes does often enough tout the methodological soundness of what he and Beeckman called "physico-mathematics," I think one must look elsewhere too. The profounder motives behind the identification of matter with *res extensa* are to be found not in method but in the philosophy of nature: Cartesian physics is, from the Aristotelian standpoint, an *anti-physics*.

In prime matter we find a philosophical concept that, if it does its job, threatens to put itself out of business. The hidden paradox in the maxim that *forma substantialis dat esse simpliciter* is that what gives being *simpliciter* must, on a plain understanding of the maxim, give it to nothing. The solution of Suárez and Toletus is to separate the metaphysical *actus* of mere existence from the physical actus of substantial existence. This does not quite beg the question, but it leaves open the question of what it is that is said to exist. The 'what', in Aristotelianism, is standardly given by designating a form. But here there is no form to designate. None, that is, unless one follows the way of Scotus and assigns to matter a form of corporeality; or takes, following Averroes and Zabarella, or Ockham, quantity to be inseparable from matter even in conception.

That was the way of Descartes also. What may be surprising is that, contrary to Descartes's confident supposition, the coincidence of quantity and extension is not a matter of course. In the Aristotelian treatment, two quite different reasons for separating them are given. From the doctrine of the Eucharist comes the Suárezian distinction between being present at and occupying a place. That distinction will again have a role in the theory of the rational soul. For now it suffices to retain the thought that the occupation of place—whose primary effect is the passive power to resist occupation of the same place by others—is not a necessary condition for having physical effects at a place. The second reason for separating quantity from extension is that 'quantity' denoted what came to be regarded as two quite different properties of bodies: *volume* and *specific density*. Even in Aristotle there are hints of such a distinction, hints made explicit by Ægidius.[64] Cartesian physics never quite met the challenge of explaining it.

64. See Hussey 1991:223, which cites *De cælo* 4c4.

Descartes's ideal was to displace Aristotelian physics from the center of natural philosophy and to put middle mathematics there instead. He saw that to do so required not merely a new method—his own, after all, only canonized the methods of existing middle mathematics—but new objects of study coextensive with those of the old physics. Hence the identification of *forma corporeitatis* with quantity, and of form with figure. The surreptitious substitution of extension for Aristotelian quantity had the salutary effect of sweeping aside superannuated puzzles, and kept, for a time, the specter of *Deus sive Natura* at bay. It also ignored the real problems I have just mentioned. But when those problems were resurrected, the context had changed: solidity and density were now *superadded* to *res extensa*.

# [5]

# The Structure of Physical Substance

So far I have examined matter and form separately. In the ordinary course of nature, neither ever exists without the other. Each has its proper office: that of substantial form is to give specific existence, to cause active powers, to individuate; that of prime matter is to subtend quantity, to be the subject of the dispositions that enable the active powers of substantial form to operate. The two thus complete one another to such a degree that, as the Aristotelians argued, to suppose that either should exist separately except by a miracle is to suppose that nature has produced it in vain.

The intimate association of matter and form raises the question of their distinctness (§5.1). Though the question becomes especially pressing in psychology, it is asked and answered first in physics. The central texts have no trouble with the separability of form from matter; on the separability of matter from form, they record controversy. The more common view was that matter could exist apart from form by divine power. One consequence is that matter and form are really distinct. The real distinction of body and soul is merely a special case. There were other more specific reasons to hold that rational souls can exist without matter; but the availability of the general argument created a presumption in its favor.

The stronger the distinction between form and matter, the more urgent the need to explain the manner in which they are united (§5.2). The predominant view was that substantial form and prime matter, when united in a complete substance, are *unum per se*, not *per accidens*. Yet it is not obvious how that could be so, if they are really distinct. The difficulty will be familiar to readers of Descartes's letters to Elizabeth. To one side, the danger was the reduction of form to a mere mode of matter, with the consequence that the soul is corruptible. To the other side, the danger was Platonism—a world in which the soul, like every incorruptible essence, stands apart from matter. The doctrine of substantial union avoids both dangers, but not, as we will see, without raising new difficulties.

At the end of §3.2 I suggested that substantial form is known not by uncovering a hitherto concealed level of structure but by its effects and causes. Among them are the accidents required for matter to receive form, which are both part of the material cause of form and, in generation, efficiently caused by it. The most fundamental is quantity. In §5.3 I look at the relation of quantity to form and the more general question of the "dispositions" required for the reception of form. Unlike the Thomists, Suárez and Toletus hold that quantity and other accidents common to material substances precede form both in generation and in the constitution of substance. In the resulting structure the quantity of a substance and the qualities that "accommodate" prime matter to particular forms inhere immediately in matter, while the active powers associated with the form inhere in matter only by way of the form. That picture is already suggestive of the division of powers in Descartes's physics.

The term *dispositio*, though never the subject of sustained scrutiny, recurs frequently in physical arguments. It denotes, roughly speaking, the arrangement of parts—spatial or other—to some end. As we will see, it figures prominently in Descartes's physics also, conveniently ambiguous as between an innocuous sense of 'spatial arrangement' and a charged sense of 'purposeful arrangement'. Such ambiguities allow Descartes nominally to agree with his opponents when in fact something quite different, and often antithetical to the Aristotelian sense, is intended. After examining the definition of *dispositio*, I will look at some examples of its use in physical argument (§5.3). The full account of a substantial form ought in principle to spell out the dispositions necessary to its reception in matter; its active powers might then be exhibited as efficient causes of those dispositions. Yet the texts here examined rarely go past showing that such and such a power or disposition must exist. At the end of §5.3 I will briefly consider why that is so.

The arguments in §3.2 on behalf of substantial form frequently referred to the active powers united by it. It is, as I have noted, those active powers that chiefly differentiate substantial forms. In §5.4 I examine the "emanation" of active powers from substantial form and the role of such powers in the generation of substances. Although the substantial forms of terrestrial substances are, by general agreement, incapable of generating their like without the aid of celestial powers, they are capable of preparing matter to receive copies of themselves; their active powers give them the means to do so. In §6 I will follow out the teleology implicit in that account: here I concentrate on the process.

## 5.1. Matter and Form Distinguished

Substantial form and prime matter are really distinct only if each is capable, at least by the absolute power of God, of existing without the other. There seems to have been little doubt among Catholic philosophers that form could exist without matter.[1] The existence of matter without form was another story. Those who, like Thomas, denied that matter had a proper existence were inclined to deny also that it could exist without form even by divine power. The central texts all diverge more or less greatly from that position. Even the Coimbrans, who hold that prime matter is pure *potentia,* distance themselves from the *Thomistæ.* Suárez and Fonseca, since they distinguish the existence of matter from its actualization by form, have little trouble arguing that God can conserve matter *in actu* even without form.

Explicitly the question turns on the actuality of matter. But a more trenchant issue is in the wings. In Genesis we read that "the earth was empty and void." Augustine and other church fathers read 'earth' as matter, and 'empty' as 'formless'.[2] God first created matter and then supplied it with "all the [species] of which this mutable world consists" (Augustine *Conf.* 12c8, p220). So matter existed naturally without form for some time. Augustine did not, of course, deny that God created form and matter from nothing. But the thought that a formless, tenebrous matter preceded all creation could not be far away. In the *Timæus,* matter is eternal; the demiurge imprints it with simulacra of the Forms, and orders it toward himself and the Good.[3] According to a certain Heraclides, disciple of Plato, there was "long ago a formless empty world," in which everything "lay weary and burdened

---

1. Suárez writes that "Aristotle and [other] philosophers perhaps could deny that [material forms] can in any way subsist separately from matter, because they judged that dependence and actual inherence in matter are essential to them. But we Catholics, who believe that God conserves accidents without their subjects, cannot doubt (although certain moderns do) that God could also conserve substantial material forms without matter, since the dependence of an accident on its subject [. . .] is greater" than that of substantial form on matter (Suárez *Disp.* 15§9¶1, *Opera* 25:532). Zabarella does not hesitate to ascribe to Aristotle the thesis that "*being in matter* is a necessary and essential condition for [material forms]" (Zabarella *De anima* 2text3, p120A).

2. "Nonnulli enim [. . .] opinati sunt in ipsa nascenti mundi origine fuisse à Deo procreatam materiam absque ulla forma, idque significari cap. 1. Geneseos illis verbis, Terra autem erat inanis & vacua, ubi, ut D. Augustinus cap. 4. libri imperfecti de Genesi ad literam ait, per terram, quæ omnibus elementis subsidet, materia, per inanitatem, eius informitas designatur" (Coimbra *In Phys.* 1c9q6q3, 1:171).

3. See, e.g., for example Calcidius *In Timæum* ¶306, 321, 329; pp307, 316, 323; Blumenberg 1988:139. Weinberg has brief expositions of Platonist, Neoplatonist and Avicennian cosmologies (Weinberg 1964:11, 21ff, 116ff). Mersenne in a letter of 1630 reports that Robert Fludd takes the Biblical *creatio ex nihilo* to be creation from formless matter: "Materiam (quam sæpissime *tenebras* vocat) esse id, quod proprie appelletur *Nihil;* ac proinde cùm Deus dicitur creare aut facere aliquid ex nihilo, intelligi creare aut facere ex materiâ." After which he asks, "if these [opinions] are not impious, what can be?" (Mersenne *À de Baugy* 26 Apr. 1630, *Corr.* 2:442).

in sad silence," until the demiurge set the heavens apart from the earth, the land from the sea, and gave the elements their proper forms (Coimbra *In Phys.* 8c2q2a1, 2:164). It is not surprising that those who propounded this theory bolstered their argument by appealing to Genesis. There is, after all, nothing in the opening verses to contradict them.

If, on the other hand, matter cannot naturally exist apart from form, then matter and form are either both eternal or both created. One could there-fore refute the eternity of matter by refuting that of forms, which was easier.[4] In showing, moreover, that matter is created along with form, one shows that matter is not, as the Manichaeans thought, evil in itself. The Church insisted as a matter of faith that all things, material and spiritual, "are produced by God from nothing according to their entire substance,"[5] and that "every creature of God is good."[6] Only an uncreated matter, therefore, could be evil; but whether reason alone could prove what faith affirmed was another question.

Aristotelians, especially Catholics, had to navigate carefully among *fundamenta* that on their face were not easily reconciled. The world had to be, on the one hand, utterly mundane, purged of divine attributes. No such doctrine as that of Stoicism, with its world soul, could be admitted; nor, certainly, could God be corporeal.[7] Nor could the existence of the world be implied in that of its Creator: creation must be a free, hence contingent, act of God. One route to enforcing the separation was to make matter and God as distant as possible. God is *actus purus,* necessarily existing; symmetry would have it that matter is *potentia pura,* having no proper existence at all.[8] Or, if it does, at least it exists independently only through miracle and is thus entirely subordinate.

4. Of the twelve arguments used by Toletus to prove that the world has not existed for an infinite time, eight presuppose that the creation was of complete substances, and not of matter alone (Toletus *In Phys.* 8c2q2, *Opera* 4:214[ff]). The Coimbrans, who devote their energies mostly to refuting the contrary view, offer four arguments, of which two presuppose that if the world has existed for an infinite time so have humans (Coimbra 8c2q3a3, *In Phys.* 2:267[ff]).

5. Denzinger 1976, no. 3025 (Concilium Vaticanum I, 1870); cf. no. 800 (Cc. Lateranense IV, 1215, against the Cathares), and no. 1333 (Cc. Florentinum, 1442).

6. Denzinger 1976, no. 1350 (Concilium Florentinum, 1442). This follows an omnibus anathema against heretics of all stripes, including Gnostics and Manichaeans.

7. See the discussion of God's body in Funkenstein 1986:42–47; on the nondivinity of the world, see Milton 1981:191[ff]. On God as matter, see §4 n.27. Certain philosophers in the twelfth-century School of Chartres identified the *anima mundi* of the *Timæus* (see Calcidius *In Timæum* ¶26, 54, etc.) with the Holy Spirit, the third person of the Trinity. One obvious objection, noted already by William of Conches, is that the Holy Spirit is "consubstantial, coequal, and coeternal with the Father"; the world soul, if it exists, must be a created thing (Gregory 1955:134[ff], 142, 149, 148n.1; cf. William's *Dialectica* [ed. Cousin], pp475–476). Giordano Bruno, according to a scandalous report written the day of his death and published anonymously in 1621, revived the identification (Mersenne *Corr.* 1:137; cf. Michel 1962:118, which cites Bruno *Lampas triginta statuarum* [1587], *Opera lat.* 3:54).

8. See Coimbra *In Phys.* 1c9q6, 1:167; Suárez *Disp.* 13§4¶7, *Opera* 25:411.

But since God's power is absolute, if it is not contradictory to suppose that matter should exist without form, then God can bring about that it should do so. The independence of matter from form is in keeping with the tendency in the central texts to mitigate the debasement of matter one finds in Neoplatonism and in Augustine. Fonseca argues that if, as Dionysius affirms, existence is the principal way for created things to participate in divinity, then matter too must exist; for "otherwise the derivation of divine *esse* would not descend through all ranks of created things" (Fonseca *In meta.* 8c1q1, 3:444aC). The readiness of Suárez and others to ascribe quantity and other accidents to matter and the increasingly elaborate theory of mixtures are other indications that matter was being granted a more independent role in natural philosophy.

We find, then, a twofold claim: matter can exist without form according to God's absolute power; matter cannot naturally exist without form.

1. The Thomists argued that matter without form would be *actu sine actu,* which is openly contradictory.[9] Soncinas, citing Averroes, writes that "matter differs from form but is never denuded of form, and indeed when it is separated from one form, it takes on another, since if it were denuded of all forms, what is not *in actu* would be *in actu*" (Soncinas *Q. meta.* 8q1resp, p180b; Averroes *In Phys.* 2c2comm12, *Opera* 4:52G). If, moreover, the existence of matter is a formal effect of substantial form, then since not even God can produce a formal effect in the absence of form, not even God can bring it about that matter should exist without form.[10]

Those were the primary metaphysical arguments. Against the Thomists, Suárez and Fonseca argue that the *actus* of existence is distinct from that of informing. Averroes therefore equivocates: to hold that matter denuded of form exists while lacking *actus* in the sense of being informed or composing a complete substance is no contradiction (see Coimbra *In Phys.* 1c9q6a4, 1:172). For similar reasons the second argument is rejected. It is false that all existence is from form, "as from an intrinsic and as it were essential cause (for this is true of complete substantial existence, but not of each partial [existence])."Though the existence of matter is naturally dependent on that of the complete substance it is part of, and thus on the formal causality of the form (which is, as we learn elsewhere, the union of matter and form), still God may intervene and preserve matter even when the other is absent and can have no effect (Suárez *Disp.* 15§9¶9, *Opera* 25:535).

9. Fonseca *In meta.* 8c1q1, 3:442aC; Coimbra *In Phys.* 1c9q6, 1:167; Suárez *Disp.* 15§9¶2, *Opera* 25:532. Cited as arguing against the separate existence of matter are Thomas (*ST* 1q66a1, *Contra gent.* 2c54–55), Capreolus, Caietanus, Ferrarius, and Hervæus—the usual list—together with Bonaventura and Durandus.

10. Suárez *Disp.* 15§9¶2, *Opera* 25:532. Since a form can be an efficient and final cause as well as a formal cause, the effect must be specified to be formal.

Since form can likewise exist without matter, the two are really distinct.[11] The manner of argument will be familiar to anyone who has read Descartes's proofs of the real distinction between soul and body. His and his opponents' reasonings, whatever their differences, share the fundamental principle that whenever God can conserve each of two things without the other, then they are indeed two things in the strongest sense. But where Descartes admits such a distinction only between soul and body, and demotes all material forms to modes of matter, the Aristotelians admit it for *all* substantial forms, material and spiritual. The difference between the soul and other material forms is established on other grounds.

In making a real distinction between matter and form, the Aristotelian version of Aristotle stands apart from many recent versions. Many commentators now would, like the *ethnici* rather condescendingly mentioned by Fonseca, deny that matter can exist without form or form without matter. Form identified as organization or disposition, or as activity or power, must be realized in a material subject.[12] In Aristotelian terms, that is to make form a mode of matter, as Descartes did. One mark of how successfully he and his like integrated their view of form into the new science is that few commentators other than those who continue the Scholastic tradition even notice the shift.

Descartes is noticeably reticent about the means by which God may conserve things that are incapable of existing apart naturally. Suárez, thorough as ever, takes up the problem. How, he asks, can God supply the effect of the form? His response is relevant not just here but wherever the Aristotelians hold that two things not naturally separable can exist by divine power without one another. The opprobrium heaped by Descartes and Boyle on the doctrine of real accidents is directed precisely against the notion that something whose raison d'être is to exist in another should nevertheless be held capable of existing when that other does not. That they take to be not

11. Eustachius, exceptionally, seems to hold that matter can exist without form but not form without matter: "Quòd si materia virtute Divinâ omni formâ exspoliaretur, posset suâ vi subsistere; quia prius naturâ est materiam esse quàm formam: at verò forma, si virtute Divinâ separetur àmateria, non posset vi suâ cohærere, sed statim (nisi virtute Divinâ sustentaretur) in nihilum rueret; quia ab eo subjecto pendet ut conservetur ex quo educitur [. . .]" (Eustachius *Summa* pt3, *Physica* 1tr1d2a9; 3:127). The difference is, however, smaller than it appears. Matter can exist without form by its own power (*suâ vi*), but can only come to be without form through God's power; form can neither exist nor come to be without matter unless sustained by God.

12. Charleton, for example, writes: "how can there be a formal aspect which is not an aspect of anything, a form which is not the form of anything?" (1987:422). On form as disposition, see, e.g., Frede 1985:21; as structure and activity, Moravcsik 1991:46. As Charlton notes, Aristotle's nondualist theory of the soul has newly become attractive to antidualist philosophers of mind, who tend to set aside as unfortunate lapses passages that intimate another view (Charlton 1987:408).

merely implausible, but unintelligible: not even by a miracle could God bring it about that quantity, say, should exist without inhering in matter.

It is worth examining, then, how an Aristotelian conceived of God's operation in such instances. I will take two: the preservation of matter without form; the preservation of accidents without a subject.

In Suárez's account, the first point is that the "dependence" of matter on form is "extrinsic."[13] Matter depends on form either as "a condition or actual disposition naturally required," or as "an informing cause [. . .] concurring and assisting *per se* in the existence of matter." The first is "quite extrinsic and *a posteriori,*" amounting only to the fact that a thing lacking such a disposition cannot naturally exist. The second, too, is extrinsic:

> Although according to this mode [of dependence] form concurs in a certain way as a cause to the *esse* of matter, it does not concur as an intrinsic cause or component of that *esse,* or as its proper subject, but only as an informing or actuating cause, and as [an] extrinsic [cause], in the sense that it is entirely distinct from the effect; and God can make up for [*supplere*] a cause of this sort as an efficient cause [*efficiendo*], even if it causes by informing. The material cause, on the contrary, is intrinsic like the formal [cause]. God cannot make up for such a cause with respect to its effect [. . .] But he can make good its causality with respect to the other component—form—even if matter, according to its genus [i.e, as the material cause] contributes to [*influat*] the existence of its form *per se* and as a true cause. And in that way God may make up for the dependence of an accident on a subject. (Suárez *Disp.* 15§9¶6, *Opera* 25:534)

Of the four Aristotelian causes, two—the material and formal—are intrinsic, since they are parts of the thing of which they are causes, while the other two—the efficient and final—are extrinsic. Not even God can make up for the absence of an intrinsic cause, since (as Suárez shows elsewhere) actual existence is necessary for an intrinsic cause to have its proper effect. But an intrinsic cause may have other, incidental, effects that in the order of nature invariably accompany its proper effect, but in relation to which it is not intrinsic. The existence of matter invariably accompanies the intrinsic effect of form (the complete substance). But matter itself is not intrinsically caused by form, nor form by matter. In a human being, which is a complete substance, God cannot make up for the absence of either matter or form. But it is possible that God should conserve the body without the soul, since the soul "causes" the body only by way of intrinsically causing the human

13. *Dependentia* and related words, used as technical terms, typically denote causal rather than, say, logical dependence; causal dependence can, of course, be understood in terms of any of the four Aristotelian causes. To say that the dependence of matter on form is extrinsic is to hold that form is an extrinsic (efficient or final) cause of matter.

being of which both are parts: "God can make up for the causality of form, not in the composite, but in the matter, not by informing it but by effecting it [*neque informando, sed efficiendo*]" (ib. ¶6, 25:534).

Suárez's position therefore is that God can, extraordinarily, bring matter into existence without the extrinsic contribution ordinarily provided by form.[14] There is no difference in God's action, but only in its circumstances. Whoever believes in the "concreation" of matter and form in complete substances must admit the possibility of the creation of matter alone. Similarly, if God allows the form of a substance to perish while preserving its matter, the miracle consists only in his having continued the action of conserving matter "in that state in which it should not be, and without the conditions that are necessary to the natural mode of existence" (ib. ¶7, 25:535). Whatever mystery resides in God's mode of operation in such instances resides also in the most ordinary instances.

My second instance, the conservation of quantity and other accidents without their subject, requires an action quite different from the ordinary. In his lengthy treatment of the Eucharist, Suárez argues thus:

(i) When the subject of quantity is ablated in transubstantiation, quantity is deprived of a "positive, real, and intrinsic mode," namely its actual inhesion or union with the subject (Suárez *De sacrif. myst.* d56§2, *Opera* 21:280). Inherence is not a mere denomination, nor is it simply presence. Hence "quantity, when it is separated from its subject, is deprived of this intrinsic mode." So it will not suffice to hold that God simply conserves quantity without its subject; some new action is needed.

(ii) When quantity is conserved separately, it is altered by receiving a new mode of being that is in fact repugnant to its former mode of inherence in a subject. Suárez admits that this conclusion is only probable, and that the contrary position (held by Scotus and Domingo Soto, among others), can be defended (¶9, 281). The advantage of his own position, which is that of Thomas, is that if the miracle by which quantity receives the new mode of being is granted, no further miracles are needed for the persistence of other accidents without matter, since they can be taken to inhere in quantity as a proxy-subject (see 282, and the passage from Thomas's *Sentences* commentary quoted at 274).

14. "Si a principio crearet [Deus] materiam solam sine forma, actio eadem omnino esset cum illa, qua Deus actu creavit materiam sub formis solumque constitisset miraculum, vel præternaturale opus in hoc, quod Deus faceret illam actionem sine concomitantia alterius, per quam induceret formam in talem materiam": "If at the beginning [God] had created matter alone without form, his action would have been the same as that by which God actually created matter with its forms, and in doing so would have done something miraculous or preternatural only insofar as God performed this action without the accompaniment of the other action by which he induced form in such matter" (Suárez *Disp.* 15§9¶7, *Opera* 25:534).

(iii)  The new mode of being entails a new action of conservation by God,
rather than a continuation of the old action. In the natural order,
God conserves a subject, together with all its forms, in a single act.
Suárez is careful to note that in this act God does not act in the role
of material cause (which, being intrinsic, cannot be substituted for)
(¶14, 284).

What does the new mode of being consist in? God, Suárez writes, "can
bring it about that an accident should in a certain manner participate in the
mode of existence of substance, by a certain mode that has formally (so to
speak) a resemblance [*similitudinem*] with the substantial mode [of exis-
tence] but materially and in being [*entitative*] is accidental, to which it
suffices that it should affect and terminate accidental being, although it is
not related to another subject" (¶13,283). The argument is largely
defensive—against the very complaint that would be voiced so many times
in the seventeenth century. It proposes to show that to call something an
accident and to suppose that it does not actually inhere in a subject is not
contradictory. Fonseca, citing the common view of Catholic theologians,
argues that the *inhærentia* essential to being an accident is not *actualis* but
*aptitudinalis*. An accident, in other words, is not the sort of thing that must
inhere in a subject if it is to exist at all. It is the sort of thing to which it is
essential that ordinarily it would so inhere (Fonseca *In meta.* 7c1q1,
3:198).[15]

The strategy should be familiar. We have seen it at work in Suárez's
account of quantity. There quantity was construed as not actual but potential extension, here the inherence essential to accidents is construed as not
actual but potential being in a subject. Are such distinctions "groundless," as
Boyle says (*Papers* 22)? They are, first of all, entirely in keeping with the
spirit of Aristotelian physics, which defines, as we have seen, change itself in
terms of the passage from *potentia* to *actus*. Reconstructions of notions like
quantity as *potentiæ* are not, it seems to me, illicit extensions of the basic
scheme. They are indeed extensions: the occupation of space by quantity is
a "change" only for someone who is inclined to regard every actuality as the

15. "Inherence is twofold, actual or aptitudinal [. . .] Actual [inherence] is that by which
an accident inheres actually [*actu*] in subjects; aptitudinal [inherence] is that by which it has a
propensity to inhere [i.e., actually]. No Philosopher either discovered or affirmed this distinction until after it was shown by faith that the accidents of the Holy Eucharist did not exist in
substance [. . .] All judged, before the mystery was divinely revealed, that there was no other
inherence than actual [inherence], and indeed that it was contradictory for accidents to be
separated from the subject substances in which they inhered" (Fonseca *In meta.* 7c1q1§2,
3:198aF). In the commentary on 7c1, Aristotle's view is said to be that "non posse ullum
accidens existere sine substantia, aut ex subjecta substantia in aliam substantiam migrare."
Both conditions are violated in the Eucharist, the first since the quantity of the host no longer
inheres in matter, the second because the other accidents, which used to have the matter of the
host as their subject, now have its unmoored quantity as their subject.

coming-to-be of something that was, or at least could have been, merely potential. Inherence, although it is not, properly speaking, a motus, is an actuality, an existing state of affairs not only for the subject but for its accidents.

There is no doubt that Suárez's argument was intended to resolve an apparent contradiction between reason and revealed truth. Since doctrines of double truth, granting to each source of knowledge its own kind of truth, had long since been ruled out, the contradiction had to be resolved, and in favor of revelation.[16] The question is not whether the shift was motivated by the desire to avoid contradiction—philosophers habitually make distinctions to that end—but whether it was entirely ad hoc. I have already suggested one way in which it could be regarded as a plausible extension of methods already proven. Another way is this. We have seen, and will see again, that the essences of things (other than God) are to be found not so much in what they actually are, or have been at a given time, as in what they potentially are at all times. Human nature is not well captured in a description of Sappho as she is now: if she is sleeping, she neither senses nor perhaps thinks; if she sits, she does not run. Yet she has the power to do all those things. Actualities come and go; only potentialities persist. Perhaps the same holds for accidents: what matters is not actual but only potential inherence.

One could argue, of course, that quantity and inherence are exceptional. But once the boundaries of the possible are marked by logical contradiction, or—what comes to the same—by God's absolute power, it becomes quite difficult to argue that the *actual* possession of any accident or mode is essential to a thing. Is it indeed contradictory to speak of 'matter without quantity'? Only if prime matter cannot but have the power to occupy space. But prime matter is that which, when joined with substantial form, yields a complete substance, and nothing more.

The case of inherence is harder. Boyle, after quoting a standard definition of inesse, and of 'accident' as *id cuius esse est inesse* (a formula cited by Fonseca), writes that if the Scholastics "will not allow these accidents to be modes of matter, but entities really distinct from it and in some cases separable from all matter, they make them indeed accidents in name, but repre-

16. The term "double truth" refers to two versions of the relation between philosophy and revelation: (i) what can be demonstrated solely by reason or experience and what is revealed are different—but all the same only one of the two, namely revelation, is true; (ii) the truths of philosophy contradict, and therefore compete with, revealed truth. The first is found in a number of philosophers in the thirteenth and fourteenth centuries, including Boethius of Dacia. It was partly in response to his *De æternitate mundi* that Stephen Templier, the Bishop of Paris, issued the condemnations of 1277. The second version, though condemned by Templier, was not, it would seem, unambiguously affirmed until the end of the fourteenth century by Blasius of Parma. See MacClintock 1956, c.4; Hissette 1977:13; Steenberghen 1991:348[ff] (on the condemnation); Maier 1955, c.1 (on Blasius).

sent them under such a notion as belongs only to substances [. . .]," a doctrine that is either "unintelligible or manifestly contradictious" (Boyle *Papers* 22). He takes it to be part of the definition of 'accident' that an accident *cannot exist separately* from the thing or subject wherein it is" (*Papers* 21). The only lexicon I have found that definition in is Chauvin's, published after the heyday of Aristotelianism. Chauvin, in keeping with Boyle's (and Descartes's) conception, defines accidents as *modes:* an accident is "a mode affecting a created substance, and so dependent on it that it cannot exist outside [that substance]."[17] But to define accidents as modes begs the question against real accidents—unless, of course, you have already accepted Cartesian physics.

The Aristotelian would insist that not all accidents are modes. Some, like figure, are modes because they depend on other accidents (§4.3). To them actual inherence is essential. Other accidents, including not only passive qualities but quantity, depend naturally only on the subject. Since actual inherence is inescapable for modes, one might conjecture that other accidents might be distinguishable from modes in that respect. Yet clearly inherence—being in a subject—is, so to speak, diagnostic of accidents: hence the move to potential inherence, which is therefore essential to all accidents; for modes the stronger condition of actual inherence is added.

In an earlier section I suggested that in Descartes and his successors self-subsistence became the primary defining feature of substance, with the consequence that the "second substances" of the *Categories* dropped out of sight, along with 'being of' (rather than 'in') a subject. A parallel simplification occurs when 'accident' and 'mode' are taken to be synonymous. Actual inherence and potential inherence coalesce; the notion of *real accident*, meaning that which is both an accident and capable of being conserved apart from substance, becomes, as Boyle quite rightly affirms, unintelligible. But that unintelligibility is not a datum of human experience, nor is it an inevitable consequent of the notion of accident. It is a by-product of the struggles of the new science against the Schools.

All the same, to conceive whiteness apart from the wall it modifies does strain the imagination. Suárez admits as much. His positive account of the existence of accidents apart from substance rests on a rather suspicious

17. "Accidens prædicamentale [. . .] est modus substantiam creatam afficiens, & ab eâdem ita dependens, ut extra eam existere non possit" (Chauvin, s.v. "accidens prædicamentale"). Chauvin's *Lexicon* is, however, a tainted witness on Aristotelian matters (Gilson 1913:v). Goclenius, a better witness, mentions no such condition; he says only that being ordered toward a subject is of the essence of accidents (Goclenius *Lexicon* 29). Micraelius states both that accidents cannot be separated from substance (*accidens non separatur a substantiis*), because its *esse* is inesse, and that an accident is called "separable" if "vel reipsa, vel mentis abstractione à subjecto separari potest." *Accidens reale* he defines simply as *forma inhærens* (Micraelius *Lexicon*, s.v. "accidens"). The Port-Royal authors, like Boyle and Chauvin, take 'mode', 'accident', and 'quality' to be virtually synonymous (Arnauld and Nicole *Logique* 47).

analogy with substances. One might regard it as what philosophers of science call a 'how possible' explanation: it defeats the claim of incoherence by showing one way in which accidents could exist without their subjects, but it does not assert that in fact they do in that way. The existence of matter apart from form is on firmer footing. That God can create matter conjointly with form will readily be granted by anyone who assents to the orthodox doctrine. If form serves merely as a condition or assisting cause in the creation of matter, then God, who can do whatever he does without assistance, can create matter apart from form, or conserve it when form perishes. It is thus rather easy to shift the burden onto those who deny that matter can exist without form.

2. The absolute possibility of prime matter existing apart from substantial form is thus established.[18] Since form can, by similar arguments, be shown to be able to exist apart from matter, the two are really distinct. What remains to be shown is that the two cannot *naturally* exist separately.

On that issue little need be said. There was no great controversy among the Aristotelians themselves.[19] The use of finality, however, in the two arguments given by Fonseca will be of interest later. The first shows that matter cannot be deprived of one substantial form without immediately taking on another. That is a consequence of a more general principle: "agents cannot intend mere privation, but rather some *actus* and perfection, to which a privation is joined." So the natures of things "are not instituted to despoil matter of any form, unless it is in order to induce another" (Fonseca *In meta.* 8c1q1§4, 3:447bC; cf. Coimbra *In Phys.* 1c9a6a1, 1:167). It follows that the corruption of a body is only the incidental effect of a cause aiming to introduce into its matter a new substantial form. Since no natural cause aims at producing matter devoid of form, there is no reason for it to come about naturally.

No efficient cause will bring about bare matter; neither will any final cause. Nature allows nothing otiose; but matter sans form would be, since "it can by itself sustain no operation." Its passivity is recompensed when it is joined with form, but not otherwise. Matter without form is matter without purpose. One might argue that it could exist for the sake of other things that do have form and are capable of acting—humans, say. But it is characteristic of Aristotelianism not to construe natural purposes entirely in relation to human ends (see §6.2). Indeed, the natural impossibility of matter

18. I will use the phrases 'absolute possibility' and 'natural possibility' occasionally as shorthand for 'possibility relative to the absolute (or ordained) power of God'. Absolute possibility thus coincides with logical possibility, but its rationale is different.

19. The authorities cited by Fonseca and the Coimbrans are patristic (see n.2 above, and Fonseca *In meta.* 8c1q1§4, 3:448bC). The only exceptions are Marsilius of Inghen and Gabriel Biel, who held that Christ's body existed without a substantial form after his death.

without form is based on the contrary claim: nothing can exist naturally that does not have its own ends.

It is instructive to contrast these arguments with those of Boyle and Descartes. Although one may show rather easily in Aristotelian physics that matter without form has no place in the natural order, it is very difficult to show that separate existence leads to absurdity. That, and not the mere task of showing that matter without form is unlikely or contrary to experience, is what Descartes and Boyle had to accomplish. It is not surprising that the job was done by changing the definition of 'accident'; little short of that could have worked. The change was not unmotivated. It was harmonious with the new physics. Rejecting potential in favor of actual inherence in the definition agreed with the general policy of banishing powers and potencies of all sorts. But from the viewpoint of the eventual losers in the debate, the arguments of the victors could not but have appeared to beg the question.

## 5.2. Substantial Union

Prime matter and substantial form are really distinct, a plurality independent of the manner in which they are conceived. Yet the complete substance of which they are parts was held to be not two things but one. The question was not whether but *how* matter and form make one substance. Since human beings are an instance of substantial union, the question bears immediately on psychology and on theological doctrines about the immortality of the soul. Among Descartes scholars, substantial union—long overshadowed by the interaction of soul and body—has recently been studied with some attention to its historical context (Hoffman 1986). But it is still unusual for the union of body and soul to be seen as what it was: a question whose terms were set not in psychology but in physics. There are, of course, particular difficulties concerning the rational soul. It alone among material forms is not "educed" from matter; it alone can exist apart from matter even without divine intervention. Nevertheless what was proved about the distinctness of matter and form and about their union could be, and was, applied mutatis mutandis to the soul. Indeed the insistence of the Church on that point hindered attempts to show by reason alone that the human soul is immortal.

Since the controversy about substantial union concerns the claim that matter and form are *unum per se,* I will examine that notion first. The question then will be whether prime matter and substantial form, when they coexist in complete substances, enjoy the highest form of unity that a composite thing can have—whether, in other words, a complete substance is *unum per se.*

1. *Unum per se, unum per accidens.* The Coimbrans distinguish several degrees of unity, from the "unity of aggregation" exemplified by a heap of stones to that of simple substances "free from commixture with matter"— angelic souls, God. In between are the two kinds of unity most relevant to the question: the *unitas per accidens* of accidents with their subjects, the *unitas per se* that results from the "composition of parts coming together in some third nature." Such unity we find in quantity, when two parts share a boundary, and—the Coimbrans add without further argument—in the composite of matter and form.

As an account of the unity of composite substances, that amounts to little more than labeling what one hopes to explain. Nor does it help much to be told that "all that arises out of distinct things without any physical and real union between them" is only an *ens per accidens* (Suárez *Disp.* 4§3¶13, *Opera* 25:129).[20] After all, there seems on the face of it to be little reason to deny that whiteness and the wall are physically united. Yet accidents, apparently without exception, are supposed to be only *unum per accidens* with their subjects.

In the Coimbrans' further remarks, however, we find two more compelling reasons to set substantial unity apart. The first is that matter, as pure *potentia,* and form, as the *actus substantialis* of matter, are "ordered first of all to each other." Alone, each is, as we have seen, an incomplete substance, and each is completed just by the other (*In Phys.* 1c9q11a2, 1:184). The second is that matter and form are united without intermediaries. Though whiteness and the wall are indeed one in a certain sense, the whiteness inheres in its subject, the complete substance, only by virtue of the substantial form that has actualized the matter of that substance. Setting aside the difficulties of the Eucharist, accidents always presuppose the presence of both matter and substantial form.

Suárez sets up the general principle that whatever has *one* essence or being [*entitas*] is *unum per se.* It will have one essence just in case it has "whatever bears on its intrinsic reason or consummation." Simple substances obviously satisfy that criterion. As for composite substances, "since matter and form *per se* are not complete, whole beings in their kind [*entia completa et integra in suo genere*], but are instituted by their nature to be composed into [a substantial nature], that [nature], which is composed immediately from them, is deservedly called, and is, a nature and essence which is one *per se*" (Suárez *Disp.* 4§3¶8, *Opera* 25:127–128). Matter cannot naturally exist without form or form without matter; in that sense, each of

---

20. *Unum per se* (*per accidens*) and *ens per se* (*per accidens*) are taken to be more or less interchangeable: "that which is *unum per se* is called [. . .] an *ens per se*, while that which is only *unum per accidens* is called [*ens*] *per accidens*" (Suárez *Disp.* 4§3¶3, *Opera* 25:126).

them, considered by itself, does not have all that is needed to fulfill its end, which is precisely to exist as part of a composite.

Yet much the same could be said of accidents and substances. As we have seen, potential if not actual inherence is essential to being an accident; and substances are said to be completed or perfected by accidents, and are thus "instituted" by nature to receive them. Why then are accident and substance not *unum per se*? The union of accident and subject "is a third genus of *entia per accidens*, which seems to depart more [than the unity of mixtures] from the first and least [genus] *ens per aggregationem*, and to approach more to unity per se, because the things from which it arises are not distinct with respect to their *suppositum* [. . .], and have a greater physical union; the [substance] is truly *in potentia* to the [accident] [. . .], and the [accident] is by its nature ordered to the [substance], and in it has its connatural perfection; in all [features] *ens* of this sort imitates that which is properly and *per se* one, although it is *simpliciter* and absolutely *per accidens*."[21] In view of those resemblances, it is difficult to understand why accident and substance are nevertheless only *unum per accidens*. The reason we were offered in the case of matter and form was that each completes the other. But, as Suárez notes, much the same could be said of substance and accident. Accidents, as we have seen, must at least potentially inhere in substances, and naturally must do so; no substance can exist naturally without the accidents appropriate to its form, either as necessary conditions or as effects.

Two more promising *differentia* emerge from questions on unity. But neither looks decisive. The first is that matter and form united constitute a single essence, although substance and accident do not. Without examining the notion of essence, one cannot evaluate the point, but at first sight it would appear that quantity at least is both an accident and part of the essence of corporeal substances. Suárez writes, "But because we hold that quantity inheres in prime matter, with which it does not make an unum per se, but *per accidens*, it thus should be added that accidental form, by the very fact that its nature is in relation to [*respicit*] a being [*esse*] of another order and category, and is not ordered so as to constitute or complete it in that order and category, does not constitute an *unum per se* with it, but an *unum per accidens*, because what is *unum per se* must be of one order and category" (*Disp.* 16§1¶13, *Opera* 25:570). The composite is a substance made up of substances. Its components are of the same category as each other and as it

---

21. Suárez *Disp.* 4§3¶14, *Opera* 25:130; cf. the similar discussion at 25:570. The term *suppositum,* applied to created things, can be interpreted as 'first substance' in the sense of the *Categories:* a "singular individual substance subsisting *per se*" (*Disp.* 34§1¶7–9, *Opera* 26:350). 'White' and 'Socrates' are said of the same individual, whereas the citizens of a republic, though united, are distinct individuals.

itself. The general principle is that a thing can be *unum per se* only with another thing of the same category. It is likely to seem ad hoc, since the only case that matters is precisely the one to be proved. Nor is it clear—especially to a modern philosopher inclined to reduce essences to essential properties—why only a substance–substance union, and not a substance–accident union, should yield a single essence. The question of the unity of complete substance has in effect been deferred to that of the unity of essence.

The second *differentia* is that accidents are always united to an entity that is itself already a composite of matter and substantial form. They are second *actus* of matter, and presuppose a first *actus*, namely, substantial form. But substantial form itself seems to presuppose certain accidents, notably quantity. In what sense, then, is it "first"? A Thomist can straightforwardly answer that prime matter doesn't even exist unless joined with substantial form; evidently no other form can precede it in union with matter. The central texts, on the other hand, mostly hold that matter has a proper existence, and some, like Suárez himself, hold that quantity sometimes inheres directly in matter (see §5.3). The priority of substantial form remains to be explained.

2. *The unity of form and matter defended.* The first and most important of the arguments refuted by the Coimbrans takes up that point generally. Either prime matter and substantial form are united by way of some connecting medium, or not. If there is a medium, then it must be an accident; but if matter and substantial form are united by means of an accident, they are not *unum per se.*

If there is no medium, then since substantial form is supposed to unite immediately with prime matter, it could, "at least by divine power, be united without the intermediation of quantity, so that from the rational soul and matter there would result a composite which was *unum per se,*" and which could only be a human being (rather than some other kind of thing). Thus we could have a human without a head or heart, which "should not, it seems, be conceded." Experience, moreover, shows that just as the generation of a substance requires a particular kind of agent, so too it requires a certain kind of matter. If form united immediately with matter, there would be no reason for this: even without a miracle the human body might take on the soul of a horse. Since that does not naturally occur, it must be that substantial form can only unite with a matter already endowed with suitable accidental forms (Coimbra *In Phys.* 1c9q11a1, 1:182–183). We are forced back to the first horn of the dilemma.

The Coimbrans deny that between two things that are mutually accommodated and proportioned, as are matter and form, there need be any medium by which they are united (a4, 1:186).[22] They choose instead the second horn of the dilemma, and the consequences of holding that matter is united immediately to form. Following Scotus they argue that, however unnatural the result might seem, Socrates' soul could indeed be joined to matter deprived of quantity. In fact, "any sublunary form can be conjoined to any portion of inferior [i.e., sublunary] matter by divine power" (Fonseca *In meta.* 5c2q14§2, 2:196bF, cited by Coimbra).

It remains to be explained why not any form can be *naturally* joined with any matter. Socrates' soul requires a body endowed with suitably arranged parts, which themselves must be composed of bone and flesh so as to subserve the various powers of the soul. It is not enough to observe that such a form *could* be joined to matter lacking suitable accidents, so that the two would then be *unum per se*. The presence of the accidents in a naturally generated substance may still be grounds for holding that as it actually is it is not *unum per se*. A ship and dock joined by a gangway are not *unum per se*, even if by welding ship and dock together we can make a thing that is *unum per se*. We find again the question raised a moment ago: are matter and form *immediately* united, even in the presence of accidents that are admitted to be naturally necessary to their union?

## 5.3. Conditions for the Reception of Form: Dispositions

That certain accidents *were* naturally necessary to the reception of even the simplest forms was universally acknowledged. Quantity at least is common to all. The Aristotelians made a standard distinction between "proximate" matter, the already formed stuff required by a particular substantial form, and "remote" matter, that which supports all the forms of a thing, substantial and accidental. We have seen that there were various opinions about the essence of prime matter. But all agreed that, however the fact was to be explained, substantial form is always naturally accompanied by a characteristic set of accidents when joined with matter.

The explication of the structure of material substances was bound up with that of generation and corruption, and of life. The active powers of a living thing have as their collective end its continued existence and the generation of its like in new matter; even sensation and intellection are subordinate to those ends for people if we consider them only as living (if we consider them

---

22. Similarly Suárez writes that the union of matter and form "depends essentially both on form and on matter [. . .] because it is their actual bond, and no other bond inserted between them is needed [. . .], since otherwise it would proceed to infinity" (*Disp.* 13§9¶9, *Opera* 25:431). The union, which is nothing other than their mutual dependence in the complete substance, is a joint mode of each, just as *motus* is a joint mode of agent and patient.

as rational, their ends are different). Experience shows that the effect of the generator consists largely, perhaps entirely, in preparing the matter of the new substance to receive its form. Since active powers are defined by their (completed) actions, and since actions are defined by their *termini*, the scientific understanding of substantial forms depends on that of the dispositions that it is their purpose to sustain and reproduce.

The simple formed matter of *Physics I* is thus unfolded into several layers. The first is prime matter. The second is prime matter together with quantity and dispositions, the proximate material cause of substance, and thus of substantial form. That relation will be studied here. The third layer consists in the active powers of which substantial form is the efficient cause. Those powers are in turn the proximate efficient cause of the dispositions of the new substance. I will examine that relation in the next subsection.

The role of accidents in the reception and sustenance of substantial form is touched on in a variety of contexts. I will divide the discussion into three parts:

(i) The rejection of preexisting substantial forms or "seeds" in matter;
(ii) the inherence of accidents, especially quantity, in prime matter independently of, and prior to, substantial form;
(iii) the notion of *dispositio* and its use in the theory of generation and corruption.

Although the first two questions become largely moot for the new science, the third retains its importance. Mechanistic hypotheses on the phenomena of life were for some time not markedly better or less ad hoc than those of the Aristotelians.[23] Their advantage consisted mostly in being mechanistic. Descartes acknowledged that his physiology was incomplete without a theory of generation; yet his attempts to devise one were at best mere promissory notes, at worst question-begging. Biology has since redeemed those notes. But we should not overlook the fact that, as Leibniz soon recognized, the arguments Aristotle raised against the reductive theories of *his* predecessors had lost but little of their force in Descartes's day.

1. *Form educed from matter.* Doctrines of preexisting form typically came up in questions about the maxim *forma educitur a materia* (form is educed from matter). The maxim was as usual not questioned; only its interpretation was subject to debate. One obvious reading is that form, if it is educed from matter, somehow already exists in it, but occultly or imperfectly. Anaxagoras "contended that everything is in everything, [. . .] so that the generation of

---

23. Two notable successes in this period—Kepler's theory of vision and Harvey's theory of the circulation of the blood—were arrived at independently of mechanistic conceptions. Descartes's theory of the action of the heart, on the other hand, and the animal spirits invoked in his psychology of perception were soon recognized to be mistaken.

a thing is only of the form," which exists in its entirety in matter beforehand. Until they are generated, such forms "are not apparent from outside on account of the indistinction which they possess in matter." Generation is simply the manifestation of a form that is otherwise not different before than after (Nifo *Disp.* 7disp11c3; 196a). Similarly Augustine argues that "for all things that are brought before our eyes, there are certain seeds concealed in the elements of the world, through which those [forms] that were hidden come out into the open" (Coimbra *In Phys.* 1c9q12a1, 1:189).[24]

The primary argument on behalf of latent forms was quite simple. All philosophers accept the principle that nothing comes from nothing. Form, as we have seen, is a thing distinct from matter. Hence "either it is something prior to generation, or it is nothing" (Suárez *Disp.* 15§2¶1, *Opera* 25:505; cf. Coimbra *In Phys.* 1c9q12a1, 1:189). If it is nothing prior to generation, then the principle is violated. If it is something, it cannot be matter itself. Indeed it cannot be other than form. But that is impossible, "whether because otherwise there would be no substantial generation"— since in effect the substance to be generated already exists—"or because otherwise contrary forms would exist together in matter."[25]

One could avert the last objection by supposing that the forms latent in matter are "inchoate" or "imperfect." Averroes, according to Nifo, held that in matter, privation was something more than the mere *potentia passiva* to form. But it cannot quite be form, since if matter already had a perfected form, it would not be fit to receive all forms. The forms that preexist in matter are therefore not merely hidden, as Anaxagoras and Augustine thought, but "imperfect" (Nifo *Disp.* 7disp11c3; 197b; cf. Averroes *In Phys.* 1com79,66,62; *Opera* 4:45C,40I,37H).[26]

---

24. Cf. Augustine *De trinitate* 3c7, p140: "Omnium quippe rerum quæ corporaliter uisibiliterque nascuntur occulta quædam semina in istis corporeis mundi huius elementis latent." Boyle found the maxim "not *comprehensible*" except if understood in some such way (*Papers* 55). But he does not therefore agree that there must be preexisting forms. Instead he argues that what the Aristotelians call "generation" is only alteration or the production of subtle from gross matter. Maignan, in his question on the eduction of form, concludes that matter, understood as pure potentia, is a "mere figment" in physics, and that the "matter" of generation consists in already formed elements (*Cursus*, Philosophiæ naturæ 1pr5, p142).

25. So Zabarella, after Avicenna: "Unless this form [of matter as such] is admitted, there will be a creation *ex nihilo*, namely, what is usually called generation will be creation [. . .] The form which is produced is produced as a whole [*tota producitur*], without any part existing beforehand [*nulla sui parte præsupposita*]. But what is produced as a whole, without any part existing beforehand, is said to be created. Therefore form is created" (*De rebus nat.* 207B). See also Toletus *In Phys.* 1c8q15, *Opera* 4:36ra.

26. The same view was ascribed to Albert (*In phys.* 1tr3c3&16; *Opera* [Inst.] 42b,70a). For an extensive discussion of his views and the responses to them, see Nardi 1960, c.2. Soncinas (*Q. meta.* 7q28, 159b) and Suárez (25:506) deny that by "inchoatio formæ" and similar phrases he meant anything other than that matter is *in potentia* to form (cf. Nardi 1960:87–93). One must admit, however, that Albert's manner of speaking is suggestive. Nardi argues that Albert did believe in an *inchoatio formæ* distinct from the *potentia* of matter (92$^{\text{II}}$).

The problem such arguments raise is not entirely obsolete. A similar dynamic operates in Thomas Nagel's argument for panpsychism. Call an *emergent* property any property of a composite whole which is not reducible—by whatever means count as reductive—to the properties of its parts. Nagel argues that mental properties are not emergent. On the other hand, they cannot come from nothing. Something like them must be inchoately present in the building blocks of matter: Nagel speaks of "protomental" properties. The Aristotelians would agree that souls cannot come from nothing. But for them having a soul *is* an emergent property, distinct from the disposition of the underlying matter. Their conclusion, as we will see, is not that matter is endowed with a protosoul, but that the rational soul is not educed from matter. The mental does not seep up from below; it slips down from above.

Any solution to the dilemma had to meet three conditions. First, it must show that there are no actual preexisting forms in matter. Second, it had to somehow interpret the maxim *forma educi a materia* so as to make it come out true. Third, it had to take account of those forms that were believed *not* to be educed from matter: the rational soul, as I have just mentioned, artificial forms, and grace (which comes to us not out of our nature but by a supernatural act of God).

To the first point Suárez argues that on pain of supposing an infinite number of mutually repugnant forms in prime matter, one cannot take the supposed preexisting forms to be just as they are in generated substances. If they are there, they must be, as Averroes thought, somehow imperfect. Generation, Suárez supposes, would then be something like an increase in intensity. But that will not work. Substantial forms do not admit of intension or remission. There would, moreover, be an "infinity of actual entities" in matter, for though they were imperfect, the "remitted" forms would not be merely possible. The higher grade of intensity of the generated form would have to be added by the agent, and so the same question arises for it (Suárez *Disp.* 15§2¶6, *Opera* 25:507). The forms educed from matter, therefore, do not exist in it in any way.

The problem now is to reconcile the fact of generation with the principle that nothing comes from nothing. 'Eduction' is defined thus: "We say that for a form to be said to be educed from the *potentia* of a subject, two conditions are required and suffice: the first is that there should be in the subject a natural *potentia* to it, since otherwise the eduction will not be natural; the other [condition] is that what is educed can neither be brought about nor endure without the aid [*sine adminiculo*] of such a subject; which [condition] they call 'depending on matter in becoming and being conserved [*pendere in fieri & conservari à materia*]'" (Coimbra *In Phys.* 1c9q12a5; 1:193; cf. John of St. Thomas *Nat. phil.* pt1q4a1, *Cursus* 2:85; Suárez *Disp.*

15§2¶15, *Opera* 25:510). Grace and artificial forms fail the first condition, since matter has no natural *potentia* to them; the rational soul fails the second, since it can be produced and endure without the body. As for the principle, if it is taken strictly, so that to come from nothing is "to come as a whole, that is without presupposing any of the parts," then substantial form does not come from nothing. It comes from matter, or rather from "a matter disposed and determined [*dispositam determinatamque*] toward forms of its sort" (Fonseca *In meta.* 5c2q4§2, 2:90F). It is, in short, generated, not created.[27]

Even if the solution to the dilemma is sound, it is incomplete. Substantial form is really distinct from matter; it is likewise really distinct from whatever accidental forms may naturally accompany it in a complete substance. Since the progenitor need not perish in reproducing itself, the form of the progeny cannot be numerically identical to that of the progenitor. That new existence, then, is left unaccounted for, even if we distinguish generation from creation.

The question 'Where does the progeny's form come from?' can be understood in several ways. One concerns the existence of the individual: 'How did the matter of this individual come to be determined (in some way or other)?' The appropriate answer is that such and such an agent acted on the matter. Another concerns the kind of thing the individual has turned out to be: 'How did the matter come to be determined *thus?*' Once the individual exists, the form itself is the answer to that question. But I am asking about the cause of the form, which is, as I have noted, really distinct from, and not already in, the matter of that individual.

One answer, favored by Boyle, which would still be favored, is that the disposition is the cause of the form, or rather that the form is nothing other than the matter so disposed. The Aristotelians, however, denied that substantial form could be thus reduced to accidental forms. A second answer is that of Avicenna. There is a celestial intelligence, "by which the forms of natural things are induced in matter disposed beforehand by a physical agent," and which is therefore called the *dator* or *datrix formarum*.[28] That

27. "Generatio enim est ex aliquo præsupposito, unde non facit totum ens, quod generatur, sed materiam perficit: at verò creatio est productio ipsius rei totius: creare ergo equum, est totam equi substantiam de novo facere: "Generation is from something given beforehand, and so it does not bring about the entire being of the thing generated, but perfects its matter. Creation, on the other hand, is the production of the whole thing; to create a horse, therefore, is to bring about for the first time the entire substance of the horse" (Toletus *In Phys.* 1c9q19, *Opera* 4:41vb). As a *motus*, generation presupposes not only a *terminus ad quem*—the thing generated—but a *terminus a quo*, that from which the thing is generated. As an action, it presupposes an object acted on, namely the matter of the *terminus a quo*. Creation has only a *terminus ad quem and requires no object.*

28. For a sketch of Avicenna's cosmology, see Fonseca *In meta.* 5c2q8§1, 2:122Cᶠᶠ. There the *datrix omnium formarum substantialium* is identified with the *motrix intelligentia* of the

answer too was rejected. Since "every agent produces what is similar to itself," only a composite substance can cause another composite substance. God is the sole exception; in him alone all form and matter exist virtually. Because "natural causes are determined by nature to certain actions," and those actions to effects peculiar to them, even angels and demons can introduce forms into matter only by bringing about the conditions under which natural causes will do what comes naturally (Coimbra *In Phys.* 1c9q12a3, 1:190–191; cf. Fonseca *In meta.* 7c7q2§2, 3:251E).

What is primarily generated, the Aristotelians emphasize, is not form or matter alone but the composite. Toletus writes that "form is not produced, or made, but the composite, in whose production form is co-produced and co-made: just as when someone bends a staff, he does not produce curvedness, but makes [something] curved, that is, when he makes the wood such, curvedness is co-produced."[29] The human form, say, is not the primary *terminus* of human reproduction: what is reproduced *per se* is not that form but a human being, composed of soul and body. The form is indeed made, but *per accidens*, as a by-product so to speak of the making of the composite. But the composite too is a new existence. To that objection Toletus responds that one cannot deny genuine activity to particular causes[30], and so one must ascribe to them a "wonderful virtue," by which they participate in the creative power of God. That, like the similarly wonderful virtue of matter to have such forms drawn out of it, "are not to be denied because they are wonderful" (*In de Gen.* 1c3q2, *Opera* 5:254vb).

We are left—as we should be, no doubt—with material substances themselves as the *datrices formarum*. But though Abel's form came from Adam, Adam's form, like his matter, must have come from God. In the beginning there was a *dator formarum*, that was also and at the same time a *dator materiæ*. It has often been remarked that Christian Aristotelianism incorporated elements of Plato's natural philosophy. One need not look far for them: Avicenna's intelligences are not rejected entirely, for example, though their role is restricted. But in the Aristotelian world there is, barring miracles,

---

lunar sphere; "natural things (even the celestial orbs), since they act by *motus,* can do nothing other than dispose matter in order that this last *intelligentia* [i.e., the lunar] should introduce substantial forms into it" (122F).

29. "Magis placet solutio Ægidii [. . .] quod non producitur forma, nec fit, sed compositum, ad cuius productionem comproducitur, & confit ipsa forma: sicut qui virgam incurvat, non facit curvitatem, sed facit curvum, id est, dum lignum tale facit, curvitatem comproducit" (Toletus *In de Gen.* 1c3q2, *Opera* 5:254vb; cf. Suárez *Disp.* 15§2¶15, *Opera* 25:511; Thomas *ST* 1q90a2). That what is made is not the form but the composite is part of Aristotle's argument against Platonic ideas (Aristotle *Meta.* 8c4, 1034a25$^{ff}$).

30. *Particular* causes, which include all sublunary agents, are contrasted with *universal causes, the heavenly substances that concur in earthly generation and that introduce substantial* forms in cases where the sublunary agent is absent, as in the equivocal generation of frogs and the like from putrefying mud (see, e.g., Fonseca *In meta.* 7c7q2, 3:250$^{ff}$, or virtually any *De anima* commentary).

exactly one "donation of forms" to the world. From that moment on, nature acts *pro deo,* by its own power though in accordance with God's will.[31]

We see here an exacting economy of the world's dependence on God. The Coimbrans write that Aristotle occupies a middle place between two extremes. Some, like Anaxagoras and the atomists, put preexisting forms in matter; some, like Avicenna, deny to sublunary things the wonderful virtue of making forms and would have them come always from above. One side threatens to make God superfluous by attributing to matter all it needs to exist; the other robs Nature, as the Coimbrans put it, of "her most noble operation"—the generation of substances (Coimbra *In Phys.* 1c9q12a4, 1:192). In Aristotelianism the world is neither so independent as to do without its Creator, nor so destitute as to require his intervention in the operations proper to it.

2. *Accidents prior to form.* Not all the accidents that invariably accompany a substantial form are conditions for its reception. Some truly "prepare and adapt" matter to form, some merely "embellish and perfect" the form (Suárez *Disp.* 14§3¶28, *Opera* 25:480). Among the latter are the active powers that, as we will see in §5.4, are said to "emanate" from form. It is reasonable to believe that such accidents inhere in matter only by way of form. Heat and lightness (*levitas*) are both invariable concomitants of fire. Yet although heat is necessary to its production, lightness might well be a mere consequence of the form. Heat is both an active power of fire and a preparatory accident in its generation; lightness is not. In more complex substances, and especially in animals, the active powers that operate in generation cannot be reduced to elemental qualities. There the forms that follow upon substantial form, and those that prepare the way to it, become increasingly dissimilar.

Embellishing accidents, then, inhere in matter by way of form; but what about those that prepare matter for it? The Thomists argued that they too inhere in matter only by way of form. Since in generation the *terminus ad quem* is a new substantial form, and since there cannot be more than one form in each substance, the old one must yield to the new. As the slogan has it, the generation of one is the corruption of another (Aristotle *De gen.* 1c3, 319a6). But then—Thomas and his followers argued—none of the accidents of the old substance, including its quantity, can be numerically identical with any accident of the new substance. Substance must be "resolved" into prime matter before it is resurrected.

---

31. Nature depends on God *in esse,* since God's conserving power is necessary to the existence of all created beings. Nature depends on God *in fieri* only insofar as God must concur in the actions of natural agents. On nature as *pro-dea,* see Alain de Lille *Anticlaudianus* 2:69–74; Speer 1991:111[ff]; and §7.1 below.

The Thomists and their opponents[32] agree that the persistence of accidents through substantial change would be a good reason to believe that they inhere in matter alone. They also agree, by and large, on the *experimenta* that I will turn to in a moment. Indeed Fonseca rests his defense of the Thomist view entirely on metaphysical arguments and finds himself obliged to explain away, or dismiss, the appearances. The real issue must lie elsewhere.[33] The positive arguments for their position reveal that the *fundamentum* is a sheaf of mutually supporting propositions already encountered in arguments about the essence of matter (§4.1):

(i) The subject of inhesion of *all* accidents is the composite of matter and substantial form (see John of St. Thomas *Nat. phil.* 3q1a6, 3q9a1, *Cursus* 2:582,755).

(ii) Matter is first of all and per se completed by and united with substantial form (ib. 583).

(iii) Matter is *pura potentia*, having no proper existence or actuality.

All existence, in short, comes from substantial form, and only that which exists can have accidents. Matter, "since it is *pura potentia* in every genus of beings, cannot of itself alone give birth, as it were, to quantity, which is an actual being, except when at the same time it is joined with substantial form."[34]

We have seen that the doctrine is troublesome, even paradoxical, where it concerns existence. Now we will see that it is troublesome in the theory of generation and corruption. Suárez, in a section of unusual length even by his standards, establishes first of all that matter has indeed "sufficient being [*entitatem*] so as to sustain this or that accident" (Suárez *Disp.* 14§3¶12, *Opera* 25:474; cf. §4.2). Even if its subsistence in a complete substance is only incomplete or partial, that subsistence will do for "accidents propor-

---

32. Thomas (*ST* 1q76a6, *In de Gen.* 1lect10), Albert, Durandus, and the usual Thomists are lined up in favor of resolution into prime matter (Fonseca *In meta.* 8c1q1§3, 3:453bF); Simplicius, Philoponus, Averroes, Avicenna, Henry of Ghent, Ockham, and "most of the nominalists," Gregory of Rimini, Gabriel Biel, and Augustino Nifo against it (Fonseca 3:453aE, Suárez *Disp.* 14§3¶10, *Opera* 25:474).

33. Theological issues weigh lightly on the dispute. Suárez mentions in passing that his position is "favorabiliorem (ut sic loquar)" to explaining the Eucharist and justifying the veneration of relics. That may have been a touchy point. If no accident of the living person is identical to any accident of her remains, the bones of a saint and the saint herself share only their prime matter; but that seems unworthy of veneration (cf. Fonseca *In meta.* 8c1q1§2, 3:451aC). If mere similarity of accidents suffices, an exact replica of a saint's scapula, though it belong to Aretino, is as venerable as the saint's.

34. "Materia enim, cum sit pura potentia in toto genere entium, non potest ex se sola quasi parturire quantitatem, quæ est ens actuale, sed simul copulata cum forma substantiali, quatenus ea forma dat esse corporei in genere substantiæ" (Fonseca *In meta.* 8c1q2§3, 3:456bB).

tioned to it," just as the subsistence, though partial, of the rational soul in a human being will do for its volitions and habits.

That much establishes the possibility of accidents inhering in matter independent of form. The more interesting part of the argument consists in showing, by appeal to the *experimenta* I mentioned a moment ago, that in fact they do.

Start with what is common property. Quantity answers to matter as quality to form. It is, as the Coimbrans put it, "the first disposition of matter," necessary to the reception both of substantial and other accidental forms. For the latter it serves as a kind of secondary subject, so that not only the substance itself but its color and heat are said to have size and figure. Though the central texts deny that quantity is absolutely necessary to matter (§4.2), they agree that it is naturally necessary (Coimbra *In Phys.* 1c4q1, 1:105).

All this suggests a structure for material substance in which prime matter together with quantity, and perhaps certain other dispositions, make up a proximate matter ordered to and receptive of substantial form. Toletus, concluding that in corruption "there does not always occur a resolution immediately into prime matter," argues that in many instances we can distinguish two sorts of accidents: those that are common to the old and new substances, and those that are proper to one. Clearly what is proper to the corrupted substance alone will not survive the change. But for those which are "common in some way, and which have no contrary in the generated substance," nothing opposes their remaining numerically the same in both (Toletus *In de Gen.* 1c4q7, *Opera* 5:262va). The clear implication is that such accidents must have matter alone as their subject, since (as all Aristotelians agree) accidents cannot "migrate" from one subject to another. Of the two incomplete substances that constitute the corrupted complete substance, only the matter persists; it alone could be the subject of accidents common to both the old and the new substance.

Experience shows that certain accidents of the old substance have exactly similar counterparts an instant later in the new one: "We experience that in the cadaver of a human just after its death the same accidents remain which were in the living human, except those faculties which are proper to living things, and which therefore effectively depend on the soul" (Suárez *Disp.* 14§3¶20, *Opera* 25:477). One may reply, as Fonseca does, that exact similarity does not prove identity, and that where reason shows that something cannot be so, the evidence of the senses is of no avail.[35] Suárez answers that

---

35. "Not by sense but by reason should one examine the numerical identity and diversity of things; as when someone sees two exactly similar eggs from the same hen: he would not be able to distinguish them numerically from each other, but someone who knows that the hen laid only two eggs, and has eaten the one that the other person saw first, will demonstrate to him

the reply "could indeed be admitted, if some reason intervened that would effectively compel us to correct our senses" and also exhibit a cause for the new accidents in the generated substance. But the metaphysical arguments are not compelling, and in many instances no efficient cause offers itself. The heat, for example, that remains in a cadaver certainly does not result from the form of the cadaver, since it quickly disappears. Nor could any external agent induce such heat, "especially when death comes about violently by suffocation, strangulation, etc." The particular cause in such instances certainly cannot, and the universal causes that are said to educe the *forma cadaveris* from the matter of the victim have no reason to do so (Suárez *Disp.* 14§3¶21, *Opera* 25:478).

The *forma cadaveris* with its scars and dwindling heat would occupy a useful though morbid chapter in the history of Aristotelianism. Here the point is that the Thomist position requires that many accidents of the old substance be exactly duplicated in the new. Some authors argue that this already offends against Ockham's razor. Worse are cases where from the new form accidents contrary to the duplicates would follow, since it cannot then be their cause. A cadaver is naturally cold; its form would not of itself produce heat. The Thomist position requires also that the quantity of the old substance be produced anew. Typically they say it is introduced at the exact instant in which the new form is introduced and the old quantity perishes. Though Suárez admits that the claim cannot be refuted, he concludes that it is "more philosophical and more in accordance with corporeal agents" to suppose that every action supposes "a quantified and divisible subject, distinct from others not only in being but in quantity [*non solum entitative, sed etiam quantitative*]" (¶25, *Opera* 25:480). More generally, since prime matter is *in potentia* indifferent to all forms, its taking on the *forma cadaveris,* say, rather than some other "requires that it should be accommodated and disposed by the agent to such a form; but this accommodation and disposition occurs only through previously existing accidental dispositions," including the quantity that supplies extension to all the rest.[36]

that they are diverse." Likewise the sun and moon appear to have the same size, yet reason shows that the sun is "five hundred" times larger. Hence even if the scars of a corpse appear to be numerically identical to those of the living person, "reason, which demonstrates the opposite, is to be believed, since to its inward tribunal belong that [question], and [all] the insensible numerical differences of things that are very similar" (Fonseca *In meta.* 8c1q2§6, 3:458bC). This passage, rather unusually, humbles the senses, thereby to exalt reason—a Cartesian strategy. But it is one of the few instances where a thesis manifestly contrary to sense is defended.

36. Although there are Thomist replies to these points (see John of St. Thomas *Nat. phil.* 3q1a6 for explicit replies to Suárez), it would be tedious to examine them. Toletus, I should note, attempts a rather uneasy accommodation of the two views (one that Fonseca rejects). He also excepts quantity from among the accidents that survive through transmutation, though his argument otherwise resembles that of Suárez. Cf. Toletus *In de Gen.* 1c4q7, *Opera* 5:263–264; Fonseca *In meta.* 8c1q2§4, 3:454.

Since quantity, at least, persists through all substantial change, it is, for reasons already mentioned, reasonable to believe that it inheres immediately in matter rather than in the composite. A leading special case is that of the human being, since the soul is a spiritual thing. Like all spirits, it is incapable of receiving quantity. But it is also incapable of being naturally joined with matter except if that matter has quantity. The quantity of a human being must inhere in the matter alone.[37]

The conclusion can be generalized. The union of a particular quantity with its matter can persist even when whatever union it has with substantial form perishes. The quantity of Caesar's body is united, we may suppose, both with his matter and with his soul. When Caesar was murdered, that quantity was still united to the same matter, but its union with the soul departs when the soul does, and it is then united with the *forma cadaveris*. Hence the union of quantity with matter and its union with form are distinct. Suárez argues that in all material substances other than the human "there intervenes no special partial union between form and quantity" (¶46–48, *Opera* 25:488[ff]). Indeed in all material substances, not just human, quantity and form are united only by way of the union of form and matter, on the one side, and that of matter and quantity, on the other. Form and quantity are not *unum per se*, but *per accidens:* a conclusion that will have its counterpart in Regius's heterodox Cartesianism.

The *differend* between the Thomists and their opponents can be traced to a dissension within Aristotle's texts themselves. The *Categories* treat individual substances as unanalyzable; accidents cannot but inhere in complete substances. The *Metaphysics* and the *Physics* add a new level, matter and form (see Furth 1988, §9). In Scholasticism that distinction was combined with the *Categories* account to yield prime matter, substantial form, and accidental forms.

The subject of inherence in the *Categories* is a concrete existing individual: Socrates is white, not his soul or body. In the *Metaphysics* and the *Physics*, on the other hand, it is at least an open question whether forms other than substantial form inhere in the composite. The Thomist position sticks with the *Categories. From the Metaphysics* it takes the distinction of matter and form, interpreting the pair as prime matter and substantial form. Only the union of the two yields a proper subject of inherence. Thereafter everything proceeds as in the *Categories*. The structure is shown in Figure 5. In generation, the complete substance perishes, and thus the subject of inherence of quantity, and so too of all other accidents (I omit those accidents that reside in

---

37. Suárez *Disp.* 14§3¶16[ff], *Opera* 25:476[ff]. Descartes expected no trouble from Aristotelians when he held that the immaterial mind is joined with the material body: far from being a reason for believing that the mind is extended, the fact that it cannot be is reason for holding that quantity inheres in the body alone.

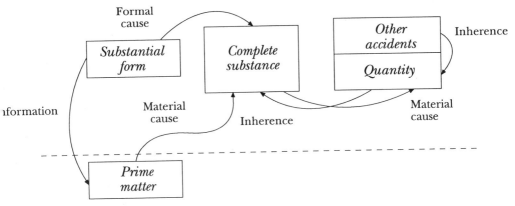

Fig. 5. Structure of physical substance (Fonseca)

the rational soul alone). Since quantity cannot naturally exist unless it inheres in a subject, it too perishes. Everything above the dotted line in Fig. 5 is replaced when the new form is introduced.

   In the theory of Suárez and Toletus, a compromise between nominalism and orthodox Thomism, the unanalyzable first substances of the *Categories* are almost entirely abandoned. Preparatory accidents inhere immediately in matter; those that are common to the old and new substances may therefore persist, along with those accidents that, although not "connatural" with the new form, nevertheless are shown to persist by experience. In particular quantity always inheres immediately in matter. The structure according to Suárez is shown in Figure 6. Below the line now is all that pertains to proximate matter. Much of it, and certainly quantity, is preserved through substantial change. The bones and scars of a dead saint are, just as the senses would have it, the very same bones and scars as those of the living saint. The role, evident to experience, of dispositions in promoting the eduction of the new form, can be explained without difficulty, since the very same accidents that "proportion" matter to its form in an existing substance existed in its predecessor.

   Suárez's solution (which is, I should note, not unique to him) has the advantage of not requiring us to deny the evidence of the senses. It has the disadvantage of raising more acutely the question of unity of complete substance. We have seen, however, that Suárez believes that the union of matter and quantity is distinct from the union of matter and form. Hence matter and form are not in fact united *by means of* quantity. Indeed form can be received by a matter lacking quantity altogether (§4.2). So they can still be *unum per se*. But matter endowed with quantity can likewise exist

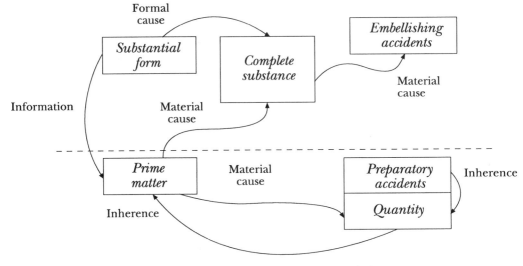

Fig. 6. Structure of physical substance (Suárez)

without form. The union of matter with form and its union with quantity are on an equal footing, a result closer to Ockham than to Thomas.

Matter has in Aristotelian physics three functions. It is the *subjectum* of substantial form, the indeterminate *being-somehow-or-other* that is specified by a particular *somehow*. It is the substrate of transmutation. It is that which all material substances have in common. In the Thomist account, as we see it in Fonseca and John of St. Thomas, the *subjectum* and the substrate are one and the same: both *pura potentia*. In fact the identification of the *subjectum* with *pura potentia* virtually forces a like identification of the substrate. Once that is arrived at, matter cannot actually be said to have those accidents that are proper to material substances; it can only be said to be *in potentia* to them.

In Suárez's account the *subjectum* is accorded a proper existence. It is not *pura potentia*. But in agreement with the Thomists he does not take any of the accidents common to material substances to be essential to matter. Since God, according to his absolute power, can deprive matter of all those accidents, what is essential to matter is only that it be *in potentia* to quantity, and thus to the occupation of space, to mobility, and so forth. The substrate of change, on the other hand, does essentially have quantity. In particular instances it may have many other accidents as well. It can be as complicated as the human body. Yet Suárez rejects the Scotist plurality of substantial forms: the proximate matter of a given form is not itself a substance. So we have not yet arrived at Cartesian body and soul.

Suárez's *Disputationes* are very nearly contemporary with the *Quæstiones* of

Fonseca; Toletus's *Physics* commentary predates them both. One cannot speak unproblematically of "development." But in light of what is to come, two tendencies may be discerned in Suárez's theory of matter. The first is that accidents, divided into the *dispositiones* that precede form and the powers that follow it, are being allotted different subjects of inhesion. Though the powers are said to inhere in the composite, the parallel with *dispositiones,* which inhere in matter alone, might incline one to regard them as inhering in the form alone. We would then have a substance with two sets of properties, one attached to matter, one to form—in animate things to the soul. The parallel with Cartesian dualism need hardly be emphasized; what will pave the way to it is the reduction of powers to dispositiones, and of *dispositiones* to the modes of quantity, and the elevation of all powers of the soul above the capacities of matter.

The second tendency is toward the reduction not only of powers but of forms to the *dispositiones* of their matter. Indeed if one could show that their so-called powers follow immediately from their *dispositiones,* the place of substantial form would become exceedingly precarious. The arguments outlined in §3.2 would still have to be addressed. But once matter is granted not only actual existence but a characteristic set of properties, an important motive behind the introduction of substantial form disappears. So the Thomists may well have thought. The dangers were not hard to discern. Matter, having been given title to the common properties of bodies, might now be free to declare its independence not only from form, but from God.

3. *The notion and uses of 'dispositio'.* The semantic field that contains *dispositio* merits attention, not least because, as I have said, Descartes takes over the term to his own ends. In the shortest chapter of the *Metaphysics* Aristotle defines *dispositio* as "the order of that which has parts, either according to place or *potentia* or form" (5c19, 1022b1), deriving διάθεσις from θέσις, meaning 'position'.[38] Disposition in place is that of a house—the roof highest, and so forth. Disposition in *potentia* is illustrated by complex virtues like prudence, "whose first part pertains to deliberation, the second to judgment, the third to command."[39] Most important for my

---

38. The *Categories* define 'dispositio' as a transient, easily mutable *habitus* or state (1c8, 8b25[ff]). In that sense heat and cold, sickness and health are dispositions, knowledge and virtue are *habitus.* Aristotle seems more to be defining a certain respect under which qualities of various sorts can be regarded than a class of qualities. The Aristotelians tend, on the other hand, to take 'quality' as a genus of which *habitus/dispositio, potentia/impotentia,* and so forth (§4.3), are distinct species, as is indicated by standard questions on whether the classification is exhaustive and disjoint. Zabarella restricts the term to animate things (Zabarella *Opera logica, Tabulæ logicæ* 120); Chauvin, though he recognizes the broader sense of the *Metaphysics,* holds that the term is "properly" used in the *Categories* sense.

39. "Formally, if we suppose that the habits of the soul are not simple qualities, but emerge out of several habits duly coordinated, so may the whole science be called a *dispositio,* because it is like a certain order of that which has parts *secundum virtutem*" (Suárez *Disp.* 42§3¶6, *Opera*

purposes is the last, disposition according to form, which Fonseca takes in its *Categories* sense of *figura*. He identifies it with what Aristotle calls the "place" or "order" of "parts in a whole," which is to say with quantity itself (see 5c13q1§4, 2:642F).

One might well ask why disposition is not distributed among other categories—place, quality, quantity, and relation. The answer is that the "order" of the parts in question is with respect to an end. It can therefore be applied even to things which have no parts but which "affect the subject for the worse or the better." Even in those that do have parts, what is foremost is that they affect the subject for the worse or the better, since only when so regarded are they qualities.[40]

Suárez's more extended treatment, well attuned to the range of the term *dispositio,* emphasizes the orientation to an end. Whatever exhibits order, ultimately, or is understood as ordered, spatially, temporally, or *secundum virtutem,* can be called *dispositio,* whether it is a *habitus* of the soul, a passive quality of the body, or health and beauty. *Dispositio* thus falls within the broad ambitus of *ordo,* which in its highest use denotes the "unifying structure of the multiplicity of being in the Universe," the *unitas universi* (Michaud-Quantin 1971:95).

In use *dispositio* is often collocated with *proportio,* a word the lexica all define quantitatively. Arguing that prime matter has a natural *potentia* toward the rational soul, Suárez writes that "matter is naturally disposed [*disponitur*] ultimately to receive the rational form," and again that "the soul is an natural *actus* proportioned [*proportionatus*] to matter" (Suárez *Disp.* 15§2¶12, *Opera* 25:509). The Coimbrans write that form is educed from the *potentia* of matter only if "its privation and dispositions preexist in matter," and later that "since a *potentia* necessarily corresponds in proportion [*proportione*] to its *actus,* matter has a natural *potentia* only to those forms that are natural to it." Concerning mixtures, Toletus writes that all parties agree that "a certain qualitative *dispositio* is necessary and proximate to the form of the mixture" (Toletus *In de Gen.* 1c10q17, *Opera* 5:303va). That *dispositio,* or *complexio,* consists in a fixed *proportio* or temperament of all four elements.

Gathering such contexts together, one sees that like *ordo, dispositio* is distributed over a spectrum extending from the de facto arrangement of parts to states or features unspecified except in relation to an end. To what extent, then, did an Aristotelian expect, or feel obliged, to produce a definite characterization of the *dispositiones* he proposed? Suárez writes that

---

25:612). Capacities analyzed into other, simpler, capacities are said to have the latter as parts secundum virtutem.

40. Fonseca *In meta.* ad 5c19 (text24): "Cum partes aliter ac aliter ordinari possunt, si certo quodam modo ordinentur, bene se habet res, si opposito, male: ut patet in exercitu, domo, complexione humorum, in artibus, scientiis, & similibus" (2:921).

any form necessarily requires an "accommodating *dispositio* from the side of matter," the necessity being as usual natural rather than absolute (*Disp.* 15§6¶5, *Opera* 25:519). The statement is plausible enough. But what is being asserted here, other than that the association of forms with proximate matters is nonarbitrary?

In the metaphysical context, the assertion is intended to remind the reader of Suárez's earlier arguments against the Thomists. *Dispositiones* are necessary conditions for forms to be causes not merely in *esse* but in *fieri*. Not only must they be present in order for a form to exist, they must also temporally precede its introduction (see *Suárez Disp.* 14§3¶28ff, *Opera* 25:480ff). Since *dispositiones* are accidents, the term also serves as a re-minder that proximate matter does not have a substantial form. The Scotist would have it that the proximate matter of the rational soul already has a substantial form, the *forma corporeitatis,* which persists after death. Suárez and the other central texts hold instead that what persists are the *disposi-tiones,* and that the *forma cadaveris* is introduced at the moment of death.

But a philosopher now would rather know what those dispositions are. It is one thing to say, for example, that conscious thought must supervene on certain physical processes, quite another to say what those processes are. One may certainly attempt to establish the first without having any idea about the second. But even committed physicalists, I'm sure, would like to be able to point a finger at the appropriate part of the brain. The inability to do so has no doubt been one reason why physicalist arguments, however cogent in the abstract, have never quite succeeded in carrying the day.

To some extent the impression that appeals to powers or *dispositiones* are vacuous rests on ignorance of other works in the Aristotelian canon.[41]

---

41. That ignorance is not without basis. By far the greater number of commentaries were devoted to the *Organon,* the *Metaphysics,* the *Physics, De generatione,* and *De anima.* In the statistics gathered by Blum from Lohr's bibliography of commentaries published between 1500 and 1650 (Blum 1988), the works rank as follows (the percentages indicate the proportion written by members of monastic orders): *Organon:* 1165 (42%); *Physics:* 1103 (40%); *De anima:* (603 (43%); *Metaphysics:* (463) 51%); *De gen. et corr.:* 286 (58%); *De cælo:* 247 (48%); *Meteorologia:* 241 (42%); *Parva naturalia:* (113 (25%); *Gen. anim., Hist. anim., Intr. anim., Part. anim.:* 25 (4%).

Though the curriculum undoubtedly had the major part in determining this distribution, the relatively small numbers of commentaries even on *De gen. et corr.* and *De cælo,* together with the near absence of commentaries on the animal books, which exhibit Aristotle at his most empirical, indicate a certain lack of interest and (no doubt) aptitude for their subjects. (I should note that members of medical faculties, who did not write Aristotle commentaries, also studied plants and animals, and that there were commentaries on Theophrastus's *De plantis* and Pliny's *Naturalia.*)

The *cursus* tell a more favorable story: in them, natural philosophy tended to grow at the expense especially of metaphysics, and to incorporate varying amounts of new information (see Schmitt in Schmitt and Skinner 1988:802ff). Nevertheless, it must be said that despite recent attempts to rehabilitate the Aristotelians, the impression given by the corpus en masse is that the character attributed to the textbooks by their enemies—the absence of experiments as opposed to authorities or commonsense generalities, the continued reliance on ancient and

*Physics* and Metaphysics commentaries are sparing in details about particular substances. But it is not their task to study them. The first descent to particulars comes with the theory of the elements in *De generatione*. There we find that among the necessary conditions for receiving the form of fire is heat to the highest degree (*summa calor*), and similarly for the other elements (Toletus *In de Gen.* 3q5, *Opera* 5:315[ff]). In *De anima* commentaries we are told, for example, that the brain must be humid enough to receive impressions, and yet dry enough to retain them. Such claims may be mistaken or primitive, but they are attempts to describe more precisely the dispositions underlying various powers or forms.

An Aristotelian will not accept offhand just any inference to a disposition or power. Sometimes Ockham's razor is invoked, sometimes the absence of any conceivable efficient cause. What is typically not done is to demand an analysis of the power in question, and, if an analysis is not forthcoming, to reject it. There seems to be, in other words, no expectation that the conclusion that such and such disposition must be supposed will be followed up by a research program that would lay bare its underlying conditions or break it down into simpler component dispositions. Even when experience is brought to bear on such problems, the facts adduced are rarely new. Almost never can one infer that a fact was witnessed with the particular intention of resolving a disputed question, or that the person adducing them witnessed them. It is far better, in fact, for the generalization to be taken from an authority, from a text that, by standing the test of time and the acid of dialectic, has become an auxiliary to common sense.[42]

Take the question of *minima naturalia*. Among the *dispositiones* required for a particular substantial form was the quantity of its matter, or rather the upper and lower bounds of the quantity it was naturally compatible with.[43]

---

Scholastic treatments—is not unjustified. The Aristotelians who *were* progressive in relation to the new science were typically not those who wrote commentaries or *cursus*. Though much more work would be needed to bear out these impressions, they do suggest that—as I will argue shortly—the barriers to innovation were not conceptual but institutional.

42. See Dear 1985:148[ff], and Dear 1987. The second essay, on Jesuit astronomy and optics, argues that the Jesuits, like the Royal Society later, began to stress singular reports whose veracity was attested by observers who had themselves witnessed them; the "objectivity" of the observers was assured by their institutional position as researchers. Of the works I have been citing, the central texts largely antedate the developments brought out by Dear. John of St. Thomas repudiates them (ignoring Galileo, he cites Thomas's arguments against mountains on the moon; he maintains that since the heavens are immutable the new stars of 1572 and 1603 were created supernaturally; see *Cursus* 2:850, 851). Arriaga, a Jesuit, and Maignan, a Minim, were to varying degrees conversant with them (Schmitt 1989, XI:223[ff]).

43. The question of *minima* concerned especially the minimum quantity possible for homogeneous substances, defined as those whose parts (or whose at-least-minimal parts) were all of the same form as the whole. Emerton argues that developments within Aristotelian theories of *minima* and mixtures laid the way for seventeenth-century corpuscularism (Emerton 1984, c.3 and 4). The central texts, it must be noted, though they affirm the existence of *minima*, show no movement in that direction.

That would seem to be a question susceptible to empirical test, even if, as it happens, the means turned out to be beyond those available to seventeenth-century science. Yet the reasons for and against *minima,* though they do draw on experience, do not arise from any investigation into that question. Here are some arguments against:

> In these [substances] there is no limit to bigness [. . .]; therefore [there is no limit] to smallness, since the reason seems to be the same.
>
> Such a minimum, if there is one, is a quantity, and is therefore divisible: it can therefore be actually divided, and then would be smaller than the minimum.
>
> An agent acts successively, and therefore produces the lesser before [it produces] the greater, and there will never be a minimum without something less being produced before it: therefore there is no limit in these things.

Here are some arguments for *minima:*

> The *actus* of active [powers] occurs in a [suitably] disposed patient, since form is not received except in disposed matter; but of such dispositions the principal one is quantity; therefore just as the rest are determinate, so too is quantity.
>
> To the action of an agent a certain quantity is necessary, in the absence of which the action will not occur; an axe can be so small that it will not cut. [. . .] A thing cannot be of a quantity such it cannot operate; for there is no being without operations. But for operation a certain quantity is necessary, with respect to smallness, and so also to being itself. (Toletus *In Phys.* 1c4q10, *Opera* 4:25rb-va)

The premises of these arguments are general; all are repetitions of statements to be found in earlier authorities; none but that of the second argument is a priori. What interests me here is that none of the premises bears immediately on the question at hand in the direct way that, say, dropping weights from a tower bears immediately on the question of whether the velocity of fall is proportional to weight. All of them could well have occurred to someone who had never heard of minima.

What could have borne directly on the question? The discovery of Brownian motion lay some two centuries ahead; instruments capable of imaging individual atoms were not available until the 1950s (Crewe 1993). On the other hand, one consequence of the existence of *minima* is that mere mechanical cutting ought sometimes to produce a new substance. *Minima* are not atoms; they could, presumably, be deprived of the requisite quantity and corrupted. Does grinding ever result in transmutation? Could flesh be cut so finely as to cease to exist? Just how small can an axe be and still work? Such experiments might not decide the question, but that would itself be informative.

Aristotelian *quæstiones* did not give birth to new "experimental" questions, answers to which would "test" answers to the original question. That particular barrenness distinguishes the new science from the old. The presumption behind my queries is that one should *do* things, make things happen with the intention specifically of resolving a dispute. The textbook authors did not take it to be part of their brief that they should, in the experimental way, *do* anything in particular. This is not to say that new questions never emerged. John of St. Thomas, having insisted that in generation the accidents of the new substance are all numerically distinct from those of the old, adds a question, not found in other authors, on "how the ultimate disposition is caused, or causes substantial generation" (*Nat. phil.* 3q1a7, *Cursus* 2:588). But his new question is simply a continuation of the argument against Suárez by the same means. It brings no new phenomena to light; it only shows that familiar generalities can be saved.

The fruitfulness of Aristotelian or Cartesian hypotheses cannot be measured by reference only to the concepts used in them. Descartes's own postulated mechanisms and configurations met with criticisms not unlike those raised against powers and qualities. It was as easy, and as informative, to suppose some mechanism or configuration that causes the phenomena we wish to explain, as to postulate a new power or quality.[44] Aristotelianism, for its part, need not have been an insuperable obstacle. There were times when Aristotelians did actively seek new experientia. This was true in the thirteenth century of Albertus Magnus and Roger Bacon. In Descartes's time the Jesuit Scheiner made extensive observations of sunspots and parhelic phenomena; Galileo's opponent Grassi emphasized his having witnessed himself the *experimenta* he published; Mersenne's appetite for facts and his devotion to experiment preceded his conversion to mechanism.[45] Even when their philosophy remained Aristotelian, their questions were no longer to be answered only through the interpretation and reconciliation of received *experientia*.

The problem with Molière's doctor is not that he appealed to a *virtus dormitiva*—which, after all, opium does have—but that the inquiry was supposed to end there. At its best, in Aristotle's theory of the soul or the theory

---

44. See Gabbey 1990, esp. 279–282.

45. See Schmitt 1981, essays 7 and 8; Dear 1987 (Scheiner and Blancanus); on Mersenne's relation to Aristotelianism, see Lenoble 1943:211–222. On Jesuit instruction and research in mathematics and astronomy, see Dainville 1978, esp. 311–354.

of mixtures, Aristotelianism did not rest with powers so obviously complex. That it did not penetrate very deeply had to do, not with the notions of *virtus, dispositio,* and potentia themselves, but with the techniques by which new questions were generated out of old ones.

## 5.4. Substantial Form and Active Powers

The arguments just laid out, and those of §3.2, suggest that in individual substances three stages of specification are built upon prime matter. The first consists in quantity together with preparatory accidents; these inhere immediately in matter, though not of course without an accompanying substantial form. The second consists in the form itself, and the third in the active powers and embellishing accidents of which form is said to be the unifying ground. I will here finish the description of the structure of material substances by examining the causal relation of form to active powers, and of active powers to dispositions.

Active powers introduce into the matter of the thing generated the dispositions required by its new form. The new form will in turn effect the powers and embellishing accidents proper to it, so that the cycle can begin anew. Thus briefly stated, the account raises a certain doubt. Actions can be defined in terms of their termini and active powers in terms of the corresponding actions. The active powers associated with form, if they pertained only to generation, would be determined by the dispositions proper to that form.[46] The form itself has no effects but by means of them and can be known only by their actions. It too, therefore, would be entirely determined by its proper dispositions, and so superfluous.

But in fact even the elements have powers with no role in generation. Though heat alone suffices to generate fire, fire is not only hot but luminous and light (*levis*). Mixtures and animate things typically have many powers with no immediate relation to generation: the medicinal virtues of herbs, the senses of animals, the human intellect. It could be argued that even such powers are instrumental, if not to generation itself, then to survival. But because they are at best among the remote causes of the *dispositiones* necessary to the reception of form, they cannot be said to be determined by those *dispositiones* in any obvious way.

In the new science the banishment of substantial forms rested on showing, or asserting, that the "actions" ascribed to powers could all be explained by appealing to *dispositiones* alone. The word 'disposition' has, in the vocabu-

---

46. On the centrality of the generative among the powers of substance, see Suárez *Disp.* 18§3¶19, *Opera* 25:621, quoted below. In *De anima* Aristotle calls generation the "most natural" of the soul's powers (2c4, 415a26), because it is common to all material forms, animate or not.

lary of a philosopher like Quine, usurped the role that was played by powers; dispositions he admits only as stand-ins for eventual reduction to microstructure. That reduction was already envisioned by Descartes and Boyle, as we will see. Descartes's mechanisms, Boyle's analogy of key and lock, were intended not only to show that the phenomena can be saved without putting any properties in matter except those it has by virtue of being extended, but also that nothing mediates between dispositions—or mere arrangements of parts—and so-called powers. Form thereby vanishes.

Even within Aristotelianism, the mediating role of substantial form called for clarification. Active powers were supposed to be grounded in it, dispositions somehow caused by it. The first relation, which was called *emanatio* or *resultatio,* did not, oddly enough, attract much attention.[47] Of the authors studied here, Suárez and John of St. Thomas alone discuss it explicitly. The neglect is odd in part because substantial forms were held to act only through their active powers. Indeed, as we have seen, demonstrations of their existence relied not on showing that they had immediate sensible effects, but on the necessity of a unifying ground of the powers that did have such effects. The indirectness of their relation to the phenomena proved to be a point d'appui for skeptics, who were happy to declare them inscrutable. I will look at the Thomist position against which Suárez argues, and Suárez's arguments for his own.

The second point touches on the activity of created substances. The central texts agree that they are effective, but only within definite limits. Unlike the first cause, God, they can in no way create complete substances, matter *and* form. They can, on the other hand, generate their like in existing matter; on that both common sense and the authorities agree. But substantial forms were believed to act only through their powers. Active powers are accidents and thus inferior to forms. Since in general the less perfect cannot bring about the more perfect, accidents should not be able to produce substantial forms, nor should substantial forms themselves be able to. Experience and theory thus seem at odds. In looking at the role of substantial forms in generation, I will examine how the clash was mitigated.

1. *The emanation of powers.* Thomas, having proved that the powers of the soul are distinct from its essence, asks whether they "flow" from its essence. His answer is that they do, since each subject is "the cause of its proper accidents." The powers of the soul are *actus* of its *potentia,* and can therefore be called its accidents, and so are caused by it. They are not, indeed, the

---

47. In *De anima* commentaries the relation of the soul to its powers is a standard question. See for example the questions from Toletus and Suárez cited below. The presuppositions of the general question, and in particular the distinction between form and powers, were not shared by all Aristotelians. The controversy over *dimanatio* occurred largely among Thomists and their closest opponents.

effects of any *motus* in the soul, properly speaking, since that would imply the soul's acting on itself. But they flow from it "by a certain natural result [*resultatio*], as color from light" (Thomas *ST* 1q77a6, Parma 1:302).

It is easy to see how form could, by analogy with its own relation to matter, be called the material cause of active powers. Some philosophers believed it was that and nothing more. But Toletus, like Thomas, holds that the soul is not only the material cause of its powers, but also "quasi efficiens, & quasi formalis," not, certainly, by effecting them as it itself is by its generator, but as a *subjectum* determines its own accidents, which depend on it and are ordered by it (Toletus *In de An.* 2c4q10, *Opera* 4:70vb). (Prime matter, of course, is a *subjectum* too. But unlike form, it is, as we have seen, *in potentia* indifferent to its accidents. Though they "depend" on it as a subject of inherence, they are not "ordered" by it. Ordering is the job of form.)

The vagueness of the language—Thomas's *resultatio*, Toletus's *quasi efficiens*—points to a puzzle that the mere coining of terms cannot solve. Either *resultatio* is an action of which the soul is an efficient cause or it is not, and the soul is at most a material cause. John of St. Thomas argues that emanation is an action, not of the form, but of the generator.[48] It is indistinguishable from the action by which the form itself is educed from matter. What gives form, gives its consequences.[49] The *terminus* of generation is not a "bare substance," but a substance equipped with whatever it needs for its proper operations. Every genuine action, moreover, "requires an operative virtue," that must be distinct from substantial form, "since otherwise the whole ground for distinguishing operative powers from substance is overturned." Form, properly speaking, has no actions at all, whether on itself or on other substances.

Emanation is an action of form only in the way that unlocking a door is an action of the key that turns the lock. When an agent produces one thing by way of producing another thing of which the first is a consequence, the second "is said to behave actively [*active se habere*]" (John of St. Thomas *Cursus* 269b, citing Thomas *Summa logica* 1c6, Parma 17:59b). A generator produces the powers of its offspring by producing its form, and only in that sense does the form "act" to produce its own powers.

Analogy has its perils. It cannot be that no intermediate link in any causal chain genuinely acts. Otherwise parents wouldn't genuinely act in producing their parents' grandchildren. What is peculiar to the instance at hand is that form, in effecting the accidents proper to it, would be acting on itself.

48. Cited in support of the view are Caietanus, Ferrarius, Soto, and the Carmelite *Cursus Complutensis.* The Scotists hold that "those passions which are identified with substance, like active powers, do not emanate from substance, but come to be through the act of generation itself. But those that are distinct from substance, like quantity [. . .] emanate from substance by a genuine action" (John of St. Thomas *Nat. phil.* 1q12, Appendix,*Cursus* 2:268a).

49. *Cursus* 2:268b; cf. Thomas *In de Cælo* 3lect7, Parma 19:160b.

Though John of St. Thomas grants that it "aids and guides" the production of those accidents, the initiator must be something else.

Suárez takes the second way. Accidents that are "immediately connected" with form, as the intellect to the soul, "are caused by substance, not only materially and formally, but also efficiently by natural result [*effective per naturalem resultantiam*]" (18§3¶6, 25:617). *Resultantia*, he admits, cannot be a full-fledged action: actions are accidents too, and so there would be a regress of accidents. By way of making the notion more precise, Suárez argues that it applies only to those accidents that are immediately connected with form. The intellect is connected thus with the soul, but the figure of the body results by way of quantity. More generally, all those dispositions which precede form in generation, and which, as we have seen, may persist through substantial change, are caused by form only by way of its active powers.

The strongest argument for supposing that *resultantia* or emanation is distinct from any action of the generator is that emanation could be "suspended" by God, so that, for example, a soul could be generated that lacked an intellect. Nor is *resultantia* the mere copresence of form and power in one substance. God could not only prevent the intellect from proceeding out of a soul, he could also at the same time immediately produce an intellect in it (¶13, 25:619). Only if having an intellect were a logical consequence of having a soul would God be unable to do so.[50] The making of a fully equipped soul, Suárez concludes, requires not one action, but two: creation properly speaking, whose *terminus* is substance, and resultantia, whose *terminus* is the accidents that complete the substance (ib., see also *De anima* 2c3¶10, *Opera* 3:582).

Even in less exotic circumstances, emanation may occur later than generation. Water condensed from hot vapor is only gradually restored to its pristine coldness; a stone may be prevented from falling until long after its maker has vanished (¶7, 25:617). John of St. Thomas replies that "in the form produced by generation [the action of] generation remains virtually, and when the impediment is removed it is again expressed in a passion and emanation" (*Cursus* 2:270). This is part of his namesake's explication of the slogan 'Everything that is moved is moved by another': the falling stone does not move itself, but is moved by the generator, whose action may continue to exist even when it itself is destroyed (see *Cursus* 2:465).[51] When impediments are present, part of that action may be delayed, just as the action of a

50. "Nam finge Deum creare substantiam animæ impediendo emanationem potentiæ ab illa [. . .]; maneret ergo tunc substantia animæ absque intellectu et voluntate auferat vero postea Deus impedimentum, reliquatque animem suæ naturæ, certe dimanarent ab illa intellectus, atque voluntas, sicut in primo instanti generationis, aut creationis" (Suárez *De anima* 2c3¶10, *Opera* 3:582; cr. also 2c1¶7, 574).

51. On Thomas's understanding of the slogan, see Weisheipl 1985, c.4, esp. 90ff.

testator may be delayed by inheritors' lawsuits. Where Suárez sees two actions—generation and emanation—John of St. Thomas sees one.[52]

If the issue seems abstruse, it is because we moderns, like the nominalists, don't believe that form and powers are really distinct.[53] The conclusion can be generalized. So long as powers are understood in terms of the physically necessary consequences of structure, they will "emanate" from it as a matter of course. Descartes's machines, as we will see, illustrate just such an understanding. In his world the issue that set Suárez and the Thomists at odds cannot arise.

2. *Forms and powers in generation.* Generation begins with corruption: since nothing can have two substantial forms at once, the matter of the new thing must be released from its old form. When water is changed to air its heat must first be raised above the latitude permitted by the form of water. That alteration at once destroys the form of water and makes its matter ready for the form of air. The moment that the work of the active powers is complete, the new substantial form emerges. Generation is completed by the emanation in the new substance of its active powers and other embellishing accidents.

Such is the gross anatomy of generation. Both substantial form and active powers appear to have some part in the process. But how is their labor to be divided? Can accidents by themselves "attain to the production of substance"?[54] In their version of the standard question, the Coimbrans first lay out what they call the two extreme views. One, which they attribute to "certain recent Philosophers," is that "the substantial form of the generator lends no active influx to generation or to other transitive actions; accidents alone, by their own force [*virtute*], and as principal causes, generate substance" (Coimbra *In Phys.* 2c7q18a1, 1:293).[55] The other, that of Scotus and the Nominalists, is that "accidents attain to the production of substance

---

52. Elsewhere Suárez grants that emanation continues and completes the act of generation, and that the soul, as the efficient cause of powers, is the instrument of the generator (Suárez *Disp.* 18§7¶10, *Opera* 25:633). But since he believes that instruments can have genuine efficacy of their own (§7.3 below), there is no contradiction.

53. The (real) identity of the mind and its powers follows, for example, if the mind is conceived to be a computer program. Hilary Putnam and Martha Nussbaum have argued that Aristotle's conception of the soul is functionalist, and so—according to the predominant brand of functionalism—analogous to a program. Mary Louise Gill compares substantial form to a "list of instructions that determines the materials and tools needed to realize a particular end" (Gill 1991:251).

54. Even though the question is stated generally in terms of substance and accidents, the only relevant cases involve substantial form and active powers. Matter and quantity cannot produce substance because they are entirely passive (Fonseca *In meta.* 5c2q6§1, 2:98C; Suárez *Disp.* 18§2¶3, *Opera* 25:599). Matter considered as substance can neither be generated nor corrupted.

55. The *recentiores* include Albert of Saxony and John Mair, a Parisian philosopher who flourished in the early sixteenth century, as well as the earlier Richardus de Mediavilla and Thomas Argentinus (Suárez *Disp.* 18§2¶14, *Opera* 25:602).

neither by their own power [*vi propria*], nor as the instruments of substance; they merely make matter apt and fit, so that in it the substance of the generator can immediately produce substantial form" (ib. 293–294). In the technical terms used to answer the question, accidents are, according to the second view, not even "instrumental" but only "dispositive" causes in generation. An instrumental cause has the same effect as the corresponding principal cause. A dispositive cause has a different effect that promotes the operation of the principal cause.[56] The utterance *Hic est calix sanguinis mei* is instrumental to the sacrament, the freshness of the wine merely dispositive. To deny that accidents are instrumental causes in generation is to deny that they are causes of substantial form at all.

The two extremes, then, deny that accidents are instrumental causes. The first makes them principal causes, at least sometimes; the other, merely dispositive always. The central texts take that favorite path of Aristotelianism, the middle way: accidents cannot of themselves bring substantial forms into existence, but substantial forms can only generate other substantial forms by way of accidents. Accidents, especially active powers, are necessary instruments in the production of substance.

That answer confers upon created substantial forms and their powers a well-defined, and narrow, rank in the hierarchy of causes. Created forms do, on the one hand, have the power to generate other forms. As the Coimbrans say, activity is "one [. . .] of the principal affections of natural things" (Coimbra *In Phys.* 2c7q18a2, 1:294), and generation is the noblest of natural actions. Forms, the noblest part of substance, ought not only to be endowed with activity but also to be the cause of the noblest action. Only those who deny efficacy to second causes altogether would go so far as to deprive form of activity.

Accidents, on the other hand, since they are less noble than forms, are incapable of producing them without assistance: "Whatever is in the effect, exists beforehand in the efficient cause, [which is] equally perfect if univocal, more perfect if equivocal: for nothing can give what it does not have either formally or eminently, as they say, that is, more excellently: but an accident, if it is compared with substance, especially with the form which is introduced [in generation] or with the composite which is generated, is much less perfect: substance is therefore not produced by the virtue of accidents [*virtute accidentis*]" (Fonseca *In meta.* 5c2q6§2, 2:99bE; cf. Coimbra *In Phys.* 1:295). That is enough to rule out the first extreme for the

---

56. Suárez, after an analysis well worth reading, concludes that "in the [. . .] most proper way [of speaking], 'instrumental cause' is said of that which concurs or is raised up to bring about what is nobler than itself, or exceeds the measure of its proper perfection and action, like heat, insofar as it concurs in the producing of flesh, and in general accidents whenever they concur in producing substance" (*Disp.* ¶17, *Opera* 25:590).

time being. But the Scotists and nominalists conclude further that accidents can in no way be said to have substances among their effects and that substances are the immediate principles of generation (Coimbra *In Phys.* 1:293, Abra de Raconis 54).

Against that conclusion the central texts urge experience, reason, and authority.[57] The elements, the heavens, mixtures, all are seen to exert their actions through accidents. For the soul, too, "the principles of accidental operations are accidents also, as is clear from vital or nutrimental heat, and from the power of gathering together and expelling excrement, and the like" (Suárez *Disp.* 18§3¶16, *Opera* 25:620). Some substances, including the heavens and the soul, appear to act at a distance. Since every immediate cause must be present wherever it acts, the heavens themselves cannot be the immediate cause of the generation of minerals (Suárez 620, Fonseca 2:102E—F). In the reproduction of animals and plants, the seed "gradually fashions through its formative power" the limbs, and so forth, even if the parent has meanwhile died (Coimbra 1:296). Although God could supply the missing efficacy of the parent (see Abra de Raconis 55), so widespread an appeal to universal causes is to be avoided, "since the right disposition of the world would have it that whatever can be done suitably and naturally by second causes is done by them" (Suárez *Disp.* 18§2¶15, *Opera* 25:603).[58]

Accidents, then, are necessary instruments in generation. But only instruments, never principals. John Major noted, though he was no doubt not the first, that hot iron can set wood afire; and we all know that rubbing two sticks together can accomplish the same end. The Aristotelians knew that a drop of water in a vat of consecrated wine will be corrupted by the accidents of the wine, even though the substance of the wine has departed. In all such cases, it seems, an accident brings about the production of a substance whose form is not present.

But we have seen that all such claims would have the less noble produce the more noble. When heat alone appears to generate fire, it must receive the aid of a superior form. If there was disagreement on that point, it concerned only the nature of the superior form: was it material? was it the celestial intelligences? Scotus and Suárez did not forego appeal even to the First Cause. Other philosophers referred the equivocal generation of frogs and worms from decaying matter to the heavens; some held that metals and

---

57. The authorities include Aristotle *De sensu* 4c4, 441b13; Averroes *In Meta.* 7com21, 31 and 12com18; along with Thomas and many of his followers.

58. Suárez adds (25:602) that the Scotist view "smacks of Plato [*sapit Platonem*]," and is similar to that of Philoponus (*Phys.* 1 *in fine*), Themistius (who writes in *De an.* 3c52 that forms "provenire ab anima mundi"), and Avicenna (see §5.3). To deprive second causes of genuine efficacy was contrary not only to the opinion of all the Scholastics but also to faith (*Disp. 18§1¶5 & 12, Opera* 25:594, 597; cf. Concilium Tridentinum, *sessio* 6, *canon* 4, in Denzinger 1976, no. 1554).

precious stones were produced in the earth by the sun; and many believed that the *forma cadaveris* must be introduced by a superior cause.

Take a case where, by our lights, superior causes are obviously not needed: the generation of fire by radiation or friction. Some philosophers explained the combustion caused by a hot iron by supposing that "some true fire has seeped into the pores of the iron, from which the [new] fire is generated" (Suárez *Disp.* 18§2¶30, *Opera* 25:609, cf. 679). Fonseca holds that the iron or the source of its heat is either fire or nobler than fire—the iron itself, being a mixture, contains all four elements virtually—and is thus suited to produce its form. The same goes for friction: "what rubs [. . .] is a mixture, in whose virtue fire is contained" (*In meta.* 5c2q6§3, 2:104bF).

Such explanations avoid an appeal to celestial causes. In other instances, according to the careful analysis of Suárez, they cannot be avoided. His account of the generation of fire and minerals starts with the observation that the heat in an "extraneous" subject "never has the intensity necessary for the form of fire."[59] Of all the elements fire has the highest degree of heat, and in any mixture the elemental qualities are always tempered. But it may happen that "although the proximate agent introduces heat only to a certain degree, there results *per accidens* the disposition of a quite different form, as when wet matter is heated by fire, and a warm wet temperament readily results, and [thus] some form that requires such a disposition" (¶30, 610). We might take this to justify not the introduction of a superior cause but the reduction of those forms to their dispositions. Suárez instead concludes that "it is not difficult to believe that the virtue of a superior agent is required [. . .], not only because the proximate agent does not have an equal or similar substantial form, but also because neither its quality nor its temperament suffices *per se* to dispose and prepare the matter" (ib.). The superior agents turn out on the common view to be celestial bodies, especially the sun.

Animal souls are the noblest of nonhuman forms. But in generation, because they reproduce by seeds, they turn out to require even more assistance than inanimate forms. The form of the seed is less perfect than that of the parent, and the mother's form cannot make up for its deficiencies. A higher cause must not only concur with the action of the seed but supplement it: "insofar as it attains to the production of the sensitive soul, not only does the seed lack power with respect to the mode of action, but also with respect to its substance (so to speak), and thus with respect to that action needs much more the actual help, influx, and concurrence of a superior cause" (¶33, 611). The slightly obscure language refers to an argument just

---

59. 'Extraneous' means 'not belonging to a thing by nature'. Iron is naturally cold. But there are other substances—air, of course, and blood—which are hot by nature. So it is not quite clear what Suárez has in mind, unless that fire is the hot thing par excellence.

made that the seed probably does not suffice even to organize the matter of the foetus. Still less, then, will it be able to educe the sensitive soul that requires such organization. Many philosophers held that the higher cause is either the celestial intelligences or the heavens working as their instruments. But they too must act through their accidents, and their accidents do not differ in kind from those of earthly things. So they are no more able to produce souls than are animals themselves. Suárez, despite the words of caution quoted above, is led to conclude that God alone is the coauthor of souls, human or not.

The oddity is that lower forms—those of elements and many mixtures— are capable of reproducing themselves without assistance. That looks to be "both an ill-ordered [*inordinata*] institution of nature and a great imperfection in the causes that generate in this way" (¶34, 611). Suárez replies that, just as it is no imperfection in man that two of his species should be needed in reproduction, so too propagation by seed, which requires the assistance of God, is "a natural condition or need" that animals "postulate by their nature." Indeed it is because living forms are so much more perfect than nonliving that they require an "exquisite and peculiar mode of generation." For that reason also they require assistance; but it remains true that nature has given to them, as to every corruptible kind, all they need to propagate themselves (¶41, 614).

Nevertheless it seems odd that the more noble and more powerful a material form is, the more fragile its mode of propagation should be, and the more dependent on assistance from God. It is as if the presumption of matter to assimilate itself to the divine had to be compensated for by a proportionate vulnerability. There was, indeed, a standard view that the more noble a substance was, the more difficult it would be to bring it into existence. Higher animals must reproduce sexually: no one individual can reproduce itself. The human soul, and all spiritual forms, could not be brought into existence at all by material forms alone. Infinite being would require infinite power to produce. Even the attenuated infiniteness of immortality, shared by the soul, could be communicated only by God himself.

The dustbin of history has long since swallowed up the Aristotelian theory of generation. Already Petrus Aureolus had complained that the use of celestial causes "is the refuge of the wretched in philosophy, as God is the refuge of the wretched in theology."[60] Their reputation would only get worse. Rather than flog a dead theory, it is more fruitful to ask what obstacle,

---

60. *In Sent.* 4d1q1a3; cf. Maier 1951:182n.29, Maier 1952:13n.10. For a brief survey of similarly unfavorable sentiments about appeals to celestial causes or God, see Hansen's introduction to Oresme *Mira.*, p50ff. Oresme too calls the use of celestial causes, demons, or God to explain marvels the "ultimum et miserorum refugium" (Prologue, p136). On the naturalization of marvels, see §6.4 below.

in Bachelard's sense, blocked a further understanding of generation. Less tendentiously put: which elements of the Aristotelians' conception induced them to bring in universal causes?

Most obvious is the principle that the less noble cannot bring into existence the more noble. Accidents alone cannot beget substances. That is bedrock: what remains is to explain apparent counterinstances. More bedrock: forms cannot act at a distance. Hence the push to the one cause that can act immediately anywhere because it is present everywhere. Scotus gave up the second principle to preserve the first. Suárez thought that doing so smacked of Plato and Avicenna.[61] Yet his own considered view of the reproduction of animals is similar. Perhaps it was inescapable: so long as forms were considered more noble than accidents, while acting only through them, they will always need a supplement; if that supplement too is a finite form, it will want a further supplement, and so forth. Since it was already firmly part of theology to hold that God concurs in every action of created things, his assistance in generation could be seen to be merely a specification of that concurrence in certain instances.

Indeed the relation of forms to their powers is not only comprised in the question of the efficacy of second causes; it is an almost exact parallel. God could act immediately to bring about change in the world; finite substances could, in principle, have been given the power to bring about accidental change (*Disp.* 18§3, *Opera* 25:621). Instead each has its proxies. The difference is that God *chooses* to operate through second causes when he does, and that some second causes are free; created substances, even those with free will, cannot choose but to act through their accidents, which in turn cannot but fulfill the ends of form. That difference aside, within each substance the hierarchy of divine and created agents is repeated in the hierarchy of substantial and accidental forms. Every substance is an image of the universe.

If the Aristotelians had countenanced a reduction of substantial forms to accidents, the ordering of forms according to nobility would have lost its force. But reduction was refused even for elemental forms. Though some philosophers, notably Alexander of Aphrodisias and Nicholas of Autrecourt, were said to have affirmed that the form of each element is its

---

61. "Posterior vero pars [the claim that inanimate forms are educed immediately by the first cause] sapit opinionem Platonis, qui dicit formas substantiales induci ab ideis separatis; nam si verum est Platonem non posuisse ideas nisi in mente divina, perinde fuit ac dicere formas substantiales effici a prima causa": The latter part smacks of Plato's view, when he said that substantial forms are induced by ideas which exist apart [from the world]; for if indeed Plato did not posit ideas except in the divine mind, it is as if he had said that substantial forms are brought about by the first cause" (*Disp.* 25:602).

dominant quality, that view was rejected as "absurd" by the central texts.[62] What is substance, Averroes argues, in one thing cannot be an accident in another. But heat, say, is found in animals too, and so it cannot be the substantial form of *fire*. Not even an essential quality can be a substantial form: if the ablation of heat destroys fire, it is not because heat is the form of fire, but because the ablation of heat removes the form, just as cutting off a man's head removes his soul (Toletus *In de Gen.* 2c2q2, *Opera* 5:309v–310r).

Since even the forms of the elements cannot be reduced to their dispositions, neither can those of higher forms. The reduction envisioned by Descartes of all material forms to modes of extension could not be admitted. Confronted with similar proposals in ancient atomism, the Aristotelians hardly felt the need for extended refutation. It sufficed to note that atomism does away with the distinction between substantial and accidental forms, and thus with the distinction between generation and alteration. The irreducibility of form was not the only reason the Aristotelians had for holding that substantial form is nobler than accidental. But it was perhaps the best physical reason.

Descartes's physics, on the other hand, by doing away with the distinction (except in the case of the rational soul), wiped away with one stroke the principal reason that drove the Aristotelians to assign celestial agents and the First Cause an essential role in generation. Or at least it promised to: he himself was unable to explain the reproduction of animals except by invoking a seed endowed with powers of organization, whose effects he could attempt to describe but whose explanation eluded him.

62. "Others would have it that elemental forms are the four primary qualities themselves [. . .] They judge Alexander and other Greek interpreters of Aristotle to believe the same. Nor does it count against [the view], that the forms of the elements should also be substances: they are referred (they say) to the category of substance, and for that reason they are substantial qualities, just as the qualities that belong to the category of quality are accidental" (Fonseca *In meta.* 8c3q3§1, 482aB). Toletus adds Nicholas to the list, and notes that John of Jandun contests the ascription of the view to Alexander (Toletus *In de Gen.* 2q1, *Opera* 5:309va). 'Absurd' is his word.

# [6]

# Finality and Final Causes

D'Alembert wrote that whatever Descartes's errors in physics, he at least showed "those with good minds how to throw off the yoke of Scholasticism" (*Disc. prél.* xxvi). Not the least of his services was to proscribe the use of final causes in physics. Final causes are not merely superfluous, not merely the "vestal virgins" of philosophy, as Bacon said, but "dangerous."[1] The appeal to ends leads to absurdity, and invites us to put ourselves in the place of the Creator, presuming to know his ends.[2]

Gratuitous examples of the sort that later critics probably had in mind are not hard to find. Abra de Raconis, for example, writes that "the two most important ends to which the sea is instituted are, first, that it should be the common domicile of fish, and second that in it there should be navigation to provide commerce and necessary goods; but to both ends its saltiness is most fit, since saltiness keeps the sea from putrefying and makes it stronger and denser so as to hold the greater weight of ships."[3] The Coimbrans write of the ocean that "when by its flooding or daily ebb and flow or the force of storms its [waters] are borne toward the shore, they contain themselves within preordained limits as if terrified of some law written to them on the sand" (Coimbra *In Phys.* 2c9q1a1, 1:224; cf. Suárez *Disp.* 23§10¶10, *Opera*

---

1. See Bacon *De augm. sci.* 3c5, *Works* 1:571. It should be noted that Bacon's jibe is directed against the use of final causes in physics—where they were in fact usually treated—and not against their use altogether. In metaphysics, ends, along with forms, are proper objects of inquiry (cf. 570).

2. *Encyclopédie* s.v. "Causes finales," 2:789. Cf. Descartes *PP*1§28, AT 8/1:15.

3. *Phys.* 384. He adds that since "God rarely effects himself what he can produce by way of second causes," the efficient cause of the sea's saltiness is probably "fiery exhalations" and solar coction (385; cf. Dupleix *Physique* 7c19, 485, which cites Aristotle, Pliny, and Plutarch). Cf. also John of St. Thomas *Nat. phil., Tract. de meteoris* 7c3, *Cursus* 2:877a–b.

25:888). The sentiment is fine, but the explanation is superfluous and based on a misapprehension of the facts. The ocean's boundaries have changed enormously over the aeons; its transient containment can be explained entirely by efficient causes. So too with saltwater fish: the sea was not made for them, we now believe; they, in a certain sense, were made for the sea.

But these are egregious instances. Descartes and those who subscribed to his polemics exaggerated the sins of their opponents, ascribing to the Aristotelians views the Aristotelians would have repudiated. Far from being confused about which things have souls and which don't, as Descartes thought, they carefully separated, as we will see, the case of rational agents, which can recognize and judge their ends, from that of irrational agents, which exhibit only the secondhand finality of instruments. While asserting the primacy of the final cause, they did not, as d'Alembert insinuates, confuse final with efficient causation. The *horror vacui*, which d'Alembert holds up to ridicule, is an instance where the Aristotelians are quite clear about the efficient causes. They also propose one or another final cause; they do use intentional language in describing how natural agents act to prevent the occurrence of a vacuum. But the use of such language does not of itself imply confusion. In Abra de Raconis's explanation of the saltiness of the sea, the end is ascribed to God, not to the sea itself. Nor must one suppose that final causes compete with efficient causes in explanation: the Aristotelians recognize both, making each a separate subject of enquiry. When Descartes and others declared that final causes were superfluous in the face of adequate efficient causal explanations, they forced the issue in a way that most Aristotelians would not have regarded as necessary.

The Aristotelians had, moreover, come increasingly to restrict final causation to the actions of rational agents. The beginnings of that development can be found in Ockham, and its culmination in Jean Buridan. Though in the period I am studying there was a return among many authors to the less radical position of Thomas, at least one author—Hurtado de Mendoza, whose *Cursus* was published in 1624—omits the standard questions on final causes from his *Physica* altogether, placing them instead among disputations about the will. That was extreme, but it reflects an increasing reluctance to take ends for "real and proper" physical causes except when cognized by rational agents.

What remains, and what Descartes especially opposed, was finality. Though the ends to which natural changes tend might not be proper causes of those changes, it was inconceivable that natural change should not have ends. The analogy implied in the use of intentional language to describe natural change retained its force. Finality, and not the valetudinous doctrine of final causes, was Descartes's real target.

The impression conveyed by Descartes of a vast distance between his physics and that of his predecessors was, therefore, not unwarranted. The very notion of natural change, defined as the imperfect *actus* of a *potentia* in the thing changed (§2.1), presupposes that the basic and most comprehensible changes in the world are directed changes, which cease of themselves after reaching their *termini,* and in which there is an intrinsic difference between the patient, in which the change occurs, and the agent, which is unchanged except *per accidens.* Other changes—corruption, say, or monsters—must be redescribed to be understood scientifically. They are not exceptions or counterexamples to the omnipresence of finality: they are, rather, necessary outcomes when natural processes meet and interfere with one another, or when a patient's matter is inadequate to receive the form specified for it by the agent.

The prototypical event of Cartesian physics—the collision of two bodies—is just the sort of event that must be redescribed. It has no *terminus ad quem.* There is no intrinsic feature in the encounter thus described that would allow us to pick out one as the agent and the other as the patient. To bring it within the scope of science an Aristotelian must find, say, that one body is a falling stone, the other a pilgrim leaving for Rome; only then can the collision, now analyzed into two independent *motus,* be referred to their ends and understood. The Cartesian condition of intelligibility is quite different: paradoxically, at once more immediate in conception and more remote in application. The falling of a stone is not a simple expression of the heaviness imparted to the stone by its generator. It is the outcome of innumerable collisions between the stone and the whirling vortex that surrounds the Earth. Each collision is subsumable under the laws of motion; but the aggregate effect can only be understood, as we would now say, statistically.

Nevertheless even Descartes, whose unbending opposition to finality was maintained after him only by Cartesians of strict observance, did not rule out reasoning from the perfection of the world as a whole.[4] Nor did he deny that people and other ensouled creatures act according to ends. To understand why and how the explanatory burden in natural philosophy and metaphysics sometimes remained with the final cause, or was shifted to the efficient cause, or even to no cause at all, it is a good idea to start with the distribution of that burden in Aristotelianism.

---

4. Sylvain Régis sustained his opposition to final causes even in the face of Leibniz's jibe that Descartes had "turned philosophers away from the search for final causes, or, what is the same thing, from the consideration of divine wisdom" (Leibniz *Ph. Schr.* 2:562). Régis answered that Descartes never intended to banish final causes from moral philosophy; as for natural philosophy, he was right to do so, since "in Physics one does not ask why things are, but how they occur" (ib. 4:334). For a discussion of the exchange between Leibniz and Régis, see Tocanne 1978:70–71.

The first item of business is to reduce to order the welter of objects said to have ends, at the same time classifying some uses of finality in physical argument (§6.1). The second is to examine the existence of ends in nature, both in the actions of individuals and in nature collectively (§6.2). Next I look at the final cause, which was for a number of reasons not easily aligned with the other three Aristotelian causes (§6.3). I conclude by studying the use of teleological principles in explanation (§6.4).

## 6.1. Varieties of End

In human affairs, a single bodily act may be seen to serve many purposes at once. I lick a stamp, intending thereby to signify that I have paid postage, and so also intending to mail a letter, which in turn repays a debt and keeps me in the good graces of the telephone company. No discussion of finality in Aristotelian physics should neglect the complexity of the ends attaching to natural change, or what I will call its *teleological structure*. Not merely the ascription of ends, but the unification achieved through the subsumption of one end to another, and of all to the ultimate end God, is what gave the Aristotelian use of finality its force. When combined with a doctrine of creation, moreover, it armed natural theology with evidence not only for the existence and power but for the intelligence of the Creator.

I will classify the ends ascribed to natural changes into three categories. The first, individual ends, includes the immediate ends of actions and of particular things. The second, collective ends, includes all those ends that cannot be achieved by one thing alone, like the preservation of the species. Individual ends are typically in aid of collective ends; as we will see, a collective end may be in conflict with, and override, individual ends. The third category, cosmic ends, consists of those ends that may properly be called the ends of nature as a whole, which for brevity's sake I will call 'Nature'. The most notable of those ends is beauty.

1. The most general definition of 'end' is 'that on account of which', *propter quid*, something is done. Aristotle holds, and the Aristotelians agree, that in any change there are two things that may reasonably be called the immediate end of the change. There is a condition, the *terminus ad quem*, the attaining of which coincides with the cessation of the change. There is also a thing for whose benefit the change occurs. That thing may or may not be the subject of the *terminus ad quem*. In the standard example of an ill person's being restored, both health and the person herself can be called "that on account of which," and so both can be called ends.[5] The condition was most often designated the *finis cuius*, the subject the *finis cui*. The terminol-

5. *Phys.* 2c3text37 (195b19); Coimbra *In Phys.* 1:230; 2c text32 (195a24), 1:229.

ogy is confusing, especially since some authors used other terms. Except in quotations, I will avoid the Latin terms. I will call the condition the *intended state* and the subject benefited the beneficiary.[6]

The intended state and the beneficiary have distinct roles in explanation. It is, first of all, trivial that every natural change should have an intended state. Natural change is defined as the imperfect *actus* of a *potentia:* it has, therefore, a corresponding perfect *actus,* its natural stopping point. What is not trivial is to decide which of the changes we see are in fact complete, and if so, what their intended states are. Descartes denied that the changes of bodies were natural in the Aristotelian sense; inquiry into their intended states became moot. Nor did the Aristotelians hold that every occurrence in nature is a natural change: monsters, which I will discuss in §6.4, are a case in point. Designating the intended state of a change, moreover, enables one to say that the change has been frustrated, or to explain why it has not yet ceased. Aristotle, in fact, points to such judgments by way of arguing that nature acts according to ends. It also, as we have seen in §2, serves to distinguish one kind of *potentia* from another, and so, since the natures of things depend on their *potentiæ,* provides the basis upon which to define natural kinds.

It is not obvious, on the other hand, that every natural change has a beneficiary. The hard case, as we will see, is generation, in which the intended state includes the existence of a new individual of a specified kind.[7] Here not just a state, but the resulting thing, is the end of the action; but the beneficiary is not that thing but something else—the owner, say. It was customary to distinguish two ends in generation: the *finis generationis,* which is the thing generated and thus the intended state of generation; the *finis rei genitæ,* or the end of the thing generated. Typically the *finis rei genitæ* is an end attainable by one or another operation of the thing generated, as restoring health is the end of medicines. The beneficiary of medicine-

---

6. "Finis Cujus dicitur, cujus adipiscendi gratia homo movetur, vel operatur, ut est sanitas in curatione; finis Cui dicitur ille, cui alter finis procuratur ut est homo intentione sanitatis; nam, licet homo curetur propter sanitatem, ipsam vero sanitatem sibi et in suum commodum quærit" (Suárez *Disp.* 23§2¶2, *Opera* 25:847). In *cuius gratia, cuius* is the genitive of the relative pronouns quis/quid, translating Aristotle's οὗ ἕνεκα or οὗ ἕνεκα τινος 'on account of which/something'. *Cui,* the dative, translates ᾧ ἕνεκα or οὗ ἕνεκα τινι 'for whose account'. Τὸ οὗ ἕνεκα, *id gratia cuius* or *id propter quid,* is Aristotle's usual name for the final cause. Cf. Aristotle *De an.* 2c4, 415b, *Meta.* 12c7, 1072b; Zabarella *De anima* 2c4text37, 437B; Coimbra *In Phys.* 2c7q2o2a2, 1:305; Fonseca in n.7 below; Eustachius *Physica* pt1tr2d2q6, *Summa* 2:143.

7. Fonseca uses the term *finis cuius* to denote not only states like health, but also products of action, such as houses, and immanent actions, such as contemplation, that issue in neither a state like health nor a product. "Finis Cuius dicitur id, cuius consequendi aliquid fit, sive id sit res aliqua, quæ per actionem fiat, veluti domus, cuius ædificandæ gratia artifex operatur, sive actio immanens, per quam nihil aliud fiat, veluti contemplatio, sive usus aut fruitio rei, ut habitatio domus, aut voluptas, quæ ex contemplatione percipitur, cum quis propter voluptatem contemplatur" (Fonseca *In meta.* 5c2q10§1, 2:153A).

making is not the medicine itself but the person who takes it. In generation, then, the intended state is the thing generated; but only sometimes is it obvious that there is something for whose benefit the thing is generated. Even if children help their parents and provide for them when they are old, still it is not clear that children are born for the benefit of their parents.

The Aristotelians held, nevertheless, that to every intended state there corresponds a beneficiary.[8] It made sense, for example, to ask not only whether the saltiness of the sea was the intended state of the changes that created it, but for whose benefit it is salty. The answer is in part that it is useful to people. Though Aristotle himself seems to have recognized the legitimacy of such questions, he does not bother much with them. Their prominence in Aristotelian discussions of ends in nature rests, as we will see, on the warrant provided by scriptural assertions of man's dominion over nature.

Supposing, then, that all natural changes do have a beneficiary, naming that beneficiary is essential to assessing intended states and the appropriateness of the means employed to reach them. Health in a human requires a degree of heat that would kill an oyster. Even among animals for whom being warm is part of being healthy, the means of heating might well be different: a carnivore should be given spicy meat, a vegetarian hot peppers. For that reason the Aristotelians said that the intended state is "subordinate" to the beneficiary (Fonseca *In meta.* 5c2q10§1, 2:154C).

Some authors thought that intended states could be assimilated to instruments or means (and thus that only the beneficiary is truly an end). Since an instrument is an efficient cause, that would reduce some final causes to efficient causes. Yet there is, Suárez argues, "a great difference between a means and a *finis cuius;* the means [. . .] is desirable insofar as it is useful for health; but health is desired on its own account, because it perfects *per se* the person for whom it is desired." He concludes that only the *whole*—a healthy person—is the "entire and adequate end of this action."[9]

That conclusion fails to do justice to instances where the intended state is a newly existing individual. Although health and the person who is healthy may be called a whole, a house and its owner may not. Though one could evade that objection by talking instead about the *finis rei genitæ*, the shelter provided by the house, it is more illuminating to arrive at Suárez's conclusion by a different route. The end on account of which a patient is given medicine is not just the patient herself, nor even her continued existence. It

---

8. "Inter hos autem fines hoc interest, quod finis Cuius semper dirigitur in aliquem finem Cui, finis autem Cui ex ratione sua non necessario dirigitur in alium finem" (Fonseca *In meta.* 5c2q10§1, 2:153C).

9. Suárez *Disp.* 23§2¶5, *Opera* 25:848; Fonseca *In meta.* ib. See Maier 1955 for a discussion of this controversy in Ockham and others.

is a certain state *of* the patient. Other states require other actions, even if the beneficiary is fixed.

In generation or in art, the teleological structure of actions has a complexity not exhausted in the distinction between intended states and beneficiaries. A veterinarian gives a horse mint to cool the blood. The intended state of the operation is the cooling, the immediate beneficiary is the horse. But the horse is treated so that the owner can plow his land. The owner is a remote beneficiary, and the horse's power to work is also an intended state. We will see a similar structure in the generation and preservation of species. Recognizing that complexity is essential, because the two kinds of end figure differently in Aristotelian physics. Questions about the causality of ends and the objects of final causality concern primarily the intended state. The intended state, rather than the beneficiary, is what "moves an efficient cause to act." The beneficiary, on the other hand, comes into its own in questions about the existence of ends in nature.

2. *Collective ends.* States of individuals and individual substances do not by any means exhaust the list of possible ends. There are also collective ends. For example, since no material individual can endure forever, a specific form—the form or part of form common to all members of a kind—would sooner or later cease to exist if it could not bring about the existence of its like in new matter. Yet all forms strive, as we will see later, to assimilate themselves to God, one of whose attributes is to exist at all times. God, for his part, has given to each kind of form the powers necessary for each instance of that kind to produce in new matter whatever is required for its like to exist. It is therefore at least possible for every material form to exist eternally, not indeed as an individual, but as a kind: "In all things there is an inborn appetite for protecting and preserving themselves. Hence the care with which they seek what is useful and healthy, and shun what is harmful [. . .] Because in that way all perishable things strive to save themselves from destruction, and if not numerically at least in species, if not *per se* at least through the species to which they belong, strive to obtain immortality; so that thus they approach, insofar as it is possible, a resemblance to the divine nature" (Coimbra *In Phys.* 4c9q1a3, 2:62).

Although the individual form of the offspring, the *terminus ad quem* of generation, is clearly the intended state of that action, neither the parent nor the offspring can rightly be called the beneficiary. The parent, although it may benefit by having offspring, need not have offspring for that reason; the offspring, although it benefits, in a sense, from being generated, need not have been the intended beneficiary of generation. What benefits in every case is the specific form, regarded as a kind.[10] Yet no one individual aims at the preservation of its kind. Preservation is a collective end.

---

10. Generation, writes John of St. Thomas, is the most natural of the soul's operations (cf. Aristotle *De an.* 2lect7, Thomas *In de An. ad locum*) because the nature of each living thing

Individual ends are typically appealed to in answering questions about particular acts or types of act. Collective ends explain rather the existence of powers in a form. The question one answers by referring to the preservation of specific form is not 'Why did this individual reproduce?' but rather 'Why do individuals with this form (or indeed any material form) have the *power* of reproduction?' Similarly, the most obvious answer to the question 'Why do souls have the power of nutrition?' is that a soul requires for its own survival certain dispositions in its matter—notably heat—that can only be maintained by eating and drinking.[11] The soul, then, is not only the efficient cause by emanation of its self-sustaining powers (§5.4) but their beneficiary.

That collective ends are distinct from, and occasionally opposed to, individual ends is well illustrated by the *horror vacui*.[12] Toletus, after affirming that God could bring it about that a certain interval was devoid of substance, argues that nevertheless a vacuum cannot naturally occur. Experience amply confirms that it cannot, and, as usual, the relevant phenomena are taken to be actualizations of a sort of *potentia*, the *horror vacui*. But for the Aristotelians as for their successors, to end the story there would have been jejune. Toletus gives two explanations of the *horror vacui*: "I believe one should say that contiguity among bodies is the most natural disposition, so that just as a continuous part draws along by its movement another part continuous with it, so too what is contiguous with another draws along what is contiguous, since no other body can follow." The contiguity of bodies is, moreover, "in accordance with the nature of the universe," since if a vacuum intervened between sublunary bodies and the heavens, their contact with the heavens would be interrupted, and the heavens could no longer guide them (Toletus *In Phys.* 4c9q10, *Opera* 4:130rb).

The Coimbrans argue more generally that in all bodies there resides a "love of mutual conjunction and society." In drops of water and in the earth, bodies form themselves into a sphere, "the greatest conciliator of union."

---

"inclines to it not on account of its own good, but on account of the common good, namely the perpetuity of the species, which is preserved by its being generated" (John of St. Thomas *Nat. phil.* 4q3a1, *Cursus* 3:86b). Because the common good outweighs the private good, generation more strongly attracts the soul than nutrition or self-preservation (ib. 87a). The reasoning resembles that of the Coimbrans concerning the *horror vacui*, which I discuss below.

11. One may wonder, if all the powers of souls inferior to the human have, ultimately, the preservation of the form in the face of corruption as their end, what end could be assigned to the powers of angels, demons, and the human soul itself insofar as it is immortal? The answer for angels, at least, is that there are other ends: the contemplation of God (which, being an immanent action, is its own end), the administration of the heavens, the revelation of divine intentions to humans (see Suárez *De angelis* 5 and 6, *Opera* 2:565$^{\text{ff}}$).

12. Aristotelian theories of the vacuum are discussed in Schmitt 1967 (Toletus is quoted at pp355, 357), Grant 1973, c.4, Grant 1981. Schmitt discusses only empirical evidence for the *horror vacui*; he does not examine the causes proposed for it; Grant is likewise much more interested in experimental evidence than in causes (but cf. n.14 below).

United and joined, they "operate much more strongly, and repel the injuries of their adversaries." The love of union brings even bodies of different kinds together, since "there is no kind of thing that subsists by itself, torn from the rest, or that if others are lacking can preserve its power or its perpetuity." In particular, sublunary forms, as we have seen, require the aid of celestial influxes to reproduce themselves. A vacuum, clearly, would "dissolve this preservative union," and impede the reception of heavenly force [*vis*], since "it cannot happen that this power should cross through an empty intervening space."

The need for union is so great that it can overwhelm what one might call the private tendencies of things.[13] The Coimbrans note that "each natural being strives resolutely on behalf of two goods, namely the common good of nature as a whole, and its own peculiar good." But the common good is the more "excellent and divine," so that sometimes a body will act to its own detriment on behalf of it. Since union is a great good, and vacuum a "most pernicious evil" to the world, we see heavy bodies move upward, light bodies downward; "even contraries, which would otherwise flee one another [. . .] come together as if they had given up their antipathy" (Coimbra *In Phys.* 4c9q1a3, 2:62–63).[14]

3. *Cosmic ends.* A collective end, then, is an end which each agent can be said to strive for but which cannot be accomplished by any one agent alone. A cosmic end, on the other hand, is an end serving nature as a whole, an intended state whose beneficiary is Nature. In their question on the ends of nature, we find the Coimbrans responding to the argument that because certain "minute animals" are "entirely useless," nature does not always act according to ends: "Nothing superfluous or without an end has been brought about by God, although to the ignorant it may at first glance seem so, just as someone might judge the tools in some craftsman's workshop to have been multiplied beyond necessity, because he is ignorant of their uses" (Coimbra *In Phys.* 2c9q1a3, 1:326). Each, being beautiful of its kind, at least adds something comely to the whole: or rather God, by allowing such a

---

13. The Coimbrans contrast the "bonum privatum" of individuals with the "bonum commune" that sometimes conflicts with the *bonum privatum.* That the political uses of such phrases are not far off may be gleaned from the references to Aristotle's *Ethics* (1c2), and to "those who manage the public good" (*Rempublicam tractantibus*) earlier in the question.

14. Since I am here interested in the final causes of the *horror vacui,* I omit discussion of Aristotelian hypotheses about its efficient causes. Scholastic authors had proposed a "universal" nature that originates from the heavens; its action consists in impressing upon individual substances the impetus that in turn moves them to fill up voids (Grant 1973:69–70). The Coimbrans affirm that "frequently nature uses a certain motive force, which for this purpose seems to be attributed, along with the peculiar virtues by which they seek their own places, to all sublunary bodies" (ib. a3, 2:64). Neither they, Toletus, or John of St. Thomas mention universal nature; Suárez, in his discussion of the *horror vacui,* refers to the "providence of universal nature, or rather of its author" (Suárez *Disp.* 18§8¶14, *Opera* 25:655), but in a discussion devoted specifically to celestial causes he mentions only their role in generation (22§5¶7ff, 25:840ff).

thing to exist, has done so. Yet adding to the beauty of the cosmos cannot be the end of any individual's actions (I set aside angelic and human arts). Like the brush strokes of Seurat, the gnats and mites of this world, unappealing in themselves, have purpose in relation to the purpose of the whole in which they are tiny but necessary parts.

The Coimbrans then answer the argument that some animals and plants are not just useless but harmful: "Nor are venomous animals superfluous [. . .], because, as St. Augustine says [. . .], they should at least be praised for reminding us to esteem that other better life [i.e., in God], in which there is the greatest security [. . .] Or finally because, as Lactantius writes [. . .], it was useful to man that he should encounter some beneficial things, and some harmful, so that in avoiding the one and striving after the other he should exercise his power of reason" (ib. 327). It cannot be any individual animal's purpose, or even that of the specific form, to be useful to humanity. They pursue their own ends. Since we do not produce them, or at least their specific forms, it is not our purpose either to have made them useful. Only in relation to the whole, in which there are both people to be reminded of a better life and venomous beasts to remind them, can that purpose be made out. Serving humanity is not, of course, itself a cosmic end. But the existence of humans, too, perfects and ornaments God's work of creation, and so whatever serves us indirectly serves that end.

Cosmic ends yield answers to very general questions about the constitution of the world. We saw a moment ago that the preservation of specific form is the end of generation, or rather of the power of generation. One could take a further step back and ask why it should matter that specific forms are preserved. The Coimbrans speak of the beauty of the world. They do not mean merely that it is beautiful to us, but that it is a fitting illustration of God's power. The multitude of actual forms testifies to, though it does not exhaust, the infiniteness of that power.

## 6.2. Existence of Ends

In Descartes's time anyone who set out to construct a physics without finality would have found an immense variety of phenomena suddenly left out of account. The range of items to which teleological reasoning was applied gave rise to an equally broad range of arguments to support the legitimacy of that reasoning.

Most fundamental is the regularity of natural change. In the second book of the Physics, Aristotle sets before his opponents a dilemma: chance or ends. What occurs fortuitously occurs irregularly; but natural changes occur "always or for the most part" (*aut semper aut plerunque*). So they must occur on account of some end. Worse yet, if there were not ends in nature, "all things would come from all promiscuously . . . men would arise from the sea, and scaly creatures from land, and winged ones burst from the air."[15]

The argument is, on its face, unconvincing. Everyone agrees that efficient causes necessitate their effects ("if the cause is given, so is the effect," writes Eustachius with his usual brevity [*Summa, Phys.* 1tr2d2q5, 2:142]—*positâ causa ponitur effectus*). So people will not emerge from the sea ever if they do not always: one does not need ends to account for that regularity. Given that we have not seen any such occurrence, and that the sea remains constant in composition, there is no reason to expect that the weird event will occur. Likewise, if people have always given birth to people, and birds to birds, and if they remain constant in composition, then there is no reason to expect that people will bear birds or birds people. So if the regularity to be explained is 'people give birth only to people, and no other kind of thing does', then an appeal to the necessity of efficient causes seems to suffice.

The possibility of a world run like a great machine by efficient causes alone did not escape the Aristotelians. They did not adopt final causes through ignorance of such alternatives. Even if "God did not concur in the actions of natural agents, but allowed them to perform their *motus* independently," still "the rock would descend, fire would generate its like, and so for the rest" (*Disp.* 23§10¶8, *Opera* 25:888; Coimbra *In Phys.* 2c9q2a3, 1:330). Suárez acknowledges that regularity would result from the operation of efficient causes alone; indeed such a world would even exhibit order. It would appear to tend toward ends: "from this impossible hypothesis it follows that nature operates in the most orderly way so as to tend toward an end, without any direction or intention of the end." But the appearance of order without the fact of being ordered is "absurd" (*Disp.* ¶9, 888). The inference is familiar, as are the agnostic retorts: nature is not in fact well ordered, and in any case "order," unless the term is used so as to beg the question, does not entail ends.

What must be added to the mere observation of regularity or even of order is either a list of phenomena that would not occur, or that would occur differently, in a world of efficient causes alone; or else reasons to believe that introducing ends, and speaking of them as causes, will yield a genuine understanding of natural change which is not to be had in consulting efficient causes alone. In 1600 there were good arguments of both sorts. Many phenomena seemed to require a coordination of efficient causes that no known efficient cause could bring about. The generation of living things was among the more striking. The sagacity of animals in some of their

---

15. Coimbra *In Phys.* 2c9q1a1, 1:323, quoting Lucretius *De rerum nat.* 1:161–162 (cf. also Thomas *Contra gent.* 3c2, Suárez *Disp.* 23§10¶3, *Opera* 25:886). Lucretius, as the Coimbrans recognize, since they quote the same passage in their question on prime matter (1c9q3a1), is arguing not that nature acts toward ends but that natural change never begins or ends with nothing. The argument of *Physics* 2c8 is discussed and defended in Cooper 1987.

actions likewise outstripped the capacities of efficient causes, and indeed seemed to entail the recognition of ends. To that the Aristotelians added the long-term stability of the world, and the usefulness of its inhabitants to humans.

The Aristotelians did not conceive of efficient causation in Humean fashion. Cause and effect are not related as antecedent to consequent in the instantiation of a law. There are indeed regularities. But they consist in the existence of kinds in which powers are united and the necessity with which the actions of power are manifested under the appropriate conditions. Efficient causation itself is blind and atomistic. Blind because it consists merely in an agent's setting into motion the *potentia* of a patient; atomistic because each agent, insofar as it is considered only an efficient cause, acts in disregard of others. Yet in many natural processes the "triggering" of efficient causes proceeds in a regular order. Those causes seem, moreover, to aim not only at their particular ends but to work in concert toward something that none could effect alone. The Aristotelians believed that guidance could not come from the efficient causes themselves, or from the matter they act on. It could be accounted for only by supposing that they act toward ends. Exactly how ends provide guidance will be seen in §6.3; here I will examine the chief instances that seemed to call for that guidance.

1. *Coordination of powers*. In §3.2 I presented arguments to the conclusion that substantial form is the ground of the unity of active powers, and in fact their efficient cause. We saw, moreover, in §5.4 that the active powers that emanate from a form serve to introduce into new matter the *dispositiones* required for it to receive a similar form. That form is the *terminus ad quem*, not immediately of the actions of the active powers, but remotely, as their joint result. Generation is one action only with respect to its remote efficient cause—the form of the generator—and its terminus—the form of the thing generated. With respect to the active powers that a form must use as its instruments, generation requires a series of alterations coordinated both temporally and spatially. In higher animals, for example, the organization of the fetus includes the "effecting of a proportionate temperament of primary qualities" and the "disposition of the members" in "marvelous order" (Suárez *Disp.* 18§2¶33, *Opera* 25:611). The particular efficient causes of those changes have so to speak no cognizance of the rest, and no reason to occur when and where they do. They will act whenever a suitable patient presents itself. Nor will adducing further efficient causes, even celestial, help unless it can be shown how they, considered merely as efficient causes, could somehow regulate the causes already existing in the fetus.[16]

---

16. A further argument to the same conclusion can be gleaned from the Coimbrans' discussion of monsters. In general, an efficient cause and its effects are similar: heat causes heat, and so forth. Each efficient cause, moreover, "strives to produce without error or deviation what is similar to itself." The instrumental powers of a seed, say, are nothing more than the

There was, as John Cooper notes, little reason in Aristotle's time to believe that the "powers of matter," that is, the efficient causal powers of the
elements and their mixtures, could by themselves have the effects observed
in generation (Cooper 1987:270). In the sixteenth century, the three principles of Paracelsus—salt, sulphur, and mercury—were no more promising
than the four elements of Aristotle had been. Celestial causes, the heavenly
*intelligentiæ*, even God himself, are of use only insofar as they provide appropriate repositories for the final causes that alone serve to explain
generation.

Just as the coordinated action of active powers in generation requires the
introduction of ends, so too does their coexistence in individuals. The same
holds for dispositions and organic parts, insofar as they provide the material
basis for powers. Suárez writes that "if matter is taken according to itself, it is
indifferent, and has no need of such dispositions or properties; but if it is
supposed already to be affected with such-and-such dispositions, already
those dispositions have been introduced on account of some end or form,
which requires them on account of its conservation or for some operation;
but that operation is in turn on account of the conservation of the species or
the individual itself [. . .] And so every connection and connate necessity
which exists *per se* in natural things, arises from being ordered to an end"
(Suárez *Disp.* 23§10¶7, *Opera* 25:887). In §5.3 it was argued that the accidents or dispositions that prepare the way for substantial form inhere immediately in matter. Matter has no reason to take on one set of dispositions
rather than another—it's just matter. It has them in order to receive the
form. But why does the form require *those* dispositions and not others? The
answer is that it requires whichever dispositions enable it to exercise the
powers of which it is the efficient cause via emanation. But now one can ask:
why *those* powers? The form itself has ends, and in particular self-
reproduction. Self-reproduction requires at least that the form be able to
produce in new matter the very dispositions that it itself needs in order to
act. Efficient causation suffices to show why, *given* the form, those powers
exist together in one individual. But it cannot, Suárez argues, show why
from *that* form *those* powers emanate. For that one must take the production
of the form to be not merely their common efficient cause but their common end.

---

qualities arising out of the *temperies* or disposition of the elements contained in it. They
reproduce, without fail, their like in the matter of the offspring. But if, as the Aristotelians
thought, the form of the animal to be generated is not reducible to the disposition of its matter,
then the qualities, if they operated independently, could not themselves produce that form
except by accident. Something else must control their operation, and produce its like through
them. That something, of course, is the form, which sets the active powers in action to the end
of reproducing itself.

In Suárez's argument we proceed from dispositions to form, from form to specific ends, and (by implication at least) from those to cosmic ends and God. At each stage we ask: why do these items exist (or why are they compresent in one individual)? The answer looks not to their efficient cause (whose existence indeed does explain theirs, in one sense) but to what will exist if they are allowed to act. It looks, in other words, to their ends. The series terminates when we reach God: provided that he is omnipotent and necessarily existent, and that an omnipotent being will create as rich a world as possible consistent with its goodness, we need nothing more.[17]

2. *Sagacity and industry of animals.* Among the most visible operations of animals are raising their young, building webs or nests, looking for food, and fighting or avoiding their enemies. For the Aristotelians, the thought that such actions could be explained solely in terms of the dispositions of animal bodies was a pipe dream. The resources of their science—or of Descartes's—were not up to the task. The independence of animals' actions from immediate stimuli, the coordination of their actions, the flexibility in their means, all told against any explanation that did not take into account the ends to which those actions are directed. The real question, which I will return to, was whether animals were guided by reason or merely by instinct. Here it suffices to note that in either case it was necessary to appeal to ends.

3. *Stability.* Under this heading I put observations not of the regularity of particular causal relations but of the persistence of certain large-scale features of the natural world. We have seen that the ocean keeps mostly to its place, and thus the proportion of ocean to land remains relatively constant. So too for the elements:

> The same is shown by the discordant amity of the four elements lying spherically in four regions, and their alternating recirculation through downfall and ruin. From earth water, from water air, from air fire, and then in the other direction air from fire, water from air, and earth from water are generated; so that in moderation, as if in equilibrium, the heavy is made equal to the light, the cold to the hot, the lower to the higher, as much as can be: to that end, namely, that they should last; while nature teaches us in great things that impartiality [*æquabilitatem*] is the nurse of concord, and concord the conservator of all things (Coimbra *In Phys.*

17. Cooper writes that for Aristotle, the permanence of animal kinds "meant the permanence of a set of well-adapted, well-functioning life forms." If, moreover, it is "*permanently* true that there *are* these given kinds of good, well-adapted plants and animals," then any physical theory must take that into account. Aristotle, Cooper believes, chose to "accept as a fundamental postulate of physical theory" that the world "so governs itself as to preserve in existence the species of well-adapted living things that it actually contains" (Cooper 1987:251). That sounds rather like an *anima mundi*, as if the world itself were an organism. The alternative is to invoke supernatural powers (271)—which is, of course, what Christian Aristotelians did.

2c9q1a1, 1:224; cf. their question on the natural places of the elements, 4c5q3a1, 2:32[ff]).

Each particular instance of transmutation could be explained by appealing to efficient causes only. What requires further explanation are certain global features—that water and water vapor are in equilibrium, or close to it, so that the air and water in the world do not eventually all get changed into earth and fire. In the Aristotelian theory of generation, there is no way to explain the rate at which one element is changed into another. So unless one brings in a final cause—the sustenance of life, say—the balance of reaction rates will be a brute fact.

Modern science has exhibited as much dissatisfaction with brute fact as Aristotelianism did. The difference lies in the means of domestication. Where efficient causes fall short, Aristotelianism turns to final causes and, as we will see, to intelligent agents capable of cognizing ends. Modern science finds ever more elaborate structures in which to combine efficient causes: clockworks, feedback mechanisms, neural nets, chaotic systems. Though it is true that very little of that machinery existed in Descartes's time, it is also true that the will to devise it may well have required the proscription of ends in natural philosophy. Although one may defend, on the grounds suggested by John Cooper, the appeal to final causes by Aristotle and the Aristotelians, one must, I think, also acknowledge that in assuaging the dissatisfaction aroused by bruteness with the proferring of ends, Aristotelianism did present an obstacle to the invention of concepts adequate for explaining, in terms of efficient and material causes alone, those otherwise brute facts.

4. *Usefulness.* To our Darwin-sharpened eyes the most glaring obstacle was no doubt the explanation of natural phenomena by reference to their convenience or agreeableness for humans. Yet here the Aristotelian would have found a happy coincidence of experience and philosophy with revealed truth.

Start with the thought that the world is filled with an "inexhaustible variety" of forms (Coimbra *In Phys.* 1:224). A world consisting only of the four elements would be boring but not impossible. Consider also that those forms are irreducible: the forms of elements cannot be reduced to their qualities, or the forms of mixtures to the proportions of elements they contain, or the forms of animals to the dispositions of their bodies. At every level something new emerges. The four-element world would remain just that, unless some power introduced into its matter the forms of higher things. Hence each form bespeaks a distinct act of creation. Why, then, would God have created so many of them?

For the Aristotelian, part of the answer is that God, given that he created higher forms, would not have failed to provide them what they need to exist and endure. In each individual the form is provided, as we have seen, with the power to produce and maintain the dispositions it requires. Exercising that power sometimes requires acting on other things, as in eating or breathing. Those too must be provided. But they must not only exist, they must be capable of being used: if our food, like Tantalus's, had the power to evade our grasp every time we reached for it, we would starve. The things an individual needs to use—where use entails alteration if not corruption of the thing used—ought to be, therefore, no more powerful than the individual itself. Conversely, what is less powerful is by that fact available. "It is," the Coimbrans write, "established by the law of nature that things of inferior grade are rendered to the more excellent, especially if they can sometimes make use of them" (*In Phys.* 2c9q2a1 1:328).

That all things in the world are thus available to humans, and their existence therefore explicable partly in terms of their utility, is thus only one case of a truth that applies all the way down: "the form of an element, the most contemptible of all, is ordained to [the use of] the form of a mixed body; the form of a mixed body to the vegetative [i.e., the part of the soul—in plants the whole—devoted to nutrition, growth, and generation]; the vegetative is possessed by the sensible; and this again by the rational soul, which embraces all ranks and perfection of forms." So "man, by the inborn right of his nobility, and the prerogative of the more eminent form, summons the whole body of nature, and claims it for himself" (ib.).

Reason here only confirms what was revealed in Genesis. Suárez, in his commentary on the six days, distinguishes three aspects of the "dominion" given to men over the rest of creation.[18] Man has by nature the capacity for dominion; by nature and by right he has power over and use of inferior creatures. He has the capacity by virtue of having an intellect and a free will. His physical power over animals consists in his being able to subject animals to himself and use them either by orders or by force. Even if some animals resist that power, "by reason, art, and industry, man conquers all [. . .]: some he instructs, according to their capacities [. . .], some he kills, either for use or to prevent harm" (*De opere 6 dierum* 3c16¶8, *Opera* 3:279). That physical power is accompanied by a moral power, a "right" (*jus*) to exercise it: "Man has this right, therefore, by virtue of his creation, and it is natural to him with respect to inferior animals, both because they exist naturally on account of man, and because no harm can be done to them in any use of them, and so if man's nature alone is considered [i.e. disregarding Original

---

18. I use the word 'men' deliberately: in every species of animal, the male form is, according to Aristotle, superior to the female; hence even in the state of innocence, Adam had power over his wife, a power that is "a certain sort of dominion, and is according to natural right, since the man is the head of the woman" (Suárez *De opere 6 dierum* 3c16¶17, *Opera* 3:282; cf. Ephesians 5).

Sin], it is licit for him to use these animals in any worthy [*honestum*] manner"
(*De opere 6 dierum* 3c16¶17, *Opera* 3:280; 'honestus' would exclude bestiality
and animal sacrifice). The argument is not that might makes right; that we
can, in fact, use animals is at best a sign that it is licit to do so. Not our
physical power alone but the possession of free will and the capacity to
deliberate about ends and means make us morally superior.[19] The physical
power, since it has the exercise of the moral power as its end, is subordinate
to it.

The moral dimension of arguments from utility merits emphasis because
it illustrates the parallel between ends and goods that will be prominent in
the next subsection. The appeal to ends makes easy the transition from what
is to what ought to be, or from what a thing can do to what it may. In natural
philosophy, on the other hand, the explanatory value of such arguments is
slight. Earlier I quoted passages from the Coimbrans answering objections
to the effect that some natural things are useless or obnoxious. The utility
they manage to impute to gnats and vipers is tenuous; it could hardly be the
total final cause of those creatures. God could have found less roundabout
ways of commending himself to us, or of making us exercise our wits. If,
moreover, utility, as opposed to actual use, rests only on the inferiority of
their forms to ours, then everything in the world except angels and other
people has utility. So general an explanation hardly does justice to the
multiplicity of natural forms, or to the workmanship with which each was
fashioned.

I close this discussion with three remarks, one metaphysical, one methodological, one historical.

Consider an acorn: it has a form of its own qua seed, but it also is *in
potentia* an oak. The form of oak is the end to which it tends, and for which it
was produced. Whether oaks actually exist or not, if there are acorns, there
are actual things in which oakness is implied. Or consider the claim that in
order to conserve itself each form must have certain active powers. If that
form exists, those powers are implied in it even if God has miraculously
prevented them from emanating from it. Similarly the end to which those
powers act is given in them even if they never act.

Let us suppose that the end is really distinct from the thing whose end it
is. Since real distinctions hold not only between complete substances but
also between, say, a soul and its powers, the scope of the problem is not
reduced much by that supposition. How, then, can a thing have a property

---

19. When God said, "Let us make make man in our image and similitude," Soto writes, it
was as if he had said, "Let us make a man who, like us, exhibits judgment, and is capable of
choice, and since he has dominion over himself and his actions, let him rule in the same way
over the animals and other things that lack reason" (Soto *De nat. & gratia* 1c3; cf. Thomas *ST*
1pt1q96).

in which is implied the existence of another thing really distinct from it? Colors, for example, have the power to produce in the eye what were called "intentional species," which are really distinct from them. Is it then somehow intrinsic to colored things, somehow implied in being a color, that there should be species and eyes? A. Mark Smith writes that "as visible [. . .] every object is defined and restricted by its capacity to express its nature though color. However, this same expression is equally restricted by what it redounds on: the visible expressions of things are realizable only in eyes, not in tongues or ears" (Smith 1981:575). Having the power to be colored in the presence of eyes is indeed a property, real in the things that have it, but dependent for its actualization on other things distinct from them, and on unproblematic relations like propinquity and simultaneity with those things. It does not quite imply the existence of those other things (save in the presence of maxims to the effect that no power will forever go unexercised), but it does imply, I would say, that there are natures such that if things with those natures exist, the powers will be manifested. In other words: no power to be colored without vision and all that vision requires, even if nothing actually has vision. Similarly, though no doubt less plausibly by our lights, if lower forms exist on account of higher forms, then the natures of the higher forms exist. Even if God had wound up creation on the fifth day, with only the birds and fish, their existence, supposing that they had the natures they now have, would have implied the natures of the higher mammals, including humans. God, after all, would not have created the means to an end without having that end in mind too.

That implication, that palimpsestic foreshadowing of the end in the means, or of the perfected in the imperfect, is in part what led Descartes to speak of the Aristotelian notion of *potentia* as 'confused'. If the natures of things do imply other natures really distinct from them, then a complete idea of a nature must include those other natures. Looking ahead to Leibniz, one might argue that the inclusion in each individual essence of all the others compossible with it is an implication of the same sort: because there is but one end—the world as a whole—in God's creative act, each individual has all the rest as its end. They are implied in it just as the oak's essence is implied in that of the acorn.

My second methodological remark begins by noting that for the Aristotelians all of the problems raised here were brought together by them as instances of the activity of ends in Nature. The fate of those problems in modern science varies greatly. The coordination of powers in generation is now explained by theories of the regulation of gene expression and the gradients that control tissue differentiation. Evolutionary theory has taken over the explanation of useful behavioral traits in animals. The stability of the atmosphere is referred to equilibrium principles and to the various geological and biological processes by which carbon dioxide, for example, is replenished or removed. The utility of animals and plants to humans falls to political economy (in its anthropological guise) and the history of domestication; its moral aspects fall largely outside science. The point is not just that the problems are treated in distinct disciplines. With due allowance for the relative absence of specialization, much the same could have been said in the sixteenth century. It is rather that the modes of reasoning about the problems have turned out to be heterogeneous, and the unification achieved by treating them as instances of final causation has turned out to be spurious. That, and not the almost equally spurious unification achieved by treating them mechanistically (see Gabbey 1990), is the lasting lagacy of Descartes and his coconspirators.

My last remark concerns the fading away of the question *An natura agat propter finem* [Whether nature acts on account of an end]. Toletus, Suárez, and the Coimbrans all devote several pages or more to that question and to others issuing from it. Later *cursus,* including those of Eustachius (1609), Abra de Raconis (1617), Hurtado de Mendoza (1624), Arriaga (1632), Poncius (1642), De Quiros (1666), and Dupasquier (1705), do not have such a question. In part this may be because the structure of the *cursus,* unlike that of the commentary, had no obvious systematic place for it. But I suspect also that the consideration of the ends of Nature was beginning to pass over into theology. There it flourished. In Butler's Analogy, the Boyle Lectures, Derham's *Physico-theology,* Bernardin de St. Pierre's Harmonies, and the *Bridgewater Treatises* written by, among others, Whewell and Babbage, the discoveries of the new science were assiduously catalogued into evidences of the hand of the Creator. Cartesian strictures, as Tocanne notes, had little effect on that industry. Unlike Descartes, most natural philosophers found no repugnance between efficient and final causal explanations.[20]

## 6.3. Character of the Final Cause

Writing on final causes in Scholasticism, Anneliese Maier concludes that in Buridan, at least, the efficient cause and the necessity associated with it had attained a primacy that presaged the eventual eclipse of the final cause in the seventeenth century.[21] Ockham had already argued, following Avi-

20. Philosophers of the later seventeenth century, "with rare exceptions, were not conscious of any opposition or conflict between the search for the efficient cause, the necessity of which was not doubted, and the search for the final cause" (Tocanne 1978:76).

21. Maier 1955, c.5, especially pp.325[ff], and Lang 1989. Lang takes Buridan's "procedure," the literary form of his work, to reflect his substantive views: "the procedure of Buridan's

cenna, that the final cause acts only by virtue of existing in the intellect of an agent; to which Buridan added that when it acts thus, it acts as an efficient cause, and that where the agent is not such as to conceive the ends by which it acts, there is no final cause at all, only efficient causes. To the argument that if there were no ends in nature, then one thing would follow from another haphazardly, Buridan replies (as we would) that efficient causes suffice. "The *regula*," Maier explains, "from which we can read off whether a natural action turns out *talis qualis debet esse* [as it should be], is no longer the goal toward which the process is directed, but the law according to which it is accomplished and according to which the same cause yields, *ceteris paribus,* always the same effect" (Maier 1955:334). In inanimate nature and even in animals, the necessity that earlier philosophers had referred to the operation of ends is referred by Buridan, if not quite to laws of nature, at least to the "univocality" of efficient causes.

The central texts present both a retrenchment in favor of more traditional views and an accommodation to the conclusions reached by Buridan. The case of rational agents, first of all, becomes the only case to which their account fully applies. Only in rational agents do ends operate straightforwardly. Animals are held to be acted upon by ends as rational agents are, but "imperfectly," because their ability to judge the goodness of things, and to deliberate about means, is limited. Inanimate things are not acted upon by ends except insofar as they are the instruments of God.

In the last of these propositions there is a significant step away from Aristotle and toward Plato. "The novelty," writes David Balme, "in Aristotle's theory was his insistence that finality is within nature: it is part of the natural process, not imposed upon it by an independent agent like Plato's world soul or Demiourgos" (Balme 1987:275). That innovation is rescinded in Christian Aristotelianism, no doubt because the Christian God provides a counterpart to the Demiurge, and because it was fitting that the nonhuman world should depend on God as an instrument on its maker (§7.2). If they disagree with Plato, it is in denying that universals exist except *in rem,* or that genuine knowledge can only be had of eternal Ideas. But when they assimilate the inanimate world to a divine artifact, and final causation to inten-

---

*Questiones* is sequential: the questions follow one after another without any clear or significant relation to a larger purpose. In this sense, his procedure is not teleological" (588). I doubt that any such relation can be borne out. The "disputed questions" and the *expositio* were standard literary forms, used by philosophers of all persuasions; they were not "procedures," if by that one means actual or ideal methods of inquiry. In fact Thomas was perfectly capable of writing a set of questions with no more structure than Buridan's (e.g., the *Sentencia de anima*), and Buridan of writing a running commentary like Thomas's on the *Physics* (see Buridan *Exp. de an.*). Lang's argument reflects, I think, a rather naive understanding of the relation between literary convention, individual authorial choice (especially in the Middle Ages), and content.

tional action, they hark back to him even as they unwittingly prepare the way to the world-machine of Descartes.

Doubts about the existence of ends in nature came largely from outside Aristotelianism. With few exceptions, Aristotelians agreed that the actions of rational and irrational agents had ends. But it is one thing to say that an action *has* an end, and another to say that it might be *caused* by its end.[22] Here various suspicions had been voiced by Aristotelians from Avicenna onward. For any genuine cause, it was thought, one must be able to answer a standard list of questions about its nature, including the *causalitas* or "causality" of the cause, its mode of existence, and the objects upon which it could act. Suspicions surrounding the final cause arose largely from the difficulty of answering the standard questionnaire, especially when the agent is "irrational"—inanimate, or animate but inferior to humans. Where for the other three causes—material, formal, and efficient—there might be controversy over details, there was broad agreement that the questionnaire could be sensibly answered. For the final cause there was no agreement.

The difficulties arose in trying to reconcile the character of ends with three conditions thought to hold of all causes. Every cause is, first of all, temporally prior to its effects (see, e.g., Eustachius *Summa* pt2, *Physica* 1tr2d1q4, 2:138). But in generation, say, the form does not exist until the action it is supposed to be the final cause of has ceased. More generally, to be a cause a thing must exist. But the ends to which people act often do not yet, and sometimes cannot ever, actually exist. A person may strive, Fonseca writes, "to make the diagonal [of a square] commensurate with its side"; yet to such a thing real existence is "repugnant" (Fonseca *In meta.* 5c2q11§1, 2:163bF). The third condition is that a cause be nobler or more perfect than its effects (Eustachius ib.). But often the ends a thing acts toward are less noble than the thing itself. Whatever end God acts toward, except himself, must be less perfect than he is. In particular the world is less perfect than God; yet it must be the end toward which he acts in creation.

In the answers to the standard questionnaire we find attempts to respond to these anomalies. The Aristotelians denied, for example, when specifying the causality of ends, that ends literally act, thus avoiding, or evading, the first. The second they answered with what at first sight appears to be an ad hoc broadening of the notion of 'real existence'. The third, and least difficult, anomaly they handled using the distinction I have already mentioned between the ends of things and the ends of their operations.

---

22. "Ut enim notat Gabriel [Biel] [. . .], finis et causa finalis non omnino sunt idem; nam finis ut sic solum dicit terminum ad quem tendit operatio, vel ad quem motus ordinantur; causa autem finalis est, quæ movet agens ad operandum" [As Gabriel Biel observes (. . .), the end and the final cause are not entirely the same; 'end' as such means only the *terminus* toward which an operation tends, or toward which *motus* are ordered, while the final cause is what moves an agent to operate]" (Suárez *Disp.* 23§9¶8, *Opera* 25:884).

1. *Causality of the final cause.* To each kind of cause, the Aristotelians believed, there corresponds a peculiar *causalitas,* a "formal reason" by which the mode of causation of each kind differs from the rest. Since every cause is a "principle that imparts being" (*principium quod influit esse:* Suárez *Disp.* 23§9¶10, 386), causality in general is "nothing other than the influx, or concurrence, by which each cause in its kind actually imparts being to its effect [*actu influit esse in effectum*]."[23] The material cause imparts *esse,* as does the formal cause, simply by being a component of substance: "matter is like a beginning, or foundation of being itself, while form consummates and completes it." The efficient cause is "like a source and principle which *per se* imparts *esse* to its effect." As for the final cause, that's a problem, Suárez admits, promising to discuss it later.[24] In the disputation on final causes, we get a precise definition. The causality of the final cause consists in a "metaphorical motion" of the will; that motion is nothing other than the real motion of which the will is the efficient cause, and which consists in being attracted to, or desiring, or loving, the end.[25]

The notion of metaphorical motion was common property. The Coimbrans write, "there are two kinds of motion, one proper, and which occurs by a true and genuine action, as when fire heats water; the other improper and figurative [*translatitiam*], and of that sort is the motion by which something is said to move which, by inspiring a love for itself, attracts and draws to itself [the soul]."[26] Why "metaphorical"? Fonseca notes that

23. Suárez *Disp.* 12§2¶13, *Opera* 25:387. I translate *influere* by the rather unsatisfactory 'impart'. 'Inflow' (meaning 'cause to flow in') is obsolete (the only quotation in the *OED* is Hobbes asking what it means). Eileen O'Neill has, however, revived the word in her useful survey of influx as a model of causation (O'Neill 1993).

24. One problem that is easily solved is how the end can be a "principle." Principles (Aristotle's ἀρχαί) are supposed to be first, yet the end is last. Suárez's answer, the standard one, is to affirm that although it is the last "in execution," it is the first "in intention"; it "moves the agent to act," and is therefore a true principle (389).

25. Suárez *Disp.* 23§4¶8–9, *Opera* 25:861. "Est ergo tertia sententia, quæ constituit etiam hanc finis causalitatem in motione metaphorica. Addit vero, hujusmodi motionem non poni in actu secundo, nisi quando voluntas in actu secundo movetur, et quando sic ponitur in re non esse aliquid distinctum ab ipsomet actu voluntatis": "There is therefore a third position, which also takes the causality of the final cause to consist in metaphorical motion. But it adds that motion of this sort is not posited in a second act, except when the will is moved in a second act, and when it is posited in the thing it is nothing other than the act of will itself." After citing numerous authorities in support of this *sententia,* Suárez adds that "nullus tamen ita clare et expresse prædictam declaravit sententiam, sicut Ocham, in 2, q. 3, a. 2, ubi ait causationem finis esse movere efficiens ad agendum [a commonplace]; illud autem *movere,* non esse aliud nisi ipsum finem amari ab agente, vel aliquid propter ipsum": "No-one has so clearly and expressly put forward the position above as Ockham has in 2, q. 3, a. 2, where he says that the causation of the end is to move the efficient cause to action; but the term 'move' means nothing other than that the end is loved by the agent, or something on account of the end."

26. "Ut quænam finis caussalitas sit scrutemur, animadvertendum est ex Divo Thoma [*ST* 1pt2q9a1] & Scoto [*In sent.* 1d1q4] duplicem esse motionem, unam propriam, quæ fit per veram, & germanam actionem, ut cùm ignis aquam calefacit; aliam impropriam, & trans-latitiam, cuiusmodi est ea, qua movere dicitur id, quod sui amorem iniiciendo allit, trahitque

some ends might well also act on an agent as efficient causes. Food will attract an animal by its smell and look. But many ends have no efficacy and yet are capable of moving an efficient cause to act: "in such a way a lower place draws a stone out of a higher, though it impresses nothing in it, and most fictitious things, which have *per se* no effective power, move many agents to obtain them" (Fonseca *In meta.* 5c2q10§4, 2:160bC). Yet there is, at least in rational agents, a genuine *motus* whose reason is given by the end. The will desires *in potentia* whatever the intellect is capable of presenting to it as an end. To desire an end *in actu* is therefore the actualization of *potentia*, or in other words a *motus*. The will itself is the efficient cause of that *motus*. But since the will is the efficient cause of all its *motus*, particular determinations of the will must have an additional reason—the end they are directed toward, or that *ceteris paribus* they would effect. In that sense, acts of the will depend not only on the will but on ends; metaphorical *motio* is that dependence.[27]

Speaking of the action of ends as metaphorical opens one avenue to solving the problem of the nonexistence of ends. If one could make sense of metaphorical action, perhaps one could make for them a reasoned exception to the rule that causes must exist to act. It also allows one to speak of items otherwise devoid of action, like places, as having effects. What it does not do, despite Fonseca's unfortunate example, is allow one to explain the action of ends on inanimate things. Metaphorical motion, whatever it may be, moves only the will. Suárez explicitly restricts his discussion to the action of ends on the wills of created agents, thereby excluding God (who cannot be acted upon even metaphorically) and inanimates. The Coimbrans, confronting the objection that "even things that lack cognition operate on account of ends [. . .] and yet they are not attracted to an end" (*In Phys.* 2c7q21a1, 1:308), reply that "natural agents [. . .] are not attracted *per se* to the end, nor do they strive toward it *per se*, but only insofar as they are directed by the first cause, in which their ends, and the knowledge and love [of those ends] pre-exist" (a3, 1:311). A natural agent, a stone say, is "attracted" to the ground not per se, not considered in itself, but only in relation to God, who has caused it to be endowed with a quality that will, as an efficient cause, impel it downward. It "strives" toward the ground only *per accidens*, as a key is said *per accidens* to turn a lock.

We may find the talk of metaphorical motion obfuscating. The real story, we think, is given in the example of an animal's attraction to food: an end

---

ad sese" (Coimbra *In Phys.* 2c7q21, 1:307). Eustachius plagiarizes this passage (*Summa* pt2, *Physica* 1tr2d2q6, 2:143).

27. Conversely, when there is no actualization of a *potentia*, there is no *motio*, even metaphorical: such is the case for divine acts. Suárez writes that in such instances "the whole manner of speaking is metaphorical" (*Disp.* 23§9¶6, *Opera* 25:883).

attracts by efficiently causing a desire that in turn efficiently causes actions that will, if the world cooperates, result in the end's being attained. Since the Aristotelians themselves agree that the cognition of an end is a *conditio* sine qua non for its being a cause, the anti-Aristotelian need only argue that the cognition is not just a condition of, but the cause of the volition that was said to result from the metaphorical movement of the will by ends, and that the cognition of an end will have been efficiently caused by the end. Though there is still reason to talk of ends, final causality is eliminated in favor of efficient causality.

2. *Mode of existence of the final cause.* The account just proposed was not only known to the Aristotelians, it was rejected by them. Avicenna and Soncinas had held that an end acts only "according to the existence it has in the soul, not outside it" (Soncinas *Q. meta.* 5q3, 55a). A standard way of stating the view was to distinguish two modes of existence: *esse intentionale* or *objectivum*, the mode of existence things have when they are objects of thought, and *esse reale*, or existence independent of thought or of the soul.[28] The claim, then, is that the end is a cause with respect to *esse intentionale*, not with respect to *esse reale*. At least one earlier philosopher, Guido Terreni, had gone on to say that the end, existing only in intention, acts by efficiently causing a desire or appetite.[29] But Soncinas did not, and the central texts do not.

Instead they insist that the end causes according to *esse reale*. Only *esse reale*, and not *esse intentionale*, satisfies three criteria. It alone fulfills the condition of being that on account of which an agent acts. The *esse reale* of the end, or its absence, determines whether the act is successful or frustrated. The means are chosen not to accomplish the *esse intentionale* of the end, but its *esse reale*. A doctor who by hypnosis succeeded only in making his patients *think* they were cured would be a quack.[30]

*Esse intentionale*, then, is not the *esse* according to which an end acts. But being thought of is a *conditio* sine qua non for final causality. Even God by his absolute power, Suárez argues, could not bring it about that the will should be drawn toward an unthought end (*Disp.* 23§7¶6, *Opera* 25:876). The

28. "Scholastici Ens Intentionale appellant Ens, quod sola intellectus conceptione & consideratione inest; seu Ens quod est intra animam per notiones [. . .], cui oppositur Reale, quod reperitur extra animæ notiones" (Goclenius *Lexicon* s.v. 'Intentionale', p256; cf. also 157). *Ens (esse) intentionale* and *objectivum* are used interchangeably.

29. In a *Quodlibetum* of 1315, Terreni argued that the "the causality of the end with respect to the movement of the agent by an appetite of the soul is efficient causality" (Maier 1955:286, citing Terreni's *Quodlibeta* 3q2). Efficient causality, he says, is that "by whose presence and motion what was *in potentia* is now formally *in actu*." But the end is "that by whose presence [in apprehension] a certain appetite that was *in potentia* is now *in actu*." It is thus the efficient cause of that appetite. I should note that none of the central texts cites Terreni or explicitly states his view.

30. "A sick person does not desire the *esse intentionale* of health, which is long since awaited in his soul, nor does he take walks and abstain from food for the sake of it, but for the sake of the *esse reale* of health" (Coimbra *In Phys.* 2c7q23a2, 1:316).

metaphorical motion of an end on the will is founded, he writes, "on a natural agreement and sympathy of the intellect and the will," in which it is necessary that the cognition of the end should precede the inclining of the will toward it (¶2, 875). Suárez and the Coimbrans both compare the "proximity" of the intellect and will to the spatial proximity necessary in efficient causation: "Just as in the efficient cause local proximity is a requirement, so in the final cause proximity in the soul, vital proximity [*approximatio quasi animalis, seu vitalis*] is a requirement" (Suárez *Disp.* 23§8¶10, *Opera* 25:881; cf. Coimbra *In Phys.* 2c7q23a2, 1:315). Vital proximity is not merely the joint presence of two powers in one individual. It is, rather, that each is necessary to the other's operation and part of its reason for being. Intellect without will would be gratuitous, but so too will without intellect.[31]

The analogy brings out what is most plausible in the idea of metaphorical motion, and helps make clear why final causality could not be reduced to efficient causality. Suárez, noting that the mere words 'metaphorical motion' do not clarify the nature of final causality, explains it thus: "by the very fact that the goodness of the end is sufficiently understood and proposed to the will, it excites [the will], and according to its force, draws the will to desire it [. . .] [This relation between cognitions of goodness and acts of will] seems to be established in experience, and its foundation seems to rest on a certain natural sympathy between the intellect and the will, insofar as they are rooted in the same essence of the soul."[32] The understanding that a thing is good just does, as a matter of fact, incline the will toward it. But it is not the thought itself that inclines the will. It is, rather, the goodness of the thing understood. Suárez, as I said, identifies the metaphorical motion of the end with the real motion by which the will produces particular desires. They are the same instance of "causing," viewed with respect first to final and then to efficient causation. Their joint effect—the particular desire,

---

31. In arguing that goodness is the formal reason of the final cause, and thus that what is bad cannot be willed as bad, Suárez writes that for the will to be moved by the bad as bad would be "repugnant to [the will] itself": "The end or institution of the will is that through it man seeks what is agreeable to himself, and flees what is disagreeable; if [the will] were to accept an inclination tending to the disagreeable insofar as it is disagreeable, it would formally and directly contradict itself and its end" (Suárez *Disp.* 23§5¶4, *Opera* 25:865). The will can intend an evil under the false impression that it is good, but not an evil as such, no more than the intellect can assent to falsehood as such. Suárez's argument, I should note, is teleological. A similar argument could be made to show why the intellect and will are in "vital" proximity. Earlier Suárez speaks of the judgment of the intellect as "bringing the end close enough that it can cause" the will to act (25:860). Such ways of talking, I take it, are intended to relieve the discomfort one might feel if one thought that ends would sometimes have to act at a distance. The recognition of an end, whether through sensation, imagination, or pure understanding, gives it an existence in the soul; and then sympathy brings it the rest of the way.

32. Suárez *Disp.* 23§4¶4, *Opera* 25:860; cf. also §5¶14, 867. Suárez eventually rejects the particular explanation of metaphorical motion that this passage is in aid of. But he does not reject the description itself.

directed toward the end—is thus immediately caused by the end, under the formal reason of goodness.

All this only makes more urgent the problem of nonexistent ends. If an end acts according to real existence, and under the formal reason of goodness, then how could anyone want jam tomorrow or try to square the circle or wish (as those in hell might) that they didn't exist? The answer is at first sight disappointing: "*Esse reale* is taken in two ways: in one way for that which is a true and positive being, and not dependent on the notions of the intellect; in the other, so broadly that it embraces any being that in some way is possible to obtain or at least is presented to appetite as possible or not impossible" (Coimbra *In Phys.* 2c7q23a2, 1:315). Fonseca, too, writes that the term should be understood to mean what is possible to be obtained, or seems so, "even if it is not truly real, but a being of reason," like being judged great, "or even negative," as when someone wretched wishes to die, or impossible (though not seeming so).[33] Though there are ways to treat some of these cases without stretching *esse reale*,[34] the Aristotelians acknowledge that esse reale must include future existence, potential or possible existence, and—most striking—apparently possible (but in fact impossible) existence.

The best way to understand their thinking is to remember that being real does not entail being actual. 'Real', in the present context, denotes what exists *in re,* as contrasted with what exists only *in anima.* Jam tomorrow, or Quine's possible fat man in the doorway, are real insofar as their being what they are does not depend on whether, or how, they are thought about. 'Mind-independent' is the closest current term. In this respect the Aristotelians were thoroughgoing realists: future things, potential things, possible things, all have real though not actual existence.

But not so thoroughgoing as to admit *impossibilia.* Here they insist only that what we pursue we pursue *as if* it had real existence and goodness. A

33. "Adverte tamen, nomine *esse realis veri* aut *apparentis,* intelligendum esse in hac materia, esse possibile obtineri, aut *quod obtineri posse videatur,* etiam si illud non sit vere reale, sed rationis, quale est concipi ab aliis, magni æstimari, et similia: aut etiam negativum, quale est miseriarum, ut ita dicam carentia, propter quam à plerisque mors eligitur. [. . .] Illud etiam adverte, *possibile* tam late quoque intellignedum hic esse, ut etiam pro eo accipi debeat, quod non videtur impossibile. Multa enim intendimus, de quibus dubitamus, num re vera obtineri queant, ut si quis dubiter, num possibile sit circulum quadrare: & tamen conetur eius quadrationem invenire" (Fonseca *In meta.* 5c2q11§4, 2:169D–F). Suárez writes that "neither genuine essence nor possible existence [*neque esse essentiæ verum, aut esse existentiæ possibile*] is necessary to final causation; apprehended being [*esse apprehensum*] suffices" (*Disp.* 23§8¶7, *Opera* 25:880).

34. The Coimbrans handle negative ends by arguing that every privation or negation (like wanting to be hungry or to cease to be) is desired only *per accidens,* as a consequence of desiring something positive. The damned wish not to exist in order to avoid eternal torment (*In Phys.* 1:316). An end, therefore, can incline the will either *toward* itself (if it is apprehended as good) or away (if it is apprehended as bad).

person who attempts, as Hobbes did, to square the circle is moved not by the concept itself, or by the existence of a squared circle in thought (which he already has achieved), but by what we would call the *content* of that thought, the squared circle conceived, though falsely, as possible.[35] Although the squared circle may not have *esse reale,* it acts as a final cause *secundum esse reale.*

Since the ends to which irrational agents act are at least possible, I will not develop the point further. What I want to retain is the thought that ends, even in the actions of rational agents, act immediately as ends, under the formal reason of goodness. Cognition of the end is a necessary condition and indeed absolutely, not just naturally, necessary. But just as the local proximity necessary to efficient causation does not constitute the immediacy with which an efficient cause operates, so too the "vital proximity" introduced by cognition does not constitute immediacy in the operation of a final cause.

3. *Objects of final causation.* The object of a cause is the thing on which it acts. To include created rational agents—human, demons, and angels— among the objects of final causation was clearly not subject to debate. They were the paradigm. Yet we have also seen that the actions, powers, and forms of irrational agents are continually explained in terms of their ends or the ends of Nature. In reading questions on the final cause, one is struck by the slippage between the practice of teleological explanation and the theory of that practice. The one occurs everywhere, the other seems to allow only for final causation in finite creatures with intellect and will. The impression is inescapable that while the ascription of *ends* may be going strong, their place among *causes* has become, as already in Buridan, tenuous everywhere except in human action.

Setting aside the case of God, there are essentially two kinds of objects to consider: inanimate things and *bruta,* or animals as distinguished from plants. The treatment in each case is what one would expect if ends could only act on rational agents. Inanimate things and their actions have ends only at second hand, by virtue of being God's instruments. The actions of *bruta* are the objects of final causation only to the extent that they possess something akin to reason. The Coimbrans grant that animals give the appearance of judging ends and deliberating on means, but they do so by sheer instinct. Even those who are inclined to a generous estimate of their capacities, like Suárez, deny that animals recognize the good as good; they are moved without knowing why.

The Coimbrans distinguish three "grades" of tendency to ends. The first is

35. Fonseca writes that if someone believes it possible that the side of a square should be commensurate with its diagonal, and strives to demonstrate it, what moves him is not "ipsa conceptio talis commensurationis, ut possibilis" but rather "ipsa commensuratio concepta, ut possibilis." Not the concept, but what is conceived, is the final cause (*In meta.* 2:170A).

that of rational agents, "who not only apprehend an end under the reason of the good and the agreeable, but also recognize the aptness [*habitudinem*] and proportion of the means for obtaining that end." Such agents alone order themselves toward ends. The second grade is "of those who at most perceive the end materially, that is, under the reason of the good and the agreeable," but who neither discern the means nor order themselves to their ends. Such are animals. The third and lowest grade is of those things "that in no way recognize ends," because they lack both intellect and sense (Coimbra *In Phys.* 2c9q2a2, 1:329).[36] Such things were sometimes called *natural agents.*

For natural agents, the only way of being acted upon by ends is to be "directed toward an end by some superior and more excellent cause, that is, by the artisan of Nature herself, who comprehends the ends of all things, and gives [to them] the propensity and powers to obtain those ends." The work of nature is the work of an intelligence, "moved by divine art and directed by ingenious reason." As the archer impresses upon an arrow the impetus by which it attains the target, Toletus writes, using a figure borrowed from Thomas, "so glorious God has given to all things a nature by which even without foreseeing their ends they are led unerringly to them."[37] Suárez, citing the same illustration, makes the point more precisely. In the actions of natural agents, "there is no final causality, properly speaking, but only a tendency [*habitudo*] toward a certain end [. . .] The adequate principle of these actions is not just the proximate natural agent, unless perhaps *secundum quid* [i.e., with respect to the order of efficient causes] [. . .], but absolutely the chief [principle] is the first cause, and so an adequate principle of such actions includes an intellectual cause intending their ends" (*Disp.* 23§10¶7, *Opera* 25:887). The analogy is familiar, its advantages numerous. An instrument clearly need not know the ends to which it is designed or used in order to be said in some sense to have those ends. God clearly has the power and the knowledge by which to order things to whatever ends he chooses. Since he is the rational agent par excellence, there would seem to be no doubt that ends could act on him by metaphorical motion. More generally, an artifactual world is the natural counterpart to a creator God, and evidently dependent on him not only in existence but in essence.

Yet the analogy raises serious problems. One concerns the action of ends on God. Although metaphorical motion is clearly not, say, alteration or change of place, it is, all the same, a change in the thing moved: an "inclining" or "drawing" of the will toward an end (cf. n.27 above). Yet God, being

---

36. See also Suárez *Disp.* 23§10¶13ᶠᶠ, *Opera* 25:589; Toletus *In Phys.* 2c9q12, *Opera* 4:74v. The three grades are found already in Thomas *ST* 1pt2q1a2 and q6a2, Parma 2:2, 30.

37. *In Phys.* 2q12, *Opera* 4:75ra; cf. Thomas *ST* 1pt1q103a1, Parma 1:395.

necessarily entirely *in actu,* can receive no impression from without. The answer, in brief, amounts to accepting the objection. Divine acts of will are not caused by ends, or indeed by anything, because they are nothing other than God's will itself, which is entirely *in actu.* Metaphorical motion, which is indeed the actualization of a potential to desire, applies only to created rational agents.[38]

The second problem is more intractable. When Aristotle defines nature, he mentions artifacts only to deny that they have natures (see §7.2). That argument is in part an argument against Plato on behalf of the independent reality and knowability of the natural world. The possibility of an *episteme* of Nature qua Nature rests on finding the principles of natural change within Nature itself, including the natures of individual things. Yet Suárez holds that an adequate principle can be found only in God.

One easy reply is that Aristotle was wrong. There are actions that cannot be sufficiently accounted for by reference "to the private properties or inclinations of particular things." The *horror vacui* or the confinement of the sea, Suárez argues, cannot be accounted for by the peculiar nature of water, but only by "the end, which rests in the perfection of the whole universe." That end "must be intended by a superior agency" (*Disp.* 23§10¶10, *Opera* 25:888). If Aristotle did not acknowledge the existence of cosmic ends that could not be accomplished incidentally in the pursuit of private ends, then his physics was incomplete.

But the reply, even if just, only accentuates the difficulty. A more cogent reply will show that some instruments, at least, inherit the ends of their users. Part of the reply will turn on the differences between natural and artificial instruments, or equivalently on the differences between human industry and divine creation. That part, I defer to §7.2. Here I consider only the relation of means to ends, and to final causes, in general.

The question, then, is whether means can really be said to share the ends to which their users put them, or for which they make them.[39] Suárez, arguing that all actions and effects of the will are effects of the final cause, writes that "when by these actions there is produced a *terminus* that has an enduring existence [*terminus permanens in facto esse*], this also is held to be an effect of the end conceived beforehand, whether in becoming as it is actually made or in existence in fact while it endures afterward. For that reason

38. Suárez *Disp.* 23§9¶3ⁿ, *Opera* 25:882ᶠ; cf. Coimbra *In Phys.* 2c7 q21a2, 1:309.

39. Philosophers now ask an analogous question about intentionality. Some objects, like thoughts, have "original" intentionality, while others, like written or spoken words, have only "derived" intentionality. John Searle, for example, insists that no computer yet made has original intentionality. The physical states of the central processor or of the video display have intentional content only by virtue of being our instruments. In Aristotelian terms, they have *intentio* only denominatively, as when we call a caricature cruel.

Aristotle said that instruments exist on account of an end, and likewise houses and other artifacts are effects of some end conceived beforehand" (Suárez *Disp.* 23§3¶19, *Opera* 25:857). That seems clear enough. Though the immediate effect of an end, as we have seen, on created agents is the metaphorical motion of the will, subsequent effects too are obviously on account of that same end. Yet he then argues that the causality of the end with respect to external actions and things is extrinsic and denominative: "But acts merely commanded by the will (and all the more so their effects) in no way are elicited immediately by the end itself, nor are they said to have an intrinsic aptitude toward the end, but are said to be ordered to the end only by an extrinsic denomination by way of interior acts [of will], as walking is merely extrinsically ordered toward health" (¶23, 858). After all, Suárez notes, if walking or some other exterior act (or, presumably, an artifact produced by such an act) occurred "on account of some other end, or by chance," "in itself and in its being [*entitate*] it would not be transmuted or have anything taken from it." Walking is walking, whether you walk for health or to buy cigarettes. The exterior act is therefore not caused *per se* by the end, but only per accidens, as a falling stone pushes air aside on its way to the ground.

Arriaga makes the same point. Considered in relation to a given end, a means may be said to be useful intrinsically, since its usefulness consists in its being genuinely effective in bringing about that end. Cutting a vein is useful to health because of its nature. But if we do not consider any end in particular, then cutting a vein is neither useful nor useless. Means, in short, are called useful or not only by "extrinsic denomination" (Arriaga *Cursus* 353).

Applying these conclusions to natural things, regarded as divine instruments, we would have to say that they too have their ends only extrinsically, and are caused by them only *per accidens*. Suárez and the rest agree, as we have seen, that even if God were as impassive as the gods of Lucretius, particular efficient causes would not cease to have the effects they now have. Suárez argues, moreover, that "if from the mutation of the cause, neither the effect nor the action is changed, that is a sign that the cause does not influence the effect [*influere in effectum*] either immediately or *per se*" (858). Natural agents, insofar as they may be characterized in terms of efficient causes and effects alone, would not differ in the Lucretian world; nor are they caused by divine ends immediately or per se.

A disturbing outcome, if you're an Aristotelian. The means of palliating it will be found, I think, only by examining the differences between nature and art (§7.3). Here I will just mention briefly one other way out. The argument proves only that if there were a Lucretian world containing individuals of the same sort that now exist, it would (setting aside collective and cosmic ends) precisely resemble the actual world. The Aristotelian may still argue, then, that although the actions of individual natural agents do not have intrinsic ends, the *natures* of those individuals do. Only if—and this would have seemed quite unlikely, as well as unorthodox, to an Aristotelian—natural kinds could be shown to emerge through the operation of natural efficient causes alone, would there be reason to doubt that natural kinds and their instances had intrinsic ends. We have seen that in fact the Aristotelians believed that sublunary agents were incapable even of reproducing their forms; still less would they be capable of creating new ones. The case for appealing to ends in natural philosophy, therefore, rests not on their being needed to explain particular events, but on their being needed to explain the existence of kinds. Cosmic and collective ends, not particular ends, will be the primary instances of finality in natural agents.

With animals, the stakes are quite different. That their activities were guided by ends, and that they to some degree recognize those ends, was amply confirmed in experience. Though there were attempts, which I will examine shortly, to assimilate the causes of their actions to those operating in inanimate things, the dominant view was that efficient causes alone would not suffice. The point of dissension was instead whether their actions were guided, like ours, by a rational intellect in proximity to a free will, or by innate instinct.

Instinct is not itself a final cause but a means, analogous to metaphorical motion in rational agents, by which ends could incite, via sense and imagination or *phantasia,* the motive powers of the soul into action. Even in humans, all actions that we accomplish "without the command or movement of the will" occur either in the manner of inanimate agents or by instinct (Suárez *Disp.* 23§3¶18 and §10¶15, *Opera* 25:856, 889–890). What interests me here is not the psychology of instinct, but rather the reasons given for introducing a third way for final causes to operate, distinct both from metaphorical motion and from the instrumentality of inanimate agents.

Suárez notes that certain unnamed philosophers, in their eagerness to avoid attributing rational souls to animals, have gone so far as to deny that animals recognize their ends at all. Instead these philosophers hold that their actions are brought about either by internal powers analogous to weight, or by external powers analogous to magnetism. Suárez immediately dispatches the view with the comment that it is "absurd, contrary to evident experience, and indeed to divine Scripture" (889).

Nevertheless Buridan came close to proposing just that view. The actions of animals depend entirely on the efficient causal powers of their own natures, the heavens, and God:

Concerning natural things I believe that the swallow, when it mates, nests, and lays eggs, no more thinks of the young which are to be generated than a tree, when it leafs and flowers, thinks of its fruit. Nor do the mating, nesting, and egg-laying of the swallow depend in their existence and order on the young. On the contrary, the young do not determine the swallow to operate thus; rather the form and nature of the swallow and celestial bodies at the appointed times and God by his infinite wisdom determine the swallow to mate, and from that follows the generation of eggs, and then, when the swallow is so disposed by its nature together with celestial bodies and God, all of them determine it to nest-building and then to egg-laying [. . .] All these [effects] issue from divine art and celestial bodies and particular agents, both extrinsic and intrinsic (like the substantial forms of natural things themselves). (Buridan *In Phys.* 2q13, p40rb; cf. Maier 1955:326)

Drawing on the familiar maxim that what is first and better known to nature is later and less well known to us, Buridan explains that although we know and intend the ulterior consequences of our immediate actions, in nature "there is no intention or appetite [. . .] aside from a *potentia* and determination to produce an effect"—namely the *first* effect, which is naturally bound to follow, and from which all later effects will come.

The Coimbrans' story differs only a little. Plants, which no one credits with knowledge of their ends, generate leaves and fruit as if they foresaw the needs of their offspring. The swallow, a stock example in these discussions (see *Phys.* 2c8, 199a26), "looks at [*intuetur*] mud; an image of the mud enters its imagination [*phantasiam*]; then judging it useful the swallow picks up the mud and carries it to this or that place," and so forth. In the progression of actions, each of which—we might say—provides the stimulus for the next, "judging" amounts only to a determinate action of the image on the motive power of the swallow, "a certain imitation of [human] judgment, which animals exercise in a crude and simple act," without the operations of composition and division characteristic of rational thought (Coimbra *In Phys.* 2c9q4a2, 337–338).

Suárez, on the other hand, grants that animals recognize the ends to which they act. But only "materially," not "formally": they do not recognize the "formal reason of agreeableness or utility, and they are not so moved that they could order one thing to another, or desire a thing formally as lovable in itself."[40] The swallow sees mud, and is inclined, willy-nilly, to pick

40. Suárez *Disp.* 23§10¶15, *Opera* 25:889, citing Thomas *ST* 1pt2q1a2 and q6a2 (Parma 2:2, 30); see also 1pt1q59a1. Earlier the Coimbrans make the same threefold distinction (Coimbra *In Phys.* 2c9q2a2, 1:329), but in their question on instinct they do not refer to it. Their primary interest is in refuting the claim, which they attribute to Jean Gerson, that instinct is a "special motion, by which God concurs with them in their works" (2c9q4a1 and 3, 1:336, 338).

it up. The mud is in fact useful for nest-building; but even though the swallow does use it to that end, it doesn't form the thought that the mud is useful, or choose it rather than some other building material.

Two conclusions may be drawn from the discussion of *bruta* and natural agents. The first is that everything except rational agents, and in a quite limited way, animals, is acted on by ends only as an instrument of God. Finality occurs everywhere, final causes only in nature's highest forms. The second is that final causality is, for the Aristotelians, always dependent on rational cognition, human, angelic, or divine. It follows that the Aristotelians are not open to the charge of animism: that charge rests on a misunderstanding of the role of analogy in their natural philosophy.[41] It follows also, however, that knowledge of the ends of nature is not, as it would have been for Aristotle, a knowledge of one sort of intrinsically natural property among others; it is instead knowledge, however deferred and uncertain, of the mind of God.

## 6.4. Teleological Reasoning

In this subsection I will detail three instances of teleological argument: the slogan 'Matter desires form'; natural limits on size; and the production of monsters. The first two are straightforward enough; the last requires a bit of preamble. Monsters themselves, or rather monstrous forms, are not themselves ends, unless perhaps of Nature or of God. Like other chance events, they are the incidental outcomes of natural changes whose ends are quite different. Part of the point in treating them is to delineate the limits of finality, as a prelude to discussing natural and violent change in §7; the other part is to illustrate in some detail the conjoint roles of material, efficient, and final causes.

Together the examples illustrate three points. First, in natural agents, the Aristotelians were inclined to treat tendencies to form as nothing other than the *dispositiones* or *proportiones* of proximate matter which were discussed in §5.3. That is not to deny that those agents have ends. The term *dispositio* is itself charged with finality. Rather it is to say that 'having such and such an end' and 'having such and such a material constitution' are alternative

---

41. After quoting Thomas's example of the arrow tending toward its target (n.37 above), Weisheipl writes that "the scholastic terminology was commonly attacked in the seventeenth century [. . .] as the expression of animism and anthropomorphism; this was due to a misconception of analogical usage—a human necessity" (Weisheipl 1985:22, n.95). Descartes did not deny that analogies were useful and even necessary to physics (see, e.g., *Principles* 4§201 and 203, AT 8¹:324, 326, 9:319, 321). That was not the issue; the issue was *which* analogies. But I agree that Descartes's serious misreading of his predecessors' views indicates that, among other things, he mistook the role of intentional idioms in their physical reasoning.

descriptions of the same accidents, one with respect to form as their normal effect, the other with respect to the combination of elemental and derived qualities which has such effects.

Second, cosmic ends like that implied in the slogan 'Nature does nothing in vain' are in certain respects functionally equivalent to laws in Cartesian science. Not only do they supply the warrant for inferences from what has been observed to occur to what could or could not occur; they also find a similar ground in the distinction between the absolute and ordained powers of God.

Finally, in examining responses to the question of whether nature intends monsters, I will uncover an interesting disparity between Toletus and the Coimbrans. For the Coimbrans, the role of monsters is the traditional one of adding variety and thus beauty to the world; for Toletus, they emerge inevitably from the proportioning of God's concurrence with second causes according to their strength. The difference, I think, is significant: Toletus's account not only avoids appeal to an aesthetic motive, but also permits an explanation of monsters to be given, if we can measure the "strength" of causes, in entirely physical terms.

1. *Matter desires form.* In this slogan from *Physics I* we see flagrant misuse of intentional idiom, and gross confusion between soul and body. The passage in Aristotle reads: "[Matter] desires [form] just as the female [desires] the male, or ugliness beauty. But [matter] is neither ugly nor female *per se,* but rather *per accidens*" (*Phys.* 192a22ff).[42] The immediate context is an argument that matter and privation are distinct, or, equivalently, that matter is not just nonbeing. Form, Aristotle argues, does not desire itself, since it itself is the desirable; privation could not desire form, since that would amount to desiring its own destruction. That, the Coimbrans say, is "most alien to the laws of nature" [*quod à naturæ legibus quam maximè est alienum*]—one of the infrequent occurrences of that phrase (*In Phys.* 1:149). Hence what desires form is neither form nor privation but a third principle, namely, matter.

The Coimbrans' explanation provides ample pretext for the accusation of animism. Beauty, they say, is "nothing other than the splendor of a thing, the sight of which attracts the soul": so Plato has defined it. But matter alone, being *nuda potentia,* has no power to produce accidents by which to be seen. Only when it has received form and with it color and other visible qualities, can matter be beautiful. On that account it desires form. Matter, moreover, can assimilate itself to the divine, from which it is otherwise "most distant,"

42. "Sed hoc, est materia, perinde appetens illud ac si marem foemina, & quod turpe est, appetat pulchrum. At nec turpe, nec foemina per se, sed ex accidenti" (Coimbra *In Phys.* 1:148). Other translations differ, but not materially (cf., e.g., Albert *In phys.* 1tr3c16, *Opera* 4.1:71). Interestingly enough, all the commentators ignore the analogy with female desire, preferring to talk only about beauty.

only by receiving form, and so too the actuality and activity of that form. On that account also matter desires form.

Matter, it would seem, can envision the benefits of form, and though it cannot act on its desire, it can (according to Scotus) be called 'satisfied' when that desire is fulfilled. But the Coimbrans know as well as Descartes that matter—and a fortiori prime matter—cannot literally be said to desire anything. The appetite of matter for form, they remark, following Thomas, is "nothing other than the inclination of matter to receive form," an inclination "deeply rooted in matter," and in fact not distinct from it (*In Phys.* 1c9q5a1, 1:160). Toletus, more precise on this point, distinguishes appetites into intellectual, sensitive, and natural. Natural appetites are the inclinations of things that can recognize their ends neither formally, as we do, nor materially, as animals do. They are nothing more than the propensities toward the good which God, who knows the ends of all things, has given to them. Matter in particular is "proportioned" to receive all forms; and by virtue of that can be said, even when it already has one, to desire those it does not have (Toletus *In Phys.* 1c7q17, *Opera* 4:38ra).

Whatever Aristotle may have meant by his dictum, it is clear that the Aristotelians do not literally attribute desire to matter. "When we say that earth desires the lowest place, or matter desires form [. . .], this is true only according to a transferred usage [*per translationem*]." Of such things it is true only that it is *as if* they had appetites in the proper sense (Fonseca *In meta.* 1c1q1§3, 1:62aE). The intentional language picks out a class of properties—which we still call propensities or inclinations or tendencies—by virtue of their analogy with a part of the class intimately known to us—our own, intellectually guided tendencies. The phenomena thus picked out in natural agents can also be described in the more neutral vocabulary of *proportio* and *dispositio*. But the Aristotelians, unlike some recent philosophers, feel no obligation to give up the analogical, "transferred," way of speaking. The analogy is stronger for the Aristotelians than for us because they do not hesitate to call that which a thing by nature inclines toward its *good*, whether that thing is human or not. To every appetite, moreover, there corresponds a state that may be called, properly or by *translatio*, the "satisfaction" of that appetite, and whose outward sign is the spontaneous cessation of *motus*. When that state is not reached, words like 'frustration' or 'miscarriage' may be used, whether the agent is rational or not (*Phys.* 199b; cf. Coimbra *In Phys.* 1:320, 339). In those respects, then, the use of intentional language to refer to features of the inanimate world highlights a genuinely common pattern of behavior. Its use would be dangerous only if one somehow forgot the all-important differences between rational and natural agents: that we know the good as good, that we have the power to refrain from following our inclinations.

2. *Limits on size.* A question on the upper and lower limits to size in material substance seems to have become a set piece in *Physics* commentaries of the sixteenth century, only to disappear in the seventeenth. I will examine only part of the question, that having to do with limits on the size of living things.

The Coimbrans' argument, once one sorts out the logical distinctions they make between upper (or lower) bounds and least upper (or greatest lower) bounds, comes to this. Every animal has a definite essence and perfection; its nutritive faculty, which controls its growth, does too; hence there is an upper bound. As for lower bounds, every soul requires organs to carry out its functions, and "the consequence is that these are determinate, and so the soul also needs a definite minimum quantity." The arguments, in other words, point out that among the perfections of the soul is that of having a quantity within certain bounds.[43] But inanimate things are thought not to have upper bounds, and yet they too have their perfections. So the question at hand seems begged.

Toletus does better (*In Phys.* 1c4q9, *Opera* 4:24vb). After citing authorities, he argues first that the manner by which living things grow imposes limits on their size. The "instrumental cause" of growth is "natural heat." Natural heat is necessary to the parts of the body; as it is lost, it must be restored by the vital heat that, generated in digestion, is distributed to the rest of the body by the liver. The larger the parts, the weaker and less intense the vital heat will be in each, and so at some point more heat will be lost than gained. Hence there is, for each species, a definite upper bound on size.

The argument has a nice feel to it, perhaps because it compares quantities and appeals to efficient causes only. In the background, however, there is a teleological principle. It is logically possible that a soul should, so to speak, be programmed to grow so large that it will inevitably die. But nothing—to cite a maxim mentioned a few pages ago—can naturally tend to its own destruction. Such a soul cannot naturally exist.

More explicit is the finality appealed to in a later argument: "that nature has never done a certain thing up to now is a sign that it cannot; but no one has ever come across men larger than mountains, or smaller than ants; nature therefore cannot do this, since if it could a *potentia* would exist in vain, unless it sometime achieved *actus*" (Toletus *In Phys.* 1c4q9, *Opera* 4:24vb). We have first a rule of thumb: what has never been known to occur

---

43. Faced with the objection that the crocodile never stops growing, they answer, correctly but irrelevantly, that each crocodile will have an upper limit to its size—namely, however big it is when it dies! They also argue that in fact it must stop growing *before* it dies, on the grounds that the nutritive faculty must fail before the moment of death. That argument too is plausible but irrelevant (Coimbra *In Phys.* 1c4q1, 1:110).

probably can't. Then an application: Brobdingnagian and Lilliputian people have never been known to occur, so they can't. To that Toletus appends
a justification for the rule. Nature, that is God, would not provide a form
with a power and then never allow that power to be exercised, because then
that power would have been created for no reason.

The argument is not a mere induction. It is not of the form 'No Brobdingnagian human has been observed, ergo no human (past, present, or
future) is Brobdingnagian'. If it were, the last step would be superfluous.
The conclusion is in fact not that no human *is* Brobdingnagian, but that no
human naturally *can* be, or that no human is *in potentia* Brobdingnagian.

For that the last step is needed. We have seen many times a *potentia*
inferred from an *actus*. Water spontaneously cools: therefore it has the
*potentia* to be cold. Such inferences presuppose that the observed state is the
outcome of a natural change, but they are otherwise straightforward if the
natural can be discerned from the unnatural without knowing what powers
a thing has. To infer the absence of a *potentia* requires more machinery.
There is no inconsistency in supposing that a *potentia* has never manifested
itself. To infer from the failure of a supposed *potentia* to manifest itself that
there is no such *potentia*, one must suppose that every one has, or has had, its
moment; and that inference will in turn rest on the economy with which
God pursues his ends.

One must suppose also that the observations made so far are an unbiased
sample of the occasions on which the supposed *potentia* might have manifested itself. That raises an interesting question. The created world yields
evidence of the magnitude of divine power. But no inference of the form
'God has never been known to do *X;* therefore God has not the power to do
*X*' can be sound. In particular the evidence, otherwise overwhelming,
against the unique events recorded in the New Testament, shows only that
those events lie outside the scope of God's ordained power. Induction retains its validity only on the assumption that God continues to adhere to his
self-imposed order; the inference from actual to potential likewise relies on
that order. If that order is altered, those inferences may no longer be sound.
No human body had the power to promote salvation until the first Communion, no bones the curative effect of saintly relics. Every statement that
proposes to circumscribe the powers of things, or list their ends, must be
implicitly relativized to an order we know to be contingent, and whose
motives may outstrip our comprehension. What we can be sure of is that
God's ends are fixed and that the means he chooses to accomplish them are
not gratuitously varied. That suffices to ensure that the powers he has given
to each thing to pursue its good will not change, and it warrants the inference from the actual to the potential.

3. *Monsters.* Chance events are an acknowledged exception to the reg-

ularity with which natural agents pursue their ends.[44] Yet monsters, at least, are not an exception to the natural order. Not only are they caused by natural agents, but their ends, although distinct from the ends to which normal births are directed, can be seen to lie within nature.

A monster, Toletus writes, is "a mistake [*peccatum*] of nature acting on account of some end, from which it is frustrated by some corrupt principle." Or, more precisely:

> A monster is a natural effect which degenerates from the correct and usual disposition consonant with its species [*effectus naturalia recta & solita secundùm speciem dispositione degenerans*]. It is called a "natural" effect, because in art there are no monsters, properly speaking, but only by analogy and resemblance to natural things. [. . .] It is said to be that which "degenerates from the correct disposition" because a natural effect is not called a monster unless there is some defect, obliquity, or deviation from what it was going to be according to the reason of that effect, and on this account monsters are called "mistakes of nature" [*peccata naturæ:* cf. *Phys.* 199a33–b4], because nature falls short and wanders in bringing about the effect. The disposition is said to be the "usual" one because what is monstrous occurs rarely and does not resemble the other effects that proceed frequently [from the same causes]. I add "consonant with its species," since what is monstrous does not preserve a specific resemblance to its proximate principal cause, so that if a dog were born of a human female, it would be a monster. (Toletus *In Phys.* 2c9q13, *Opera* 4:75rb)

The outward signs of monstrosity are rarity and lack of resemblance with other, more frequent, effects of the same causes. A monster, I should note, differs from its normal kin not by its form (it is still a member of the same species) but by its *dispositio*—the arrangement of its parts, or perhaps the temperament of its elemental qualities. The inward *ratio* of monstrosity is a departure of the *motus* of generation from the *actus* implied in the powers

---

44. What follows is intended only to address questions about finality. A survey of textbook questions on monsters would be a sizable undertaking, although they all draw on Aristotle, Plutarch, Pliny, Galen, and Albert. The standard work on philosophical theories of monsters in the Renaissance is Céard 1977; cf. also his edition of Ambroise Paré's *Monstres.* Daston and Park 1981 discuss learned and popular belief; Darmon 1977 details a number of seventeenth- and eighteenth-century cases. Paré, Daston and Park, and Darmon include contemporary illustrations. Reisch's chapter on monsters includes a well-known depiction of five more or less credible instances (*Marg. phil.* 8c19, p345; also at 405, in the chapter [9c40] on human deformities). Goclenius's entry, which treats the term in its broad sense of 'prodigy' or 'portent', presents an exhaustive classification of *monstra* in every department of nature (s.v. 'Monstrum', 708ff). The *locus classicus* in Aristotle (*Phys.* 199b4ff) is examined briefly by Lerner (1969:169f). In his introduction to Nicole Oresme's *De causis mirabilium*, Bert Hansen gives a lucid presentation of the relation in Aristotelian thought between marvels and natural order; on monsters in particular cf. p62 and notes 37, 38. Oresme himself devotes a long section of the *De causis* to monsters and their causes (*Mira.* 3§6, 229–248).

that produce it, whether that is to be ascribed to the principal cause itself—
the seed—or to the concurring causes and conditions sine qua non.

The Coimbrans list five causes or conditions which must concur in generation (Coimbra *In Phys.* 2c9q5a2):

> (i) the formative virtue of the seed, which "delineates and builds the whole body limb by limb";
> (ii) the matter "from which the fetus coalesces";
> (iii) the womb or *receptaculum;*
> (iv) the primary qualities "on whose temperament the health and integrity of animals depend";
> (v) extrinsic factors like the "salubrity or inclemency of the region, the influx of the stars," and so forth.

Of these five, (i) and (v) are efficient causes, the rest are aspects of the material cause; I will refer to all but (i) as "impeding" causes. Deficiencies or abnormalities in these conditions will produce characteristic defects. If the seed is crude or sluggish, for example, the offspring may have confused, useless parts, or revert to a lower form, as when a man is born with a ram's head. In such cases "nature, since she cannot attain a perfect finish, contents herself with an inferior grade" (Coimbra *In Phys.* 1:341).

Sometimes the matter of several seeds from different species is promiscuously confused, and half-men or half-beasts are born. This happens in African rivers when the water is low (Toletus says it happens in India). Sometimes there is too much matter, or too little, and dwarves or giants result. If there is too much heat in the womb, the offspring may be born with a beard (which has, the Coimbrans add, "occurred in our own Portugal"); too little, and it will age prematurely. The heavens, whose crucial role in generation was studied in §5.4, can affect the parts of the offspring in many ways. Sometimes, the Coimbrans write, "when those stars dominate, that most favor the generation of men, it happens (if we believe Albert in his *De mineralibus,* and [John of] Jandun in question 24 of this book [i.e., in his *De mineralibus*]) that the animals generated then, and even the stones that congeal, partly imitate the likeness of man" (1:342).[45] One last cause is the parents' power of imagination, "which occasionally makes the formative faculty wander from its target, and imprints upon the fetus absurd or alien

---

45. Paré cites from a French translation of Aristotle's *Problemata* a story of Albert the Great. In a certain village a cow gave birth to a calf that was half human. "The villagers, dubious about their pastor, brought him to judgment, and meant to burn him and the cow together." Albert persuaded them that the monster had come about "by a special constellation," and so the pastor was "delivered and purged from the imposition of so execrable a crime." Paré for his part doubts that Albert was right (*Monstres* 68; for the source, see 172n118). Whatever its detractors may have said, it is clear that Aristotelian science was not *entirely* useless.

figures." Such are the *marques d'envie* that Mersenne asked Descartes to explain.[46]

Thus far the story contains nothing but efficient and material causes. Finality enters when one asks why there should be monsters, or, as the Aristotelians put it, whether Nature "intends" monsters. In the Coimbrans' response, it becomes clear that the sense they give to the question is 'what part of Nature in particular intends monsters?' There are, as it turns out, four possibilities: the principal cause, the impeding cause or causes, the principal and impeding causes together, and God. The principal cause does not intend the monster: "it strives to produce without error or vice an effect similar to itself" (*In Phys.* 2c9q6a1, 1:344). But a monster, as we have seen, is by definition dissimilar to its principal cause. The impeding cause, too, does not intend the monster, for similar reasons; by itself, it would produce nothing of the sort. Only the two taken together—call this the total cause—can be said to intend the monster, or rather what the Coimbrans call its *fundamentum,* the respect in which a thing is a monster, like two-headedness or six-fingeredness. That Nature should intend some effect "is nothing other than that it should incline toward it"; the total cause does incline toward the monstrous effect, and in that respect so does Nature.

That, however, leaves the initial question unanswered. It is true that the total cause, here as elsewhere, tends toward its effect. How could it not? But the question was rather: why should such total causes, having the characteristic effects they do, exist at all? The answer, as one might expect, consists in showing that God intends that they should. With respect to his ends, nothing, the Coimbrans affirm, happens "fortuitously or by chance." The frustration of the action of a particular cause is not the frustration of a divine end, because God "does not absolutely will the ends of all particular causes." Some he promotes; others, "by reason of their weakness," he allows to fail. His ends, which are universal, are never frustrated.[47]

And what are those ends? The only one mentioned is the beauty of the whole. The theater of the visible world is ornamented by "the variety of images not only of elegant things, but also of the deformed," just as the charm of pictures is augmented by adding dark colors to light (a2, 1:346).

46. To Mersenne 29 Jan. 1640, AT 3:20; 30 Jul. 1640, AT 3:120; cf. *Diop.* 5, AT 6:129. For a grotesque example, see Paré *Monstres* 36, fig. 28 (= Darmon 1977:176, fig. 41). This is a child born in 1577 (Paré says 1517, but Céard corrects it to 1577) with the face of a frog, because its mother conceived it while holding a live frog to cure her fever. Darmon lists a number of similar instances.

47. A passage from Thomas's *Summa* cited by the Coimbrans here reads: "Since God is the first and universal cause not merely of one kind of thing, but universally of all beings, it is impossible that something should occur outside the order of divine governance; but if something seems to escape, when considered according to one cause, from the order of divine providence, it is necessary that it should have been sliding back into that order according to another cause" (Thomas *ST* 1pt1q103a7, Parma 1:398).

Toletus answers the two questions rather differently, the second in rather less Panglossian fashion. What intends the monster is the impeding cause. While it is true that it is not the principal cause of the monster, it is that part of the total cause from which the monstrous effect frequently follows. The conjunction of the two may be rare, but given that the first is operative, if the second acts, the same deformity will necessarily occur (Teletus *In Phys.* 2c9q13, *Opera* 4:76rb). That, for Toletus, suffices to say that the impeding cause "intends" its effect. Unlike the Coimbrans, he does not insist that what intends the effect should resemble it, or that it should issue in that effect for the most part unconditionally. Although a closer study would be needed to establish the point, it does seem that Toletus, in ignoring resemblance, and in requiring the 'for the most part' condition to hold only conditionally, has a more modern view of efficient causation.[48]

His answer to the second question strikes me in a similar way. Toletus asks how God could "concur with both the impeding cause and the generating [i.e., principal] cause." It would seem that the one concurrence impedes the other, and that God is frustrating his own action. We have seen that the Coimbrans answer, in effect, that God subordinates particular to universal ends. Toletus's answer is that although God concurs with every cause, he concurs with each according to its nature, so that "with stronger causes he concurs more strongly, and with weaker causes more weakly." Hence the one concurrence impedes the other "not from the side of the concurrence, but from the side of the causes." If there were, he adds, "a supreme Emperor, who had beneath him various Kings, who would seek arms and troops, and receive them according to their condition and manner; and when two Kings contended, would overthrow the one who was weaker," that Emperor would be like God, who, though he assists all things, does do according to their nature and power and, when two contend, allows the stronger to win. Though the Coimbrans note in passing that God allows the weaker cause to fail, their emphasis is on the universal and the particular, and on the beauty of the whole. Since in doctrinally similar texts like these, emphasis yields most of the differences from one to the next, the shift is worth remarking.

No one who has read Deleuze's Nietzsche could fail to be struck by this

---

48. Two recent discussions of Aristotle's phrase 'always or for the most part' and its contrary 'rarely or never' differ on a similar point (Mignucci 1981, Judson 1991). Mignucci holds that 'always or for the most part' should be taken to indicate a high absolute probability: normal births happen most of the time, monsters only a few times. Judson argues that "always or for the most part" should instead be taken to denote a conditional probability: given that an event is an instance of bovine generation, the likelihood of its being an instance of the generation of a normal calf is high. The Coimbrans, then, side with Mignucci: what is "intended" by a cause follows from that cause most of the time, whatever other causes are operating. Toletus, on the other hand, holds that an effect is intended provided only that it occur most of the time given that the other causes are also operating.

passage. Though Toletus speaks of providence, the justice of his world perilously resembles that of Thrasymachus. The beauty that the Coimbrans, following Augustine, take to be God's leading motive is nowhere to be seen. Divine concurrence, moreover, is nothing other than a ratification of relations intrinsic to nature: as Toletus puts it, nature's outcomes are determined *ex parte causarum*, not *ex parte concursus*.

Perhaps God has arranged it so that the more noble, the more lovely, is also the stronger. The struggle of kings might then issue in beauty. But only *per accidens*, at least from our standpoint. In his question on fate, Toletus defines fate, following Boethius, as the "inherent disposition of *mobilia* [sc., *corpora*, natural things] according to which divine providence binds them to its order" (*In Phys.* 2c6q11, *Opera* 4:69va). That disposition, considered with respect to the divine mind that arranges it, is called "providence." With respect to the things themselves, it is called "fate." All things under the heavens are ruled by fate—"all movements, and mutations, and other passions." Human bodies are, as we have seen, subordinate to the heavens, especially in generation, which is of all the soul's powers the most distant from the intellect and will. Are we too, then, ruled by fate? No: the will is only inclined by the body, never compelled; only God rules it. Yet if the heavens can impede every bodily movement, or at least every movement not immediately caused by the will, then the sovereignty of the will is restricted, as Descartes would later hold, to the soul itself. We escape fate, and are stronger than the stars, but only in what we will, not in what we happen to accomplish.

Whatever their differences, Toletus, Coimbra, and the other central texts agree that every event that looks like a counterexample to finality can be analyzed in one of two ways.

The event may be the outcome of two causes acting simultaneously on the same subject, or in the same place, each of which, were it to act alone, would fit the scheme of natural change. Here the model is the encounter of two people at a market. Their meeting is not the end of either one's actions; but it can be analyzed into two actions each of which does reach, or would have reached, a natural *terminus*. The generation of a monster that results from the mixing of seed, or from the intervention of celestial causes, can likewise be analyzed into the frustrated, but still directed, actions of each kind of seed.

Or the event may be the outcome of a single cause acting upon matter that is not fit to receive that action. The quantity of seed is not an active power, yet an overabundance of seed can frustrate the action of the *vis formatrix* embodied in it. What opposes the active power is not another active power, but the absence of a corresponding passive power. The action of that one active power is, clearly, directed to the same end it would have attained had the matter been fit.

In neither case is there an action that lacks an end. There are only actions that fail to attain their ends. The treatment of monsters here is of a piece with the treatment of chance events generally. Chance, Suárez writes, is not "a peculiar cause instituted *per se* to such an effect, but can be any created efficient cause, insofar as by accident and without intention [*præter intentionem*] there is conjoined with its *per se* effect another rare and fortuitous effect" (Suárez *Disp.* 19§12¶5, *Opera* 25:743). No effect is a chance effect with respect to God, who foresees and intends all that happens; only with respect to particular things can there be chance events. In monstrous births, Toletus holds, the generator fails to achieve its end; the celestial influence, if it is present, does achieve its end, which is to dispose terrestrial matter as it sees fit. That it makes the seed deviate from its usual path is an outcome that happens to be conjoined, because of the simultaneous action of the generator on the same matter, with its intended effect (Toletus *In Phys.* 2c6q9, *Opera* 4:67rb).

I said at the outset of this subsection that ends in Aristotelian physics supply a functional equivalent to laws in later physics. I will conclude by making that claim more precise. Let us suppose that in both the earlier and the later science, experience, whether it is the common experience appealed to in many Aristotelian arguments or the experimental results used in modern science, warrants not merely singular claims but counterfactuals of the form 'Whether or not *S* occurs, if *S* were to occur, *T* would occur'. I take it that in many instances, Descartes and his predecessors would have agreed that such counterfactuals—call them phenomenological regularities—are true. But thereafter they diverge.

The Aristotelian looks for an agent—sometimes a specific nature, sometimes Nature as a whole—and a description of *S* and *T* such that *T* would be an intended effect of a certain operation characteristic of that kind of agent, given that *S* has occurred. The generality with which a phenomenological regularity is credited derives from the fact that agents of that kind will all act according to the same ends, whether these are the purely formal ends of §6.2 or ends peculiar to that kind. The necessity implicated in the counterfactual form of the regularity derives from the fact that the effect, given that the agent has triggered an appropriate efficient cause, that no impeding cause overcomes its action, and that the matter it operates on is adequate, will be bound to follow. For example: the birth of a calf, given that its parent has the end of reproducing its form and has set the seed in motion, that the heavens do not intervene in an unusual way, and that the matter operated on is not excessive, and so forth, is bound to occur. That is the kind of warrant the Aristotelian would offer for the commonsense belief that if Bessie were to give birth, she would have a calf.

The Cartesian looks for a description of *S* and *T* which will permit a derivation of the conditional 'If $S$ occurs then $T$ occurs' from the laws of nature together with assumptions about the natures of the things involved. (I leave further details for Part II.) Where in Aristotelianism the nature of a thing entails that it shall have certain ends, according to which it will act—of necessity if it is an irrational agent—, in Cartesianism the nature of a thing entails that certain characteristic effects must follow if the thing or certain of its parts is moved in certain ways. The necessity of the effects is inherited from that of the laws they instantiate. Fire consists of small particles moving with great speed; placed next to wood, those particles must break off particles of similar size and speed from the larger particles of wood they collide with. The necessity of that effect is inherited from that of the laws they instantiate. Thus the Cartesian justifies the commonsense belief that if a flame were put next to something wooden, that thing would burn.

The claims of the opponents of Aristotelianism that its appeal to ends was superfluous, or that it must have arisen from a confusion of matter with soul, are, it seems to me, warranted only after one has adopted the Cartesian way of explaining phenomenological regularities (or something similar). Considered from within Aristotelianism, there seems to be neither superfluity nor confusion in the appeal to ends. One has indeed the option, if one is explaining the actions of inanimate agents, of construing talk of ends in terms of material dispositions and what follows from them. So we saw in Toletus's treatment of the dictum 'Matter desires form'. That option seems to have been increasingly favored by the Aristotelians, although one does not find them pursuing it as far as some of their Medieval predecessors had. But to insist, as later philosophers did, on choosing that option to the exclusion of the appeal to ends which is also available would be to give up the unification of inanimate, animate, and rational agency which the appeal makes possible, and to abjure the Aristotelian notion of nature altogether.

# [7]

## Nature and Counternature

Nature, as everyone knows, comes in two sizes: one size fits all, and extra large. The first, standardly glossed as 'essence' or 'quiddity', is that which defines each individual substance. In Aristotelian physics, it is the principle of motion and rest—provided that we understand by motion *natural* motion, and that the individual is correctly described. The second, glossed as 'world', 'universe', or 'cosmos', is the system of things which have natures in the first sense. It is a system, a unity rather than a mere aggregate, by virtue of efficient causal relations, like the relations between celestial powers and terrestrial souls which were studied in §5.4, but more significantly by virtue of the teleological relations studied in §6.

In the first part of this section I will examine the various senses of the words 'nature' and 'natural' acknowledged in the central texts, and certain of the contrasts in which the natural participates. Their complexity was a topos for philosophical lexicographers almost from the start. Out of that motley, two themes, pervasive in Aristotelianism, emerge: the theme, which we have already seen in several guises, of order; the theme of intrinsicness, which I interpret in terms of explanatory autonomy.

I then turn to the contrasts of 'natural' and 'supernatural', and of 'natural' and 'preternatural', 'contranatural,' 'violent'. I show that the first contrast is ambiguous: there are events that, though their causes lie outside nature, are "owed to nature." Nature enjoys a certain autonomy with respect to them. There are, on the other hand, truly miraculous events whose explanation lies, but for the fact that they occur to natural things, entirely outside nature.

Preternatural, contranatural, and violent changes all have causes within nature. The contrasts between them rest, as I will show, on the ways in which a thing can fail to be autonomous with respect to the changes that occur to it: in the contranatural, with respect to final causes; in the violent, with respect to efficient causes. The preternatural, that to which a thing's nature is indifferent, is violent but not contranatural. Although preternatural changes neutralize, as I will put it, the difference between "according to nature" and "against nature," and are therefore not subject to evaluation in terms of ends, the exception they yield to the directedness of natural change is always subsumed under a higher end.

In §7.2 I turn to the definition of nature, and its relations with essence and finality. Nature, as Aristotle defines it, comprises matter and form, a passive and an active principle. But in keeping with the importance of agency in the Aristotelian conception of natural change, the active principle predominates. One immediate consequence is that essence and definition depend primarily upon what a thing does, and not on what can be done to it. If we look forward to Cartesian physics, we see that because agency has no role in the nonhuman world, Descartes must find other ways to account for the obvious fact that the world includes distinct natural kinds. I then ask to what extent finality does and must enter into the definition of natural kinds.

In defining nature Aristotle contrasts it with, and yet thinks it in some ways analogous to, artifactual form. Turning to that analogy in §7.3, I show that, by virtue of their mode of production, which consists chiefly in the spatial rearrangement of their material, artifacts have, in the proper sense, no form—and, therefore, no nature. The absence of nature in artifacts, which for Descartes was the key to their use in the understanding of natural things, renders problematic the time-honored analogy between human art and divine creation, and the conception, which in §6 we saw was essential to giving ends a role in physical explanation, of natural agents as instruments of God. A reconciliation of that conception with the fundamental belief in genuine natural agency was indeed achieved. But for anyone who insisted, as Descartes did, on the absolute freedom of God's will, the underlying tension, and the alternative of employing the analogy to erase natural agency, would have been apparent.

## 7.1. The Uses of Nature

Virtually every philosopher who attempts to define 'nature' or its sister words in other languages remarks on the variety of its uses. Some attempt to regiment that variety by picking out one use as central. Some, like Boyle, urge its elimination, suggesting other terms thought to be less equivocal.[1] Yet the word has remained, its complexity undiminished. A century after

1. Boyle, "A Free Enquiry into the vulgarly received notion of nature" §2, *Papers* 179[f].

Boyle's "Free Enquiry," the *Encyclopédie* retains his eight senses and adds several more for good measure. Since then, things have only gotten worse for the fastidious. Raymond Williams calls 'nature' "perhaps the most complex word in the language," adding that a complete history of its uses "would be a history of a large part of human thought."[2]

Complexity, and no doubt complaints about complexity, have accompanied 'nature', *natura*, φύσις, and their kin all along. One need not search far for the reason. Nature was, as G. E. R. Lloyd has argued, not waiting to be discovered by Xenophanes or the authors of the Hippocratic corpus. It had to be invented, and once invented, it was "repeatedly contested" (Lloyd 1991:418,432).[3] Though the medical authors and their rivals are long gone, notions of the natural, and the numerous contrasts in which 'nature' participates, have proved to be of permanent utility. At the same time they have proved to be permanently contentious, a point nicely underlined in Lloyd's essay by his references to the environmental movement.

Aristotle's definition of φύσις in the second book of the *Physics* is, therefore, more the fresh gambit of a new hand at an old game than a deliverance of empirical inquiry or conceptual analysis. So too, though they have the air of well-rehearsed, and fixed, matches, are the *quæstiones* devoted to that definition in the commentaries. But before I turn to them, I want to survey the territory of the natural of which these texts are but the most learned province.

The diligent Goclenius, at the outset of his entry on *natura*, lists ten significations, virtually all of which can be found at the heart of one or another controversy.[4] Slightly systematized, they are:

(i) "the essence of each being, whether it be substance or accident,"

---

2. Williams 1985:219, 221. On the roots of the natural–supernatural distinction, see Lubac 1946, Auer 1964.

3. One mark of an invention is that not everyone has it. Rolf Schönberger remarks that "anyone who looks at the texts of Judaism must observe that a Semitic equivalent for the concept 'nature' is altogether absent" (1991:216).

4. Goclenius *Lexicon* s.v. 'Natura', p739. The *loci classici* for the various senses of 'nature' were Aristotle *Meta.* 5c4 (1014b17[ff]) and Boethius *Dua. nat.* p1341[ff]. With these precedents to justify it, listing the senses of 'nature' becomes a standard part of arguments on nature. The Coimbrans, saying there are many more, list just six (*In Phys.* 2c1q1a1, 1:203), collapsing Goclenius's senses (i) through (v) into their second, his (ix) and (x) into their third, omitting his (viii) (no doubt because it is not immediately relevant to physics), and adding two new ones: in the first place "the divine mind, maker and founder [*parens*] of all things," and in the last place "the generation of living things, which is called *nativitas*," which is included to motivate the etymology from *nascendo*. Abra de Raconis lists *natura* in the sense defined in the *Physics*, which corresponds to Goclenius's (ii) and (iv), from *natura* in the sense of *essentia*, which is applicable not only to corporeal substances but to God, mathematical objects, and accidents (Abra de Raconis *Phys.* 67). I am not sure what the significance of these differences is, or even whether they are significant—in some instances we may have simply variation for its own sake.

This is the site of an old contest indeed. The Coimbrans link the use of *natura* to denote essence with the theological dictum that "in the three divine persons there is one nature," which was made dogma by the Council of Chalcedon in 451.[5] In such contexts *natura* is allied with the more technical term *subsistentia* (which is found almost exclusively in metaphysics).[6]

    (ii)  "substance," as when Pliny says "Nature is the maker of things";
    (iii)  "both substance and its qualities, affections, operations of nature";
    (iv)  "the matter or form of a natural body separately";
    (v)  "the form of inanimate natural bodies only, as when nature is opposed to the soul, and distinguished from it."

These are the senses with which the definition given by Aristotle in the *Physics* is most directly concerned. The last of them, which is often the subject of a separate question, impinges immediately on the immortality of the soul, since all that is complex and entirely natural is subject to corruption.

    (vi)  "the quality and affection of a natural body, as when a magnet is said to attract iron by nature."

Among those qualities is the "vital heat" of animals. This was the sense of 'nature' appealed to in the controversies among Hippocratic, temple and folk medicine detailed by Lloyd.

    (vii)
        "natural efficient causes"—Goclenius's example is the slogan *Natura nihil facit frustra;*[7]
    (viii)  "the state of a man not reborn [through baptism], as opposed to the state of grace."[8]

This last sense is again the site of battles both old and new, grace being an instance of the *non*natural with urgent and fundamental interest for the

5. *In Phys.* 1:203; cf. Denzinger 1976, nos. 300–303 (Conc. Chalcedonensis, Actio 5). The fourth Lateran Council (1215), in a *Definitio contra Albigenses et Catharos*, uses the form cited by the Coimbrans: "tres quidem personæ, sed una essentia, substantia seu natura simplex omnino" (ib. no. 800, Conc. Lateranense IV, c1; cf. also no. 1330, Conc. Florentinum, Bulla "Cantate Domino" [1442]).

6. See Suárez *Disp.* 34§2–4, *Opera* 26:353–379, where the three terms *natura, suppositum,* and *subsistentia* are defined and distinguished. In seventeenth-century *cursus* one chapter in the otherwise much-reduced metaphysics segment is typically devoted to *subsistentia* and *natura.*

7. Similarly, the Coimbrans record the sense "natural causes, insofar as they act according to an ingrained propensity," and list a number of *dicta* of the same sort.

8. On this sense, see Auer 1964:333. There is an ambiguity in this use of nature as between the "pure" nature with which we—in the person of Adam—were created, and the debilitated "nature" that we, through Adam, acquired in Original Sin.

descendants of Adam and Eve.[9] I will return to the question of nature and grace below.

In each of these first eight significations, *natura* is said of an individual, whether as totality, essence, form, or accident. I have a nature (or rather several), you have a nature, the rocks and the earth they are made of have natures. On the other hand, *natura* applies to all the individuals of the class otherwise designated by terms like *mundus, universum,* and *cosmos:*

> (ix) "the world or παν, that is, the totality [*universitas*] of things that the world consists in," for which usage Goclenius's example is "In nature nothing is idle and ἄτακτον [sc., disorderly]" (cf. *Phys.* 8c1, 252a11);
>
> (x) "the order of natural things ordained by God."[10]

Nature in these inclusive senses I will denote by the capitalized word 'Nature'; for the other, individual senses, I will use 'nature'.

Although it may be futile to look for a unique core sense of *natura,* one key is to be found not in what the senses share but in their collocation. Although not all of them explicitly assert the existence of a cause or principle of a thing's actions, to speak of a thing's nature rather than its essence or substance is to bring the entity referred to into the field of cause or principle, as opposed, say, to that of existence or definition. More important, it is to imply that the cause is intrinsic.

This is clear in the physical uses of the word. The presupposition of the use of 'nature' is that among the actual or possible changes associated with each thing, some are recognized to belong to it, some not. Those changes, unlike extraneous changes, find their explanation in the thing itself. More precisely: though they may be elicited or triggered by another, the way they proceed is determined from within. In a box I confine an object; at some moment I open all sides of the box. The object, if predominantly earthen or watery, will fall; if predominantly airy or fiery, it will rise. The fall or rising are set off, allowed to manifest themselves, by something outside the object. But thereafter the object itself determines what will happen.

For each thing, then, there is a set of at least possible happenings for which a sufficient explanation of the course of those happenings (though not perhaps of their inception) can be given by referring only to its features. I will say that the thing has *explanatory autonomy,* or that it is *autonomous,* with

---

9. "By the term 'nature' we designate the natural faculty of free will to choose the good, and by the name 'grace' the assistance [*auxilium*] and gift infused in us by God for those tasks to whose undertaking and completion nature does not suffice" (Soto *De nat. & gratia* 1c2, 4v).

10. Both the Aristotelians and their successors take pains to assure their reader that by 'Natura' they do not mean anything like a person or an *anima mundi.* On Alain de Lille's personification of *Natura* in the sense of "cosmic principle," see Speer 1991:111ff and Chenu 1957, c.1.

respect to that set. The features that do the explaining, or some more fundamental feature that gives rise to them, are the thing's *nature;* and the autonomously explicable happenings are the *natural* happenings.

Typically those explanations will at some point advert to the *kind* that the thing belongs to. This animal stores food because, being an ant, it is provident; that animal flees because, being a hare, it is timid (see Coimbra *In Phys.* 2c9q3a2, 1:334). Goclenius's senses (i), (iv), and (vi) all concern nature in the sense of kind: a lodestone attracts iron by nature, which is to say, by *its* nature, by virtue of being the kind of thing it is. As Montgomery Furth notes, it is a datum for Aristotle that each individual, especially biological, is "permanently endowed with a highly definite specific nature [. . .] which it shares with other individuals" (Furth 1988:72; though Furth here refers only to animates, and among the Aristotelians the datum holds universally). What lodestone does by nature and what gold does are different; attraction is natural to lodestone, not to gold. Natural kinds and natural acts presuppose one another.

Explanatory autonomy, applied to the world, is a restricted version of what is sometimes called "naturalism."[11] It holds that there is a delineable subset of worldly goings-on whose causes, efficient or constitutive, are to be found in the world itself. There may be other happenings whose causes are only to be found in God. Aristotle argues that the origin of *motus* must be traced to a prime mover; but in all other respects he supposes the world to have explanatory autonomy. In §5.4 we saw that many Aristotelians were convinced that the generation of humans, and perhaps even of all animate beings, is not explicable by natural causes, while the generation of inanimates is explicable.

The world is self-sufficient only to a rather limited degree; but to some degree it must be if physics is to be more than descriptive. It is not trivial to hold that the world of sensible things enjoys a degree of explanatory autonomy. Anyone who denies that second causes are genuinely efficacious is thereby denying that the fundamental level of explanation is to be found within nature. One might admit that at another, less profound—call it phenomenological—level, some sort of explanation is to be had. But if one holds, as Malebranche did, that all genuine causation is mediated by God,

11. 'Naturalism' has, unfortunately, as many senses as 'nature'. Keith Hutchison defines it as the view that "virtually all events in the material world, and many of man's moral attitudes as well, can be accommodated by human reason without appeal to the supernatural." Newton, believing as he did that the solar system needed a nudge from God every so often to remain stable, was not a naturalist; Laplace, who did away with nudges, was (Hutchison 1983:297, 298). Naturalism, I think, is better stated as a claim about the causes of events and features in the sensible world than as a claim about the scope of reason. If it is true that physics can do without the hypothesis of a divine governor, that is because the causes of natural change are themselves all to be found in nature, or, more circumspectly, that the *differentiating* causes are found there.

then one is denying radically the sufficiency of any merely physical account of natural change. The Aristotelians, though they believed that all second causes require the concurrence of God, did not deny genuine efficacy to them. They were to that degree more naturalistic than some of their successors.[12]

The applicability of 'nature', whether to individuals or to the world, rests on there being changes that belong to a thing, on its explanatory autonomy with respect to those changes. Nature is, therefore, a relative concept, relative to the class of changes picked out as natural. To the objection that *natura* denotes something absolute but is defined as a relative term, Buridan responds that it is not, in fact, absolute: "the term 'nature' signifies matter and form and supposes them [*supponat pro eis*] but it does not signify them absolutely; rather [it signifies them] with respect to natural motion and rest" (Buridan *In Phys.* 2q4, 21fva). Even though 'nature' denotes ("supposes for") something that can be defined absolutely, it is itself a relative term, like 'element'.[13] It is, moreover, relative to natural motion and rest (and not, e.g., violent or miraculous motion). Contrary to what one might think from the order of definitions in the *Physics,* the concept of nature presupposes, and does not give rise to, a demarcation of natural change.[14]

That demarcation is pretheoretical, or at least prephysical. Its significance derives in part from what is not included: for the world as a whole, the *supernatural;* for individual things, the *preternatural.*[15]

The supernatural was, in Aristotelianism, no simpler a notion than the

12. Hutchison emphasizes the naturalism of "radical" Aristotelianism (cf. Steenberghen 1991:325–335), as opposed to the "mitigated" versions proposed by Thomas and most other Aristotelians. But because Hutchison underestimates the reinforcement that Thomism received in the latter part of the sixteenth century, at least in Catholic schools, and the well-honed ability of the central texts to achieve a prima facie consistent position on the role of the supernatural or to push problems offstage, he exaggerates the extent to which seventeenth-century natural philosophers found it needful or useful to oppose the bad Aristotle of Averroes and Pomponazzi.

13. Buridan *In Phys.* 2q4, 21va; cf. *In Meta.* 2c4, quoted in Schönberger 1991:217, n.1. Buridan's argument was not new. Toletus (*In Phys.* 2c1q1, *Opera* 4:46va) cites Albert (*In phys.* 2tr1c2) and Thomas in support of the claim that *natura* is a *relativum.*

14. The point can be reinforced by two observations about the Aristotelians' defense of Aristotle's definition, which will be discussed later. The first is that several authors argue that when Aristotle calls nature the principle of rest and motion, only *physical* motion is meant, and not that of which angels are capable, for example, the motion of the will in desire or of the intellect in contemplation (cf., e.g., Abra de Raconis *Phys.* 69). The second is that as Toletus, following Albert, points out, the *quies* referred to in the definition is *natural* rest, not the enforced rest of something prevented, say, from descending to its natural place.

15. As Schönberger puts it: "As a rule concepts are determined, insofar as they are demarcated vis à vis others, and thereby defined [i.e., in the etymological sense of having their boundaries fixed]. What is designated by natural being can only be determined if one sets it against others" (1991:217). Whether or not this structuralist homily suits all occasions, for a concept like nature, part of whose bite comes from its being *denied* of some things, understanding the various modes of *not* being natural is indeed essential to understanding the natural itself.

natural. Woven into it were at least three strands: the immediacy of divine action in certain events, the extraordinariness of the effects of such action, and its portentousness, its significance as a revelation of divine intentions. The interplay among them is complicated. To understand it, I turn for a moment to the supernatural event par excellence, the infusion of grace by God into his elect.[16]

Soto, explaining why the attainment of grace requires supernatural assistance, begins by delimiting those acts that are "owed to nature": "The general influx and concurrence of God, although it is a spontaneous and voluntary [act] of divine majesty [. . .] is enumerated among natural causes, because God has agreed to concur with the nature of things, so that [. . .] they may perform their natural operations, as if this were owed to nature [*naturæ debitus*]. [. . .] On that account, when the philosophers and theologians say that the sun by nature shines, the heavens naturally turn, fire burns by its own nature, and the like, they include under the name of 'nature' the general efficacy of God (Soto *De nat. & gratia* 1c2, 4v). If one considers God's mercy alone, one will regard all creation, this "machine of things" (5r), as a work of grace. But once God has created a certain form, he would frustrate his own intention were he not to concur in that thing's operations. In that sense one might say that God, though he remains of course free, has bound himself to Nature.[17]

In purely natural things, which in humans includes all that can be attained by reason, God concurs with whatever operations are needed to accomplish them (Soto c3, 7ff). The effects of concurrence, one should note, are determined entirely by the natures of the agents with which he concurs. It is not at all surprising that God should concur, for example, in the growth of our bodies, which is needed for our survival and reproduc-

---

16. Henri Lubac argues that the word *supernaturalis* achieves currency only with St. Thomas and is inscribed into dogma only at the time of the Council of Trent (Lubac 1946:327). *Supernaturalis* applies, on the one hand, to what exists in nature, but is beyond its power to produce; and, on the other, to what exists, as God does, outside of nature: the "miraculous" and the "transcendent." Grace is both, because it is neither implied in human nature nor obtainable through natural action, and because it "contains a reality that already [i.e., in this life] renders man a participant in divine nature" (401). My concern here is chiefly with the first sense; the second will turn up in §7.2. What binds the two together is that what lies beyond Nature's power can be effected only by something above Nature.

17. Suárez too argues that when God voluntarily concurs with second causes, his volition is "not entirely absolute [. . .], but is accommodated to the natures of things, and as if in accordance with a debt of just distribution [. . .] As also by God's will they have the power of acting, but not by a volition entirely superadded and as if by grace adjoined to the volition by which he willed their existence; instead the two volitions are connected according to a natural debt and connection of things themselves. So God's having willed that fire should heat, and water cool, is no bare volition but in its way owed [to them], on the supposition that he wanted to create such things" (Suárez *Disp.* 18§1¶10, *Opera* 25:396). It is not hard to see how someone who insisted on the absolute freedom of divine acts would be inclined to deny that one act of God could induce any sort of obligation to another. Cf. §7.3 and n.39 below.

tion. Since it is a factor common to all such events, concurrence can be omitted from their explanation. One doesn't need to say, once the theological setting is understood, that the paper caught fire because it was heated and because God concurred in its being heated. Nature enjoys, in short, explanatory autonomy with respect to what is owed to it.

But humans, unique among creatures, are endowed with free will; and Adam, our representative and type, chose, as the Bible says, to disobey God. That sin, for reasons that Soto expounds at length, has tainted us all, and cannot be remitted even if a person acts justly according to reason. Its "disorder and deformity," whose most obvious outward manifestation is death, has "remained as a habit, affixed to our nature" (c9, 30v), but not— since it resulted from a free act of Adam, and not from God's creation of the human form—genuinely natural.

Since our natural ends do not require that we be released from the disorder of original sin, it cannot be said that God, merely by creating us as we are, has bound himself to supply us the means to release ourselves. Acting in accordance with reason or following the dictates of natural justice can never suffice to achieve grace, although it may be a necessary condition. If, when a person is baptized, the disorder is removed, that is not simply by virtue of God's concurrence in whatever natural acts take place. God concurs, for example, in the movements of the priest. But those movements would not suffice to remove the disorder of Original Sin unless God added to his usual concurrence a special "assisting" act. That assisting act is *supernatural.*

Since second causes by themselves could never effect the remission of Original Sin, God himself must not merely concur in remission but himself effect it. One mark of the supernatural is the *immediacy* of divine action. But the fact that an act is brought about by God without the aid of second causes does not in itself suffice to distinguish supernatural acts. Suárez holds that God must "supplement" the power of celestial agents in the generation of animals (*Disp.* 18§2¶37, *Opera* 25:612; cf. §5.4 above). Even if God's act can still be described as concurrence, it is, nevertheless, of a different order than his concurrence in the generation of the elements. There natural causes suffice entirely, and concurrence factors out; but in the generation of animals, natural causes do not suffice. Nature is not autonomous with respect to those actions.

But generation is not usually thought of as supernatural. The reason, I suppose, is that even though in generation divine concurrence not only allows but supplements the action of natural causes, still that concurrence, just because it is necessary, is "owed" to them. It will arrive whenever the natural causes are set to act, just as weaker kinds of concurrence do. Even human generation, where the soul, because it is not educed from matter, must be immediately created, is not supernatural. We can explain even that creation in terms of the appropriate configuration of second causes, since God's action follows as a matter of course: it would be a miracle if a human soul were *not* created at the right moment. In that sense nature is autonomous even with respect to those actions.

Thus the supernatural, when it does not amount merely to God's action in those instances where he alone can produce a certain effect, is set apart also by *extraordinariness*. Thomas argues that on the one hand "every work that God alone can perform can be called miraculous." But in a stronger sense, the miraculous requires that the "form induced [by God's action] is beyond the natural power of the matter [in which it is induced]," and that there should be "something outside the usual and customary order of causes and effects" (Thomas *ST* 1pt2q113a10, *Opera* [Parma] 2:454). God concurs with, and even supplements, the powers he has implanted in created forms; but those are just the powers of which Aristotle said that they act always or for the most part. Even human generation, though it requires the creation of a soul, is part of the natural order. Since concurrence is the exercise of God's ordained power, we have what by the seventeenth century was becoming a standard identification: that which occurs according to *potentia ordinata* is the *ordinary*. It is not surprising that some philosophers should have begun speaking of *potentia ordinaria*.[18]

But that inclination, though understandable, is misleading. It is true that when God does what is owed to nature, the acts will not be surprising; their occurrence is, though not caused by natural things, explicable in terms of their powers and deficiencies. But events supernatural in the stronger sense may also be perfectly ordinary and are entirely to be expected. Priests and parents did not doubt that, if the ceremony were performed correctly, the taint of Original Sin would be washed away by baptism.

Immediacy and extraordinariness, then, are interdependent but distinct strands in the notion of the supernatural. On the third, portentousness, I can be brief: though it was much discussed in theology and in popular literature, it does not figure greatly in the central texts. The supernatural, in its guise of the extraordinary, was an invitation to exegesis. Not just monsters but prodigies of all sorts—comets, meteors, earthquakes, floods, frogs pouring from the sky—were examined for what they might reveal both about the future and about the intentions and sentiments of God.[19] The central texts,

18. See Oakley 1984:57[ff] on the shift from *ordinata* to *ordinary*.

19. On the interpretation of prodigies, see, e.g., the views of Cornelius Gemma (*De naturæ divinis Characterismis . . . .*, 1575) analyzed by Céard. Gemma proposed a new science, an *ars cosmocritica*, which would be to the disorders of nature what symptomatology or semiology was to the disorders of humans (Céard 1977:370[ff]). It is to be noted that events with perfectly respectable natural causes, like eclipses, were also held to be portents: for some, but not all, authors on marvels, the availability of a natural explanation did not close off the possibility, or

though they do not explicitly discourage the exegesis of prodigies, do not encourage it either. Their steadily naturalistic standpoint may well have been intended to inoculate students against too credulous a reception of popular practices of divination and against ascribing strange events to the actions of demons (whose existence they do not doubt).[20]

The contrast of the natural and the supernatural cannot be reduced to a straightforward opposition between what is caused by natural things and what is caused by God alone. Nor can it be reduced to the opposition of the usual and the rare. Still less was it what it became in the seventeenth century, and perhaps is even now: a simple contrast between what is according to the laws of nature, and therefore explicable naturally, and what violates those laws, and is therefore, from the standpoint of physics, essentially surd. A long road must yet be traveled to reach Hume.

Unlike the supernatural, the contranatural is, in its definition, straightforwardly derivative from the natural. Natural change is according to the nature of a thing, contranatural change "against its nature and inclination" (Toletus *In Phys.* 4c8text57, *Opera* 4:126rb). What is not so clear is just which changes are against a thing's nature or inclination. Since the immediate context of Aristotle's longest discussion of contranatural change in the *Physics* is a classification of the ways in which changes can be contrary to one another, one might expect that contranatural change just is change contrary to one or another of the changes that follow from a thing's nature.

But two problems, one raised by Aristotle himself, the other by his commentators, reveal an unexpected complexity.

1. Since nothing aims at its own nonexistence, but instead tends to preserve itself, it would seem that corruption is always contranatural. In particular, death is against nature. But Aristotle himself holds that although generation and corruption are contraries, neither is natural or contranatural. In fact corruption, since it is the inevitable outcome of the natural process of aging, is natural. Then he adds: "But if what is violent [*quod vi est*, i.e., what exists by force] is præternatural [translating Aristotle's παρὰ φύσιν], a corruption will also be contrary to a corruption, namely [a corruption] which is natural [will be contrary] to one which is violent" (*Phys.* 5c6, 230a29[ff]; Coimbra *In Phys.* 5c6text57, 2:157). The violent, *quod vi est*, was introduced earlier in his arguments against the void: "First of all, every *motus* is effected

---

the need, for interpretation (see 488; for the contrary view, see 340, 443, 451, 473[ff]).

20. Daston and Park argue that "for the educated layman [. . .] and even more for the professional scientist of 1700, the religious associations of monsters were merely another manifestation of popular ignorance and superstition" (1981:24). The central texts, which reflect of course the view of the well educated in the late sixteenth and early seventeenth centuries, exhibit an earlier stage of that development: their view is that monsters have entirely natural causes, though they can, perhaps, *also* be signs.

either by force or by nature. But it is necessary, if there should be violent *motus,* that there is also natural *motus,* since in fact the violent is præter-natural, and what is præternatural is posterior to the natural" (*Phys.* 4c8, 215a1$^f$; Coimbra 4c8text67, 2:52). Stated only conditionally in *Phys.* 5, but asserted in *Phys.* 4, is that violence is preternatural, "alongside" nature. (I set aside for later the claim that the natural is prior to the violent.) Now the sense of 'violent' is, as Thomas puts it, that a violent act is "from a extrinsic principle when the patient does not contribute any force [*a principio extrin-seco vim passo non conferente*]" (Thomas *In Phys.* 8lect7, *Opera* [Leonina] 2:388). The patient, in other words, "does not have an inclination" to the act (John of St. Thomas *Nat. phil.* 1q9a4, *Cursus* 2:188a).

The characterization of violence is ambiguous, as John of St. Thomas notes, between a "negative" interpretation of 'does not have an inclination', according to which the patient neither helps nor opposes the extrinsic principle, and a "positive" interpretation, according to which it does oppose the extrinsic principle. If the patient neither helps nor opposes, then the violence is preternatural, properly speaking. If it opposes, then the violence is contranatural. Some Aristotelians doubted that there were any preter-natural actions: that is the second problem, which I will come to shortly.

Now back to corruption. It seems clear that corruption is not just preter-natural but contranatural. How could a thing not oppose its own destruc-tion? And yet Aristotle holds that some corruptions are natural, others violent and therefore contranatural.

Recall that a class of changes was earlier said to be natural for an agent if it enjoyed explanatory autonomy with respect to them. Violence is evidently nonnatural on that definition, since the force behind it is extrinsic. So violent corruption will be nonnatural, and corruption that starts from within will be natural by contrast.

Yet it is difficult to understand, in the face of the general principle that things tend to preserve themselves, how corruption could be natural. I think we can make explanatory autonomy do the work here too: what is con-tranatural about corruption, whether it comes from within or not, is that it goes against the good of the individual. An individual can enjoy explanatory autonomy with respect not only to a certain class of changes but with respect to a certain kind of cause. Violent acts are those for which a thing fails to be autonomous with respect to efficient causation (I take 'vis' to denote the efficient cause alone). "Contrafinal" acts are those for which a thing fails to be autonomous with respect to final causation. Corruption from within—which in aging humans is brought about by the gradual loss of vital heat—has no final cause within the organism. One could tell a story about how individuals must die to make way for their descendants, or some larger story about the fallenness of the material world; but whatever story one tells it will not, and cannot, according to the principle that nothing intends its own demise, include the ends of the individuals themselves.

Earlier I said that Nature was autonomous not only with respect to those events whose efficient causes lie within it, but also with respect to those events whose causes lie outside it, but which are "owed" to it. Those acts are owed to natural things that, given that they exist and have the natures they have, and thus the characteristic ends they have, must be performed if they are to fulfill those ends. (Descartes's use of the thesis that God cannot be a deceiver to vindicate our clear and distinct perceptual ideas of corporeal substance is one instance of an act owed to Nature.) Such acts are, with respect not to their efficient causes, but with respect to their final causes, entirely natural. Corruption, on the other hand, even if it serves a collective or cosmic end, is always contranatural.

2. The remaining problem has already been stated: are there, in addition to natural and contranatural changes, preternatural changes as well? The problem comes up in questions on the motions of the heavens. Aristotle argues that because no violent motion can be perpetual, the motion of the heavens, which is perpetual, must be natural.[21] It also seems to have come up in discussions of divine action, and especially of the punishments of demons in hell (John of St. Thomas *Nat. phil.* 1q9a4, *Cursus* 2:189b). Beyond those specialized questions is another: what scope was there already in Aristotelianism for a concept of change, and thus of *ens mobile*, the object of physics, that would neutralize the otherwise all-embracing framework of activity, nature, and finality? The preternatural—that to which natures are indifferent—is one part of this new physics, and its corresponding *potentia*, the *potentia neutra* or *obedientialis* of matter to the manipulations of art is another. Though there were doubtless a number of avenues to the new physics of the seventeenth century, including the mechanics based in the pseudo-Aristotelian *Problemata*, and the middle sciences, whose practice was indfferent to many of the distinctions elsewhere essential to Aristotelianism, it is significant that without departing from that tradition one could already find, in its interstices, some elements of what was to come.

That the preternatural is a kind of neutralizer of key contrasts is apparent when one considers the chief objection raised against it. Zabarella reports that his Paduan predecessor, Marcantonio Zimara, in his *Theoremata* "entirely denies" that there is any middle ground between natural and violent motion, where 'violent' is understood to mean 'contranatural' (Zabarella *De*

---

21. Zabarella *De motu ignis* 2, *De rebus nat.* 292A–D; Aristotle *De cælo* 2text17, 269b6. Since the question of the existence of preternatural movements does not obviously arise in the *Physics* (though an able commentator would have no trouble finding pretexts for raising it), and since it is therefore absent from most *Physics* commentaries and *cursus*, a detailed discussion is outside the scope of this work. My purpose here is to lay out the problem raised by the preternatural and to lay the ground for referring to it in the Part II of this work.

*motu ignis* 4, *De rebus nat.* 293C^ff). Aside from citing texts where Aristotle treats the division as exhaustive, Zimara points out that the argument of *De cælo* mentioned above is invalid if there is a third alternative. One can infer from the fact that the motions of the heavens are not violent that they are natural only if natural and violent cover all cases.

John of St. Thomas reports that one of his colleagues at Alcala, a certain Ioannes Ramirez, argues that every *motus* is either natural or violent in the sense of not being from an intrinsic principle. That in itself does not, as we have already seen, preclude the possibility of change which is neither natural nor contranatural. But Ramirez goes on to argue that all bodies have an internal propensity to move toward or away from the center of the universe; if they "recede" from that place, they are being moved violently. The sense in which "recession" is to be taken is not clear: but if it includes any motion that is not toward the natural place of a thing, then every motion will be either natural or contranatural.

Both Zabarella and John of St. Thomas, who would, I think, have agreed on little else, argue that there can be preternatural change. Zabarella offers a variety of arguments, of which the most interesting here is this: "[The motion of fire around the heavens] is not natural if one takes the natural to be that which is brought about by an internal principle; but it is not so apart from nature [*præter naturam*] that it opposes or injures nature; instead it nourishes nature and preserves it. It is rather, similar to both [natural and contranatural motion]" (ib. 8, 300E). Zabarella is contrasting *motus* whose efficient cause is not contained in their subject with *motus* whose final cause is not contained in the subject, which is the contrast I drew a moment ago in explaining how dying of old age could be at once natural and contranatural.

John of St. Thomas, quoting his namesake (Thomas *In de Cælo* 1lect4, *Opera* [Leonina] 3:16) makes a similar point (ib. 195b). The circulation of fire and air in the heavens is not caused by an intrinsic principle, but neither is it contranatural; it is, instead, "in a certain way above nature, because such a *motus* inheres in them through the impression of superior bodies, whose *motus* fire and air follow according to perfect circulation, while water [follows it] according to imperfect circulation"—whence the tides. John of St. Thomas adds that it is not against nature because moving in a circle whose center is the center of the universe is neither moving toward nor away from one's natural place, whatever that may be.

Circular motion, then, is preternatural because it neutralizes the contrariety of natural places. John of St. Thomas then notes that on this account, a revolving wheel might be said to move preternaturally, and that since only contranatural motions are resisted by a thing's nature, and thus subject to quick decay, the revolution of the wheel, "once impressed [. . .] would never stop, unless it were held back from outside" (196a). He answers that the wheel would (like any projectile) be moved by an impressed "impulse" that because it is extraneous to its subject will eventually vanish. The forced choice between natural and contranatural is rejected, only to be replaced by an equally forced choice—so it would seem—between extraneous and native or connate qualities.

The neutralization of contrasts is only temporary, *aufgehoben* by a new contrast. A similar pattern is exhibited in John of St. Thomas's arguments on behalf of the claim that when God acts as the "principle and root of every *motus*" he cannot be acting violently (the punishment of demons is another story). It is evident in experience, first of all, that some *motus* are not violent though they result from an extrinsic principle. Such are the tides (whose motion we have already seen described as preternatural) and the illumination of the air by the sun. Contrary to what Aristotle thought, not every such motion is violent (188b).

John of St. Thomas adds to that a posteriori argument an a priori reason. The coordination, he says, of superior and inferior causes is itself natural. Indeed an inferior agent inclines yet more to a *motus* coming to it from a superior agent than to its own *motus,* "because it inclines more to the conservation of the whole than to its own." If, as everyone would agree, its inclination to its own conservation is natural, "all the more so will its inclination to the conservation of the whole be natural." But the conservation of the whole depends on the connection of the parts, so that if an inferior part disobeyed its betters it would threaten the whole.

We saw a particular case of what one might call the communal spirit of natural things, and their submission to higher ends than their own, in the Coimbrans' explanation of the *horror vacui* (§6.2; cf. John of St. Thomas's own discussion in the present question, 190b, and his vacuum question, 1q17a1, 362b). Bits of matter will move, even against their natural inclinations, and even at risk to themselves, to ward off any threat to their spatial continuity.

It is tempting to read off a politics from such arguments. I will content myself, however, with pointing out the coincidence of two orders. The first is the order of power: God's power infinite and unparalleled, that of the angels greater than any material thing, ours greater than the animals', and so forth. The second is the order of perfections: the part is subordinated to the whole because the perfection of the whole is superior to that of its parts. The two orders are linked: the more powerful an agent is, the more fully it takes the perfection of the whole, its own contribution to which is proportionate to its power, as its end.

Here as before the preternatural, as well as the locally contranatural, are *aufgehoben* through the intervention of a new contrast, this time between part and whole. There is no place in Aristotelianism for a change that is simply, and without further explanation, "outside" nature. Even if, more-over, a change is preternatural with respect to an individual, it will not be preternatural if viewed with respect to the ends of something else, whether they be those of the substantial form of a thing—the ground upon which the extraneous and the innate are distinguished—or those of a whole or a higher agent to whose ends the individual's own ends are subordinated. It will therefore, in a broader sense, have an end, and will therefore terminate when that end is reached. The preternatural, like the contranatural, is therefore always transient.

## 7.2. Individual Natures

Aristotle, having in the first book of the *Physics* settled on matter, form, and privation as the principles of natural things, proceeds in the second to define 'nature' itself. The nature so defined is the nature of individual things. Though the *Physics* acknowledges the use of 'nature' to denote the system of natural things, it does not explicitly define that sense of the word.[22] I will first examine what the commentaries have to say about the definition itself, and certain issues that arise immediately from it. These include alternative concepts of nature rejected in favor of Aristotle's, the differences among animates, inanimates, and artifacts, and the applicability of the notion to the soul. Following that I will enlarge the scope of the discussion to consider the relation between nature and essence, and the finality inherent in the classification of natural kinds.

    1. *The definition of nature.* Expositions and questions on Aristotle's defini-tion of nature can be seen to have two purposes. One is to make the concept precise through successive parings-down. The *terminus a quo* is the genus 'principle', the *terminus ad quem* matter and form, with primacy accorded to form. The other is to establish among all things, natural or not, a hierarchy, and to ascertain the position of animates and of the human in that hier-archy. Fulfilling those two purposes yields the object of physics—*ens natu-rale,* according to the dominant view—and gives it a position in the order of knowable things, below divinity and above art.

    In the usual translation, Aristotle's definition reads: "Natura motionis, & quietis, eius rei, in qua est, primum, & per se, & non ex accidenti, prin-cipium quoddam sit, & caussa [Nature is a certain principle and cause of movement and rest in which it exists first, and per se, not accidentally]"

---

22. See, e.g., *Phys.* 8c1, 252a13 (Coimbra *In Phys.* 8c2, 2:271): "For nature is the reason of the order of all things." It is actually not easy to find an occurrence of φύσις in the *Physics* which must be interpreted in this sense; nor is it very common elsewhere, although in many instances it is not easy, perhaps not possible, to decide (Bonitz 1870, s.v. φύσις).

(Coimbra *In Phys.* 2c1 explanatio, 1:199).[23] Aristotle says 'principle *and* cause', the Coimbrans note, so as to exclude privation, which is a principle but not a cause. "Nothing else, except matter and form, is acknowledged by Aristotle to be (a) nature" (2c1q2a1, 1:205). Nature is called the principle of motion and rest so as to exclude mathematical forms, which are not causes of motion, and nonphysical causes of motion—separated substances like God and the angels.

Such causes are excluded by the phrase *in qua est,* which signifies that nature is an *intrinsic* principle. That there are intrinsic principles in Nature is, as we have seen, no idle assertion. One alternative, briefly noted by the Coimbrans, is the view that nature is a *universal* principle, manifesting itself, to be sure, in individuals, but properly belonging to none. Albert writes that certain philosophers hold that "absolute" nature is a "form and power issuing from the first cause through the motion of the [outermost] sphere, which, after it comes forth, is diffused into all natural things and becomes in them the principle of movement and stasis."[24] Such theories, which were not without their analogues among Renaissance philosophers, did not strike the Aristotelians as serious competitors, because they effectively denied efficacy to second causes.[25] But they are worth mentioning, if only because the combination of a denial of individual natures, especially active, and an affirmation of a universal force proceeding from God, finds an echo in Descartes, where the quantity of motion imparted to matter by God at the outset is the only source of activity in the world.

The next part of the definition, *per se & non ex accidenti,* serves to exclude both art, in the sense of skill, and its products. Since I discuss artifacts in particular in §7.3 below, I will not dwell on them here. Medicine is a skill and a cause of the actions of those who practice it. But medicine in the

---

23. As usual, the cited text varies slightly from one commentary to another, or even within a commentary. The version that the Coimbrans analyze in their *quæstio* on the correctness of the definition is "Natura est principium & caussa, ut id moveatur & quiescat, in quo primò, per se, & non ex accidente inest" (2c1q2, 1:205), a close paraphrase of the translation into better Latin. Their Greek text coincides with the Loeb edition.

24. *In phys.* 2tr1c5, *Opera* (Inst.) 4.1:83, see also 49. As authorities for the view, which he refutes, Albert cites the Pythagoreans, Plato, and Hermes Trismegistus. His immediate source, however, which is also mentioned by the Coimbrans, is Avicenna (*Suff.* 1c7, 17vb). Toletus writes that for Plato, "nature is life, or a certain force suffused through bodies, their governor and director," a "substance proceeding from God, and divided amongst lower things," yet in itself one substance (*In Phys.* 2c1q1, *Opera* 4:46ra).

25. Among others were Francesco Patrizi da Cherso (Ingegno 1988:257) and Robert Fludd, with whom Mersenne engaged in an extended controversy. Mersenne reports that according to Fludd, "God, while he is a certain light diffused through the whole world, does not enter anything unless he has first assumed like a cloak an ethereal spirit [. . .] God forms a composite with this ethereal spirit. He resides with it especially in the Sun, from which he radiates outward to generate and vivify all things. In that way God [is] the form of all things and does all things, so that second causes do nothing per se" (*À de Baugy* 26 avr. 1630, *Corr.* 2:441).

doctor who treats herself is not her nature or part of her nature. She only happens to be both the subject in which the skill resides and that which is changed through the exercise of the skill. But when the soul acts on the body, or when active powers emanate from substantial form, the object of such actions cannot but be the body or the substance itself in which the soul or the form resides.

The relation between an artifact qua artifact and the natural changes that happen to it is similar. Artifacts do undergo change. Sculptures erode, paintings fade, saws get rusty. But it is not, or so Aristotle argues, by virtue of being a *saw* that the saw rusts. It is by virtue of being made from iron. A saw only happens to be that which rusts; iron rusts by nature. The principle of that change does not reside in the saw, but in the stuff it is made from, its *materia ex quo*.

So far we have it that natures are intrinsic per se causes of natural change. The goal, as I said, is to show that matter and form alone are nature in the truest sense. Hence the word *primo* in the definition. The Coimbrans, like other commentators, construe it as an adverb modifying *in qua est*. A nature must exist in its subject *primo*, which the Coimbrans gloss as "not by another, or so that nothing is in [the thing] beforehand." We have seen that use of primo in the definition of substantial form, the first form joined with matter, or at least the first that yields a complete substance. Here the standard doctrine that substantial form and prime matter are alone "first" in their respective constitutive roles yields the conclusion that they alone are natures.

But which is *the* nature of a thing? One way to raise the issue is to ask whether 'nature' applies to active principles, passive principles, or both. In the commentaries, that query was prompted by an oddity in Aristotle's definition. Nature is the principle not of *moving* but of *being moved*. In the Greek, it is the ἀρχή τοῦ κινεῖσθαι, with κινεῖθαι a middle/passive infinitive; in the usual Latin translation, κινεῖσθαι becomes *motio*, glossed in the Coimbrans' paraphrase by *ut id moveatur*, a passive form.[26] The definition looks incomplete, since form, which is active, is surely nature if anything is. Aristotle himself says it is nature more than matter (*Phys.* 193b7). Why, then, the passive verb?

This is no mere quibble. As it stands, the definition suits matter rather than form. But the Coimbrans note that prime matter, being indifferent to

26. On the forms of κινέω, see Waterlow 1982:161ff. She argues that κινεῖθαι "is passive as to its grammatical form, but not necessarily passive as to its meaning." It does not, in other words, presuppose an agent. Nature could be a principle of intransitive change, an internal principle of the procession of phenomena in a thing—not unlike the essence of a Leibnizian individual substance. But since (see §2.2) for Aristotle every intransitive change is at the same time a transitive change, κινεῖθαι should here be taken—as all the Aristotelians take it—to be passive.

all forms, is nature with respect not only to natural motion but to violent motion, "and so there is no *motus* which is not natural with respect to matter" (Coimbra *In Phys.* 2c1q2a3, 1:208–209; cf. Abra de Raconis *Phys.* 75). If nature were matter alone, all change would be "natural"—a good outcome for Descartes, but not so good for an Aristotelian, since then directedness and naturalness would be radically distinct.

The Coimbrans defend the definition by making a threefold distinction. Some things, they say, are above nature: God, the angels, the *intelligentia*. Though angels are capable of local motion, they can only be moved from outside; they do not move themselves. Some things are below nature: artifacts, "which are so ignoble that they have neither a principle of being moved themselves or of moving others" (1:208). The middle place is occupied by "physical things, which are neither so abject as not to lay claim to a principle of any motion, nor so perfect that they cannot be moved." They can move others, for example, in generation, and themselves, locally or by growth. In that they resemble their betters. But they also have a principle of being moved, and in that they resemble neither their betters nor their inferiors. That is why Aristotle mentioned only a principle of being moved in his definition, and not of principles of moving. He did not intend to deny that nature includes active principles. He only intended, as the immediate context of the definition shows, to distinguish natural things from artifacts.

There is, however, a complication. If all natural things, animate and inanimate, have an active principle as well as a passive, then how do animates differ from inanimates? Aristotle, after all, says that the *differentia* of the animate is that it moves itself; but if heaviness, say, is construed as an active power, then heavy things, animate or not, move themselves. One answer, that of Averroes, was to suppose that animate things alone have the power not only to act, but to initiate their actions (Toletus *In Phys.* 2c1q2, *Opera* 4:47rb). Against that are not only a number of Aristotelian texts, but the basic principle that nothing acts on itself. Animates can move themselves, as Aristotle says elsewhere, only because they are inhomogeneous. Under scrutiny their apparent self-movement turns out to be the movement of one part by another (47va). Inanimate things, being extended, have potential parts. But the "bits" of one individual are not different enough to satisfy the condition of the basic principle. So they cannot move themselves even in the somewhat Pickwickian sense in which an animal can.

Toletus concludes that Averroes's stratagem is superfluous: "Neither the form nor the matter of inanimates is an active principle of their movement, but only a passive [principle], and in that respect such forms differ from the forms of living things, which are active principles of their own movement: so that when a stone is borne downward, or fire upward, they are not actively moved either by their form nor by their matter" (47vb). That solution does not contradict the claims made earlier (cf., e.g., §4.3, and Toletus *In Phys.* 2c6q10, *Opera* 4:61ra) that heat, cold, and various other qualities are active powers. They are, but only as extrinsic causes (cf. also Coimbra *In Phys.* 6c1q1a2, 2:224$^{ff}$). The heat of fire does not act on the fire itself (which is, in any case, as hot as it can be). It acts only on other things. Weight is a more difficult case, since it follows from the substantial form of a thing, and appears only to act on the thing itself. Without entering into detail, I will note that in the standard treatment of falling bodies, the generator of a heavy body is the principal efficient cause of falling, and the heaviness of a body only an instrumental cause (Toletus 4:48ra). The basic principle is preserved, and with it the denial of active powers to inanimate things.

We arrive at a fourfold classification of substances according to the presence or absence of active and passive principles, as in Figure 7. The classification is at the same time a ranking. Artifacts, at the lowest, the most "ignoble" level, have no nature at all: "no artifact, as such, has in itself a determinate and certain principle of its own motion, either active or passive." Since a statue, say, is clearly not animate, it has no active principle. It has no passive principle, either, qua statue. It tarnishes or rusts because of its material; and since the same kind of statue can be made out of different kinds of matter, some of which may rust and some not, proneness to rust belongs to the material it happens to contain, not to the statue itself (Coimbra *In Phys.* 2c1q2a2, 1:211). At the next rank is the inanimate, which, having only a passive principle, is least among natural things. The animate occupies a middle station. It is superior to the inanimate by virtue of having an active principle of *motus,* but it is inferior to separated substances by virtue of having also a passive principle. Passivity, in short, is ignoble, but lacking power altogether yet more ignoble.

What is left is to ascertain the place of the human soul in the hierarchy. That is no surprise. The soul is in part natural, in part "transnatural," as the Coimbrans put it (2c9q4a2, 1:213). It is indeed, like other animals' souls, the principle of vital functions, like growth, nutrition, and sensation. But other actions, like grasping intelligible things by divine illumination, which "befit [the soul] when it is separated from the body and outside the dregs of matter," are tacitly excluded from the definition of nature. The soul, to the extent that it is the principle of those actions, is not a nature. Some, notably Avicenna, had argued that it is not a nature at all. But true philosophy and the Church agree that the soul—all of it, and not just its lower parts—is the substantial form of the body, and therefore its nature (2c9q4a1, 1:212). So long, at least, as it remains joined with the body, it is, therefore, part of the natural world.

2. *Nature and essence.* Virtually all lists of the senses of *natura* include *essentia* and *quidditas* among them. The relation of nature to the other two is

| Substance | | Principles | |
| --- | --- | --- | --- |
| | | Active | Passive |
| Supra naturam | God, angels | + | − |
| Habens naturam | Animate | + | + |
| | Inanimate | − | + |
| Infra naturam | Artifacts | − | − |

Fig. 7. Classification of substances

succinctly stated by the Coimbrans: "essence implicates [*importat*] an order-ing toward the existence of a thing, quiddity toward its definition, and nature toward its operations."[27] But since nature includes both form and matter, and since it was a disputed question whether form alone, or both form and matter, were comprised in the essence of material things, it is worth examining the question more carefully. After doing so, I will argue that *natura,* which corresponds reasonably well with what philosophers now call the 'natural kind' of a thing, must be defined in relation to ends. That is not surprising, but it is worth bringing out because it helps set the terms for the Cartesian project of a physics without finality: such a physics will have to persuade its audience either that natural kinds are not needed in explana-tion (as some recent philosophers have argued) or that they can be simu-lated by other means.

The Coimbrans note that some of Aristotle's predecessors, who were not fortunate enough to have discovered form, held that essence consisted in matter alone (*In Phys.* 1c9q5a1, 1:162). To my knowledge no one within the Aristotelian tradition agreed with them. The entire question, therefore, turns on whether matter should be included in essence. If it is, then essence and nature coincide; if not, then they will diverge, and the role of nature, as opposed to form alone, in the classification of substances will be correspon-dingly diminished.

The question was an old one. It had already drawn the attention of the early Greek commentators Themistius and Alexander of Aphrodisia. Be-cause it pertains to doctrine on the Incarnation, it also shows up, in various forms, in patristic sources. By the time the central texts were being written

27. *In Phys.* 2c1q1a1, 1:203; cf. Suárez *Disp.* 15§11¶4, *Opera* 25:558; 2§4¶6, 25:89.

the arguments had, as usually happens when a question is both well aged and theologically sensitive, become quite complex.[28]

Nevertheless they fall into one of two patterns. The first is this: matter can change while essence remains fixed; if matter were included in essence, that would be impossible, so it is not. The most obvious instance is the growth and nutrition of animals: "if matter was contained in the essence of Socrates, for example, that singular matter would be of his essence, which is proved to be false. First, because the matter of Socrates is perpetually dispersed by the action of heat as it consumes moisture: but the essence of a thing must be fixed and stable" (Coimbra *In Phys.* 1c9q5a1, 1:162). We don't consider Socrates to have become a different kind of individual just because he loses a pound or two. Yet losing and gaining weight are changes in matter, or so it would seem. Another instance, less obvious to us, is resurrection: "if that singular matter were of the essence of this man [Socrates], the same man could not be called back to life when bodies are resurrected, unless he assumed the same body; but that seems false, since the same matter, at least partially, can belong to many [people], as is clear among those who feast on human flesh." The resurrected Socrates will have the same essence as the Socrates of Athens only if their bodies are identical. But that will be true also of Caesar, say, and Charlemagne, whose matter, we may suppose, incorporates some bits of Socrates' body. If the essences of those three individuals include their matter, then their resurrected bodies will have to contain the same parts they did on earth, including the bits that have successively been incorporated into all of them. The picture is unsettling, to say the least.

A third, extreme case that would likely not occur to us is this: "A worm is sometimes generated out of the species of the Eucharist. Such a worm is certainly a natural being. But it is not constituted from matter, since the matter of the bread vanishes after consecration, and thus no matter is the substrate of [*subsit*] the generation of the worm. Therefore not every natural being is constituted from matter, from which it follows that matter does not pertain to the essence of a natural being" (ib.). If a thing can have worm-essence without having any matter, then clearly worm-essence does not include even the property of having some matter or other, let alone a particular matter.

The second pattern of argument against including matter in nature proves the converse: essence can change while matter remains fixed. So-crates the man and Socrates the cadaver share the same matter. But they differ in essence: one is rational, the other not. More generally, matter is all

28. Defenders of the exclusion of matter from essence include Averroes (*In Meta.* 7comm21, 34, arguing against Avicenna), John of Jandun, Nifo (*Disp.* 7disp13); Soncinas, rather surprisingly, does not take a stand, offering arguments and refutations for both sides (*Q. meta.* 7q26, p157r–v).

of one species (§4.1). People and horses, since they are different species, must differ in form alone.[29]

The Coimbrans stage their response to these arguments in three *conclusiones:* that matter is a part of the essence of natural things; that "singular" matter is included in the essence of singular things; that the included singular matter is not "indeterminate and vague," but determinate and identical throughout the existence of the thing.

What is theologically at stake in the question is that the human being, including the incarnated Christ, should be in the fullest sense an *ens naturale*, a natural being. Suppose that matter were not included in the essence of human beings. Then "the soul would be an *unum per se*, and complete in itself," and the union of soul and body, contrary to received doctrine, would not be (see §5.2). It would follow that since the soul is immortal, and the union only an *ens per accidens*, no substance is destroyed in death, and human beings are in every way immortal. If, furthermore, the essence of the human does not include specific matter, the essence of an individual will not include its singular matter, where 'singular matter' denotes the individual bit of stuff informed by individual form. Were this so, then 'human', which designates an essence consisting of both matter and form, could not truly be affirmed of 'Socrates' (*In Phys.* 1c9q5a2, 1:163).

To reach their final conclusion, the Coimbrans start by defining intrinsic and adventitious unity. A thing has intrinsic unity simply by virtue of existing: as the commonplace puts it, 'being' and 'one' are mutually convertible. Adventitious unity is conferred on a plurality by something else, roughly in the way that a certain unity is conferred on a plurality of causes by an end toward which they all tend. The first is always the same, the second is "various and fleeting" (164). Some philosophers, notably Peter of Lombard, overlooking that difference, found it necessary to hold that the matter that a person receives from his parents persists in him all his life. But the Coimbrans believe that a unity of "continuation" suffices, by which they mean that although parts are continuously being destroyed and replaced, at no one time is Socrates' matter replaced in toto.[30]

---

29. Since it is not clear why Socrates the man and Socrates the cadaver cannot share part of their essence, namely, their matter, the argument is incomplete as it stands. What needs to be filled in is that individual form corresponds to individual or specific difference and that 'essence' is to be taken as that by which a thing is first constituted as an actual individual (see Suárez *Disp.* 2§4¶6, *Opera* 25:89). Socrates' corporeality and his animality do not distinguish him from other things; nor does the species of his matter; only his individual form does so. That individual form, moreover, is the first form (here the working assumption seems to be the Scotist theory of a plurality of forms; cf. §4.1) by which the individual Socrates is an actual existing thing. It alone, therefore, is his essence.

30. These arguments do not suffice to handle the Eucharistic worm. For that the Coimbrans, following Thomas (*ST* 3q77a5), have recourse to a special act of God by which at the very moment the worm is generated either new matter is created, or existing matter translated from elsewhere, or the matter of the host re-produced (ib. a3, 166).

3. *Nature and ends.* Nature, then, includes both form and matter. But they are not of equal weight. Aristotle argues that "that by which a natural being is *in actu* is nature more" than matter (*Phys.* 2c1, 193b6; Coimbra *In Phys.* 2c1 *explanatio*, 1:202). The underlying reason is that *actus* is the specification of *potentia* (§3.1), and thus closer to defining a thing than *potentia*. As Aristotle says, a thing is *said* when it is *in actu*: we normally say 'the dog' or 'the statue', not 'the body' or 'the bronze': 'said' in such usages means 'designated according to its definition'. Definitions are perspicuously exhibited essences. But in individual substances, form is the *actus*, and therefore the specification, of matter; and so form is nature—in its guise of essence—more than matter.

The relation of nature to ends has, therefore, two aspects: nature as form, nature as matter, the first having greater weight. The question I will be asking here is whether the Aristotelians in fact appeal to ends in definition, and to what extent they must. How deeply, in other words, does teleology penetrate into the workings of Nature?

Though every material form has among its ends its own sustenance and the generation of its like in new matter, those are purely formal teleological ends that do not serve to distinguish one material form from another. By 'purely formal' I mean that for every form, one can enter its name in the blanks of '—— acts so to sustain itself' and '—— strives to reproduce itself', without knowing anything at all about the form, except perhaps that it is a material form. If such ends were all that obtained in the world, teleology would serve only to establish in broadest outline the character of material substance; specific natures would be defined in other ways.

What we are looking for is ends whose definition will not lead us back to the form itself. There are two ways that could happen. One is if there are ends independent of the formal ends just mentioned. The other is if there are ends that, though instrumental to the formal ends of self-sustenance and reproduction, can be defined independently. I will present samples of each.

In *Metaphysics* commentaries one often finds, at the beginning, a question elicited by the opening sentence of that work: "All men by nature desire knowledge." In showing that Aristotle was right, the commentators typically conclude that not only do people desire knowledge, but that "metaphysics is most desirable to man insofar as he is man, both by a natural appetite and by a rational appetite ordered in the best way" (Suárez *Disp.* 1§6¶21, *Opera* 25:63–64). A thing can be described as having an appetite for something either properly, if it desires that thing through the recognition of its goodness or apparent goodness, or improperly, if it merely has a propensity to the thing (Fonseca *In meta.* 1c1q1§3, 1:62; cf. §6.3). Among appetites properly speaking, some are natural, in one of several ways. An appetite may simply be given in a thing's nature, without, as Suárez puts it, being effected by any action of that thing (¶4, 54). It may exist potentially in a thing by nature, but actually only under certain conditions, like hunger—the contrasting term in such cases is *adventitious* or *extraneous*. It may simply agree with a thing's nature, like a desire to listen to music, in which case the contrasting term is *contranatural*.[31]

Clearly only the first two kinds of natural appetite pertain to the definition of a thing. But people do not always actually desire knowledge, nor are they incapable of turning it down when it is offered—the Aristotelians do not have so rosy a view of human nature as to overlook our occasional desire to be secure in our ignorance. On the other hand, the desire for knowledge cannot but be elicited by the appropriate objects in a person who is neither distracted nor physically impaired nor depraved; it will never knowingly be hated, although one may freely refrain from acquiring it (Fonseca §6, 1:66B–E; Suárez ¶8, 55).

The reason that a desire for knowledge will be elicited is that knowledge will necessarily be seen as good. Indeed, for reasons that do not need to be spelled out here, the contemplation of truth is held to be the highest good, and its discovery the most excellent operation of the soul. Knowledge is useful, of course, but Aristotle, with the wholehearted concurrence of the Aristotelians, holds that it is desired most of all for its own sake. Our natural felicity, he says in the *Ethics,* is the contemplation of God and separated substances (cf. Suárez ¶34, 63; Fonseca §6, 69C).

The soul is that which has the power to live, sense, and think, says *De anima.* But since thinking is the most excellent of its operations, and since each thing exists on account of its operations, the soul exists on account of thinking. That thinking, moreover, itself has an end: knowledge, of which the highest form is knowledge of God. Could one then define the human as that form whose end is knowledge of God? Not quite. Other species have that end too—the angels, perhaps God himself. The human is that whose end is knowledge of God, but to define us specifically one must add something about the means. That, as we will see, requires reference to the material part of our nature. What matters here is that the contemplation of God, though insufficient to define human nature, is not a purely formal end.

There are also ends that, though instrumental to purely formal ends, can be defined independent of them. The end of sensation is to bring species of

31. Four technical terms are used here. An appetite is *innate* if it exists actually or potentially without any previous cognition of the goodness of its object, and *elicited* otherwise. An appetite is said to be necessary *quoad exercitium* if it will always arise under the appropriate circumstances and, if once it is elicited, the will cannot but actually be moved by it; necessary *quoad speciem* (or *quoad specificationem*) if it will always arise under the appropriate circumstances, and the will is free only to be actually moved by it, or to withhold its consent, but not to be averse to it. See Fonseca *In meta.* 1c1q1§4–5, 1:62–64; Suárez *Disp.* 1§6¶5, *Opera* 25:54.

sensible things to the judgment, or the *vis æstimativa* that in animals corresponds to the human power of judgment, so that things sensed may elicit desire or aversion in the soul. Although it is true that ultimately sensation aids self-sustenance, there seems to be no need to refer to that end in defining sensation.

Having sensation is in fact part of the definition of 'animal', as contrasted with 'plant'. Among animals, the possession of all five senses, whose ends can be defined in terms of their proper *sensibilia,* distinguishes higher from lower. To the extent that the senses are defined teleologically, so too the soul will be. Yet having sensation does not distinguish one species, or even one genus, from another. For that we must advert to matter. And if we ask why sensation, rather than some other operation or even none at all, should be a means to self-sustenance, again we must advert to matter.

All roads, then, lead to matter, and prime matter first of all. Suárez, in his question on the relation of matter to essence, argues that matter "in some way contributes to the operations of substance," since if it didn't, the soul would not join with it to make an *unum per se* (*Disp.* 36§2¶11, *Opera* 26:485). Its contribution, however, is quite limited: matter "is of the essence of material substance as a certain beginning [*inchoatio*] and foundation of that essence, which is completed by form" (¶12, 485). It seems, therefore, to contribute to the essence of material substance no more than what follows from being a material substance—corporeality, which consists in having dimensions, and occupying space.

Now one could say that the corporeality of material substances is for the sake of their forms. Certain of the operations that belong to those forms by nature require that the form be "incorporated": the singular matter to which an individual material form is joined would be for the sake of those operations. In that rather lax sense the material part of the thing's nature would be defined in terms of the thing's ends. The human body exists so that the soul may have sensations, so that it may thereby know the world and eventually God. But there is a twist in that account, which will be best brought out by turning to proximate matter.

Proximate matter is the disposed, quantified matter to which a given specific form, if it is to exist naturally, must be joined. We have seen that the relation of dispositions to form was controversial. But everyone agrees that matter, even proximate matter, cannot exist naturally without a unifying, end-supplying, form. (The "organized body" of which Aristotle says the soul is the form is not a single thing until it has form; experience confirms that after the soul departs, the body, lacking its principle of unity, rather quickly falls apart.) Now it is true that the existence in *this* body of such and such a disposition is explained by the operations of its soul. The roots of a plant are beneath so that it may receive nourishment from the ground. But the disposition itself, so it seems, can be defined apart from any consideration of those operations.

Yet the question is not quite so simple as that. Consider a giraffe, which is twelve feet tall. 'Being twelve feet tall' makes no reference to an end. But for that very reason, an Aristotelian might argue, it is not the most scientifically informative designation. If we call that quantity "tree-grazing" height we understand better why giraffes have that height and not some other: giraffes, being the sort of animals that nourish themselves by eating the leaves off trees, will achieve that end only if they are of a certain height.

So we have got from height to eating. The end of eating is rather more obvious: animals eat in order to preserve their vital heat and to grow. And what is vital heat? As the name implies, it is the active quality that powers, so to speak, the various operations of the soul. Even memory and imagination require the circulation of animal spirits in the brain, and animal spirits cannot be produced but by heating.

There is a certain pattern to these answers. They all propose that the operation or feature we want to explain is for the sake of some operation: but they also seem inevitably to refer to particular facts about the matter of the animal. John Cooper has characterized Aristotelian teleological explanations as appealing to what he calls 'hypothetical necessity' (Cooper 1987). The "hypothesis" is that some operation is to be performed, and the necessity consists in its being the case that given such and such material means, the operation can only be carried out thus. Given what light is and how colors work, if you want to see colors you had better have a transparent sense organ, since anything itself colored would not be capable of sensing all colors. The transparency of the crystalline humor of the eye is explained jointly by the end, vision, and by the nonteleologically defined properties of its materials.

It would seem, then, that some features of the proximate matter of a substance can and should be defined teleologically. But if Cooper is right— the examples seem to bear him out—then there will always be a nonteleological component in the definition of any material natural kind. Aristotle himself would not be troubled by that conclusion. But a Christian Aristotelian, convinced that the world included purely spiritual creatures and dedicated to making teleological sense of our existence, would have a further issue to worry about. Our highest aim, which we share with the angels, is the knowledge of God. Having a body, being a partly material form, is clearly not needed to achieve that aim, since the angels enjoy a vision of God to which only the most saintly among humans can aspire. Our bodies do provide instruments by which to know God. The senses provide evidence of his power, intellect, and benevolence; yet they are prone to error and, because of their link with the passions, to leading us astray. So even if God has given us what is due to us by nature, a person might still, however unjustifiably, envy the angels and question the justice of the God who gave us so inadequate a nature. Original sin goes some way to explaining our condition. But Adam was, after all, human: he knew God through his senses just as we do.

The only justification I can see is that God in the plenitude of his goodness and power will have chosen to create natures in every degree of perfection. That *I*, this individual human substance, should quarrel with the nature he has bestowed on me would be to demand of God that he should omit the human altogether from the ranks of created things, since every person may make the same claims. I have no more cause to quarrel with my status than do angels to revel in theirs.

The question at the outset was to what extent natures are defined in terms of their ends. Purely formal ends will not do, because they take this form: whatever this thing is, it will act in order to preserve itself. Such ends do not distinguish one thing from another: what distinguishes things is the means by which purely formal ends are accomplished, and defining the means seems to presuppose a nonteleologically defined material nature. Among human ends that are not purely formal, the knowledge of God is the highest. Yet that end, although it distinguishes us from every other material nature, does not distinguish us from the angels. Again the means must be referred to: the senses and imagination by which, barring divine illumination, we come to know God. Sense can hardly be defined except in reference to our material nature. The question then becomes this: to what end should there be things that can indeed recognize their Maker, but only by way of matter? The answer seems to be: to no other end than that God should have manifested his power by creating all possible forms.

## 7.3. Artifacts, Human and Divine

Art, in the broad sense of human industry or the products thereof, is at once a model for the understanding of nature and a paradigm of the nonnatural.[32] In Aristotelianism the comparison serves to glorify and justify our

32. On the analogy and disanalogy of art and nature in Aristotle, see, among many others, Fiedler 1978, c.5; on Renaissance versions, Close 1969 and Schmitt 1983, c.5. A wide-ranging examination of the shifting balance between imitation and creation in the understanding of art is found in Blumenberg 1957. Goclenius, I should note, lists three primary senses for the word *ars*, of which only the first is relevant here: "Habitus ποιητικòς (factivus) seu μηχανικòς, ποίημα seu μηχάνημα aliquod post se relinquens nobile vel ignobile, ut Ædificatoria." A *forma artificialis* is "quæ fit ab arte in materia" (Goclenius, s.v. 'ars' and 'artificiale', p125–126).

labors, and at the same time to emphatically call to mind the infinite difference not just between our powers and those of the Creator, but between our powers and the powers in Nature to whom God has, so to speak, delegated responsibility for the natural order. That order we can emulate, and even bring to fruition where it falls short, but we cannot alter it in any essential way. In what follows I will first examine the products of art, their causes and kinds, showing why artifacts, though they resemble in certain respects the works of God and of our own natural powers of reproduction, fail even to be substances. That conclusion was foreshadowed in the hierarchy laid out earlier: artifacts, having neither active nor passive powers, have neither form nor matter to call their own.

I then turn to the analogy and disanalogy of art and nature, first with respect to the artifacts and then with respect to the act of production. It will become clear that the Platonic denigration of art is not entirely absent from the Aristotelians' manner of thinking. Human industry is, in their view also, secondary and superficial. There is, on the other hand, in the conception of art as a supplement to nature, the germ of a quite different view, one that came to predominate in Descartes's thinking.

The twofold analogy between God's labors and ours, and between God's works and ours, though treated as benign, leads to a certain difficulty that was fully exploited by Descartes. The trouble can be brought out by considering the effects of reversing the order of the two members of the comparison. One can say—and this is what Aristotle meant—that art is like nature, whether by resembling it in form or in its production. That seems harmless enough. But if art is like nature, then so too nature is like art. Yet artifacts are bereft of any intrinsic principle, active or passive, of movement. If nature were indeed like art, nature too would lack such a principle: it would have, in short, no nature. That is what Descartes meant in his version of the well-worn adage (*PP* 2§23, AT 8/1:53). At the end of this subsection I will examine the Aristotelian response to that argument.

1. *The causes of art.* Artifacts in the strict sense result only from human industry. Humans are the only natural agents fully capable of recognizing their ends (which requires, as we have seen, a formal recognition of their goodness). That restriction has two important consequences for the classification of artifacts and the relation of artifactual to natural kinds.

The first applies to those artifacts that "imitate" nature by virtue of an intended resemblance to natural things. In natural production, the effect resembles the cause by virtue of being specifically the same form as some form that exists really in the cause. In art, the resemblance must be mediated by thought. But resemblance mediated by thought can only be a distant, attenuated version of resemblance in form. Toletus, echoing the *Metaphysics,* notes that of sensible accidents, many are common to several kinds of things—colors, tastes, and so forth. Only figure is "diverse in diverse species," and even in diverse individuals. He continues:

> Man, the king of this world, and for whose sake [natural things] are made, depends in his understanding of them on the senses, and with the senses he perceives substances only through their accidents; truly he would always be deceived and would not be able to distinguish forms and substances, if there were not some accidents by which substantial forms are designated: such are the figures and exterior forms of things [. . .]
>
> Figures, therefore, follow substantial forms [. . .] They are so conjoined with them, and attendant upon them, that many judge them to be the substances themselves [. . .] But in fact they are not substantial forms, they are only indices [of form]. Hence a perfect similitude of two things is most noticeable in their figures, so that those things seem to be of the same species that participate in the same figure, and are named by the same name. (Toletus *In Phys.* 2c2q6, *Opera* 4:54va)

It is for that reason that arts like painting and sculpture, which aim at likeness, aim at likeness in figure above all. There is a natural relation of designation already in the relation of figures to forms; that designation can be imitated in the artificial designation of that which has both the figure *and* the form of a lion by something that has only the figure (54vb). The latter, the artificial designation, is anticipated in the common sense or the imagination, which like art, can designate forms only by figures. When the resemblance of the final cause to its effect is mediated by sense or imagination, the most secure ground of resemblance, if not the only ground, is figure.

That restriction, though telling, would not be decisive were it not for the second consequence of the human origin of art. Human activity is quite limited in its effects. Cutting, shaping, and joining change only the figure of things, painting or gilding only their surfaces, moving only their place. Even though we can, however imperfectly, come to conceive the forms of things, we have not the power to introduce those forms into matter. "Art," the Coimbrans write, "never forms true things, like a true tree, but imitations of the true, and, as someone rightly said, truly false [*verè falsas*]" (Coimbra *In Phys.* 2c1q5a2, 1:216). Resemblance mediated by thought, even if it were not limited in thought itself, would be limited by the means of production available to us.

These two limitations in the cause of art will serve to demarcate artificial forms from natural. But before I develop the Aristotelian notion of artificial form, I should mention three kinds of exception. Only the last raises serious difficulties.

Some arts have no product. They aim, as Toletus says, only at actions:

singing, dancing, navigating. Such arts cannot be said to imitate nature by representing it. Their practitioners are said, as we will see, to imitate nature, or the Creator, in their mode of acting.[33] In any case, such arts will not be relevant to an account of artificial form.

Certain other arts "accomplish nothing except to apply natural agents, so that upon their being applied, nature produces some effect, as from the mixture and distillation of bodies, or in the production of certain animals" (Arriaga *Phys.* 2d6§2, *Cursus* 319a). In all such instances, the action is entirely natural. The human contribution consists only in bringing about the necessary proximity of the agent to the matter it acts on, and the form produced is not different from the form that would have been made had the agent and the matter come together without our help.[34] Again nothing new is added to the stock of forms.

Some arts, finally, rather than imitate preexisting natural effects, imitate those "which ought to have pre-existed" (Toletus *In Phys.* 2c2q6, *Opera* 4:55ra). They "supply those things that nature lacks, as in the art of making houses, clothes, and similar necessities of life." That might seem to be stretching the notion of imitation (cf. the quotations from Cusa below, n.41). In what sense ought houses to have existed, and how could the houses we build resemble houses that merely ought to exist? The answer, it seems to me, can be gleaned from an argument that turns up in questions about the ends of nature. If nature acted according to ends, she would treat us, her most noble members, the most beneficently. But she does not; "she treats animals like a mother, us like a stepmother. [. . .] nature clothes beasts, but casts man naked upon the shore as if from a shipwreck. To strong and pugnacious beasts she gives arms, to timid ones cleverness, quickness, and cover: but man she sends to fight in the arena weaponless, slower than the swift, weaker than the well-armed" (Coimbra *In Phys.* 2c9q1a2, 1:325). The reply, of course, is that Nature has given us the best weapon of all, intelligence, and the *organon* of instruments, the human hand. But here the point is not how Nature made up for casting us naked upon the shore. It is

33. If such arts have a product, it consists in a certain skill or *habitus* acquired by the artist. We saw that in the definition of nature such skills, because they are only principles *per accidens* of being moved in the thing that has them, are excluded. Skills are not active powers. They only modify the actions of active powers: "the movement of dancing, which is performed by the motive faculty inhering in the limbs, owes to art the fact that it is performed elegantly, symmetrically, and rhythmically" (Coimbra, 1:218). Arts of skill merely introduce adverbial modifications of operations we were already equipped to perform. They do not add new powers to those we possess by nature. If one were to consider the dancer herself to be a kind of artifact, human material formed by teaching as a bed by the carpenter, the artificial form *dancer* would be as inefficacious in its own right as a bed is.

34. The most spectacular example of applying natural agents, mentioned in several texts, is the invocation of demons through amulets or astrological figures (see Coimbra *In Phys.* 2c1q7, 1:217ff).

that some such things were owed to us, both by Nature—that is, God the Creator—and by nature—according to what is naturally required for us to flourish. Nature *ought* to have given us coats. Instead she gave us the ability to imagine them, and hands to make them.

The arts that supplement Nature, then, imitate the instruments she has given other forms, but not to us. That account helps to remove a certain difficulty that would otherwise mar comparisons between artifacts and natural things. Clément Rosset argues that for Aristotle "the artificial object possesses the paradoxical privilege of having more to tell us about nature than any natural object" and that Aristotle's conception of nature is therefore anthropomorphic (Rosset 1973:241–242). Aristotle does say that "according as a thing is done, so it comes about naturally" (the Latin reads: *ut quidque agitur, sic natura aptum est,* the last translating Aristotle's πέφυκε). But he also immediately says, in a passage Rosset omits, "according as a thing comes about naturally, so it is done [sc. by humans: *ut natura aptum est* [. . .], *agitur*]" (*Phys.* 199a8ᶠᶠ). There is a symmetry here that Rosset, eager to press the charge of anthropomorphism, overlooks.

Nor is art Aristotle's only model for the understanding of nature. In the biological works, comparisons of nature to art do occur, but Aristotle is far more interested in comparing animals to each other. More significantly, although it is true that, in keeping with the precept to start with what is better known to us, one may explain the workings of the lungs by reference to a bellows, the workings of bellows will have been, on the argument I am presenting, originally modeled on those of some natural pump, perhaps even the lungs themselves. What we know better about bellows is how air moves through them, or what ends might be served by them, and that we know because we have more experience with them. Aristotelianism is no more anthropomorphic in its understanding of nature than physiomorphic in its understanding of art. If art truly does imitate nature, even where she has left us wanting and forced us to use our intellect, then to use artifacts as an analogy is simply to use a familiar bit of Nature to understand an unfamiliar bit.

The products of art, then, imitate what Nature has done or what Nature ought to have done. Imitation, as we have seen, consists in the first instance in likeness of figure; in the second, in likeness of function. A fur coat is "made" not so as to resemble the coat of an animal in size or shape, but to keep us warm as the animal's coat keeps it warm. Here it is significant that the analogies of nature to art which do occur in Aristotle are not from, say, animals to pictures of animals (which would be gratuitous), but from organs to artifacts of the supplemental sort. Those are, so to speak, the extrinsic organs we make for ourselves. If we could not conceive them by looking at their natural prototypes, and fashion them by cutting and joining and so forth, they would, because owed to us by nature, have been made for us by nature, just as the heart and lungs are.

2. *Artificial forms*. All art, whether imitative by figure or by function, is secondary, superadded. The Aristotelians explicate that secondariness, and justify it, in several ways. I will start with the one cause I have not yet examined, the material cause, move to the efficient cause, and then to the formal cause.

The central texts agree that the material of artistic production, unlike the matter of natural generation, is already a complete substance. (I will use the word 'material' to describe the matter of artificial forms, which, unlike the matter of substantial forms, is a complete substance.) The Coimbrans institute a threefold comparison between divine creation, natural generation, and artistic production: "As art supposes nature, so nature supposes God. In other words, just as art effects nothing unless a Physical composite [substance] is available to it, in which it arranges an artificial form; so nature generates nothing without an underlying matter created by God, in which it induces a natural form. And so through these grades [of existence] things come to light: God produces from nothing, nature from potential being, art from perfected being; God by creating, nature by generating, art by bringing together or arranging" (Coimbra *In Phys.* 2c1q5a1, 1:214). Art is so far like nature. But just because the "forms of artifacts supervene on actual being, on an already perfected and absolute thing," namely, composite substance, they bring no alteration to that thing's nature. They cannot add to or remove its powers. They can only modulate them.

Two examples will illustrate the point. The Coimbrans consider whether art can undertake the works of nature. The instances adduced on behalf of the affirmative include automata, of which they give a half-page list of stock examples from antiquity, and astrological images that are said to have magical powers not accounted for by their material. In their response, they argue that in both a natural power is simply being modified or elicited. Automata "are not moved by their forms, nor by art, but by nature," as in clocks driven by weights. Art does not effect those movements, it only "modifies and tempers" them (Coimbra *In Phys.* 2c1q7a3, 1:219). Astrological images and other such magical instruments, to the extent that the marvels attributed to them are real, arise "from the industry of demons," who for nefarious reasons of their own wish to perform them (a2, 219).[35]

The reasons for this limitation will become clear if we consider the movements by which artifacts are made and the forms that result. Arriaga, affirming that artificial form is not distinct from the parts and locations [*ubica-*

---

35. Reisch (*Marg. phil.* 7tr2c20, p316) likewise argues that the names inscribed on such images "have no power [*virtus*]"; they are "signs of an occult pact with demons." For some detailed descriptions of figures used in divination, see Agrippa *De occulta* 500$^{\text{II}}$, and lib. 4.

*tiones*] of the artifact, argues first that to understand *which* picture we see in a painting, it suffices to understand its colors "with such and such a location on the canvas" (Arriaga *Phys.* 2d6§2,*Cursus* 319b). Similarly, to understand *which* word has been written on a page, "nothing else need be understood but that the ink is located this way or that."

What is significant in those arguments, and in rest of the question, is the terms *ubicatio, ubicatum,* and so forth, which I have translated by 'location', 'located', and so on. These are learned formations, originally coined in discussions of the category *Ubi* or 'Where'. *Ubicatio,* 'where-ing', is, like many nouns in *-tio,* used to denote both the action of the corresponding verb *ubicare* and the result of that action. To 'ubicate' something is to change its *Ubi,* that accident of a thing which consists in its being *here* or *there* (Suárez *Disp.* 51§1¶14, *Opera* 26:976; cf. §4.2).

Artificial forms are produced through *ubicationes* of existing substances. In the making of a wooden or marble statue, Arriaga writes, "through the cutting away of superfluous or impeding parts [. . .], the others that remain, without any new form preserve the distance and expressive proportion of a man, and effect a statue" (319b; cf. Suárez *Disp.* 16§2¶18 and 17§2¶11, *Opera* 25:580, 588). Hence there is no artificial form "distinct from the parts of the wood thus located, and the absence of the redundant parts." Cutting away the redundant parts can produce no new substantial form in those that remain; after all, if other parts had been cut away at random, the action would have been the same, and no one would say that a new substantial form had been made.

More precisely, the artificial form is not distinct from the relative locations of the parts. If the whole statue is moved, in the absolute sense, the *Ubi* of every part is changed, but the distances between the parts are not. So "morally speaking," the figure is not changed (Arriaga 320). The Elgin Marbles are not different sculptures than the ones Lord Elgin packed away in Athens. (Though one could imagine a conceptual artist creating a work called 'Five Thousand Two Hundred Eighty Busts of Homer' by moving the same bust one mile, a foot at a time.)

The action of the artist consists in nothing more than moving existing parts here and there. For an Aristotelian that is not the sort of change that could ever yield a new form. While it is true that, as Aristotle says, every natural change presupposes a concomitant change of place (*Phys.* 8c7, 260a28^ff), including generation and corruption, generation and corruption cannot be reduced to the concomitant changes of place. The action of the artist yields at most a new figure, a new *dispositio* of the material. Artifactual kinds, therefore, or specific artificial forms, are nothing other than types of figure. Though we may denominate a work according to its subject, whose figure it resembles, that denomination is no more than a convenient way of designating a certain figure-type, of which both the work and the model are tokens.

Figure, being a mode of quantity, is entirely inert (§4.3). "Just as quantity by its own character [*suopte ingenio*] is idle and inert, and given to nature, as if it were a second matter, to sustain accidents, so the form induced by art receives no efficacy" (Coimbra *In Phys.* 2c1a6a2, 1:217). Since, as we have seen (§7.2), the nature of a thing depends primarily on its substantial form, and on the active powers attendant on that form, figure is perhaps as far from nature as an accidental form could be. A classification, moreover, of individual substances according to their figures would only accidentally be a classification having any use in physics; and even though each species is marked, as it were, for our benefit by a characteristic figure, a taxonomy of figures would be articulated not on the nature of things, but on one of its last and least concomitants. One could as usefully classify things by the first letter of their names.

The tenuous place of figure in the natural order, and the corresponding exiguity of our ability to affect that order, is underlined by the relation of artificial form to its matter. The material of art, as we have seen, consists in substances perfected already by nature; the matter of an artificial form is the matter of the substance or substances it is composed out of. Suárez asks whether artificial forms can, like other forms, be said to be educed from the *potentia* of matter (see §5.3). His answer has two parts. The first is that when a person produces an artifact, he does so "by a natural motive force," and thus the induced artificial form does not exceed the natural *potentia* of the material. Even if the movement needed to produce the artifact is violent, as lifting a stone would be, still the form intended in that movement, the *Ubi* of the stone, is educed from an inherent *potentia obedientialis* of the material (§2.1). The artificial form, on the other hand, because it is intended only by the maker, and not in the end of any of the movements of the material, "comes about merely as a result," and so is not, properly speaking, educed from matter—a distinction it shares with the rational soul, but for quite different reasons. So there is, Suárez concludes, no natural *potentia* of matter or material to artificial form per se, "for although the intention of the maker tends toward them *per se,* and thus in such manner directs through art his actions [. . .] still that action, by which he executes his design, does not terminate *per se* and immediately in such a form, but in some other mode, from which such a form results" (*Disp.* 16§2¶18, *Opera* 25:580). The artificial form is intended only in thought. What is intended in nature, or in other words, the ends per se of the movements that go into making an artifact, is at most the various *ubicationes* of the parts.[36] The form "results"

36. "In the instruments of art it sometimes seems that by the form [i.e., the shape] of the instrument the form of the effect is proximately achieved, as when the seal imprints wax or money with a figure similar to its own, although then also in fact figure is not brought about *per*

from them, as the shape of a constellation may be said to result from the motions of the stars it contains, but it is not the actualization of any natural *potentia* in its material parts.

Indeed it is doubtful that any figure, artifact or not, is intended per se in the inanimate world. Even the spherical shape of the earth, though evidently the result of natural movements, was not intended by any of the *mobilia*. It came about *per accidens* from the movements of individual bits of earth, each of which strives to reach the center of the universe. So too the globular shape of a drop of water might result *per accidens* from the "love of conjunction" of its particles (Coimbra *In Phys.* 4c9q1a3, 2:62; cf. §6.1), each of whose movements aims only at achieving the highest degree of unity with its companions.

Artificial form, then, is nonnatural several times over. It is superadded to existing complete substances by God so that their forms may be designated to us in sensation. It is neither an active nor a passive principle, serving at best to modulate the action of such principles, and is therefore never part of a thing's nature. It is never intended per se in natural action, but only in the thought of rational agents; toward it, matter has no natural but at most an "obedient" *potentia*. Artifactual form, which is to say figure, is the perfect substitute for natural form if, like Descartes, you don't believe there are substantial forms, natures, natural ends, or *potentiæ*.

3. *Divine art.* The parallel between human art and divine creation, and between artifacts and creatures, is not without its dangers. If human artifacts are devoid of activity, and fail to constitute genuine natural kinds, then to conceive of God as an artificer puts in jeopardy the activity and substantiality of his artifacts.

There are indeed a number of similarities between human and divine modes of operation. Human art, as we have seen, is accomplished according to a cognized form. The substrate material has only a *potentia neutra* toward that form. The form itself is no more than figure imposed upon an already complete substance. The resulting artifact lacks all spontaneity: what it does it does only denominatively, as a key is said to turn a lock.[37]

In divine production, too, an idea of the thing precedes its production. Though in creation matter and form are produced simultaneously, the idea is of the form, and prime matter, like the material of art, has only an undifferentiated *potentia,* natural no doubt but "neutral" in the sense of not tending to one form rather than another. If we restrict our attention to the

---

*se,* but rather place [*Ubi*] [. . .] Therefore more generally it is said that while it is true that these instruments do not achieve *per se* figure or artificial form, still that is not because they concur instrumentally, but because the form [i.e., shape] is not *per se* producible in any case" (Suárez *Disp.* 17§2¶11, *Opera* 25:588).

37. "Formæ naturales sunt actuosæ, & quasi vivæ; formæ verò artefactorum tanquam stolidæ, & emortuæ, nullam effectricem vim habentes" (Coimbra *In Phys.* 2c1q5a2, 1:217).

production of inanimate things, or even to irrational (material but not human) things, divine artifacts, though they do not lack all spontaneity, lack the highest degree, free will. There is, finally, some reason to consider natural things to be instruments by which God accomplishes his ends (see §6.2). All created things "strive toward the end which the first agent has proposed for himself," and because he operates all things on account of himself (Corinthians 1c12), they, too, all regard that end.[38]

In what respect, then, do creatures, especially inanimate creatures, differ from artifacts? In particular, since artifacts have no active powers per se, in what respect can inanimate creatures be said to have such powers?

The last question impinges on the broader question of the efficacy of second causes.[39] From that complicated question I take only the conclusion regarding inanimate things: though their active powers lack the untrammeled spontaneity of will, they too can perform those acts that are proportionate to them, just as our senses tell us they can. For although they are inferior to spiritual things, and act always of necessity, they are not "maximally distant from God [. . .]; according to their form bodies have a greater perfection and similitude to God [than prime matter, which does lack all efficacy], and so according to [their form] they can have effective power" (Suárez *Disp.* 18§1¶14, *Opera* 25:597). Since not every action requires infinite power, and since it is not repugnant to material forms to have finite power, it is possible that God should have "communicated an active perfection" to them in addition to the perfection of existence (¶8, ¶10, 595–596). Indeed if he had not, not only would the variety of forms and qualities have been superfluous, it would have been only apparent: "for if fire does not heat, but rather God does in its presence, he could equally naturally heat in the presence of water; and therefore from that action [of heating] we can no more conclude that fire is hot than that water is" (¶6, 595). And if heat were not in the nature of fire, nothing would be: fire would have, in effect, no nature. Since it is obvious to sense, and not at all repugnant to reason, that the world is populated by things having genuinely different natures, it must be that those things are genuinely capable of acting as well as of being acted on.

---

38. "Sicuti agentia secundaria nihil absque influxu primæ causæ efficiunt, ita ad eum nituntur finem, quem primum agens sibi propositum habet; sicque ut agens primum semetipsum tanquam ultimum suorum operum finem intendit; quia universa propter se operatur; ita eundem finem secundaria agentia spectant" (Coimbra *In Phys.* 2c9q2a1, 1:328).

39. Suárez's arguments, along with many others, on the efficacy of second causes are treated in Freddoso 1988. On the argument in ¶6, see esp. p.109: "To Aquinas, Molina, Suarez, or any other robust Aristotelian, denying active causal power to an entity amounts to nothing less than denying that entity that status of being a substance." To deny that, Freddoso observes a bit later, is "tantamount to denying that there are any material entities at all and a fortiori to denying that scientific knowledge is possible."

That, then, is bedrock: inanimate things do have the active powers they seem to have. What remains to be explained is how they can, nevertheless, be called instruments of God, and how God's action in producing them differs from the action of a human artificer.

Suárez lists five ways in which in which principal causes might be contrasted with instrumental causes (*Disp.* 17§2¶7–17, *Opera* 2:587–591):

   (i) the principal cause is "that to which the action is properly and simply attributed";
   (ii) the principal cause is that which "by its own virtue flows into the effect";[40]
   (iii) the principal cause is that which "proximately and by its own influx flows into the effect," or, in other words, that from which the effect receives its being immediately and per se;
   (iv) an instrumental cause is that which "only acts insofar as it is moved by another," and the principal cause that which "*per se* and without the motion of another has the force to operate";
   (v) an instrumental cause is that which "concurs in effecting or is raised up [*elevatur*] to effect something nobler than itself, or beyond the measure of its own perfection and action," as when heat concurs in the making of flesh.

According to the first four of these ways, instrumental causes do not have any efficacy of their own. The influx of *esse* characteristic of true causes, or in the fourth, the *motus,* comes from the principal cause alone. Anyone who held that irrational agents were instruments of God in any of those ways would be denying genuine efficacy to them.

Suárez, not surprisingly, favors the fifth way of contrasting instrumental and principal causes. Doing so allows him to deny that creatures are divine instruments in any sense that would deprive them of efficacy. They are indeed subordinate to God, serving his aims even at the expense of their own, but they are principal causes with respect to all that is within their own powers to accomplish. Since we have seen that it is not repugnant or contrary to experience to hold that some actions *are* within their power, it follows that irrational agents do have genuine efficacy.

What remains is to show how God's action in producing them differs from those of human artificers. Art, we have seen, either imitates Nature in the sense of depicting or designating natural things, or supplements Nature, bringing forth through the deliberate collocation of natural causes what Nature ought to have made. It is clear that supplementing Nature to supply

---

40. 'Influx' or 'flowing in' is Suárez's most general term to designate the causality of causes, or the proper reason according to which they are causes (*Disp.* 12§2¶10 & 13, *Opera* 25:386, 387).

needs that our own natures do not suffice for, like clothing or shelter, is irrelevant to divine action. God has no needs. His art is instead like that which issues from exuberance rather than exigency, the unwritten song that springs spontaneously to the lips of someone fortunate. Like that song it has no prototype. The exemplar of a created thing that existed eternally in God's intellect before creation is not "expressive or indicative of the thing"; on the contrary, like the spoon imagined by Cusa's *idiota*, it is regarded in intuition and imitated *by the thing*.[41] "But the expression of things themselves is attributed to divine ideas in a peculiar way: because the divine essence which grasps the definition of the idea, insofar as it is known by God to be imitable by created things, contains the man, for example, eminently, of whom there is an idea, insofar as it is imitable by him, and similarly other things, and perfectly represents them" (Coimbra *In Phys.* 2c7q3q2, 1:247). God's ideas "express" created things because God knows that such ideas can be brought forth in material forms ("imitable" by Nature). The imitative relation is exactly converse to that which obtains in human art: nature imitates the divine intellect that eminently contains it. Since God is perfectly capable of realizing those forms that are suited to matter, his ideas represent them "perfectly."

Human art, as Plato said, is doubly imitative: first of nature, of which it can realize only the figures (color, which is also used in art, is imitated only in its arrangement: the red of a painting is not an artificial quality, but the thing itself applied to a different object), and then of the divine mind. What is more important, however, is that God alone has the infinite power required to realize, in the full sense of that word, the substantial forms of which he has the original ideas. Created things, except for the most inferior, lack, as we have seen (§5.4), sufficient power even to reproduce their own forms. Animals require the assistance of celestial powers, humans that of God. But even if humans or animals had the power to reproduce their forms, they do not have the power to produce them afresh. Only by God, one of whose names is causa sui, are the forms of things not only reproduced but originated. Divine art differs from human art in having the capacity to produce autonomous instruments, in whose actions it will indeed concur, but which it does not need continually to operate.[42] The

---

41. "This art [i.e., handicraft] is indeed an imitation, but not an imitation of nature; it is an amotion of the *ars infinita* of God himself, and like that art originary, primitive, creative, though not insofar as it has created the world. *Coclear extra mentis nostræ ideam [aliud] non habet exemplar* [The spoon has outside our mind no other exemplar: Cusanus *Idiota de mente* 2, *Werke* 5:240]. The spoon, no high product of art, is still something absolutely new, an *eidos* not hidden in nature. The mere 'layman' [*idiota*] is the one who brings it forth: *non enim in hoc imitor figuram cuiuscunque rei naturalis* [for in this I do not imitate the figure of any natural thing]" (Blumenberg 1957:268). Not the figure, indeed: but it can imitate the function.

42. "Formæ naturales sunt actuosæ, & quasi vivæ; formæ verò artefactorum tanquam stolidæ, & emortuæ, nullam effectricem vim habentes" (Coimbra *In Phys.* 2c1q5a2, 1:217).

closest we can come to that capacity, apart from generation, is when, by putting a projectile into violent motion, we temporarily impress upon it a motive quality not unlike the heaviness that God has permanently impressed on earthy and watery things. But that quality, or impetus, is evanescent, bound to be overcome by the natural powers that oppose it. In all other instances the most we succeed in doing is to modulate the activity of the natural powers that only God or Nature can impart to matter.

Aristotelianism reached an accommodation of sorts between two strongly held *fundamenta* whose harmony was less than certain: the omnipotence of God and the experienced activity of natural things. A few earlier philosophers had found them incompatible, believing that "to whatever degree efficacy is granted to creatures, to that same degree it is taken away from the divine power of the Creator" (Suárez *Disp.* 18§1¶2, *Opera* 25:593). Any efficacy in things would indicate an insufficiency in God's power. Against that Suárez argues that it is through no insufficiency that God allows second causes to act in his stead, but "through a voluntary and most prudent application of his power": it would be wasteful were infinite power to be employed where finite power will do (¶10, 596). The omnipotence of God is preserved, since he not only could have chosen to apply his power directly, but can at any time prevent second causes from operating by withdrawing his concurrence, as he did when the companions of Daniel walked unscathed through the furnace of Nebuchadnezzar (Daniel 3, a frequent example in questions on second causes). At the same time the manifest activity of Nature is acknowledged and given a rationale.

It should be clear that the analogy of nature and art, as a tool for the scientific understanding of nature, requires delicate handling. It will not do even to make instrumentality the ground of the analogy, since artifacts, having no proper efficacy, cannot even be instrumental causes per se according to Suárez's definition.[43] Artifacts cause nothing per se, and are caused by nothing per se except the intentions of their makers. The comparison of natural things and artifacts can yield no information about their natures, but only about the manner in which those natures express themselves.

---

43. Suárez says that "in all orders of things there can be instruments of this sort"—that is, according to the fifth way of contrasting instruments and principals—and lists natural, supernatural, and artificial forms. But what follows makes it clear that what he means is that artificial forms can have instrumental causes, not that they can be instrumental causes themselves (Suárez *Disp.* 17§2¶17, *Opera* 25:590).

PART II

# Bodies in Motion

# [8]

## Motion and Its Causes

Descartes's physics was largely discredited by the beginning of the eighteenth century. Except among the sect of Cartesians, which included, ironically enough, professors of the sort that Descartes had once numbered among his opponents, the predominant view was that of Huygens, who in 1693 told Bayle he could find "almost nothing I can approve of as true in all the physics, metaphysics, and *météores*" of Descartes (*To Bayle* 26 Feb. 1693, Appendix, *Œuv.* 10:403). Huygens treats his early "preoccupation" with Cartesian philosophy in terms not unlike those that Descartes had used of *his* preoccupation with Aristotelianism: a childhood disease, in which verisimilitude was taken for truth.

Even at the time of the *Principles,* some of Descartes's correspondents were doubtful, especially about the fourth collision rule, which seems manifestly contrary to experience. According to that rule, a smaller body in motion can never so much as nudge a larger body at rest, however quickly it moves. The Latin *Principles* already includes an elaborate explanation intended to show why the rule does not hold of bodies immersed in fluid. In the French *Principles* the remarks following the rule itself are amplified in an apparent attempt to lay to rest the misgivings of Clerselier and others.

The falsity of the rules has preoccupied historians; their response has often been to cast about for some obvious physical truth that Descartes somehow overlooked: the relativity of motion, or the failure, already noted by Leibniz, to satisfy the requirement that to continuous variation in the cause there should correspond a continuous variation in the effect. Although it is true that a person who did not recognize those truths *could* have committed the mistakes Descartes did, the failure to recognize them hardly suffices to explain why he made just *those* mistakes and not others. Descartes's successes—notably the first and second laws of motion—are of a piece with his failures. The demonstrations scarcely differ; nor, for that matter, does the kind of empirical evidence adduced in illustration of their conclusions.

I start with Descartes's definition of motion (§8.1). Descartes, explicitly marking his distance from the Aristotelians, acknowledges only *motus localis*. But even that does not quite capture the magnitude of his departure. Local motion, for the Aristotelian, when it is natural has, like any natural change, a definite *terminus*. A body out of its natural place is *in potentia* toward being in that place; when it reaches its natural place, it stops. Violent change too comes to an end, because it is always opposed by a tendency to some contrary natural change, as when a heavy body is thrown upward. Only a few preternatural motions are exempt from the requirement, and it is to them, and to the incessant motion of the heavens, that one must look to find the persistence that Descartes ascribes to every motion.

For Descartes, motion is always entirely actual, the instantaneous rupture of a body from its neighbors. It is, moreover, *from* a definite place, but never by nature to a place: it has, as I will show in the discussion of the second law, a direction but no *terminus*. Nevertheless Descartes, like the central texts, holds that motion is a genuine mode, though a peculiar one, of the *mobile*. It is peculiar because it is a *joint* mode of the two bodies that are separated by it, or, in Descartes's own words, *reciprocal*. Many commentators have misinterpreted that claim, and taken Descartes to hold that motion is relative while at the same time insisting that rest and motion are genuine contraries. Like Daniel Garber, I think that Descartes's view is consistent; passages where he seems to be acknowledging the relativity of motion turn out to be passages in which he rejects the vulgar notion of "place" on which the supposed relativity is founded.

The discussion of the laws of motion and the rules of collision I divide into two parts. The first (§8.2) concerns the mathematical description of motion, or what Kant called *phoronomy,* and the second concerns the causes of motion, in particular the role of God. In a discussion of the first two laws, I show that what is conserved according to them is not *motus* itself, which Descartes defines as the instantaneous mode of "rupture" between a body and its surroundings, but the modes of that mode—its direction and speed. I then examine in some detail the third law, the rules of collision, and Descartes's conception of quantity of motion. I emphasize three points: first, that the archetypal *phænomena* to be saved in the rules were reflection and the loss of motion by a body in a resisting medium, of which the leading special case was refraction. The notorious Rule 4 was arrived at in part because Descartes realized that pure reflection—in which a reflected body changes direction without transferring motion to the reflector—is impossible under the rules of collision he had formulated before he began work on the Principles. The second point is that the classification of physical situations in the rules of collision works quite differently than in classical physics; the conservation of motion serves only to determine how motion is transferred, and not what I call the *outcome* of the collision. Unlike his onetime collaborator Beeckman, Descartes never treats speed as a signed algebraic quantity, or rest as a limiting case of motion. Contrariness of direction and contrariness of rest and motion, determine outcomes independently of speed and volume. Finally, I argue that the geometric representation of speed, which Descartes continues to conceive in the Aristotelian manner as an intensive quantity, made natural a "hydrostatic" model of the transfer of motion in collisions.

In §8.3 I turn to the causes of motion and the proofs of the three laws of motion. Like Hatfield and other recent commentators, I emphasize the metaphysical underpinnings of Cartesian physics, turning to Aristotelian accounts of the three kinds of divine action Descartes appeals to in his proofs: creation, conservation, and concurrence. The principal issue among commentators has been whether force pertains to God, to bodies, or—in the most recent interpretation by Garber—to nothing at all. There is, I argue, a way to reconcile the alternatives. What has made the issue difficult is that in conservation and concurrence, as in the actions that Soto describes as being "owed to nature" by God (§7.1), there is a *determination* of change in one body by the features of another, but no *causation* except by God.

## 8.1. The Definition and Mode of Existence of Motion

Descartes's conception of *motus* has perplexed commentators, not least of all because in the view of many it is internally inconsistent. The difficulties in understanding it are several. There is, first, a significant change in Descartes's conception from the early works to the *Principles*. After vigorously defending against Morin the rather unorthodox claim in the *Dioptrics* that *motus* and *actio* are one and the same, Descartes effectively conceded the point in the *Principles,* criticizing his own former view.

The second difficulty lies in the discussion of place in the *Principles*. Descartes, in good disputational style, argues his own definition in opposition to two others: explicitly he opposes a relational notion of place, while implicitly he opposes the Aristotelian notion of *Ubi*. Supposing that *motus localis* is to be defined as change of place, neither of those notions of place will do. Instead place is to be defined as the surface common to a body and those immediately contiguous with it.

The third difficulty arises out of the definition of *motus* itself. Many commentators, mistaking the *reciprocal* translation of a body out of its neighborhood for the relative motion of a body with respect to a reference frame, have argued that Descartes cannot both define *motus* as he does and hold that it is a mode of bodies. Although that criticism, as Garber has argued, is misguided, still the logic of the mode in question, which seems to belong to two bodies at once, and its temporality raise serious doubts.

Contrary to the Aristotelian conception, and indeed to Descartes's own treatment of it elsewhere, *motus,* defined in the *Principles* as the rupture of two bodies, is not successive. Reciprocal translation presents the strange aspect of no sooner beginning than it must end. For all that the definition tells us, two bodies that are already ruptured cannot be said to move with respect to one another *or* to be at rest. I will show that Descartes handles the successiveness essential to motion by an appeal to the notion of *trajectory.* That notion, it should be noted, is not relative to a reference frame; speaking anachronistically, one might liken it to the much more recent notion of a *world-line.*

The peculiarities of motion as it is defined in the *Principles* should put the reader on guard against reading classical conceptions into his physics, even if one takes them to be present only *in ovo.* In particular the notion of reference frame is, I believe, simply not pertinent to his thinking. Far more important is the task Descartes's definition was meant to perform, which was both to provide a substitute for the Aristotelian definition of motion, with its inadmissible machinery of actus and *potentia,* and to introduce an instantaneous notion of change to serve as the subject of the laws of motion.

1. *Vulgar conceptions.* In the second part of the *Principles,* after showing that "all variation in matter, and all the diversity of its forms, depends on motion," Descartes turns to the nature of *motus.* The nature of *motus,* he advises us, must be carefully distinguished from its causes, of which the first is God.[1] The "vulgar" conception, according to which motion is the "action, by which a body migrates from one place to another," must be replaced by a conception that issues from the "truth of the thing" (§25, 8/1:53).

Somewhat unusually, the vulgar conception is not that of the Philosophers but of his own earlier self. In the *Dioptrics* he had called light "a certain movement, or a very quick and lively action" (AT 6:84). Defending this use of 'action' against Morin, Descartes responds that "the signification of the word *action* is general, and comprises not only the power [*puissance*] or the inclination to move, but also movement itself."[2] In his response, Morin accuses Descartes of equivocating between motion and its cause, or between *potentia* and *actus.*[3] Descartes stands his ground, and replies simply that "movement is the action by which the parts [of luminous bodies] change

---

1. 2§23 and 36, AT 8/1:52, 61.
2. *To Descartes* 22 Feb. 1638, 1:543, *To Morin* 13 Jul. 1638, 2:204.
3. *To Descartes* 12 Aug. 1638, 2:291, reading 'forme' for 'force', as Descartes did.

their place."[4] The exchange ends there, with Morin believing that Descartes can't tell *actus* from *potentia*, and Descartes believing that 'action' is still the proper term by which to designate what is transmitted from the sun to our eyes.[5]

The Descartes of the *Principles*, who has recently refreshed his knowledge of Aristotelian terminology, writes that "I have always held that it is one and the same thing that, when referred to the *terminus a quo*, is called 'action', but when referred to the *terminus ad quem* or *in quo recipitur* [in which it is received], is called 'passion' " (*To *** [Hyperaspistes]* Aug. 1641, 3:428). A few months later he advises Regius in similar terms: *actio* is *motus* designated with respect to the mover.[6] *Motus* is "always in the *mobile*, not the mover" (*PP* 2§25, 8/1:54); it must therefore be distinguished from its causes, which are typically distinct from the *mobile*. None of those claims would dismay an Aristotelian (cf. §2.3). They too were careful to distinguish between *motus* and the causes of motus.

Though the sources on which Descartes drew in formulating his early views are not obvious,[7] the motivation seems clear enough. Descartes wanted a single word to cover both actual motion and what he would later call the tendency or *conatus* of a body not actually moving toward motion. In the *Dioptrics* the particles through which the tendency he identifies with light is transmitted are taken to be at rest; later he took them to be in actual circular motion, which according to the second law of motion gives rise to a *conatus* directed outward. The earlier view had forced him to assert that motion and tendency to motion, or "action," were subject to the same laws and, in the letter to Morin, to an assertion about the signification of the word 'action' which to an Aristotelian could only seem an expression of ignorance.

4. *To Morin* 12 Sep. 1638, 2:364.
5. The relation between motion, tendency to move, and action in Descartes's earlier work is discussed in Prendergast 1975. Prendergast appears not to believe that the definition of motion in the *Principles* amounts to a repudiation of the earlier view, in part because he takes 'force' (or "the power causing motion," as in the letter to More of April 1649, AT 5:404) and 'tendency' to denote the same entity.
6. *To Regius* Dec. 1641, 3:454[f]. Later he adds that some might call the "force [*vim*] by which [a hard body] admits the movement of other bodies" an action, but that to do so is incorrect, since then the action would be in the patient and the passion in the agent (455). Though the Thomist Soncinas believed that resistance to change is an *actio* of the patient contrary to that of the agent (§2.4), I know of no one who called the mere reception of *motus* an action, for the very reason that Descartes gives.
7. See Garber 1988:346n.6. Goclenius notes that *actio* can be used to denote *motus*, but calls the usage "improper"—a solecism for *actus* (Goclenius *Lexicon* s.v. 'actio', p44). But he also records a classification of *actiones* under which local motion is an *actio creaturarum naturalis nonviolens*. Toletus attributes to the twelfth-century author Gilbertus Porretanus the view that *motus* belongs to the category of action (*Opera* 4:83ra), but does not take the view seriously. Henry More, on the other hand, writes that unlike Descartes he judges *motus* to be "that force or action, by which the bodies you say are moving are pulled apart from each other" (*More to Descartes* 23 Jul. 1649, AT 5:380)—so perhaps the "vulgar" conception had some currency.

Yet Descartes's concessions on *motus* and the identity of *motus* and *actio* did not preclude his continuing a more fundamental disagreement with Aristotelianism. The other vulgar conception Descartes rejected was the conception enshrined in the definition of *motus* as the *actus* of a being *in potentia*. Although Morin was not obtuse in taking the "action" of light particles to be an Aristotelian *potentia,* the corresponding notion of *conatus* in the *Principles* is clearly not a being *in potentia* (see §8.2).

Nor is *motus* itself the *actus* of a being *in potentia*. Descartes, not without a certain slyness, suggests that his revision of the Aristotelian theory of natural change retains local motion while rejecting the other kinds of motion. In fact what he retains is at most an extensional equivalence with the Aristotelian concept. More precisely: a body will be changing in place according to Descartes at every moment in which it is changing in place according to the Aristotelian definition; the converse will also hold provided that motion in a vacuum is excluded. That equivalence, however, belies the underlying difference in concept.

2. *The definition of motion.* Descartes begins his study of motion in the *Principles* in good Aristotelian style: first a definition, and then an explanation of its parts. What one should understand by *motus,* "so that a determinate nature is attributed to it," is

> a translation of one part of matter, or one body, from the vicinity of those bodies, that immediately touch it and are regarded as being at rest, to the vicinity of others. Where by *one body* or *one part of matter* I understand all that which is transferred at once, even if again that [body] itself should consist of many parts that have other motions in them. And I say it is a translation, and not the force or action that transfers [the body], to show that it is always in the *mobile,* not in the mover, because these two are usually not precisely enough distinguished; and it is merely a mode [of the body], not some subsistent thing, just as figure is a mode of a figured thing or rest [a mode] of the thing resting. (*PP* 2§25, AT 8/1:53–54)

The "translation" of a body is not the action that produces or stops movement: "*motus* and *quies* are nothing other [. . .] than two diverse modes of the *mobile*" (2§27, 8/1:55). Yet it is said to be a translation with respect not just to any other body, but with respect to contiguous bodies "regarded as resting [*quiescentia*]." With respect to those bodies, it is "reciprocal": "one cannot understand a body *AB* to be transferred from the vicinity of a body *CD* unless at the same time *CD* is understood to be transferred from the vicinity of the body *AB,*" and the same force or action is required for both translations. Hence, "if we wish to attribute an entirely proper nature to *motus,* and not [a nature] related to something else [*ad aliud relatam*], when two contiguous bodies are transferred, one to one part, the other to another, so that they are mutually separated, we should say that *motus* is just as much in one body as in the other" (2§29, 8/1:55–56). But that would offend against usage. We don't say when we walk that we and the earth both move, but only that we move. After explaining why usage often treats *motus* as proper to just one of the two things it really belongs to, Descartes concludes that "everything that is real and positive in bodies that move, and on account of which they are said to move, is found also in the other bodies which are contiguous with them, but which we merely regard as resting" (§30, 8/1:57).

Descartes then argues that although a body can "participate" in innumerably many *motus* by virtue of being part of other bodies, there is a "unique *motus* of each body that is proper to it." Though it may be useful to divide a *motus* into parts, as Descartes himself does in his proof of the rule of refraction, still "absolutely speaking, one should count but one [motion] in each body" (§32, 8/1:58).

Descartes's definition and his explanation of it thus raise two questions. The first is more logical than physical: can *motus,* if what is real and positive in it must be attributed equally to two things, be a mode of either? Leibniz didn't think so: "if nothing else inheres in motion, except that respective change, it follows that no reason is given in nature why motion should be ascribed to one thing rather than to others. Consequently motion is nothing real. And so in order to say that something moves, we require not only that it should change position with respect to others, but also that the cause of change—the force or action—should be in it" ("Animadversiones," ad 2§25, *Ph. Schr.* 4:369). 'Nothing real', as the context makes clear, means that the attribution of motion depends on an arbitrary choice of the body regarded as being at rest. The distinction between motion and rest is a distinction in reason only.

The key step in Leibniz's argument is that since translation is reciprocal, there is no reason to say that it belongs to one of the two bodies that are separated rather than the other. Instead it belongs to neither. The ground of that inference must be that nothing real can belong to two really distinct things at once. Leibniz therefore reverts to the view that the basis in reality of ascriptions of proper motion is the *vis* or action that causes it.

Descartes seems to hold that it is a mode of both. That position, if Descartes had chosen to make it explicit, would have required explaining; I will later attempt to provide one. But even if we grant it, a second question presents itself, this time to the claim that a body has one motion at a time. Many commentators have found it inconsistent to hold both that motion is a mode of *res extensa* and to attribute to it the kind of relativity exemplified by passages like this:

And indeed, in order to determine the position [*situs*] of a body, we must do so in respect of certain other bodies, which we regard as immobile: and insofar as we do so in respect of various [bodies], we can say that the same thing at the same time both moves and does not move. So that when a ship travels on the sea, a person sitting in the stern remains always in one place [*locus*], if one defines that place by the parts of the ship among which he keeps the same position, and the same person continually changes place [*locus*], if one defines it by the shorelines, since he continually recedes from one and approaches the other. (*PP* 2§13, 8/1:47, cf. §24, p53)

Change of place defined in this manner will not be unique, nor will it be genuinely contrary to rest: a single body can "move" in contrary directions, or be both at rest and in motion, depending on which bodies are taken to be at rest. Yet Descartes, forgetting, as it were, what he admits in 2§13 and 2§24, defines in 2§25 a nonrelative conception of motion. His misunderstanding of the relativity of motion not only vitiates that definition but also, according to the critics, leads him to formulate empirically mistaken rules of collision. Indeed it is not hard to show that the laws yield inconsistent results if equivalence of situations is defined as it is in classical mechanics. But although Descartes's third law of motion and the collision rules derived from it do not accurately describe what happens when bodies collide, Descartes's recognition of a certain relativity in some conceptions of motion is not inconsistent with the definition of 2§25. Cartesian motion, as Garber has argued, is reciprocal but not relative.

I will argue further that the notion of *frame of reference*, which is central to the classical conception, has no application in Cartesian physics. I will offer a reading of the crucial phrase 'regarded as being at rest' (*tanquam quiescentia spectantur*) to show that Descartes is not in fact taking the bodies so designated to be a local frame of reference in relation to which motion is defined.

3. *Place and vicinity.* One reason later philosophers have failed to understand what Descartes intended in defining *motus* is that they take him to conceive place and space in classical terms. He does not. Since the definition of *motus* rests on that of place, any error in understanding Cartesian place will redound on the understanding of Cartesian motion. Descartes's view is indeed a revision of the Aristotelian conception; but it is not, for all that, the conception of Huygens and Newton. I will start by briefly reviewing the Aristotelian position, and then develop Descartes's.

The Aristotelians allotted to three terms of art the senses in which a thing might be said to be at or in a place: *Ubi, locus,* and *situs. Ubi,* as we have seen (§4.2), is that by which a thing is said to be here or there. It alone is a genuine mode of bodies. *Locus* is defined in the *Physics* as the surface of the bodies surrounding a thing. The third term, *situs,* defined in the *Categories* as the orientation or posture of a thing, is of little importance in physics.

The Coimbrans, to establish that the *terminus* of local motion is *Ubi,* consider three alternatives. Some philosophers hold that it is "a relation of distance to the poles, [or] of the thing contained to the surface that contains it." The Coimbrans reject that out of hand. There is no *motus* in the category of relation: relational properties are always acquired by way of nonrelational *fundamenta.* Other philosophers take the *terminus* to be an "entity resulting in the *mobile* from the surface of surrounding bodies," or *locus* regarded as a mode of the *mobile.* The Coimbrans reject that on the grounds that genuine local motion in a vacuum is possible. What remains is *Ubi,* or "existence in space," which is just "the quantity of the *mobile,* according as it exists in this or that part of space, or a certain mode which quantity takes up according as it coincides now with this, now with that part of space, genuine or fictitious."

Space, in their view, includes not only what actually contains bodies, but the "imaginary" space beyond the heavens, in which no body, but only God, is present (*In Phys.* 8c10q2a4, 2:369$^f$). In imaginary space, there is *Ubi* but no *locus.* Similarly, angels can undergo change of *Ubi,* but since they do not occupy space, they cannot undergo change of *locus.* Although a new *locus* will be acquired in the local motion of bodies (except in a vacuum) along with a new *Ubi,* it is acquired only per accidens (3c3q2a2, 1:352). Local motion, therefore, is change of *Ubi,* not change of *locus.*

Descartes, on the other hand, holds that corporeal substance and extension or quantity coincide. Substance and space differ only as the singular or specific from the generic: "when a stone is removed from the space or *locus* in which it exists, we suppose its extension also to be removed, insofar as we regard it as singular and inseparable. But meanwhile we judge that the extension of the place, in which the stone existed, remains and is the same, although now that place is occupied by wood, water, air, or some other body." What we call the same "place," generically, is nothing but a volume, situated among other volumes, which can be filled with different individual bodies (2§11–12, 8/1:46). It is not a mode of anything, except insofar as a body happens actually to occupy a generic place. Singular place, or *locus,* to which I will return in a moment, is a mode of the "singular" extension of a thing—*its* extension, as opposed to extension in general—, and thus a mode of the thing.

Simplifying a bit, I take Descartes to introduce two terms of art, *situs* and *locus,* to replace the Aristotelian terms. Only *locus,* as it turns out, is relevant to the definition of *motus.*

*Situs* is the space, singular or generic, that a body occupies, designated with respect to "certain other bodies, that we regard as immobile." We can designate a *situs* without defining its boundaries precisely; in that loose sense, the same "place" may be occupied by bodies having different figures and magnitudes, and the same body may occupy differently designated "places" at the same time.

*Locus,* on the other hand, is "the surface that immediately surrounds the thing placed [*superficie quæ proximè ambit locatum*]" (2§15, 8:48). That surface is "nothing other than a mode" of the surrounded body, if by 'surface' we mean the entire surface common to the surrounded body and those that surround it, and if we consider it to be the same surface so long as it keeps the same magnitude and figure. If a ship in a stream is held up by the wind, one may "easily believe that it remains in the same place, although all the surrounding surface [i.e., the combined surfaces of the surrounding bodies] changes" (49). That point will be crucial to establishing that the earth is at rest.

So defined, *locus,* like *Ubi* and unlike the *receptaculum* of certain Platonists, is not a substance but a mode of substance. But unlike *Ubi* it is implied in the very idea of determinate extension (which Descartes always understands to be actual, not potential). Since a vacuum—including the supposed imaginary space outside the heavens—is not just naturally but absolutely impossible, Aristotelian arguments for distinguishing place from *Ubi* carry no weight. Wherever there is motion, there is change of place, and conversely; *Ubi* is superfluous.

*Situs,* too, is rejected, not because it is superfluous but because change of *situs* is for several reasons not a suitable idea of motion. The first is that *situs* is not a mode of bodies, but only a way of designating their extensions. If motion is to be a mode, it cannot be change of *situs.* The second reason is that *situs,* unlike *locus,* is not unique (that is the point of 2§13). If *motus* were defined as change of *situs,* each body that moves would not have a *unique* motion, and motion in one direction would not be genuinely contrary to motion in the opposite direction. The third is that if place, properly speaking, could be designated by relative locutions like 'outside of', then place and extension would be distinct. As the Coimbrans note, those who deny the existence of imaginary space do so on the grounds that "a distance or interval capable of receiving body, but without body, cannot be understood," and for the same reason they deny that a vacuum can be produced, even by God's absolute power (*In Phys.* 4c9q3a1, 2:68). That is precisely Descartes's position: no vacuum even by God's absolute power, no imaginary space, no place without body.[8]

---

8. Denying the existence of imaginary space allows Descartes to dismiss the problems raised by Buridan for Ockham's account of motion (see §2.2). Ockham had denied any real distinction between *motus localis* and the *mobile;* Buridan argues that the outermost heavens can move through imaginary space, or rotate, and that *motus* is therefore not just a mode of the *mobile,*

The definition of motion refers not to *locus* itself, but to the closely related notion of the "vicinity" (*vicinia*) of a body. A body is in the vicinity of another if the place of the one coincides in part with the place of the other—if, in other words, they are contiguous. Descartes explains his use of that term, rather than *locus,* in the definition of motion: "I have added, moreover, that translation occurs *from the vicinity of the bodies contiguous to it to the vicinity of others,* and not from one place to another: because, as I explained above, 'place' is taken in various ways, and depends on our thought: but when by 'motion' we understand the translation that occurs from the vicinity of contiguous bodies, then since only one set of bodies can be contiguous [to the *mobile*] in the same moment of time, we cannot attribute several *motus* to the *mobile* at the same time, but only one" (2§28, 8/1:55). It is essential that what counts as one body be "all that which is translated at once." If it were not, then in Figure 8 below the putative body *BC,* as it slid along the hollow cylinder *KL,* would not be translated from the vicinity of *KL, AB,* and *CD* to that of another set. But since *AB* and *CD* are translated with it, the *mobile* is actually *AD* or a body that contains *AD.* Once that is noted, *motus* appears to be unique, since the set of bodies touching the *mobile* is at each moment unique, and is different just in case the *mobile* moves.

That would settle the question of the relativity of *motus,* but for two problems. The first is that Descartes insists (2§29) that the translation occurs out of the vicinity of not just any contiguous bodies but "*those* only, *which are regarded as resting.*" The 'regarded as' would seem to introduce the relativity to thought that Descartes intended to exclude from the strict sense of *locus.*[9] The second is that what Descartes calls the *determination* of motion cannot be uniquely fixed merely by reference to the vicinity *out of* which a body is translated, and the others *into* whose vicinity it is translated can be uniquely defined only after the motion is done. That problem, which is crucial to understanding what becomes of the directedness of motion in Cartesian physics, will be treated in the discussion of the second law in §8.2. Here I concentrate on the relativity of motion.

4. *Translation.* The absolute fact by which mutual motion and rest are distinguished Descartes calls *translatio.* Two bodies *A* and *B* have been translated from each other's vicinity during a certain interval of time if at the beginning of that interval they are contiguous and at the end they are not. 'Being contiguous' and its negation are symmetric relations; so too is translation (2§29, 8/1:55), and there is no reason to regard it as inhering in one

---

but really distinct from it. In Descartes's view the world can neither be translated nor rotate, since it does not, properly speaking, have a place at all.

9. In addition to the commentators cited in Garber, see Prendergast 1972 in Moyal 1991, 4:104. Prendergast holds that "if we are to take this text seriously the reality of rest and motion is destroyed." That, as we have seen, was Leibniz's argument.

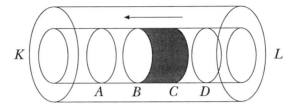

Fig. 8. Vicinity

body rather than the other. It inheres, independently of our manner of conception, in both if it holds in either.[10] Descartes calls it "reciprocal."

Both the logic and the temporality of translation are peculiar. Let the mode in A which consists in being contiguous with B be denoted c, and the mode in B which consists in being contiguous with A be denoted c'. Then presumably even God cannot bring it about that one should exist while the other does not. So c and c' are not even modally distinct. The same holds of that mode s of A which consists in being apart from B and the corresponding mode of B. The same too must hold of translation, which for A consists in having c and then s (which is contrary to c). Yet Descartes holds that in general modes, among them *motus,* which belong to really distinct substances are themselves really distinct (1§61, 8/1:30). It is no wonder Leibniz decided that *motus à la mode Descartes* is nothing real.

The logical peculiarity shows up in a somewhat different way in an example that More took to be a kind of *reductio* of Descartes's view.[11] After arguing that motus is reciprocal, Descartes considers why we don't ordinarily say the earth moves when we lift our feet: "The principal reason for this is that *motus* is understood to be of the whole body that moves; nor could it be thus understood to be of the whole earth on account of the translation of certain of its parts out of the vicinity of smaller bodies to which they are contiguous: since often several such translations, mutually contrary, may occur in it" (*PP* 2§30, AT 8/1:56). It's not hard to come up with examples. In Figure 9, the central body B must, with respect to the left-hand body A, be said to move from left to right, although with respect to C, it must be said to move from right to left. On the face of it those are contrary motions that a single body cannot undergo simultaneously. The resolution of the paradox

10. "Motion and rest differ truly and modally, if by 'motion' is understood the separation of two bodies from each other, and by 'rest' the negation of this separation. But when only one of two bodies that are separating from each other is said to move, and the other to be at rest, in this sense motion and rest do not differ except in reason" (AT 11:657; "Cartesius," a collection of notes gathered together by Leibniz, was probably written in 1642, cf. 11:647). Garber has emphasized the importance of this passage (Garber 1992:167).

11. "This article [i.e., 2§30] seems to contain a most evident demonstration that translation or local motion (unless it were merely a respect of bodies) is not in any way reciprocal" (*More to Descartes* AT 5:385). Unfortunately Descartes never responded to this part of More's letter.

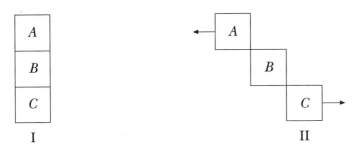

Fig. 9. Contrary motions

lies in considering *which* motions those directions are the directions of. Direction, as we will see later, is a mode of *motus*. The *motus* that B has by virtue of its mutual translation with respect to A and the *motus* it has by virtue of its mutual translation with respect to C are distinct modes of B. The directions are thus directions of distinct *motus,* and are therefore not contrary.

It may seem that the proposed resolution flies in the face of Descartes's insistence that each thing has but one *motus* proper to it (2§31). One answer suggested by the passage quoted above is that the two *motus* might be regarded as belonging not to the entire body B but to its parts—the top and bottom halves, say. But that is unsatisfactory: B is solid, and its parts are at rest with respect to one another. A second answer is suggested by a remark at the end of §31. Descartes considers the motions of the wheels of a watch carried by a man walking on a ship at sea. Each wheel has, he says, one motion proper to it—that which corresponds to the change in its singular place, or, in other words, to its translation as a whole out of the vicinity of certain bodies touching it. It also has other motions by virtue of being "adjoined to the man walking," and to the ship, and to the sea, and finally to the earth, "if indeed the whole Earth be moving." "All these motions," he concludes, "truly are in these wheels; but because so many are not easily understood at once, nor can even all of them be known, it would suffice to consider in each body that unique motion, which is proper to it" (2§31, 8/1:57). Since the translation by which motion is defined need only be with respect to some contiguous bodies, not all, "regarded as resting," perhaps neither A nor C is suitable for defining the proper motion of B, even though the reciprocal motions in B defined by them are truly in B. I suspect that ultimately Descartes would, if presented with situations like that of Figure 9, or others yet worse, simply deny that B has a proper motion, which is to say: B is at rest.[12]

12. That solution is not satisfactory either. The situation of Fig. 9 occurs all the time in the circularly moving concentric vortices that surround each star and planet. Take A, B, and C to lie

The temporality of translation is just as problematic. Translation, though it would seem that it can take place only *at* an instant, can never occur *in* an instant. There is no medium between touching and not touching, and in that sense the change from one mode to the other can only occur *at* an instant. Nevertheless, after it has occurred, the two bodies will be a finite distance apart, with other bodies intervening. Since every body that moves through a finite interval of space does so in a finite interval of time, the translation could not have taken place *in* an instant. Descartes is cognizant of that peculiarity, I think, when in the proof of the second law of motion he writes that although "no motion occurs in an instant," still we can "designate" the instants at which a body is moving (2§39, 8/1:64).

The odd thing is that, if we follow Descartes's way of thinking, we can in fact only pick out *intervals* of time in which we know that at some instant or other two bodies now separated had the reciprocal mode of translation. That mode no sooner appears than it disappears; its nature, it would seem, is to be "borne toward its own destruction," to quote Descartes's erroneous criticism of the Aristotelian concept of *motus* (see the discussion of the first law in §8.2). Although what one might call a *completed* translation can never occur in an instant, the translation referred to in 2§25, which is in a certain sense never a completed motion, can *only* occur in an instant.

5. *Succession.* Although the *motus* defined in 2§25 is a temporally point-like *translatio,* Descartes elsewhere clearly takes *motus* to be successive. To understand how he thought he could generate successive change from a *definiendum* in which succession is lacking, it is worth considering first how the Aristotelians treated succession. We have seen that in Aristotle's definition in *Physics* 3c1 *motus* is understood to be a process that—if change of substance be set aside—occurs through time. More precisely, it is a mode of the forms taken on successively during a change, consisting in the tendency of each to lead to the next by way of being on the way to the *terminus ad quem.*

At least one Aristotelian, Fonseca, characterizes that tendency in terms of the instantaneous introduction of forms one after the other. The intermediate form reached during a change, considered in itself, "will not obviously satisfy the definition of *motus* or of acquisition, but only that of a form of a certain perfection just acquired." But "when [the form] is taken, not just insofar as it has been introduced, but also insofar as it is immediately to be introduced, now it will satisfy in every way the true and relevant definition of *motus*" (Fonseca *In meta.* 5c13q9§2, 2:718–719). Since in local motion the

_____

in adjacent layers of the vortex, A in the innermost, B in the next, C in the outermost. The layer containing A moves faster than that containing B, and that one in turn faster than the layer containing C. If at a certain moment A, B, and C all lie on a line through the center of the vortex, as in Fig. 9 (I), a short time later they will be as in Fig. 9 (II). Yet B cannot be said to be at rest, since if it were, it would have no *conatus* to move outward from the center.

"form" is the place attained at each instant in motion, we have it that the mode of the *mobile* with respect to which it will satisfy the definition of *motus* is that of being just about to take a certain place.

But for the Aristotelian *motus* is essentially successive. As the Coimbrans put it, we cannot conceive the way and tendency to a form without succession, since it is "the *mobile* gradually attaining the perfection of the form," where 'gradually' means "part after part" (Coimbra *In Phys.* 3c2q2a2, 1:336). What unites the successive acquisitions of partial forms into a single *motus* is clearly the perfected form, the *terminus ad quem* common to all those acquisitions. A partial form that is, in Fonseca's words, "immediately to be introduced," is so because it is the next step on the way to the *terminus*.

Setting aside for the moment the *conatus* or inclination to move which Descartes introduces with his laws of motion, and the conserving power of God, Descartes has no such means by which to unite one instantaneous change with another. There are no incomplete *actus*, no partial forms, in his physics. There are only the actual joinings and disjoinings of bodies at various times. Or indeed, since in each instant two bodies that touch are as one, there are only actual bodies at various times. Change consists in the rearranging of boundaries within the one big body, the indeterminate *res extensa* we call the world.

It is not surprising that the *definiendum* in Descartes's definition of motion should be a punctual event and not an essentially successive entity like Aristotelian *motus*. What is surprising is the untoward consequence I have mentioned: motion with respect to something is defined only in that instant when two bodies are rupturing. Yet he clearly thinks of some motions as continuing through time. One of the examples he uses to illustrate the claim that each body has but one motion is the following: "If the line *AB* is carried toward *CD*, and at the same time the point *A* is carried toward *B*, the straight line *AD*, which the point *A* describes, depends on the two rectilinear motions—from *A* to *B* and from *AB* to *CD*" (*pp*2§32, 8/1:58, see Figure 10). The motion of the point *A* toward *B* or toward *D*, and the motion of the line *AB*, cannot be brought within the scope of Descartes's definition except by extending it.

Now I think Descartes's definition can be extended consistently with the claim that each thing has but one motion. The examples in 2§32—the one given above and the example of a point on a rolling wagon wheel—show that the successive loci taken up by a point continually moving define a geometric figure such that to each instant of motion there corresponds exactly one point of the figure. That holds independently of how the motion is conceived.

To make things simple, consider two spheres *E* and *F*, which are at first contiguous and sometime later not contiguous. That situation must have

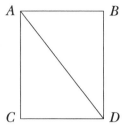

Fig. 10. Combined motions (*PP* 2§32, AT 8/1:58)

arisen as a result of successive ruptures of *E* or *F* or both with the bodies surrounding them. We consider the bodies defining the place of *E* at each instant to be at rest, not only with respect to *E*, but with respect to each other. (If *E* is immersed in a fluid, for example, the bodies it is in the vicinity of must be moving with respect not only to *E*, but to each other.) Abstracting from whatever other motions they happen to be involved in at that moment, we can consider them to be one body, call it *G*. There is no reason not to include *F* in that body. *E*, as it is translated out of the vicinity of *G*, is thus translated out of the vicinity of a body of which *F* is part. It will leave behind a series of places, in the generic sense, which joined together constitute a tubelike virtual body, which I will call the *physical trajectory* of *E*.[13]

The physical trajectory of *E* is in fact independent of *F*, which serves only to determine the *first* vicinity included in the trajectory. Because it was defined in abstraction from all movements but those of *E*, the trajectory is nothing other than a potential part of the one big *res extensa*. In Figure 10, the diagonal line *AD* is the physical trajectory of the point *A* through the interval of time represented in the figure, and the whole rectangle *ABCD* is the trajectory of the line through that interval. Similarly the "very intricate" line traced by a point on the circumference of a rolling wheel is the trajectory of that point, if the point be regarded as distinct from the wheel. If we imagine the trajectory to be generated by a succession of translations, each of them adding its increment, then the trajectory, like its parts, will be defined absolutely.[14]

13. Trajectories, as readers acquainted with Special Relativity will note, are like worldlines. In both Galilean and Einsteinian kinematics, the fact that two bodies have separated is invariant under the admissible transformations. The distance between two points (or space–time points) is likewise an invariant. For periodic motions, like the revolution of a point on a wheel, the statement in the text would have to be modified, since in some reference frames the trajectory will be a finite closed curve, and in others it will be an indefinitely long open path.

14. It is tempting to think that in the example of the wheel, Descartes is implicitly appealing to a frame of reference in which the earth is at rest. But there is no *t* axis in the diagram; there is only the line *AD*, which represents as simultaneous parts of one body the spaces successively occupied by the point–body *A* as it moves. The diagram is not intended to represent *A*'s

'Moving with respect to' can now be defined more generally. Suppose that *F* is contiguous with the trajectory of *E*. Then *E* will have left the vicinity of *F* at some time, and traversed a certain amount of space. Not only will it have moved with respect to *F* in the narrow sense described earlier, but in a broader sense it will have moved with respect to *F* even after it ceased to touch *F*, by virtue of traversing the spaces intermediate between the space contiguous with *F* and its present locus. The locution 'moving with respect to' continues to denote a joint mode of *E and F*, since a similar argument shows that *F* has been moving through the same period with respect to *E*.

The phrase 'regard as resting' is a red herring for those acquainted with later developments in kinematics. It does not mean 'regard as a *rest frame*', not even in the eccentric sense in which, according to Garber, a moving thing's rest frame is constantly moving with it (1992:171). It is true that Descartes, perhaps to accommodate the Aristotelian conception of local motion as an *exitus* from one locus to another, allows his reader momentarily to regard the *motus* as inhering in the *mobile* alone, and not in its vicinity also. But the nonrelational facts upon which judgment about motion are based are reciprocal facts of touching and not touching. Descartes wants to define motion in the Aristotelian manner as the successive acquisition of place—but without adverting to *potentia* and *actus*.

Descartes, as he was elaborating his succedaneum to the *cursus* of Eustachius, aimed at a particular sort of rapprochement with the Aristotelian definition, to which the phrase 'in the singular instants which can be designated while it moves' is the key. He takes the Aristotelian definition to be a way of designating a thing that moves at just those times it is moving. That Aristotle does so by invoking the machinery of *actus* and *potentia* is irrelevant. What matters is to find something which can plausibly be called a mode of extension and which will exist in bodies when and only when they are moving. That something is translation. Though he calls it *motus*, it resembles the object Aristotle was trying to define only in that respect.

There were, of course, other motives. Many critics, seeing what they regard as a sophistical application of the definition of *motus* to demonstrate that the earth has no proper motion, have held that the definition was devised ad hoc to yield that result. Like Garber, I am inclined to think that Descartes was not led to his definition merely by the desire to reconcile his physics with the Church's rejection of Copernicanism.[15] But neither was he

---

motion with respect to a reference frame in which *A*, *B*, and *C* are at rest, but rather *A*'s motion simpliciter, the "world-line" of the point-body *A* as it moves from the generic place *A* to the generic place *D*.

15. Though Descartes was aware that the decision of the Court of the Inquisition against Galileo was not "immediately thereby an article of faith," he nevertheless treated it as if it were (*To Mersenne* Apr. 1634, AT 1:285). The light of faith he treated as superior to the light of

simply working out problems internal to the physics he had developed up to the point of embarking on the *Principles*. What he would have discovered or rediscovered in his reading of Eustachius, Abra de Raconis, and perhaps other central texts in the interval between the Meditations and the *Principles* was that a host of problems, of which the controversy with Morin had already given him a taste, attended the concept of motion. Those problems he had ignored when he was writing *Le Monde*, either because he had forgotten them or because he thought that he could appeal to an immediate intuition of the nature of local motion (*Monde* 7, AT 11:39–40). Some of them he continued to ignore, like the debate about the interpretation of *potentia*. Others, like that of defining the instant of motion or of determining whether succession is essential to motion, had to be confronted, whether because his own physics demanded an answer or because an Aristotelian audience would expect one.

Descartes should not be treated as doing ineptly what Huygens and Newton later did well. He is trying to do better what in his view the Aristotelians had done poorly: to explain what is true of a thing at each moment of its *absolute* motion. Excising *potentia* from the body of physics, however, leaves him with nothing but Fonseca's 'form about to be introduced', the instantaneous, punctual *translatio* from the vicinity of one set of bodies to another, and the thought that by gluing together the spaces successively occupied by a body one can construct a physical trajectory to which the relative conception of place is irrelevant.

## 8.2. Persistence, *Conatus* and Quantity of Motion

The causes of motion are two: "the general cause of all the motions that exist in the world," and "the particular cause, by which it happens that single parts of matter acquire motions that they did not have before" (*pp*2§36, 8/1:61). The general cause is God himself. The "particular and secondary" causes of diverse motions in single bodies are "certain rules or laws of

---

reason, and since the pronouncements of the Church on matters of faith were authoritative, its expressed view on the movement of the earth functioned in Descartes's physics as an undoubted truth, similar in that respect to the identification of matter with extended substance.

A comparison with the theory of *Le Monde* shows that the materials for formulating the theory in the *Principles* are already there: the main vortex of the sun, which carries the planets with it, the subsidiary vortices of the planets, the containment of each body always within its own vortex (Descartes *Monde* 10, AT 11:69–70). The only missing ingredient is the definition of *motus* that Descartes did not find it necessary to formulate (*Monde* 7, 11:39). It is undoubtedly no coincidence that the definition he does formulate in the *Principles* should allow him to assert that he has "withdrawn all motion from the Earth more truly than Tycho and more carefully than Copernicus" (*PP* 2§25, AT 8/1:53–54). But the definition also serves ends unrelated to that of managing a prudent measure of agreement with the Church.

nature" (2§37, 8/1:62). In this subsection I treat those laws and the conceptions of *conatus* and quantity of motion which accompany their proof and application. In §8.3 I will examine the general cause, and the relation between divine action and force.

1. *Persistence: the first law.* The first law, in both its early and later versions, applies not just to motion but to any "state," including figure and size. I give the versions in *Le Monde* and the *Principles:*

> Each part of matter, in particular, continues always to be in the same state, as long as an encounter with others does not constrain it to change.[16]

> Each thing, insofar as it is simple and undivided, remains, *quantum in se est,* in the same state always, nor is it ever changed [*mutari*] unless by external causes.[17]

The later version differs from the earlier in two respects. Its scope is not restricted to parts of matter, and it replaces the phrase *en particulier* with the phrase *quantum in se est,* which I have left untranslated.

Although the substitution of *res* for *partie de la matière* seems not to have had any particular significance in the parts of the *Principles* Descartes managed to write, the phrase *quantum in se est* is of some significance not only for Descartes but for Newton's response to Descartes. Alan Gabbey, drawing on I. B. Cohen's interpretation, argues that the phrase has two connotations.[18] The first is "the limitation of the body's power to remain in this or that state"—the fact that it must interact with others. The second is equivalent to *sua sponte, ex natura sua,* and *sua vi,* where the stress is rather on what is natural or innate to a thing.

In Descartes's thinking, as in that of his predecessors, persistence can be regarded under a negative or a positive aspect. Negative persistence is the permanence of a state *in the absence of* external causes. I will call this simply *persistence.* Since in Aristotelian terms change by an external cause is violent change, persistence will have, if it has a natural cause at all, a cause arising from a thing's own nature. Persistence will in that sense be the natural condition of the thing. If, on the other hand, a thing were to cease to exist even though no external cause acted on it, its perishing, on the same reasoning, would likewise be natural. Descartes, in an argument I will

16. "La première [règle] est: Que chaque partie de la matière, en particulier, continue toujours d'être en un même état, pendant que la rencontre des autres ne la contraint point de la changer" (*Monde* 7, AT 11:38).

17. "Unamquamque rem, quatenus est simplex & indivisa, manere, quantum in se est, in eodem semper statu nec unquam mutari nisi à causis externis" (*PP* 2§37, AT 8/1:62).

18. Gabbey 1971:32n.91; Cohen 1964:148.

discuss at the end of this subsection, holds—erroneously—that the Aristotelians take *motus* to perish by its own nature.[19]

Positive persistence is the permanence, or the tendency to permanence, of a state *in the face of* external causes. This I will call *resistance*. Although Descartes's first law asserts only that a thing will persist, in his second and third laws it is clear that a thing will also resist certain kinds of external cause. Resistance, as we have seen, was construed by some authors to be passive, by others as active, or, more specifically, reactive. That ambiguity remains in Descartes's physics, and, as I will argue at the end of §8.3, its underground survival helps to explain why he formulated the rules of collision so as to create a marked asymmetry between motion and rest.

In §2.4 I discussed *potentia resistendi*, the power each thing has to remain as it is and to resist being changed. Zabarella takes it to be the consequence of an active *nisus* or striving of each thing to preserve itself as it is: "Resistance is, properly speaking, the repulsion [*propulsatio*] of an action apart from any consideration of reaction; such repulsion has no other being than privative, since it is the non-admission or the non-undergoing of action, and follows upon the nature of the form striving to conserve itself" (*De rebus nat.* 440F). Striving to preserve oneself is primary; the nonadmission of the action of external causes is, as Zabarella later makes clear, only one means of self-preservation. Galileo, on the other hand, construes resistance passively. In his treatise on the elements he writes that resistance is "permanence in a proper state against a contrary action," which he does not distinguish "from the thing's very existence whereby it endures" (Galileo *Early Notebooks* 244). That is in certain respects very close to Descartes's view. The *persistence* asserted in the first law is, once God's role in conserving created things is understood, nothing other than their existence with respect to the continuing action of God according to his immutable will.

That things persist in their natural states, and resist being deprived of them, was not a new idea. The Coimbrans, in their question on divine conservation, agree that "to have an inborn propensity not to exist is contrary to the nature of created things, since all incline to the opposite, and have [. . .] an ingrained desire to perpetuate themselves, insofar as that can happen" (Coimbra *In Phys.* 2c7q10a3, 1:269). Elsewhere they speak of desiring one's own destruction as "most alien to the laws of nature" (*In Phys.* 1:149). We have seen that in the forms of corruptible things the primary expression of the tendency to self-perpetuation is generation, the reproduction of forms in new matter, by which they achieve a simulacrum of the immutability of God.

*Motus*, too, at least under certain circumstances, was thought to persist, or

19. See §8.3 on the role of self-preservation in arguments about divine conservation.

at least not to cease of itself.[20] Considering whether the circular motion of the heavens could have been going on eternally, the Coimbrans note that "it is not a contradiction, either from the side of the *mobile*, or from the side of the *motor*, to suppose that motion in a circle should exist from eternity." The reason that the motion of the heavens could not have been going on eternally is not that it must have ceased of itself in a finite time, but only that there is a "repugnance" in supposing, for example, that there have been infinitely many days and nights (8c2q7a3–4, 2:284[ff]). In a few quite special cases, even terrestrial bodies might be thought to persist indefinitely in motion. The Coimbrans argue that the continuing motion of a projectile is owed to a *vis* impressed upon it by the projector, and not, as Aristotle thought, to the impulsion of the medium (6c2q1a8, 2:246[f]). They hold, as we have seen, that nothing tends by nature to its own destruction, or equivalently that a thing will persist if not acted upon by others. They argue that a body can move in a vacuum, that such a movement can be successive rather than instantaneous (4c9q4–5), and that a vacuum is absolutely possible (*In Phys.* 4c9q2, 2:67[f]); in fact they use such a motion to argue for the *vis impressa*.[21] It would seem to follow, though the Coimbrans do not say this, that if a projectile moves in a vacuum, then the *vis impressa*, since it is opposed by nothing and since it will not tend to its own destruction, should not decrease. The projectile should therefore continue to move indefinitely.[22]

Abra de Raconis gives the argument a thorough airing, only to reject it:

> This quality [i.e., the *vis impressa*] cannot be corrupted. It would either be corrupted by its contrary, but it does not appear to have a contrary, since it is well conjoined with gravity and levity, and these alone, it seems, could be contrary to it. Or else by the absence of the projector, on which it would depend for its conservation; but that is not so, since if the projector is

20. For late Aristotelian views of the persistence of *motus*, see Maier 1951, esp. pp. 295–304, Clavelin 1968:112[ff], and Wolff 1978:212[ff]; for Galileo's view that circular motion persists, see Maier 1967:488[ff], Clavelin 1968:239, and Wallace 1981:323. Wallace's chapter 15, a discussion of Maier's essay on the impetus theory in Galileo, includes a detailed examination of questions on projectile motion by several of Galileo's predecessors at the Collegio Romano.

21. The argument is also found in Abra de Raconis: "It can happen, as in a vacuum given by divine power, that an Angel or someone else should hurl a missile, and this motion would not be by something urging it on from behind, since there is no such thing; it therefore comes from elsewhere, namely from an impressed *impetus*" (*Phys.* 248).

22. Mutius Vitelleschi, one of the professors at the Collegio Romano studied by Wallace, argues on similar grounds that a preternatural motion, namely the circular motion of fire around the heavens (see §7.1), could go on forever (Wallace 1981:335, 337). I am not suggesting, of course, that Descartes knew of Vitelleschi's unpublished notes, but that a version of Descartes's first law would not have been alien to one strand of Aristotelian thought. Descartes would more likely have come across such reasoning in Galileo's *Dialogo sopra i due massimi sistemi del mondo* (1616; cf. Maier 1967:488); but it is with Beeckman that a fully attested connection can be made.

absent or even immediately destroyed, the impulse impressed on the projectile can remain and bring about its motion. Or else on account of the resistance of the medium, but that too is false, because otherwise, since in a vacuum there is no resisting medium, and since in it there could occur the movement of projectiles and impressed impulses, an impulse of that sort would never cease; therefore if that impulse were produced in a projectile, it would persevere in it perpetually. (*Phys.* 249)

Referring to Suárez, he argues that nevertheless the "impulse" would cease, "namely from the repugnance of the body in which it is impressed" (250).[23]

The Aristotelian account thus postulates a force or *impetus* which is given to a projectile by the projector. Some philosophers held that the force is under certain circumstances opposed by nothing, and therefore will continue to cause motion; others that it would, since it is extraneous, be opposed under all circumstances and would always cease to act.

It is by way of opposing the very idea of such a force that Descartes's sometime mentor Beeckman first formulates his version of the first law. Commenting on an argument of Scaliger, he writes:

> [Postil: *Coelum semel motum semper movetur.*] On Scal[iger], *de Sub. exerct.* 68,1 [Scaliger *Exer.* 68¶1, 105r–106r].—It seems that one should hold that the heavens are moved not by the Intelligences, nor by a continuous volition [*nutu*] of God, but rather by their own nature and that of their place, they can never by themselves come to rest, once moved. (Beeckman *Journal* 18 Jul. 1612, 1:10)

The argument is that since the heavens—by which he means the heavenly spheres—are homogeneous, and everywhere equidistant from the center of the universe, "there is no reason why they should be said to be able to come to rest per se." They have no reason to rise or fall, and they are, so long as they merely revolve, always in the same place. It follows that, if somehow they are moved, they will continue to move forever. Sometime after June 1613, five years before his encounter with Descartes, Beeckman generalized the conclusion. "No thing," he now writes, "once moved, ever comes to rest unless on account of an external impediment" (1:16). So the sun will continue to move, even if it is not "bound" to a heavenly sphere. More signifi-

---

23. Abra de Raconis's reference is inaccurate, at least in relation to the modern editions of the *Disputationes*. The passage he probably meant is *Disp.* 21§3¶27 (*Opera* 25:801). But Suárez is there attempting to explain why the impetus of a body *thrown upward* disappears, a somewhat different question. Nevertheless his conclusion suits Abra de Raconis's purposes: "because this quality is in a contrary subject that resists its activity, on whose account alone this quality is impressed by way of an instrument, it does not require always to be conserved there, and because otherwise the subject always resists it and its action, the nature of such a thing requires that gradually it should cease from being conserved."

cantly, a stone, "propelled in a vacuum, would move perpetually; but the air, which always strikes it anew, hinders it, and thus brings it about that its motion is diminished" (1:24). Beeckman now included not only all bodies, celestial or not, but both varieties of motion, circular and rectilinear.

The Aristotelians never took the absence of external causes to be a sufficient account of persistence. Beeckman does. Against Scaliger, who had argued that because the heavens are always in their natural place, their movement has no natural cause and is therefore voluntary,[24] Beeckman replies that if we set aside the cause of their *beginning* to move, the cause of their *continuing* to move is simply that there is no reason for them not to.[25] One need not suppose that God or the celestial intelligences constantly intervene to keep them going. Beeckman cites the Ockhamist maxim, by now common property: *quod ergo fieri potest per pauca, male dicitur fieri per plura* (*Journal* 1:10). Although the Aristotelians suppose that continuing motion requires the continuing presence of a *vis impressa*, Beeckman holds that it requires no cause at all.

Descartes, for his part, was careful to deny that *conatus* denotes anything like a vis or power. Such language is to be interpreted as a kind of shorthand for certain counterfactual claims: "When I say that globes of the second element strive to recede from the center around which they turn, one should not suppose that I impute any thought to them, from which that striving would proceed; but only that they are so located, and incited to motion, that they would really go in that direction, if no other cause impeded them" (*PP*3§56, 8/1:108). In the next paragraph he adds that "frequently many diverse causes act at once on the same body, so that some impede the effects of others." Indeed, since there cannot be a vacuum, any body that moves is not just frequently but of necessity acted upon by others. But even if it is not clear how the antecedent in the counterfactual should be interpreted, the intention is clear: to construe *conatus* counterfactually serves to eliminate active powers, especially voluntary, from corporeal nature.

24. "What moves, moves so as to come to rest. It cannot come to rest except in its place. It will therefore not move from its place, if motion has already occurred in order to attain that place. So if a part of the Heavens moves by a purely natural motion, it is not in its place. But all parts of the Heavens are in their place, and so they do not move so as to occupy their proper place. Hence it is necessary that they should be moved by a motion other than a purely natural motion" (Scaliger *Exer.* 68, p105).

25. In *Le Monde* Descartes denies, in language not unlike Beeckman's, that one need "give a reason for the fact that a stone continues to move for a certain time after it leaves the hand of the person who throws it" (*Monde* 7, AT 11:41). Nevertheless he also believes that every existence, continuing or not, needs a cause. To that extent it is Beeckman who has the more modern view. But in other respects, notably in their view of the naturalness of circular motion, Descartes looks to be the more modern, in part just because he ascribes the conservation of motion to God, whose action, he believes, is simple. The contrast illustrates well the difficulty, or rather the imposition, in the notion of scientific advance or "modernity."

Yet Descartes does not quite believe that the persistence of a body in its present state needs *no* cause save the absence of reasons to change. *Every* existence, whether new or merely a continuation, needs a cause at every moment of its existence—the sustaining power of God, a voluntary cause par excellence. The proof of the first law rests on that necessity. But since the cause in question is supernatural, he agrees with Beeckman in denying that continued motion requires anything like a *vis impressa*.

One last point. Descartes in *Le Monde* 7 says the Aristotelians believe that motus, "contrary to all the laws of Nature, strives of itself to destroy itself" (11:40). He liked this argument well enough to repeat it in the *PP* (2§37, 8/1:63). The Aristotelians agree that a self-destructive tendency would be "most alien to the laws of nature" (Coimbra *In Phys.* 1:149; cf. §6.4). They say, moreover, that motus tends toward *quies,* which is to say that it tends toward a *terminus* whose attainment coincides with the cessation of *motus* ("*motus* of itself tends toward the *quies* of the *terminus ad quem,* at which it remains").[26]

But it would be surprising if such an outright contradiction had escaped the scrutiny of the Aristotelians or their Medieval predecessors. In fact it did not. Quies is twofold: there is the not-yet-changing *quies* of the *terminus a quo;* there is the already-changed *quies* of the *terminus ad quem.* Only the first is genuinely contrary to *motus.*[27] A seed that has not yet germinated has a *quies* contrary to the *motus* of growth, or, in other words, the privation of the *motus* to which it is naturally inclined. A full-grown plant, on the other hand, though it is no longer growing, does not have the privation of growth, but only the negation of growth. In §3.1 we saw that not just any negation of a property is a contrary of that property: this is a case in point. The *quies* toward which *motus* tends is not its contrary, but only its negation.

The point can be put another way: *motus* tends *per se* to its *terminus,* and to quies only *per accidens.* What is intended in change is a certain end, whose attainment is typically accompanied by the cessation of change. But in one instance at least, the end is attained precisely by the perpetual continuation

26. Toletus *In Phys.* 5c6text54 (229b29$^{\text{IT}}$), *Opera* 4:163rb; cf. Goclenius *Lexicon* s.v. 'quies', p944. Some modern critics of Aristotle would agree that motion tends to its own destruction. Kosman writes that motion is "auto-subversive, for its whole purpose and project is one of self-annihilation" (Kosman 1969:57; Waterlow 1982:106, 123). The Aristotelians would regard that as a confusion between the *per se* tendency of *motus* to its *terminus* and the *per accidens* tendency to rest.

27. "*Motus* can be called a kind of quieting [*quietatio*], that is, a way toward *quies,* which partly coexists with the *motus* itself. What moves is partly coming to rest with respect to part of the *terminus ad quem* and partly moving toward the remaining parts: it is therefore not contrary to *quies* itself: for a contrary does not tend to its contrary" (Toletus *In Phys.* 5c6text54, *Opera* 4:163rb). John of St. Thomas distinguishes between a *quies* that consists in the conservation of an acquired state, and to which *motus* is ordered per se, and a *quies* that consists in the privation of *motus* (*Nat. phil.* 1q1a1, *Cursus* 2:16, 172; cf. Coimbra *In Phys.* 1:33, where the distinction is credited to Scotus and Durandus).

of change: the motion of the heavens. Their end, which is the imitation of God and the promotion of the well-being of sublunary substances, is accomplished in the motion itself (Coimbra *In Phys.* 2c7q22a2 and c9q2a1, 1:314, 327). In general *motus* does not, simply by virtue of being *motus,* tend toward rest. If a certain kind of motus does naturally issue in rest, that is only by virtue of intending a specific end, not rest itself.

Descartes's first law was no novelty in the physics of the period, not even in Aristotelian physics. Though there certainly were Aristotelians, notably Suárez, who would have disagreed with it, there was, especially in the treatment of preternatural motion, ample precedent. Nor was it a novelty to treat *motus* as a state. Many Aristotelians took *motus* to be a mode either of the form acquired in passing or, in local motion, of the *mobile* itself (see §2.2).[28] What is new is Descartes' denial that motion is in any sense an incomplete entity; his application of the persistence of motion to *every* physical situation, so that a body in motion, whether its motion is constant or not, always has a tendency, determinate in direction and speed, to continue its motion; and finally his insistence that the cause of any change in motion must be external to the *mobile,* so that in Aristotelian terms no change in motion is natural.

2. *Direction, tendency, and the second law.* It is in the second law that Descartes begins to depart both from his predecessors and from his contemporaries Galileo and Beeckman. That law takes the following forms:

> When a body moves, even though its movement occurs most often in a curved line and though it cannot even make any [motion] that is not in some way circular [. . .], still each of its parts in particular tends always to continue its own [movement] in a straight line. And so their action, that is, their inclination to move, is different from their movement.[29]

> Each part of matter, considered separately, never tends in such a way that it would continue to move according to any curved lines [*lineas obliquas*],

28. John Wild writes that for Descartes motion "does not have the structure of from–to, but is a fixed mode or quality, like figure which is either present or not present." He regards that conception as a mistake that "may be traced back to [Descartes's] youthful inability to understand the Aristotelian descriptions of potency, as mediated by the late scholastics with whom he was familiar" (Wild 1941 in Moyal 1991, 4:32). But if to call *motus* a mode is a mistake, then not only Descartes but virtually all the central texts make it—and Aristotle, who sometimes calls *motus* a passion, a quality, or a quantity, does too. Descartes, though he jokes about the incomprehensibility of Aristotle's definition, did not misunderstand the notion of potency. He understood it quite well, and rejected it. Garber, too, seems to think that Descartes's treatment of *motus* as a mode distinguishes him from the Aristotelians (Garber 1992:196). But although Descartes certainly rejects the conception of that mode as a *via* or *fluxus,* by nature incomplete and tending to a determinate terminus, the ontological status assigned to *motus* is precisely the same.

29. "Lorsqu'un corps se meut, encore que son mouvement se fasse le plus souvent en ligne courbe et qu'il ne s'en puisse jamais faire aucun, qui ne soit en quelque façon circulaire [. . .], toutefois chacune de ses parties en particulier tend toujours à continuer le sien en ligne droite" (*Monde* 7, AT 11:44).

but only according to straight lines, even though many [parts] are often made to turn aside through their encounter with others, and though [. . .] in every motion a certain circle is made by all the matter which is moved at the same time.[30]

Although the first law applies to every state of a thing, the postil to 2§39 makes it clear that the second law is concerned to spell out just what God conserves when he conserves motion. "Every motion," it says, "of itself is straight" (63), to which one should add that since every straight line has a direction, every motion of itself has a direction. Determining that direction is no problem: if a particle actually moves in a straight line, the direction of its motion is at every instant the direction of the line. The direction in which it "tends" to move is presumably also the direction of the line. The purpose of the second rule is to show, by way of defining what God conserves at each "instant," that there is such a tendency even when it is *not* manifested without interference.

We have seen that the rupture or separation of two bodies, which is the absolute fact underlying ascriptions of motion, cannot but occur in an instant. I understand 'instant' here in the standard Aristotelian sense: an instant is the boundary of a determinate duration or interval of time.[31] The instant of rupture is the boundary between the interval during which two bodies are joined (or the duration of the one body compounded of the two) and the interval during which they are apart. It is in that sense that rupture occurs "in" an instant—not because an instant is akin to an infinitesimal *interval* of time but because rupture, like an instant, is a boundary between durations—that of the compound body, and those of its separated parts.[32] Descartes, as we have seen, treats the boundary as a mode, which in that respect is analogous to the surface shared by contiguous bodies. That mode

30. "Unamquamque partem materiæ, seorsim spectatam, non tendere unquam ut secundùm ullas lineas obliquas pergat moveri, sed tantummodo secundùm rectas; etsi multæ sæpe cogantur deflectere propter occursum aliarum, atque [. . .] in quolibet motu fiat quodammodo circulus, ex omni materiâ simul mota" (*PP* 2§39, AT 8/1:63; cf. 9/2:85).

31. "Instans physicis est idem, quod non tempus, seu extremum temporis, quod improprie dicitur tempus momentaneum, & indivisibile, seu individuum" (Goclenius *Lexicon* s.v. 'instans', p245; cf. Fonseca *In meta.* 5c13q10§2, 2:730A). Time was typically treated as a continuous quantity. Just as a line segment was said (though not always by the same philosophers) to be both bounded by and composed of indivisible points, an interval of time is both bounded by and composed of indivisible instants (needless to say, there were serious difficulties in devising a coherent doctrine of *indivisibilia*). See Suárez *Disp.* 45§5¶1, *Opera* 26:551, 559ᶠᶠ (on points) and §9¶1, ¶9, 26:586, 589 (on time as a quantity).

32. All the difficulties that attended contemporary attempts to understand how a line could be composed of indivisible points would, of course, also attend any attempt to understand how a motus enduring through time could be composed of instantaneous ruptures. It is not clear to me that Descartes ever confronted the problem (which is understandable, given his hostility to the paradoxes of the infinite and to infinitesimals in mathematics). I do think, however, that Descartes was not committed by his view to "discontinuous" time, and still less to the continual re-creation of the world at each instant (cf. n.88 below).

is, in Descartes's way of thinking, itself capable of having modes.[33] One of them is direction, denoted by expressions like *ad illam partem;* the other, which figures in the third law, is speed.

Descartes calls that mode a *conatus* or *inclinatio*. The reason, I think, is this. Suppose a body ruptures first from the vicinity of a certain collection of bodies *N*, and later from the vicinity of another collection *N'*. The mode of being joined-with-and-then-separated-from *N* cannot be numerically identical to the corresponding mode with respect to *N'*. Descartes explains to More that "there is one mode in the first point of a body *A* which is separated from the first point of a body *B;* and another that is separated from the second point; and another that [is separated from the third], and so forth" (*To More* 30 Aug. 1649, AT 5:405). It is not clear what sort of physical situation Descartes has in mind—either a sliding motion or else the simultaneous rupture of one surface from another conceived as infinitely many simultaneous point-ruptures. In any case, it would surely follow that if *A* ruptures first from *B* and later from another body C, a fortiori the two modes will be different. The first law tells us, then, not that numerically the same *motus* (defined as in 2§25) will persist, but that the same *kind* of *motus* will persist. Presumably that would be a *motus* having the same direction and speed. To have, at a certain instant, a *conatus* in a certain direction, then, is to have in that instant the mode of rupture in that direction; the word *conatus* connotes the persistence of that mode according to the first law.

What the law asserts of *conatus* is quite strong: at every instant the *conatus* of a moving body has a unique direction in the direction of the tangent to the trajectory of the thing at the point corresponding to that instant. It asserts, moreover, that the manifestation of *conatus* in a body not acted on by others would be rectilinear motion in that direction. What is proved, however, in his presentations of the second law is rather less.

There is, first of all, no argument to show that the direction assigned to the inclination should be that of the tangent. A look at two well-known Cartesian textbooks shows that the gap was noticed. Régis and Rohault start by observing that if a body *A* is determined at a certain instant to move toward a point *B*, and if its determination does not change until it reaches *B*, it will describe a straight line from its starting point to *B*. If a body has, on the other hand, traveled in a square, "one must conclude that at the four angles where it changed its determination, it was deflected by encounters with

33. In a slightly different context (a response to the objection of Hobbes that determination cannot be an accident of *motus*, since *motus* is itself an accident), Descartes writes that "there is nothing disagreeable or absurd in saying that an accident is the subject of another accident, as when one says that quantity is the subject of the other accidents [of corporeal substance]" (*To Mersenne* 12 Apr. 1641, AT 3:355; on quantity as the subject of accidents, see §5.3 above). Gabbey argues on the basis of that passage that speed and determination are modes of the mode *motus* (1980:257).

certain other bodies that resisted its movement and determination" (Régis *Cours* 1pt2c14, 1:337; cf. Rohault *System* 1c13¶5, 1:79). A circle can be regarded as an infinite-sided polygon; a body that moves in a circle therefore "undergoes a continual violence through encounters with other bodies," since otherwise it would not change its determination at every instant (Régis 1:338; Rohault 1:80).

According to Rohault, "the only Determination that is natural to a Body in Motion" is a straight line (1:79). The problem then is to determine which line the body will follow if released; that line will also represent its inclination even when it is not released. Rohault claims that the line will be the tangent. Unlike Descartes, he gives a reason: the tangent is that line that makes the "least Angle" with the curved path that the body was traveling on until now.[34]

Régis's argument is, to my mind, more Cartesian in spirit. If the body *A* in Figure 11 moves around the circle to *F*, it must be deflected at all the points between. But suppose that at *F* it encounters no obstacle. Though *A* will have taken on many different determinations as it moves, "the later [ones] destroy the earlier," so that at *F* it will have only the "last" determination. "But this carries it toward *G*, because it must take the inclination that the curved line has at the point *F*, which is measured by the tangent *FG*" (1pt2c14, 1:338).

Descartes, though he knew that mathematicians sometimes treated the circle as an infinite-sided regular polygon, rejected the use of infinitesimals in mathematics. On the other hand, he does treat the motion of falling bodies as if it consisted in a series of segments of uniform motion, each at slightly greater speed than the last, and each arising from the impact of a particle on the falling body. Circular motion, too, arises either because a body "is retained by something which obliges it to keep always at the same Distance" (Rohault *System* 1:80), or because it is at each moment resisted by others along its path. Although it would be incorrect to suppose that Descartes believed that there are infinitesimal motions, it would not be entirely misleading to say that *conatus,* or the "first preparation for motion" as he once called it (*PP* 3§63, AT 8/1:115), is treated *as if* it were an infinitesimal motion.

The arguments I have just examined presume that inclination, or "determination," as Rohault calls it, is rectilinear; Régis's argument in particular presumes that motion in the small must be in a straight line. That may seem

---

34. The argument is circular if the angle between a line and a curve meeting in a point *P* is defined as the angle between the line and the tangent to the curve at *P*. But one can show at least that for any other line through *P* besides the tangent at *P*, the tangent will be between that line and the curve. On the measurement of angles, see Hobbes *Critique* 23§2ff, p270ff, and the references cited there, as well as *De corpore* 2c14§7ff, *Opera* 1:159ff.

obvious now. It is built into the differential triangle used by later authors to calculate derivatives, and thus instantaneous speed and direction. But some of Descartes's contemporaries did not find it so. During his first collaboration with Descartes, Beeckman wrote that

> *what is once moved, in a vacuum always moves,* either according to a straight line or a circular line, either on its own axis, like the Earth in its diurnal motion, or around a center, like a ring. For since any minimal part of the circumference is curved, and curved in the same way as the whole periphery, there is no reason why the circular motion of the Earth should abandon this curved line and proceed on a straight line. For a straight line has no more natural and regular a nature and extension than does a circular line, because the parts of the circumference stand to the whole as the parts of a straight line to the whole. (Beeckman *Journal* 23 Nov.–26 Dec. 1618, 1:253; cf. 256)

Oddly enough, Beeckman had earlier argued that a stone set on a revolving wheel and suddenly released will not continue to move in a circle but "in a straight line toward the place toward which it was directed in the moment it was released" (*Journal* 16 Mar. 1618, 1:167). Though Descartes used such examples to argue that motion in the small is rectilinear, Beeckman evidently either did not consider them in that light or denied their relevance. Like Galileo, he continued to believe that circular motion is as "natural" as rectilinear motion, and that motion in the small can be either curved or rectilinear.

Descartes, then, could not take it to be obvious that motion, as it is "comprised in an instant," must be rectilinear. In *Le Monde* he offers this proof:

> God conserves each thing by an uninterrupted action; consequently he does not conserve it such as it may have been some time before, but precisely such as it is in the same instant he conserves it. But of all movements the straight line is the only one which is entirely simple and whose nature is comprised in an instant. To conceive it, it suffices to think that a body is in action to move toward a certain direction, and that is the case in each of the instants that can be determined during the time it moves. To conceive circular motion, on the other hand, or any other that might exist, one must consider at least two of its instants, or rather two of its parts, and the relation between them. (*Monde* 7, AT 11:44[f])

The nature of the straight line "comprised in an instant" is simply to be *toward* a certain direction. By 'direction' I mean the relation that holds between a point *A and any other point B* on a straight line through that point.

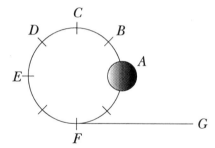

Fig. 11. Movement along a tangent (Régis *Cours* 1pt2c14, 1:338)

There are many paths from *A* to *B*. Of those paths only the line itself is given merely by giving the point *A* and the direction. Any other path—the circular *minimum* imagined by Beeckman, say—requires more information, an additional point *C* at least (see Figure 12). The directions from *A* to *C* and from *C* to *B* will be different. It is in that sense that only a straight line is "entirely simple."

In the *Principles* the proof is this: "The cause of this rule is the same as that of the preceding, namely the immutability and simplicity of the operation by which God conserves motion in matter. For he does not conserve it, unless precisely such as it is in the very moment of time in which he conserves it, without regard for that which may have occurred just before. And although no motion occurs in an instant, still it is manifest that whatever moves is determined, in the particular instants that can be designated while it moves, to continue its motion toward some part, according to a straight line, and not according to any curved line" (Descartes *PP* 2§39, AT 8/1:64). The simplicity here is not that of the motion but of God's "operation," and Descartes merely claims that it is "manifest" that what is determined to move toward some part will be determined according to a straight line. In the example that follows, Descartes adds only that none of the "curvedness" of the stone's motion prior to the instant at which it is released can be understood to remain in it. But the point is not whether the determination of motion in one instant includes its determination at earlier instants; it is whether the instantaneous mode of rupture could include, among its modes, a mode according to which it would move on a curve. The whole weight of the argument therefore rests on identifying the conservation of the mode of rupture at each instant with a simple operation of God; and the simplicity of that operation must rest not merely on the fact that motion in an instant has a *direction*, but that its natural path *along* that direction is a straight line. We are thus led back to *Le Monde* and its criterion of simplicity.

The Aristotelians employed different criteria. In *De cælo* Aristotle argues

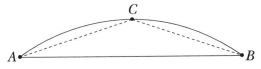

Fig. 12. Simplicity of rectilinear motion

that circular motion is simpler than rectilinear because its *terminus a quo* and *terminus ad quem* are one: "Circular motion is perfect, while rectilinear motions are imperfect, because they do not return to their starting point, like circular motion, but rather they have a *terminus* which is most distant and contrary to their starting point: and for that reason, as at the starting point they begin to move, so, when they are at their *terminus,* they begin to come to rest" (Aristotle in Thomas *In de Cælo* 2lect1, *Opera* [Parma] 19:78). When the naturalness of place is understood as its relation to the center of the universe, the heavens—or any body moving in a circle around the center— though they are moved locally, do not move from one natural place to another. Their motion, once begun, can be understood without reference to any other place but the one they occupy. In that sense it is simpler.

A second criterion of simplicity, this time in the sense of unity, is found in the Physics. Aristotle there defines the unity of *motus* in terms of the unity of the mobile, the unity of the time through which the *motus* occurs, and the unity of the form. In local motion, the form acquired in motion is place, and a local motion is therefore one if it is to one place (Coimbra *In Phys.* 7c4q5a2, 2:151; see also q4, 2:147[ff]). The motion of a falling body, therefore, is one motion, as is the motion of an animal toward its food. But the motions of the animal's parts, considered in relation to their elements, are typically not one, since the animal moves its parts both upward and downward. More generally, when the motion of a body is neither upward nor downward, its unity does not depend on the place it tends to but on the end with which the agent sets it in motion. For inanimate things, the only natural local motions unified with respect to those things themselves are the fall of heavy bodies and the rise of light bodies.

Descartes denies that any local motion in itself has a *terminus.* Even the tendency to move does not have a *terminus.* It only has a direction. Nevertheless, if we consider the direction to be operating in the role of a *terminus,* then Descartes's criterion of simplicity—a simple motion is that which has one direction—appears to be a transposition into his physics of the Aristotelian criterion of unity. Neither criterion, I should note, tells us anything about the *path* of motion. But if a thing moves in one direction only, and if direction is understood, as I suggested earlier, in terms of the relation of points on a straight line, then clearly motion in one direction only will be rectilinear. The tendency to *simple* motion will be a tendency to *rectilinear* motion.

The interpretation I have given of Descartes's proofs of the second law shows why Cartesian *conatus* is not *potentia* in disguise, but rather a substitute for it. The difference, despite Descartes's insinuations in 3§56 and elsewhere, is not that to attribute *potentia* to a thing is confusedly to attribute thought to it. I have said enough in Part I to show that the Aristotelians were not committing any such error, and that Descartes's descriptions of them are tendentious. The difference is rather that Cartesian *conatus,* though determinate in direction (and in quantity), is directed toward no *definite* place. In this it resembles not *potentia* but *impetus,* which is, as we have seen, a kind of bare striving forward with no natural *terminus.*

3. *The third law: contrariness and quantity of motion.* The third law and the rules of collision that derived from it have received far more attention than the first two laws. On the face of it there is an obvious reason, succinctly put by Paul Tannery in his notes to the French translation of the *Principles:* "While the two preceding laws are today considered to be truths scientifically acquired, the third has been rejected [*ruinée*] from the seventeenth century on through the work of Huygens on the impact of bodies. It is on this point that the principal error of Descartes' physics bears" (AT 9/2:86, note c). The falsity of the third law infects the rules of collision as well, and all subsequent explanations that depend on them.

But the clear sense of many historians that Descartes's adoption of the third law calls for explanation in a way that his adoption of the first two does not arises, I think, from a dubious method of interpretation. The method is to start from the true laws, or Newton's laws, and to understand Descartes's laws as deviations from them, as a movement of thought that fell short of its *terminus ad quem.* It is exemplified in the appendix by Tannery added to the French translation of the *Principles.* Tannery begins with the classical laws of the conservation of the movement of the center of gravity and the conservation of kinetic energy. He then shows that the rules of collision lead to deviations from those laws (9/2:328$^{ff}$). Numerous subsequent treatments have followed the same path, explaining Descartes's views in terms of what one might call the *aberrational* pattern. The pattern reflects an almost Cartesian sense that error results from the perturbation of a faculty that would otherwise track the truth.

Not only does Desmond Clarke, for example, following an almost universal practice, rewrite the rules in modern notation (instead of using proportions, as Descartes does—the point of insisting on the difference will become clear later). Clarke makes it his primary task to uncover the "deep conceptual confusions" that led Descartes astray, and to show that, however painfully, he was "genuinely groping his way towards a satisfactory dynamical account of familiar natural phenomena" (Clarke 1982:213, 226). He explains the "unsatisfactory character" of Descartes's rules by taking them to be "tentative efforts to formulate a dynamics." Efforts they were, but tentative they were not, at least in presentation.[35] There were, in any case, many other ways Descartes *could* have gone wrong, so the tentativeness of the rules tells us little about the way he *did* go wrong.

The aim of the analysis that follows is to restore some of the context in which Descartes worked. It will, I think, illuminate Descartes's thinking in a way that no analysis for which Newtonian physics is a *terminus ad quem* has done. I will analyze the law and its corollaries in terms of a hierarchy of contraries: *state, direction, volume, velocity* (or *total speed*), and *quantity of motion*. The reader familiar with the third law will note that quantity of motion is at the end, not the beginning of this list. I will then examine the contrariety of rest and motion and the Cartesian treatment of quantity of motion, showing that the rules of collision are to be understood to a large degree in the nonquantitative setting of the Aristotelian notion of contrariety, and that even the quantitative part of the rules is best understood in terms of the Aristotelian treatment of quantity of motion as an intensive quantity.

It is here that Descartes's use of proportions becomes significant: his term "degree of speed" (*gradus celeritatis, degré de mouvement*) typically denotes the aliquot parts of a *total speed* that is never specified. Hence the use of proportions rather than algebraic formulas like Huygens's $ax + by$ in his determinations of the exchange of quantity of motion.[36] The conclusion will be that the neglect of the relativity of motion, or of the continuity of outcomes that Leibniz expected, is not a *conceptual* failing on Descartes's part, as if he had overlooked an aspect of motion that would have been obvious.

The third law (in *Le Monde*, the second law) is stated in the following terms:

> When a body impels another, it cannot give it any motion without losing at
> the same time the same amount of its own motion; nor take from it any,

---

35. Descartes consistently took the difficulty that others found in understanding the rules to be a sign not that they were incorrect, or only approximately true, but that they were not sufficiently clear (see *To Clerselier* 17 Feb. 1645, AT 4:183; *Entretien avec Burman* ad 2§46, AT 5:168). Gabbey, I should note, has argued that the rules of collision were added very late in the writing of the *Principles* (Gabbey 1980:262f). Costabel concludes that the rules are "only a sketch," the principal part of an "attempt at translation [. . .] that aims to put simple and true instruments at the disposal of the average person" (1967:249). But they were not, of course, *presented* as mere attempts—a point Costabel himself underlines.

36. See the "Troisième Partie" of the 1652 draft of Huygens's *De motu corporum ex percussione* (*OEuv.* 16:99). The published *De motu* adheres to the proportional mode of presentation. I know of no passage where Descartes uses a formula like $ax + by$ to denote the total quantity of motion of two colliding bodies.

without augmenting its own by the same amount. (Descartes *Monde* 7, AT 11:41)

The third law of nature is this: where a body that moves meets another, if it has less force to continue according to a straight line than the other [has] to resist it, then it is deflected in some other way, and while retaining its motion gives up only its determination; but if it has more [force], then it moves the other body with it, and however much it gives [the other] of its own motion, it loses the same amount. (Descartes *PP* 2§40, AT 8/1:65; the French does not differ significantly)

The explanations after the rules bring out two points. The first is that the primary *expériences* upon which Descartes bases the rules comprise two essentially different outcomes of encounters among bodies. The first is *reflection;* the second I will call *absorption,* of which refraction is a special case. Since the rules tend to be treated in isolation from their application, and since preserving the phenomenon of reflection was crucial to the formulation of the problematic Rule 4, it is worth dwelling on these two archetypal outcomes.

Reflection consists of the reversal of that part of a thing's motion which is normal to the reflecting surface. The motion *AB* of a particle toward a surface *CD* is divided into a component *AF* parallel to *CD* and a component *AE* normal to *CD* (see Figure 13). Only *AE* is opposed by the surface *CD*. That surface, one should note, is part of a body which is supposed to be immovable (*inébranlable,* says Régis); otherwise the law of reflection will not hold precisely. The vertical component, successfully resisted by the immovable *CD,* is turned into its contrary *BF.* The horizontal component *AF,* which is unopposed, persists, so that in the interval of time *t* succeeding and equal to the interval in which the particle moved from *A* to *B* the horizontal component of the particle's motion will be *BH.* Since the entire motion of the particle must in the interval *t* bring it to the circumference of the circle CAGD, the vertical component of its motion can be constructed by drawing a vertical line upward from *H* to *G.* That segment is equal to *BF* (Régis *Cours* 1pt2c18¶7, 1:352f; cf. Descartes *Diop.* 2, AT 6:97, 102–104).

Reflection, then, is the archetype of the first part of the third law.[37] The point of dividing the motion *AB* into the components *AE* and *AF* is to remove from oblique reflection that which distinguishes it from perpendicular reflection, thus preparing the way to a straightforward application of

37. Garber credits John Schuster with remarking on the importance of reflection in Descartes's understanding of collision (Garber 1992:36on.40). Descartes, as I will argue later, discarded an earlier rule of collision in which the situations of Rules 4, 5, and 6 received uniform treatment because, according to that rule, a body colliding with another body at rest would never be reflected.

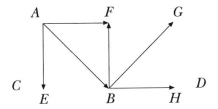

Fig. 13. Reflection (after Régis *Cours* 1:352)

the first part of the law. The reflecting body, assumed to be immovable, has all the force it needs to resist the reflected particle, and so the particle changes direction without changing speed. Since it gives up none of its motion to the larger body, and does not change in volume, quantity of motion is trivially conserved in reflection, just as the third law would have it.

The archetype of the second part is the movement of a solid body through a resisting fluid. In *Le Monde*, Descartes turns the tables on those who worry about an impressed force to carry projectiles along: "Having supposed the preceding [rule], we are exempt from the difficulty that the Doctors [i.e., the Aristotelians] find themselves in when they want to give a reason for the fact that a stone continues to move itself some time after leaving the hand of the one who threw it: for one should rather ask us why it does not continue always to move. But the reason [for that] is easy to give. For who could deny that the air in which it travels puts up some resistance to it?" (*Monde* 7, AT 11:42). Certainly not the Aristotelians; for some of them, in fact, the absence of resistance would entail that the movement was instantaneous. Descartes adds:

> But if one fails to explain the effect of [the air's] resistance according to our second Rule, and if one thinks that the more a body can resist the more it would be capable of arresting the movement of others [. . .]: one will right away have a great deal of difficulty in giving a reason why the movement of the stone dies away rather when it encounters a soft body, whose resistance is mediocre, than when it encounters a harder body, which resists it more. And also why, as soon as it exerts a little effort against [the harder body], it turns back immediately as if to retrace its steps, rather than stopping or interrupting its movement on that account. (ib.)

The second part of the third law, then, explains why motion in a fluid *continues but is slowed down*. That, in Aristotelian and Cartesian physics, is the only condition under which local motion in terrestrial bodies can naturally occur. Refraction, the object of Descartes's first successful work in mathematical physics, is a special case of motion in a fluid medium: it results from the *momentary* slowing down of a particle moving across the interface between media of differing hardness.

In accordance with the two archetypes, the rules of collision have effectively two kinds of outcome:

(i) *reflection:* reversal of direction, without loss of motion (Rules 1, 4, and the second part of 7);

(ii) *absorption:* continued motion in the same direction with transfer of quantity (Rules 2, 3, 5, and the first part of 7).

Rule 6 is exceptional in combining both reflection and transfer of motion, but that, as we will see, is because its conditions are a perfect compromise between those that give rise to (i) and those that give rise to (ii). In what follows I will first quickly lay out the hierarchy of contraries that I will use in my analysis, and then give a brief account of each law, after which I will take up the contraries rest and motion.[38]

The hierarchy is summarized in Figure 14. The first two entries, *state* and *direction,* are two-term polarities. Even though Descartes takes any direction to be contrary to any other, in the rules the only contrariety that matters is as between one direction and the opposite direction. In keeping with his terminology I call the opposites 'right' and 'left'. The last three rows are continuous quantities. What matters in the rules is whether one body, designated *B,* has a greater, smaller, or identical quantity. It is to those relations that the contest model delineated by Gabbey applies.[39]

I have arranged the contrarieties according to two principles. The first is that the rules of collision can be perspicuously presented in terms of a series of "tiebreakers" that successively determine the outcome of contests. According to that model the "stronger" of the two contestants wins—which is to say that it manages both to persist in its condition as far as possible and that it forces the other to change in such a way as to make its own persistence possible. This accounts for the order of the last three rows. The second is that Descartes treats changes of state, direction, and velocity (volume is fixed, and quantity of motion is therefore proportional to speed) as qualitatively different kinds of change.

To the list above I add one more pair of contraries, *union* and *disunion.* When the outcome of a collision is the continued motion of the two bodies

38. The longer version of Gabbey's essay lays out a "two-fold contrariety model" of the rules; a close reading of the account shows that there are three contrarieties: rest and motion, direction, and quickness or slowness of speed (Gabbey 1980:260$^f$). The latter, though it is mentioned in 2§44 as a kind of contrariety—slowness being to quickness as rest to motion—is, in Rule 7, absorbed into a more complicated comparison of quantities of motion.

39. See Gabbey 1971:16$^{ff}$ and Gabbey 1980:243$^{ff}$, citing Herivel 1965:49$^f$.

| Contrariety | Terms | Symbol |
|---|---|---|
| State | Rest, motion | $s, s^*$ |
| Direction | Right, left | $d, d^*$ |
| Volume | <, >, = | $M$ |
| Velocity (total speed) | <, >, = | $v, v^*$ |
| Quantity of motion | <, >, = | $Q$ |

Fig. 14. Contrarieties in the rules of collision (symbols with a * denote the condition after collision)

in the same direction, they will in fact be moving with the same speed and will therefore be at rest with respect to one another according to the definition in 2§25. But that amounts to saying that they become *one* body or, more pregnant, that the loser is forced to give up its integrity and to become part of the winner. Descartes, in a paragraph I will examine more closely later, writes that each body—one should say "notional" body—has, if joined with another, a force [*vis*] to resist attempts to separate it from the other, and that if it is disjoint from another, it has a vis to remain disjoint. But being joined with another is nothing other than being at rest (in the strict sense of 2§25) with respect to it, and being disjoined or ruptured from another is being moved with respect to the other. The overcoming, therefore, of a body's state of rest, and the dissolution of its integrity, are the same occurrence viewed under different aspects. But because union and disunion are rather the effects than the causes of change, I have left them out of Figure 14.

For the sake of convenience I now give the rules in full, as they appear in the Latin *Principles* (2§46–52, AT 8/1:68–70).[40]

> [*Rule 1*] If two bodies, say *B* & *C*, were entirely equal, and moved equally speedily, *B* from right to left, and *C* toward *B* from left to right, then when they met each other they would be reflected, and afterward strive to move, *B* toward the right and *C* toward the left, giving up none of their quickness.

40. For the French, which incorporates significant additions by Descartes, see 9/2:89–93 and Alq. 3:196–204. I translate *velox* and its derivatives by 'speedy' and so forth, *celer* and its derivatives by 'quick', and so forth. Picot uses *vitesse, vite,* and so on, for both.

[*Rule 2*] If *B* were even a little greater than *C*, and the rest as before, then only *C* would be reflected, and both bodies would move toward the left with the same quickness.

[*Rule 3*] If the volume [*mole*] were equal, but *B* moved even a little more quickly than *C*, not only would both strive to move toward the left, but also there would be transferred from *B* to *C* half of the quickness by which [*B*] exceeded [*C*]: that is, if there were at first six degrees of quickness in *B*, and only four in *C*, after their collision each would tend toward the left, with five degrees of quickness.

[*Rule 4*] If the body *C* were entirely at rest, and were slightly larger than *B*, then however quickly *B* moved toward *C* it would never move *C*; but would be repelled by it in the contrary direction: because a body at rest resists a greater quickness more than a lesser quickness, and this in the ratio of the excess of the one over the other; and so there would always be a greater force in *C* to resist, than in *B* to impel [*C*].

[*Rule 5*] If the body *C* at rest were smaller than *B*, then, however slowly *B* moved toward *C*, it would move [*C*] with it, namely by transferring to [*C*] part of its motion, so that both [bodies] would afterward move equally quickly: namely, if *B* were twice as large as *C*, it would transfer to [*C*] a third part of its motion, because that one-third part would move the body *C* as quickly as the two remaining parts of the body *B* which is twice as large. And thus after *B* met with *C*, it would move one third part more slowly, that is, it would require as much time to move through two feet of space as before to move through three. [. . .]

[*Rule 6*] If the body *C* at rest were most precisely equal to the body *B* which is moving toward it, it would be partly impelled by it, and partly repel it in the contrary direction: namely if *B* came toward *C* with four degrees of quickness, it would communicate to *C* one degree, and be reflected with the remaining three toward the opposite direction.

[*Rule 7*] If *B* & *C* were moving in the same direction, *C* more slowly, *B* following it more quickly, so that it caught up with [*C*], and if *C* were larger than *B*, but the excess of quickness in *B* were greater than the excess of size in *C*: then *B* would transfer only as much of its motion to *C*, as [would be required] for both afterward to move equally quickly and in the same directions. If, however, on the contrary the excess of quickness in *B* were less than the excess of magnitude of size in *C*, then *B* would be reflected in the contrary direction, and retain all its motion. And these excesses are computed thus: if *C* were twice as large as *B*, and *B* did not move twice as quickly as *C*, it would not impel [*C*], but would be reflected in the contrary direction; but if it moved more than twice as quickly, it would impel [*C*]. Namely, if *C* had only two degrees of quickness, and *B* had five, two degrees would be taken from *B*, which, being transferred to *C*, would effect only one degree [of quickness], because *C* is twice as large as *B*: and by that it would result that the two bodies *B* and *C* would afterward move with three degrees of quickness.

To keep matters simple, I concentrate on the Latin *Principles,* leaving aside the French additions and revisions, and the reformulated proof of Rule 4 in a letter to Clerselier.[41]

The rules fall first of all into two groups according to the contraries rest and motion. Rules 4, 5, and 6 treat the cases where C is at rest and B in motion; 1, 2, 3, and 7 the cases where both are in motion.[42] In the first group, with C at rest and B in motion, contrariety of direction is inapplicable, and we have three cases: $M_B < M_C$, $M_B > M_C$, $M_B = M_C$, corresponding to Rules 4, 5, and 6, respectively. I use the word *outcome* to denote the states and directions of the two particles after the collision, whatever the quantities of motion may be. The outcomes, according to the contest model, are as follows:

(i) [*Rule 4*] If $M_B < M_C$ then *B* loses. The direction of B after the collision, $d^*_B$, will be contrary to its direction $d_B$ before, since that ends the conflict between *B*'s state and *C*'s state, which will not change since *C* wins. *C*, it should be noted, maintains its integrity.

(ii) [*Rule 5*] If $M_B > M_C$ then *B* wins. *C* must change its state from motion to rest. The question then is how much motion *B* gives to *C*. Descartes holds that *C* not only changes its state, but loses its integrity as well. It therefore must move equally quickly as *B*.

(iii) [*Rule 6*] If $M_B = M_C$, then since there is no comparison of motion with rest, the remaining tiebreakers—speed and quantity of motion—are irrelevant; with no tiebreakers left, the only possibility is that both bodies lose and both win. *C* is forced to take on the contrary state, that is, to move, while *B* is forced to reverse its direction. Both retain their integrity.

It is essential to notice that in these three rules—the "rest" rules—the *velocity* of *B* is irrelevant to the outcome, where by 'outcome' I mean change or lack of change in *s*, *d*, and integrity. Descartes subsumes Rule 4 under the third law by supposing that the resistance of *C* toward being moved is always greater than the force of *B* toward moving it, on the (insufficient) grounds that *C*'s resistance is proportional to *B*'s speed.[43] I will return to that claim later.

41. See *To Clerselier* 17 Feb. 1645, AT 4:183–187. An analysis and translation of the French rules and the letter to Clerselier are given in Garber 1992:248–262. See also Clarke 1982, appendix 2 (reprinted in Moyal 1991, 4:110–122).

42. The following analysis is closest in spirit to those of Costabel and Gabbey, although I have taken more liberties with the order of the rules. Costabel subordinates the contrariness of rest and motion (which he calls a "contrariété de vitesse" on the basis of 2§44, see Costabel 1967:243) to the contrariness of direction, which allows him to take up Rules 1–3 before Rules 4–6.

43. The grounds are insufficient because they apply equally well to the situations of Rules 5 and 6. Garber suggests that the force of resistance is proportional also to the volume of the resisting body, so that *C*'s force of resistance will be equal to $M_C v$ (Garber 1992:240, 358).

In the remaining four rules, where both bodies are in a state of motion, contrariety of direction is applicable, and so they in turn fall into two groups. In Rules 1, 2, and 3 the directions of $B$ and $C$ are contrary. The symmetry of right and left allows the cases $M_B < M_C$ and $M_B > M_C$ to be regarded as one. We therefore have two cases: $M_B > M_C$ and $M_B = M_C$. When the volumes are equal, the next tiebreaker is speed, and again symmetry allows the cases $v_B < v_C$ and $v_B > v_C$ to be combined.

    (iv) [*Rule 2*] If $M_B > M_C$ then $C$ loses. It therefore changes direction and is joined with $B$. Descartes considers only the case where $v_B = v_C$. In that case the velocity of the resulting body $CB$ will be $v_B$. My suspicion is that the other two cases would have been treated analogously to Rule 5.

    (v) [*Rule 3*] If $M_B = M_C$ then speed is the tiebreaker. If $v_B > v_C$ then $C$ loses. It changes direction and is joined with $B$.

    (vi) [*Rule 1*] If $M_B = M_C$ and $v_B = v_C$, then there is no tiebreaker (the quantity of motion will be the same for $B$ and $C$). At least one of the two bodies must change direction, since they cannot interpenetrate. But $B$ does not resist the action of $C$ more than $C$ resists the action of $B$. So they both change direction, and neither yields any motion to the other. Nor, of course, does either lose its integrity.

Quantity of motion, it will be seen, has played no role in determining winners and losers. Its role is limited to determining, in those instances where a transfer of motion is mandated, the amount to be transferred. Any arbitrariness is removed either by supposing that the two bodies move as one, or that (in Rule 6) the outcome of a tie is a mixture of the two ways in which the tie could have been broken.

Only in the last rule is quantity of motion explicitly a tiebreaker. In Rule 7, both the state and the direction of the two bodies are the same. Both are moving, and moving to the left. Clearly if $B$ is to collide with $C$, it must have a total speed greater than that of $C$; but in the only cases considered under the rule $C$ has a greater volume than $B$. Quantity of motion is therefore the only tiebreaker left.[44] There are three cases: $Q_B < Q_C$, $Q_B > Q_C$, $Q_B = Q_C$. The third case is taken up only in the French *Principles*.

---

Since $B$'s force of motion is equal to $M_B v$, and $M_C > M_B$, $C$'s force of resistance will always overcome $B$'s force of motion (Gabbey has a similar account but with a different estimate of $C$'s force of resistance; see Gabbey 1971:27$^f$). That is clearly the intent of the French version of the proof of Rule 4; the Latin is not so clear.

44. Descartes does not in fact state the condition explicitly in terms of quantity of motion in Rule 7. He writes: "if the excess of quickness in $B$ were greater than the excess of size in $C$." 'Excess' denotes a ratio greater than one, so we have $v_B > v_C$, $M_C > M_B$, and $v_B/v_C > M_C/M_B$. That is equivalent to $Q_B = M_B v_B > M_C v_C = Q_C$.

(vii) [*Rule 7a*] If $Q_B > Q_C$, then $C$ loses its integrity, while $B$ gives up sufficient motion so that the two bodies can travel at the same speed.

(viii) [*Rule 7b*] If $Q_B < Q_C$, then $C$ keeps its integrity, while $B$ is forced to change direction; the speed of each body remains what it was.[45]

The analysis so far is summarized in Figure 15. What is striking is that, contrary to what one might expect from the third law, quantity of motion has only a minor role in determining the outcomes of collisions. It is the determinant per se of the outcome only when both bodies are in motion and in the same direction. In Rules 4, 5, and 6 quantity of motion has no role in determining the winner of a contest; it serves only to determine how motion will be allocated when there is a transfer.[46]

The rationale of the rules has two heterogeneous components. One, qualitative, serves to classify physical situations. Unlike a classical physicist, Descartes takes the situation of Rules 1, 2, and 3, in which the directions of motion are contrary, to be physically distinct from that of Rule 7, in which they are not. More egregiously from the classical point of view, he distinguishes the rest cases in Rules 4, 5, and 6 from the others.

The other component, which is quantitative, serves to determine a precise amount of motion to be transferred when there is transfer. The third law merely tells us that the amount by which the motion of one body is decreased (or remitted, to use the proper Aristotelian term) and the amount by which the motion of the other is increased (or intensified) will be equal.

I will now examine first the contrariness of state, and then the concept and use of quantity of motion.

*State.* A physicist now will write $v_C = 0$ to denote the state of the body $C$ in the rest cases (Rules 4–6).[47] There are innocent anachronisms, no doubt, but this is not one of them. Instead it conceals the fact that Descartes did *not* incorporate the relatively new conception of rest as a limiting case of motion into his physics.

In Aristotelian physics, as I have mentioned, there are two quite different ways of being without motion. When a thing is *in potentia* such and such its *quies* is a genuine contrary to motion. But when a thing that was *in potentia* such and such and is now entirely *in actu* such and such, its *quies* is only the

---

45. The three cases of Rule 7 are clearly meant to parallel the three rest rules (4, 5, 6), with 7a corresponding to 5 and 7b to 4. If we work backward from 7a to 5, and from 7b to 4, then the counterpart to $Q_C$ in 4 and 5 should be a quantity greater than $Q_B$ if $M_C > M_B$ and less than $Q_B$ if $M_C > M_B$. The French *Principles* make it clear that the resistive force of $C$ is equal to the quantity of motion that *would* have been transferred had $C$ and $B$ both moved off to the left at equal speed. I return to this question below.

46. See Clarke 1982:220ff.

47. So, for example, Clarke 1990:212 and Jammer 1991:316.

| State | Direction | Volume | Speed | Q | Winner | Outcome | Union | Rule |
|---|---|---|---|---|---|---|---|---|
| $s_B$ contra $s_B$, $s_B$ = motion, $s_C$ = rest | | $M_B < M_C$ | | | C | $d^*_B$ contra $d_B$ | ○ | 4 |
| | | $M_B > M_C$ | | | B | $s^*_C$ contra $s_C$ | ● | 5 |
| | | $M_B = M_C$ | | | — | $d^*_B$ contra $d_B$, $s^*_C$ contra $s_C$ | ○ | 6 |
| $s_B = s_B$, both in motion | $d_B$ contra $d_C$, $d_B$ from right to left | $M_B > M_C$ | $v_B = v_C$ | $(Q_B > Q_C)$ | B | $d^*_C$ contra $d_C$ | ● | 2 |
| | | $M_B = M_C$ | $v_B > v_C$ | $(Q_B > Q_C)$ | B | $d^*_C$ contra $d_C$ | ● | 3 |
| | | | $v_B = v_C$ | $(Q_B = Q_C)$ | — | $d^*_B$ contra $d_B$, $d^*_C$ contra $d_C$ | ○ | 1 |
| | $d_B = d_C$, right to left | $(M_C \geq M_B)$ | $(v_B > v_C)$ | $Q_B > Q_C$ | B | — | ● | 7a |
| | | | | $Q_B < Q_C$ | C | $d^*_B$ contra $d_B$ | ○ | 7b |

Fig. 15. Outcomes of collision (asterisk * indicates the value of a state after collision; symbol ● denotes union or the loss of integrity in the loser, while ○ denotes the absence of union and the maintenance of integrity; conditions in parentheses are not explicitly mentioned)

negation, not the contrary, of the *motus* that led up to it. Though both kinds of *quies* are qualitatively, not quantitatively, distinct from their corresponding *motus*, the first is a privatio, the second a natural *cessatio*.

For Descartes, since *motus* has in general no *terminus ad quem*, the distinction between the two kinds of *quies* has no pertinence. At each instant a body is either being moved or not, either undergoing rupture from the vicinity of neighboring bodies or not. If it is not, then it is in fact united with them. It is only a notional part of a larger body. The difference, for a given piece of *res extensa*, between being at rest and being in motion is as between *potential* and *actual* distinctness. Potential here is not to be understood in the Aristotelian physical sense of *in potentia*. That would imply that a potentially distinct body had actual distinctness as its *terminus*. What is potentially distinct is merely that which, by God's absolute power at least, can be made to exist separately. This use of *in potentia* one finds already in Aristotelian discussions of the metaphysics of indivisibilia like points and lines.[48]

Cartesian *quies* is therefore not a limiting case of Cartesian *motus*. There is, as Leibniz noted later with disapproval, a *saltus* between the two. That *saltus* one can trace back to Descartes's attempt to define *motus* as a *respectivum*, as a joint mode of two bodies. He is committed thereby to constructing, from properties admissible in his physics, a scheme that will yield a kind of absolute change that can plausibly be called change of place. Since he admits neither Aristotelian natural place nor anything like Newtonian absolute space, the alternatives are few. If one really conceives of body as *res extensa*, the only temporally varying relations among bodies are distance and adjacency. Relations of distance, however, require a choice of reference points from which to measure distance, and there is, as Descartes himself argues, no reason to choose one set of reference points rather than another. Only adjacency remains.

An immediate consequence of Descartes's definition of motion is that the *inception* of motion is always accompanied by the joining and disjoining of parts. When in Rule 5 the smaller body at rest is set into motion by the larger, and joined to it, the smaller body must be at rest with respect to some collection of neighboring bodies, and in fact part of them. In the letter to Clerselier, Descartes writes that "by a body which is without movement, I understand a body that is not in action to separate its surface from those of the other bodies that surround it, and, consequently, that forms part of another hard body which is larger" (*To Clerselier* 17 Feb. 1645, AT 4:187). The remark, whatever its success in vindicating the fourth rule, shows that

---

48. Suárez distinguishes potential *existence* from potential *isolation* (namely, by division). All of the infinitely many points in a line exist *in actu*, but no point exists in isolation, divided from all other points, except perhaps by God's absolute power. See Suárez *Disp.* 40§5¶29, 33ᶠᶠ, 44, *Opera* 26:559, 561, 563.

*setting a body in motion* and *separating one body from another* are coextensive. It should not be surprising that in the *Principles,* resistance to rupture and to being moved are mentioned together: "But here one must carefully note what the force of each body to act on another or to resist its action consists in: namely in this alone, that each thing tends, *quantum in se est,* to remain in the same state it is in [. . .] For this reason what is conjoined to another has some force to impede its being disjoined; that which is disjoined, to remain disjoined; that which is at rest, to persevere in its rest, and consequently to resist all those things that can change it; and that which moves to persevere in its motion, namely, in motion with the same quickness and toward the same direction" (*PP* 2§43, AT 8/1:66). Later Descartes argues that there is no "glue" stronger than rest by which to hold the parts of a hard body together: "No other [mode] can be more opposed to the motion by which these particles might be separated than their rest. And aside from substances and their modes, we acknowledge no other kind of thing" (2§55, 8/1:71). The "force of rest," to which I will return in §8.3, and the "glue" that binds together the potential parts of one body are one and the same. That point is often overlooked because in the rules of collision Descartes takes the bodies *B* and *C* to be "so divided from all others that their motion would be neither impeded nor helped by the others around them" (2§45, 8/1:67). Yet in Rule 4, at least, as the letter to Clerselier shows, he cannot entirely forget that *C* is united with a larger body.

Rest is not motion *in potentia.* Aristotelian *potentia,* which in Descartes's view is fatally infected with finality and obscurity, is supplanted by the colorless *potentiality* of the parts of quantity. Every part of a body is itself a potential body; all that is needed to make it so is the "action to separate its surface" from the whole. The contrast of union and disunion, and so of rest and motion, is of a different order than the contrast between one degree of motion and another. It resembles what is now called a *phase transition*—from solid to liquid, or liquid to gas—rather than the continuous increase of velocity suggested by the physicist's $v = 0$. Like a phase transition, the transition from rest to motion requires a *quantum* of force greater than that needed merely to bring about change within a phase. But unlike any phase transition, it sometimes cannot be accomplished by any *quantum* of force, however large.

Yet the suggestion that a body at rest resists the attempt of another to move it by virtue of being part of some larger body cannot be the whole story. The outcome, as we have seen, is continued rest only if the body *C* at rest is *larger* than the body *B* moving toward it. Then it will remain at rest no matter how quickly *B* is moving. If, on the other hand, *C* is not larger than *B*, then *C* will be moved no matter how slowly *B* is moving. But *C* will still be part of some larger body.

The problem is not just that the reason offered cannot account for the difference in outcomes. It is that when $C$ is at rest, there is no obvious reason to pick it out as *the* body with which $B$ collides rather than some larger or smaller part of the whole that $C$ belongs to. In the simplest case, where $B$ and $C$ are brick-shaped objects meeting face on, one might suppose that only those points of $C$ which lie along the line of motion of some point of $B$ are the part with which $B$ actually collides. Call that part $C^*$. What Rules 4 through 6 would tell us, if this is right, is whether $B$ will break off $C^*$ and absorb it, or else be reflected by $C^*$, so that $C$ remains whole.

I said that Descartes cannot entirely forget that $B$ and $C$ have other bodies around them. The point can be put more sharply: Descartes *cannot* regard the bodies surrounding $C$, when it is at rest, as neither impeding nor assisting its motion. Not, at least, if he wants potential bodies mutually at rest to cohere. Later authors have often chastised Descartes for "overlooking" the relativity of uniform motion. That is incorrect: he recognized that defining motion in relative terms was possible and rejected it (2§15, 28). The difficulty is rather in the handling of the conditions under which the laws would hold.

A look at Descartes's response to an *expérience* offered by Mersenne will clarify the issue. Mersenne, sometime in late 1641, has written to Descartes that a ball $A$, colliding with two balls $B$ and $C$, with $B$ as large as $A$, and $C$ smaller, will "push the small one $C$ by means of the large one $B$, almost without making $B$ move [*sans faire quasi mouvoir B*]" (*To Mersenne* 17 Nov. 1641, AT 3:452). The expérience is a counterexample not to the collision rules in the *Principles,* but to an earlier rule: when a body $B$ collides with a body $C$ at rest, it *always* communicates to $C$ whatever motion is needed for the two bodies to proceed with equal speed in the direction of $B$'s motion.[49] I will call Descartes's earlier rule, which is, as an early annotator of the manuscript letter noted, "contrary to his principles," the *1639 Rule,* although Descartes may well have arrived at it before 1639.

Mersenne takes the 1639 Rule to entail that in Figure 16, $B$ and $C$ should move off to the right with equal speed. But in fact $B$ stays almost at rest and $C$ alone moves. Descartes's answer is this: "Although at the first moment when the two balls $B$ and $C$ are touched [by $A$], they no doubt move with equal speed, still, because $B$ is heavier than $C$, it is much more hindered by the inequalities of the plane on which they roll, and it is those inequalities that stop the ball $B$ and are not capable of stopping the ball $C$; even if the 2 balls were of the same size, $C$ could go faster than $B$, since all the inequalities of the plane that resist it also resist $B$ following it, and they use their forces

---

49. *To Mersenne* 25 Dec. 1639, 28 Oct. 1640, AT 2:627; cf. *To Debeaune* 30 Apr. 1639, 2:543 and *To Mersenne* 28 Oct. 1640, 3:211.

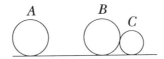

Fig. 16. Collision (AT 3:452)

conjointly to overcome them; but what resists *B* does not hinder *C*, which for that reason can immediately move away from [*B*]" The conditions of Mersenne's *expérience*, in other words, are not ideal. Had the plane been perfectly smooth, *B* and *C* would have rolled off to the right together just as they were supposed to.[50]

Perfect smoothness, or perfect inelasticity, is a condition that could be fulfilled, at least by divine power. But the condition of Rules 4, 5, and 6, which is both that a body be at rest and that its movement be unimpeded by others, cannot be, not if mutual rest is also supposed to explain the cohesion of the parts of a simple body.[51] If Descartes gives up the assumption that the body at rest is unimpeded, then he has no obvious way to determine its quantity. But if he gives up the equivalence of mutual rest and cohesion, he cannot explain why the outcomes in Rules 4, 5, and 6 should differ.

The dilemma is striking not least of all because it is absent from his earlier thinking. The 1639 Rule, the one to which Mersenne objects, yields the same outcome whatever the relative sizes of the body at rest and the body in motion. The two will always move off with equal speed in the original direction of motion. The rule is still incorrect, but it is at least less counterintuitive than Rule 4, if one is to judge from the objections to that rule.

There are, it seems to me, two reasons for what from our point of view is a step backward.[52] The first is that only Rule 4 (and the related Rule 6), and not the 1639 Rule, yields *reflection* rather than *absorption* when one of the two bodies is at rest. From the time he was writing the *Dioptrique* onward, Descartes always takes the reflecting body to be at rest; but the 1639 Rule would yield absorption.[53] The second is that, perhaps at the same time he realized

50. In his response to another counterexample, this time to Rule 5 (*To Mersenne* 23 Feb. 1643, AT 3:634ᶠ), Descartes appeals both to the imperfection of the plane on which the balls roll and to their elasticity. In a later letter he actually attempts to estimate the contribution of elasticity to the outcome (*To Mersenne* 26 Apr. 1643, 3:652ᶠ).

51. Gabbey notes the dilemma, adding that "coherence seems not always to have been Descartes' strong suit" (Gabbey 1980:314n.162). I am suggesting that although the position may have been physically incoherent, it had serious motivations.

52. The interpretation given here differs from that of Clarke, who takes Descartes to have used already a version of the "least modal change principle" in formulating the collision rules in the Latin *Principles* (Clarke 1982:223ᶠᶠ). I am inclined to take the letter to represent a later stage in Descartes's thinking.

53. In a letter to Mersenne from 1640 Descartes notes that the reflecting surface must be "hard and immobile," and that in nature "the reflection of an ordinary ball never occurs exactly in equal angles, nor perhaps that of light" (*To Mersenne* 28 Oct. 1640, AT 3:208; cf. also

that the 1639 Rule would not work, he was reading Aristotelian treatments of motion. Although he rejected the finality implicit in the Aristotelian scheme (§3.1), he incorporated into the contest model, already present in *Le Monde,* the contrariety of rest and motion and of direction. He was able, therefore, to adapt to his own purposes a language familiar to the Philosophers, whom he hoped would adopt the *Principles* in teaching. The result is a hybrid, in which the contests that determine change of state and change of direction are prior to the comparison of quantities of motion; the conservation of the quantity of motion does not determine outcomes but serves only to determine the allocation of motion when motion is transferred.

But it is not evident that the world could *not* be as Descartes describes it. His laws are not, indeed, invariant under Galilean transformation from one rest frame to another moving uniformly with respect to it.[54] They are invariant only under the mirroring of right and left, which he uses to reduce the number of cases to nine (Rules 1 to 6 together with the three cases of Rule 7).[55] Nor do they satisfy Leibniz's continuity condition: to the continuous passage from $M_B/M_C = 1 + \varepsilon$ to $M_B/M_C = 1$ there corresponds a passage from absorption (Rule 2) to reflection (Rule 1), a gross disparity in outcome (Leibniz *Ph. Schr.* 4:378; cf. also Jammer 1990:317). That, however, is what the contest model would lead one to expect, when the stakes—the direction of *C*'s motion—are discontinuous. I am not sure why God could not have created a world in which the laws of physics were *not* invariant under Galilean transformations (in fact Maxwell's laws are invariant only under Lorentz transformations), or in which continuous change of physical quantities yielded discontinuous outcomes (as in the photoelectric effect). Perhaps he didn't; but not because he *couldn't.*

*Speed and quantity of motion.* The third law tells us that whenever motion is transferred, the total quantity of motion does not change. But it does not tell us directly how much will be transferred. Nor does it tell us how quantity of motion is to be measured. All that the *Principles* tell us in general is that the

---

To Mersenne for Hobbes, 21 Jan. 1641, 3:289, and *Diop.* 2, AT 6:94–95). Both Mersenne and Bourdin ask Descartes why in oblique reflection the vertical part of the body's movement is changed while the horizontal part is not (cf. *To Mersenne* 29 Jul 1640, 3:107[f] and 111, and *To Mersenne* 30 Jul. 1640, 3:129), to which his answer is that the reflecting surface opposes only the vertical determination of the movement of the reflected body.

54. A *Galilean* transformation has the form $x^* = x - vt$, $y^* = y$, $z^* = z$, $t^* = t$, where the * denotes coordinates in a frame $F^*$ moving at velocity $v$ along the $x$-axis of the original frame $F$ (Rindler 1979:3). A law is *invariant* under Galilean transformations if it remains true when $x^*$, etc., are substituted for $x$, etc. A *Lorentz* transformation is the counterpart of the Galilean transformation in special relativity (ib. 31).

55. He leaves out the case where both bodies are in motion, in opposite directions, $M_B > M_C$, and $v_B \neq v_C$. The case $v_B > v_C$ could presumably be regarded as analogous to Rule 3. The case $v_B < v_C$ cannot be decided except perhaps in the manner of Rule 7, by comparing $Q_B$ and $Q_C$. I find it significant that he did not do so: it supports the claim that in Rules 1 to 3, where directions are contrary, the contest is between volumes and speeds rather than directly between quantities of motion.

force with which a body perseveres in rest or motion "should be judged sometimes [*tum*] by the magnitude of the body in which it exists, and the surface according to which that body is disjoined from another; sometimes [*tum*] by the quickness of the motion, and the nature and contrariety of modes by which diverse bodies meet one another."[56] I will set aside the problem of measuring the force of resistance of a body at rest and consider only the force of a body in motion. That force, Descartes tells us, is directly proportional both to speed and to volume or *moles:* "When one part of matter moves twice as quickly as another, and the other is twice as big as the first, then the same *motus* is in the smaller as in the larger; and by whatever amount the motus of one part becomes slower, the *motus* of some other [part] becomes quicker" (*PP* 2§36, AT 8/1:61). Provided we know how to individuate bodies, the nature and measurement of volume present no serious difficulty. Speed, on the other hand, is, like motion, not obviously a mode of extension, nor is its measurement simple. In what follows I will, by comparing the Aristotelian and Cartesian treatment of speed, show that Descartes remains closer to his predecessors than is usually recognized. I will show also that quantity of motion was conceived in geometric rather than in algebraic terms, and that the rationale of its allocation in the collision rules becomes much clearer once the geometric conception is worked out.

Descartes treats the terms 'quality' and 'mode' as having the same denotation. They differ only in connotation: when we consider the substance affected or varied by a mode or quality, we call it a *mode;* when we denominate it by way of the variation, we call it a *quality* (*PP* 1§46, AT 8/1:26). I know of no passage in the Principles where Descartes calls *motus* a quality rather than a mode. In *Le Monde*, however, he writes that rest and motion are both qualities, distinguishing them not from qualities *simpliciter* but from "real" qualities, a term used to denote sensible qualities as they inhere in things (*Monde* 7, AT 11:41; cf. Goclenius *Lexicon* 912). This is in keeping with the definition in 2§25 of *motus* in the strict sense of rupture: rupture or the absence of rupture is not itself a quantity but a mode of the surfaces of the things ruptured.

Earlier I left it unresolved whether Descartes could define speed and direction for *motus* in the strict sense. I suggested only that by appealing to the notion of physical trajectory one might be able to define them without shifting to motion in the vulgar sense—motion relative to an arbitrary frame of reference. The first thing to note is that Descartes inherits from the Aristotelians a twofold notion of speed. The first is *average speed,* measured

---

56. "Visque illa debet æstimari tum à magnitudine corporis in quo est, & superficiei secundùm quam istud corpus ab alio disjungitur; tum à celeritate motûs, ac naturâ & contrarietate modi, quo diversa corpora sibi mutuò occurrunt" (*PP* 2§43, AT 8/1:67). The force of *tum . . . tum . . .* is ambiguous: it can indicate either alternative or coordinate members of an enumeration.

by comparing how far two bodies will travel in a given time. In Rule 5, for example, Descartes writes that "after B collided with C, it would move one-third slower than before, that is, it would require the same time to move through two feet of space as before it required to move through three" (2§50, 8/1:69).[57] The second notion, alone referred to in the collision rules, is *instantaneous speed*. Instantaneous speed, whether alone or combined with volume into quantity of motion, is what determines the outcome of collisions in the rules where both bodies move; in Rules 5 and 6, volume determines the outcome, but the allocation of motion after a collision depends on the instantaneous speed of the moving body.

In the Aristotelian tradition, instantaneous speed was held to be an intensive quantity—the intensity of *impetus*, which was typically taken to be a quality analogous to heat.[58] Extensive quantity is quantity of bulk [*quantitas molis*]. Intensive quantity of virtue [*quantitas virtutis*] is that, "according to which one [virtue] is more perfect than another [. . .] Similarly, one quality is more intense than another, or weaker: and according to that there is also proportion. For we say that one heat is twice as intense as another" (*In Phys.* 7c5q5, *Opera* 4:205ra). The chief difference between intensive and extensive quantity is that the parts of an intensive quantity, unlike those of an extensive quantity, cannot be separated to yield smaller quantities of the same kind. But one intensive quantity can—so the Aristotelians believed—be said to stand in proportion to another, whether of the same quality and subject, or of others. In particular, an intensity can be divided into proportionate parts or *degrees*, and smaller or larger intensities measured according to those degrees.[59]

The same quality can have both an extensive and an intensive quantity. The heat of a warm body, for example, has an extensive quantity, the volume of the body, and an intensive quantity that may vary from part to part. For a body having but one spatial dimension—a line, in other words—Nicole Oresme represented the extensive quantity by a horizontal line segment and the intensive quantity at each point by another line perpendicular to the first line at that point. The *area* of the resulting figure Oresme treats as a quantity that may be compared with other quantities (*De config.* 3c5, p404[ff]). One could call it the *total quantity* of a certain quality in a subject.

<hr />

57. The measurement of speed so defined does not require measurements of elapsed time, which could rarely be made with sufficient accuracy. It requires only that certain events—the inception of motion, the passing of a mark—be observed to be simultaneous. The measurement of speed in rectilinear motions by distance traveled in equal times comes from Aristotle (Toletus *In Phys.* 7c4q4, *Opera* 4:203rb).

58. See Maier 1958, c.3 and Murdoch & Sylla 1978:231–241 for surveys of Medieval treatments of motion as an intensive quantity.

59. The third book of Mersenne's *Nouvelle pensées de Galilée* illustrates the geometric, rather than algebraic, manner in which relations between time, speed, and distance traveled were treated (Mersenne *Nouv. pens.* [167[ff]]).

One application of that method of representation is to motion. In a simple body, motion will have a uniform intensity over all parts of the body. The only extensive dimension along which its intensity varies is time. Oresme represented motion by diagrams in which the horizontal axis is the duration of the motion and the length of the perpendiculars is its intensity, which is to say, its speed. Descartes had some knowledge of the technique. Diagrams similar to Oresme's appear sporadically in his works, notably in the 1629 proof of an incorrect law of falling bodies, and in a correct law included in a letter to Mersenne in 1634.[60] Descartes also used such diagrams to represent what we now call *work* as *une force à deux dimensions*.[61]

The instantaneous quantity of motion in a body, which is the measure of its impulsive or resistive force, can be represented by a rectangle whose base is a line proportional to volume, and whose height is proportional to the speed of the whole body. The third law tells us in effect that the sum of the areas of the two rectangles representing the quantity of motion of the two interacting bodies before the collision will be equal to their sum after the collision. It does not tell us what they will be.

We have seen that the outcome of a collision is either a reflection or an absorption. The last six rules can be divided accordingly into three groups. In the first group, Rules 4 and 7b, there is *pure reflection,* that is, change of direction of the loser with no exchange of motion (Rule 1 yields a reflection of both bodies). In the second group, Rules 2, 3, 5, and 7a, there is *pure absorption,* that is, the two bodies move in the direction of the winner with the same speed. It follows that there will be an exchange of motion between

60. *To Mersenne* 13 Nov. 1629, AT 2:72, *To Mersenne* 14 Aug. 1634, 2:304; cf. also the early *Cogitationes privatæ* 10:219 (a correct proof of the law, in response to a problem set by Beeckman). The diagram in Beeckman's version (*Journal* 1:262) differs from an Oresmian diagram in making the vertical axis represent time and the horizontal ordinates speed. In the letters to Mersenne, Descartes, rather confusingly, makes the vertical axis represent both the total distance traversed (a quantity that in Beeckman's diagram is represented by the *area* under the curve representing velocity) and the speed of the motion at instants during the motion, represented by the horizontal axis.

Even in 1634, when he states the law correctly, he still gets the diagram wrong (Mersenne, on the other hand, drawing on Galileo, both states the law correctly and gets the diagram right in his *Harmonie universelle* [2pr2, Mersenne *Harm. univ.* 1:89], which was completed in 1634). Only in 1639, in a letter to Mersenne on the motion of strings, do we find a diagram in which intensity (the "force" applied to the string) is plotted against time, not distance, and distance represented by areas (*To Mersenne* 30 Apr. 1639, 2:535). Later Descartes uses a similar diagram to illustrate the speed of descent of falling bodies (*To Huygens* 18 or 19 Feb 1643, 3:620; on this letter see Nardi 1986). In other less carefully drawn figures it is impossible to tell what the vertical axis is supposed to represent (e.g., *To Mersenne* 11 Mar. 1640, 3:36).

61. In letters to Mersenne on the relation between *force à une dimension* and *force à deux dimensions,* the "one-dimensional" force required at each point of a body's motion to lift it is plotted against the path traversed; the area of the resulting rectangle represents the "two-dimensional" force or the work required to lift the body through the whole distance (*To Mersenne* 12 Sep 1638, 2:357, 359; cf. Carteron 1922:245, 252). Beeckman had earlier used the Flemish word *cracht* 'force' to denote the product of weight and distance moved (*Journal* 1–8 Oct. 1628, 3:93; cf. De Waard's note on Beeckman in Mersenne *Corr.* 2:123).

them. In the third group, which includes only Rule 6, there is reflection and transfer of motion; Descartes characterizes the outcome as a compromise between the outcomes of Rules 4 and 5.

Given that the outcome includes a transfer of motion, the conservation of total quantity determines the allocation of motion after collision only in combination with the supposition that the outcome should be absorption. I repeat the relevant portions of Rules 5 and 7:

> if B were twice as large as C, it would transfer to [C] a third part of its motion, because that one-third part would move the body C as quickly as the two remaining parts of the body B which is twice as large. And thus after B met with C, it would move one-third part more slowly, that is, it would require as much time to move through two feet of space as before to move through three.

> [. . .] if C had only two degrees of quickness, and B had five, two degrees would be taken from B, which, being transferred to C, would effect only one degree [of quickness], because C is twice as large as B: and by that it would result that the two bodies B and C would afterward move with three degrees of quickness.

There are, clearly, allocations of motion that satisfy the third law and leave B moving less quickly than C. Only the third law together with the requirement that after the collision $v_B{}^* = v_C{}^*$ determines how much motion should be allocated to C.

One could immediately represent the outcome in Rule 5 by

$$v_B{}^* = v_C{}^* = \frac{M_B}{M_B + M_C} v_B.$$

But it seems to me to be more in keeping with Descartes's thinking to proceed in the following way. In Rule 5, he writes that one-third of B's motion will move C, which is half the size of B, as quickly as the remaining two-thirds of B's motion move B. Represent B's original quantity of motion $Q_B$ by a rectangle R of arbitrary base and height, and represent C's original quantity of motion, which is nil, by a line half as long as the base of R, as in Figure 17. Descartes tells us in effect that "degrees" of the motion of B can be converted into degrees of the motion of C in proportion to their volumes, so that in Rule 5 one degree of B's motion yields two degrees of C's motion. In Rule 7a, on the other hand, C is twice the size of B, and so two degrees of B's motion yield one degree of C's. The rationale of the conversion is easy to see if one imagines converting areas on the base ab in Figure 17 into areas on the base cd, as if one were pouring fluid from one graduated cylinder to another.

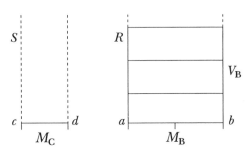

Fig. 17. Calculating motion after collision (Rule 5)

The problem of determining how much motion should be transferred to $C$ then becomes the problem of transferring a certain degree $x$ of the original motion $v_B$ of $B$ to $C$ according to the conversion factor $M_B/M_C$ so that the new motion $u_B - x$ of $B$, or the level of fluid in $R$, will be equal to the new motion $(M_B/M_C)x$ of $C$, or the level of fluid in $S$. A little calculation shows that

$$x = \frac{M_C}{M_B + M_C} v_B.$$

The new motion of the two bodies will be $(M_B/(M_B + M_C))\, u_B$, just as before. Similarly, in the example used by Descartes in Rule 7a, we have initially the situation in Figure 18(a). The motion of B is divided into five degrees, of which two convert into one degree of motion in C. Since C has by assumption two degrees already, diminishing B's motion by two degrees and transferring them to C will leave both B and C with three degrees of motion. We end up with the situation in Figure 18(b).

If Descartes had adhered to the 1639 Rule mentioned earlier, in which motion is always transferred when a body at rest meets a body in motion, then Rule 7a could be reduced to Rule 5 by taking the initial velocity of C to be zero and that of B to be $u_C - u_B$. That rule, however, has the fatal defect of making it impossible for reflections ever to occur. Instead we have the roundabout comparison of $u_B/u_C$ with $M_C/M_B$, and the division of cases according to the obscure thesis that relative slowness "participates in the nature of rest" (PP2§44).

Taking minor liberties with the term, I call this the *hydrostatic* model of the transfer of motion. In it a simple body is treated as a graduated cylinder with base proportional to its volume, into which or out of which degrees of motion can be poured. The level of the liquid in the cylinder corresponding to one body can be compared with that in others; in practice this means

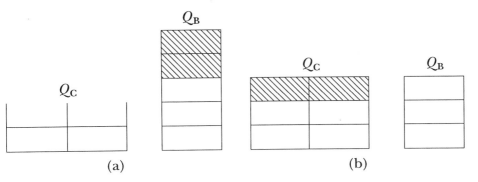

Fig. 18. Calculating motion after collision (Rule 7a)

comparing the "degrees" or fractional parts of one body's total speed with equal levels of liquid in the cylinder corresponding to another. Since degrees need never be converted into absolute units, one need not take the variable $v$ to be in units like meters per second, any more than in a geometrical diagram one need take the lengths of lines to be measured in meters or yards.

The hydrostatic model yields some insight into Descartes's comments on "natural inertia" in a letter to Debeaune in 1639. After having denied in an earlier letter that there is any "natural inertia or slowness [*tardiueté*] in bodies" (*To Mersenne* Dec. 1638, AT 2:466), Descartes now explains that "since if each of two unequal bodies receives as much motion as the other, this same quantity of movement does not give so much speed to the larger as to the smaller, one may say in this sense that the more matter a body contains, the more *Natural Inertia* it has; to which one may add that a body which is large can better transfer its movement to other bodies than a small [body], and that it can be moved less by them. So that there is a sort of *Inertia,* which depends on the quantity of matter, and another [sort] that depends on the extension of its surface" (*To Debeaune* 30 Apr. 1639, 2:543–544). The "other" sort, which is clearly related to the "force of rest" described in the Principles as depending both on magnitude and on "the surface according to which this body is disjoined from others," I will return to shortly. The inertia that depends only on the quantity of matter is, in effect, nothing other than the conversion factor $M_B/M_C$, the relative size of the cylinder into which or from which motion is poured. 'Large' means 'relatively large'; one can more easily fill a small cylinder from a large one than a large cylinder from a small one. So although there is nothing in any body that opposes motion, either in the sense that a body, considered in itself, cannot be moved by even the smallest force, or in the sense that it has of itself some quality contrary to motion, the transfer of a certain amount of motion to a large body will result in fewer degrees of speed than the transfer of the same amount to a small body.

That much is consistent with supposing that body is extension and nothing more. The force of rest, on the other hand, seems to imply that a body can have a mode, namely rest, which is not only contrary to but actually opposes motion. It still remains puzzling how the resistance of rest to motion, which in Rule 4 must be taken to be indefinitely large, can be coherently fitted into the system just outlined. On the face of it, that system would be far more congenial to the 1639 Rule than to the rules Descartes arrived at in the *Principles*.

Here I think one should not disregard, as commentators tend to do, the reference to the surface area of the resting body. I suggested earlier that a body genuinely at rest is in fact part of a larger body (which need not be at rest), and that the change from rest to motion is at the same time a change from union to disunion with the larger body. Even in the letter to Clerselier, where Descartes is improvising an alternative explanation of Rule 4, he notes that "a body that is not in action to separate its surface from those of the other bodies that surround it" is therefore "part of another hard body which is bigger," referring Clerselier to the definition of motion at 2§25.

How big is that bigger body? As big as it needs to be for Rule 4 to hold—in other words, indefinitely large. In material added to the French *Principles*, Descartes works out, in accordance with the hydrostatic model, the amount of motion that would have to be transferred to the body $C$ if it were to be moved as quickly as $B$: "So, for example, if $C$ is twice as large as $B$, and $B$ has three degrees of movement, it cannot impel $C$, which is at rest, unless it transfers [to $C$] two of its degrees, namely one for each half [of $C$], and retains for itself only the third, since it is no larger than each half of $C$, and can go no faster afterward than they do" (2§49; 9/2:91). The third law does not forbid the transfer of more than half of a thing's motion. Indeed it would seem to enjoin the transfer in situations like that of Figure 19, if the two bodies move afterward with the same motion. But a body at rest exhibits a double resistance: the resistance described a moment ago, which belongs to every body, whether at rest or not, and a second resistance specifically to the change from rest to motion. The second resistance is proportional to the quantity of motion the resting body would receive from the moving body if there were a transfer of motion according to the hydrostatic model. That quantity—two degrees in the example above—is effectively all that the moving body can give (if it gave more, say the amount denoted by the rectangle *aa* in Figure 19, and moved $C$ no faster, there would be a net loss of motion, which would violate the third law). It is, in other words, the measure of the force that the moving body possesses to impel the body at

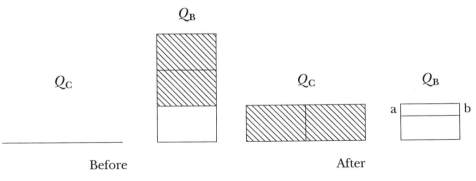

Fig. 19. Calculating motion after collision (Rule 4)

rest to move. Since it is, in fact, not greater than the force that the resting body has to resist it, the moving body must instead turn back, retaining, as the first part of the third law tells us, all of its motion.

The additions in the French *Principles* were not Descartes's first attempt to clarify Rule 4. In the letter to Clerselier, who apparently recognized that the outcome in Figure 19 is not ruled out by the third law alone, Descartes had invented an entirely new principle to explain Rule 4: "when two bodies that have incompatible modes meet, there must indeed occur some change in those modes to make them compatible, but this change is always the least that can be, that is, if, when a certain quantity of these modes is changed, they can become compatible, no larger *quantity will change*" (*To Clerselier* 17 Feb. 1645, AT 4:185). Motion and rest are incompatible modes; they can be made compatible either by pure reflection or by absorption. The body in motion must either be reflected or give to the body at rest sufficient motion so that the two will afterward move as one. But to do so it must give up more than half its motion, and (Descartes adds) more than half its "determination to move from the right to the left, insofar as this determination is joined to its speed." That change would be more than the change of all its determination, and so it changes all its determination. The result is therefore reflection.

Though some writers find this proposal an advance upon the doctrine of the Latin *Principles*, I am inclined to think that it is, like the original, a kind of hybrid, this time of the hydrostatic model and what Pierre Costabel calls a "principle of economy" to which Descartes is otherwise unfriendly (Costabel 1967:248). Like many hybrids, this one proved to be sterile. Since the quantity of determination in the direction of motion seems to be just the speed of the motion, speed effectively enters into the calculation of the "total change" *twice*. To say, as Costabel does, that "no other passage is as close to the vectorial conception of speed" is too charitable.[62] The very fact that Descartes treats determination and speed as distinct quantities capable of addition shows that he is far from a vectorial conception as ever. He was closer to the truth—if that matters—in his answer to Bourdin's objections against the proofs of the laws of reflection and refraction. There he writes that "in speaking of the determination to the right I mean the whole part of the movement that is determined toward the right" (*To Mersenne* 3 Dec. 1640, AT 3:251). The determination to the right and the determination downward together make up the entire determination. Even so, in the proof of the law of reflection itself, Descartes does *not* derive the motion after reflection by adding together the unchanged horizontal component and the new vertical component. He derives it by construction from the horizontal component and the circle at whose circumference the reflected body must arrive if its total motion is unchanged.[63] If having a vectorial conception requires the *composition* as well as the *resolution* of motions, then Descartes does not have one.[64]

The hydrostatic model, then, was retained by Descartes through every stage in his thinking about the laws of motion. But in both the *Principles* and in the letter to Clerselier it was combined with heterogeneous conceptions. The *Principles* subordinates the transfer of motion to the contrarieties of rest and motion and of direction, while the letter attempts to unite the transfer of speed and determination under a "principle of least modal change." Garber takes the letter to be a step forward, on the grounds that the principle enables the outcome of collisions to be calculated "*directly*, without a detour through force for proceeding and force of resisting and the balance of forces in an impact contest" (1992:247). Indeed, because of the double role of speed in determining the quantity of motion and the quantity of

62. Costabel, however, remarks on the "strange declaration that says that determination is joined with speed and that permits Descartes to speak of 'half' in relation to a modality that seemed not to be subject to quantification" (Costabel 1967:248). Gabbey, on the contrary, argues persuasively that *determinatio* is a quantity (Gabbey 1980:258). The determination in the direction *e* of a body moving in the direction *d* is, in modern terms, the projection of its velocity vector onto the direction *e* (ib. 259). In Fig. 3, for example, the horizontal determination is measured by *AF*, and in the vertical direction by *AE*. The quantity of its determination in the direction *d* itself is just its speed.

63. See Gabbey 1980:256. The version of the proof from Régis cited earlier in this section is essentially Descartes's.

64. In a response to Hobbes, Descartes writes that even though he could have spoken of the composition of two motions, he did not, "for fear that it might perhaps be understood that the quantity of these velocities thus composed, and the proportion of one to another, would remain" (*To Mersenne for Hobbes* 21 Jan. 1641, AT 3:288). The ensuing exchange reveals little about his understanding of the composition of motions. The only other examples I know of in which motions are composed occur in letters to Huygens (18 or 19 Feb. 1643, AT 3:624), and to Mersenne (23 Mar. 1643, 3:640ᶠ), both concerned with the trajectories of jets of water. In the letter to Huygens, he writes: "In the water drop, two movements are composed, without the quickness or slowness of the first changing anything in the second."

determination transferred to the body at rest, one can regard the principle as an attempt to make the hydrostatic model determine not only the allocation of motion but the outcome of collision. But as Garber himself carefully notes, the principle does not cover all the situations envisaged in the rules of collision. Nor does Descartes apply it to changes in the state of the resting body, since he still has no way of quantifying the modal change from rest to motion.

The path to classical mechanics was strewn with difficulties, some of which can be recognized only if we forget where it was headed. Descartes's first law, often heralded as the discovery of inertia, was in fact something else: it was the claim that the negative resistance of a thing to change—the persistence of its states in the absence of external causes—needs no natural cause. That was not entirely new, though the application to motion was relatively unusual. What distinguishes Descartes from some of his predecessors, and in particular from the Aristotelians, is the claim embodied in the second law: instantaneous *rectilinear* motion, or *conatus,* alone persists.

The immediate consequence of that claim is that any motion that is not rectilinear must have a cause. Since Descartes admitted only local motion, and denied that any body moves itself, that cause could only be contact with another body whose state is incompatible with the first. Here we find what is perhaps the most deeply compromised part of Descartes's theory of motion. The *negative* resistance, or persistence, argued for in the first and second laws does not entail a *positive* resistance to change. Yet virtually everyone would have agreed that efficient causal change requires *power* to overcome the positive resistance of an otherwise persisting state. That inference—from negative to positive resistance, and then to the requirement of an overcoming power in efficient causes—underlies the use of the contest model.

Descartes did his best to eliminate active and passive powers in Nature. But he was only partly successful, even in the heart of his physics. The success lay in his understanding of "natural inertia," represented by the conversion factor $M_B/M_C$ in the hydrostatic model of the transfer of motion. Such inertia offers no positive resistance to change and requires no power merely to be overcome. As he says in the French version of Rule 5, and had earlier maintained even for the situation in Rule 4, even the smallest force suffices to put a body in motion. The failure was in his understanding of rest. A body at rest offers, in addition to its natural inertia, a positive resistance to change, quantifiable only ex post facto. In the situation of Rule 4, it turns out to have been however much was needed to resist the moving body $B$; in the situation of Rule 5, it turns out to have been equal to natural inertia. In either case the body at rest is still conceived to be acted on by the body in motion; the asymmetry of the agent-patient relation in Aristotelianism is carried over into the asymmetry of the force of rest and the force of motion.

Consider again the paragraph of the *Principles* in which the *vis cuiusque corporis ad agendum vel resistendum* (the force of each body for acting or resisting) is more nearly specified, after having been introduced in the third law. The two *vires* are said to arise "from this one [thing], that each thing tends, *quantum in se est*, to remain in the same state in which it is, according the law stated first" (*PP* 2§43, AT 8/1:66). But that common origin is belied by the difference between them. The *vis ad agendum* can only be the force of persevering in a motion "of the same speed and in the same direction." All the other forces mentioned, including that of persevering in rest, are resistive. Though each body in a collision can be neutrally described as persevering in its present state as much as possible, given the incompatibility of its state with that of the other, Descartes still thinks of the body at rest as receiving or refusing the action of the body in motion. The body at rest is to that extent the patient in the collision. In keeping with the Aristotelian conception of transitive change, moreover, there is no reaction of the patient on the agent, except *per accidens*. Reflection, the key instance, is not a reaction of the reflecting body on the reflected body. It is the continuation of the reflected body's state as far as possible given that its present direction of motion is incompatible with the persistence of rest in the reflecting body.

The fact that, from a later point of view, Descartes's laws are not invariant under transformation from one reference frame to another in uniform motion with respect to the first, or that they do not satisfy Leibniz's condition of continuity, is only symptomatic of his retention of the asymmetry of agent and patient. Disregard of the obvious, even if Descartes were convicted of it, would hardly suffice to explain the rules he did devise. Careful attention to the Aristotelian physics he did not entirely leave behind, and to the parallel efforts of contemporaries like Beeckman, yields a fuller and more credible account.

## 8.3. Natural and Divine Agency: The Problem of Force

Perhaps the most vexing issue in Descartes's physics is the significance of force.[65] The Latin *vis* and the French *force* preside at the center of an extraordinarily complex network of concepts and denotations. They present difficulties beyond those offered by technical terms like *velocitas* and

65. The study of force in Descartes's physics has been advanced by a number of historically and philosophically distinguished contributions, among which Carteron 1922, Gueroult 1934, Gueroult 1970, Westfall 1971, Hatfield 1979, Gabbey 1980, Grosholz 1991, and Garber 1992 have been the most important to my thinking on the questions discussed here.

*determinatio*. Aside from the problem of sorting out the variety of contexts in which *vis* and its cognates occur,[66] there is the fundamental paradox emphasized by Martial Gueroult. Descartes insists that in body there are no modes except those that follow from extension, its principal attribute, or from its being a substance. The modes of extension are figure, size, and motion; those that follow from being a substance are duration, order, and number. But he also attributes to some bodies a *vis agendi,* and to others a *vis resistendi*. Since in collisions the contest between the *vis ad pergendum* of the "active" body and the *vis ad resistendum* of the "passive" body determines the outcome, the quantity of those forces has genuine effects. It would thus seem that force must be yet another mode of body. But it is, evidently, neither figure nor size nor duration. As for motion, Descartes has taken some pains to ensure that motion as he defines it will not be confused with the "force or action" by which motion is transferred.[67]

It would appear that force, since it has genuine effects on bodies, must be either a body itself or a substance apart from bodies, or a mode of bodies. Descartes ridicules what he conceives to be the Philosophers' view, according to which force, like weight, is a quasi substance, a "little soul." But force, although it has no obvious place among the modes of extension, cannot, it would seem, be dismissed entirely in the way that he dismisses weight or occult qualities. Call this the *problem of force.*

The solution of the problem is intimately bound up with the role of God in bringing about natural change. That role is characterized most precisely in the sections of the *Principles* leading up to the laws of motion, and the proofs of those laws. Descartes there adapts the view predominant in Aristotelianism: every created thing depends at each moment on God for its existence. He uses an analogy with the conservation of the quantity of matter, which was more likely to gain assent, to argue that divine immutability entails the conservation of the total quantity of motion in the world as well.

---

66. A useful *vademecum* is found in Westfall 1971, appendix B. Unlike Westfall, I don't think that an "incompatible chaos of meanings" is attached to the word 'force' and its cognates. Like most terms not owned by a dominant theory or assigned standard criteria of application (like chemical names, when used to denote samples of reagents whose purity is established by standard methods), *force* and *vis* were context-dependent. Their ambiguities were resolved in the way competent speakers usually do—by reference to surrounding words, the subject matter, the speaker's identity, and so forth. No doubt *vis* had in some contexts a sense it cannot have in others. To that degree its senses were incompatible. But I am not convinced that they were chaotic. Merely to criticize, moreover, the manner in which the word was used may well lead one to overlook an important development—the development in the seventeenth century of new methods in scientific discourse for stabilizing the senses of key terms.

67. *PP* 2§25, AT 8/1:54. Gueroult concludes: "There is indeed a paradox, since forces are all entirely reduced to modes of extension, while, on the other hand, they are defined, insofar as they are 'forces', in opposition to [modes of extension]"—in opposition because forces cause modes (1970:88).

Since the outcome of collision is said to depend on the relative strength of the force of acting in one body and the force of resisting in the other, commentators have looked to divine action as the key to solving the problem of force. Some, like Gary Hatfield, have identified force with divine action. Others, like Martial Gueroult and Alan Gabbey, have taken force to have a double aspect: as a *cause* of change, it is a mode of divine action; as the *effect* of change, it is a genuine mode of extended substances. Most recently, Daniel Garber has argued for what one might call a "nominalist" view: to say that a body has a force of acting or of resisting is nothing other than to say that God will conserve that body in accordance with the first law; to say that in collision there is a contest of forces is just to say that God resolves incompatibilities in the manner prescribed by the third law, and allocates motion so as to conserve its total quantity.

It seems to me that since a close reading of Descartes's own texts has not been decisive, it will be worthwhile to approach the problem of force obliquely, by considering the manner in which God was thought by Descartes and his contemporaries to conserve created things, as well as the relation between divine and natural action. Though Descartes does not explicitly bring out the connection, the problem of force, insofar as it is a problem of locating the causes of natural change, is one part of the vast question of the role of second causes. The connection does become clear later, when certain aspects of Descartes's thought are taken up into the occasionalism of Malebranche and others.

1. *The first law and immutability.* Descartes prefaces the exposition of the laws of motion in the *Principles* with a paragraph on the causes of motion:

> The nature of motion having thus been examined, it is necessary to consider its cause, which is twofold: the first, universal and primary, which is the general cause of all the motions that exist in the world; and then the particular [cause], by which it happens that single parts of matter acquire motions they did not have before. Concerning the general [cause], it seems manifest to me that it is not other than God himself, who in the beginning [the French adds: by his Omnipotence] created matter along with motion and rest, and who now, by his ordinary concurrence alone, conserves as much motion and rest in the whole as he then put [in it]. For although that motion is nothing other in the matter moved than its mode, still it has a certain and determinate quantity, which we easily understand to be able to be the same in the whole universe of things, even if it changes in particular parts. [. . .] We understand also that it is a perfection in God that he should be not only immutable in himself but that he should operate in the most constant and immutable way: so that with the exception of those changes that evident experience or divine revelation render certain and which we perceive or believe to occur without any change in the Creator, we should suppose no other [changes] in his works, to avoid

attributing some inconstancy to him. From which it follows that it is most agreeable to reason that from the very fact that God moved the parts of matter in diverse ways when he first created them, and now conserves that matter in just the same way and according to the same reason that he first created it, we should suppose that he also conserves always the same amount of motion. (*PP* 2§36, AT 8/1:61; 9/2:83)

The argument of §36 starts with an uncontroversial thesis. God is the first cause of all motion.[68] The Creation was not of a quiescent matter to which motion was added, or of a matter capable of moving itself, but of matter and motion together.[69] Since Descartes, as we will see shortly, holds in agreement with the central texts that the action of conservation differs only in reason from the action of creation, it follows that if God created motion then he conserves it as well. Now "evident experience" tells us that some individual parts of matter change their motion. So God's operation in conserving motion must change to that degree. But the *total motion* shared by the parts of matter can still be conceived to remain constant.

Descartes builds a parallel with a more widely accepted principle of conservation. The Aristotelians believed that prime matter could be created and annihilated only by God. In the absence of reasons (like the received doctrine of the Eucharist) to believe the contrary, it is most in keeping with God's immutability to suppose that God, once having created a certain quantity of matter, would conserve that same quantity.[70] Although the quantity contained in a particular individual may vary, the *total* quantity of prime matter shared among individuals will not. But (Descartes insists) matter and motion are here on a par; and since motion, though it is only a mode, does have quantity, its total quantity too is conserved.[71]

68. That God is the ultimate cause of all that exists was unquestioned. There was, however, controversy about the identification of the Prime Mover, whose existence is demonstrated in *Physics* 8 and *Metaphysics* 12, with God, since some characteristics attributed by Aristotle to the Prime Mover—in particular that the Prime Mover does not have infinite power—are incompatible with those of God (cf. Coimbra *In Phys.* 8c10q3, 2:370ᶠᶠ).

69. Since 'part of matter' and 'sharing the same motion' are correlatives in Descartes's physics, one could regard the cocreation of matter and motion in Cartesian physics as a transposition of the Aristotelian view that matter and form were created together (see §5.1). Gueroult has emphasized that the doctrine of the divine tweak [*chiquenaude*], setting into motion a previously immobile, and therefore featureless, matter is no part of Descartes's view (Gueroult 1970:125ᶠᶠ).

70. Since the central texts do not accept the identification of matter with quantity, the common view that matter can be neither created nor annihilated entails nothing about the conservation of the *quantity* of matter. Toletus, however, argues that "there is only as much matter as there was at the beginning of the world, and will be at the end"; the argument seems to be that in generation or corruption there can be division into parts or joining of parts, but not change of quantity (*In Phys.* 1c7q16, *Opera* 4:37va).

71. Toletus concludes, in his question on creation, that "the World, together with its *motus* and time, was made by God," and later that "to the production [of the world] there corresponds a unique instant, which was the last instant of the non-being of *motus*, because then [i.e., at that instant, conceived of as the backward limit of time] there was no *motus* and

The components of the argument are: (i) the immutability of God; (ii) the thesis, implicit but necessary to the argument, that creation and conservation are only distinct in reason, so that in particular they have the same object and—barring evidence to the contrary—are subject to the same laws; (iii) the experience of change in motion; (iv) the characterization of motion as a quantity, summable across individuals. It is (i) and especially (ii) that interest me.

The immutability of God has in the derivation of the laws of motion a double role. It ensures that they have the character of *laws:* they hold at all times and places and for all bodies (or, in the first law, all things). Descartes clearly feels no need to argue that the "ordinary concurrence" of God in natural change will, given his immutability, indeed have a lawlike character.

The second role of immutability is more remarkable. In §36 Descartes argues that since it is a perfection in God that he should operate always in the same way, we should regard his action as changing only if revelation or evident experience require us to do so. Now the first law says that a simple undivided body "never changes unless by external causes [*nec unquam mutari nisi à causis externis*]" (*PP* 2§37, AT 8/1:62). This will follow from the proposition argued in §36 if "evident experience" does not require us to postulate any internal principles of change in bodies. But experience requires us only to suppose that body is *res extensa,* and extension contains no principles of change. We then have no reason to suppose that God's action will change, and his immutability as a reason for supposing it will not.

Immutability returns in the proof of the second part of the third law. Since I discuss the proof in more detail later, I note one detail. At the end of the proof Descartes adds: "And so the continual mutation of creatures is itself an argument for the immutability of God [*Sicque hæc ipsa creaturarum continua mutatio immutabilitatis Dei est argumentum*]" (*PP* 2§42, AT 8/1:66). What leads to that fillip is an argument to the effect that changes of speed will be necessary if the total quantity of motion is to be conserved; that such changes do occur is, paradoxically, evidence for the absence of change in their cause.

It is a nice trick to make the continual *mutatio* of things an argument for divine *immutatio.* More than that, it is a trick the Aristotelians thought could not be pulled off. Suárez, citing Scotus, concludes that even from the perpetual movement of the heavens one cannot prove God is immutable, since the mover of the heavens could itself be movable in space, if not actually moved (*Disp.* 30§8¶9, *Opera* 26:115). He begins his own proof of

---

immediately after there was" (*In Phys.* 8c2q2, 212va, 213rb). Since *tempus* is that which is measured by the regular motions of the heavens, the creation of time is the creation of *motus,* and Toletus assumes, not without reason, that time existed as soon as the world did.

the immutability of God with the statement, "I do not believe that [this attribute] can be evidently proved from any immediate effect, but from the other attributes it can be demonstrated *quasi a priori*" (¶2, 113). Descartes, who was not averse to beating the Aristotelians at their own game, may have had such claims in mind when he put forward the paradoxical argument.[72]

The question to which that argument addresses itself is ancient. Supposing that a perfect, and therefore immutable, being is at every moment the conserver or creator of all created things, how can there be change in those things? Would that not imply change in their ultimate cause? The Aristotelians typically take up the question in relation to the biggest change of all, the creation of the world. As Fonseca puts it: "How could God, when before the foundation of the world he was not a cause *in actu* of things, have begun to be their cause *in actu* without any change?"[73] Since it was part of revealed truth that the world is not eternal, one obvious answer—that the world is coeternal with God, who is therefore always its cause *in actu*—is ruled out from the start.

The standard answer has two parts. First of all, since *motus* is in the *mobile*, not the mover, action does not change the agent per se (see §2.3). This principle applies to creation as well as to those actions in which the object of the action exists beforehand. As Toletus puts it: "Even in inferior agents, an agent may pass [*pervenire*] from a proximate active *potentia* to *actus* without any change in itself [. . .] In the same way God from not producing becomes a producer, since the production is not in him but in the product" (*In Phys.* 8c2q2, 4:212vb). Fonseca, who considers the question at some length, argues first that the ground upon which a cause is said either to be causing or not causing—the *ratio causandi*—is a "concurrence" of the cause with its effect. The concurrence of fire with the heating of water "is really the same as the form of fire, or heat, which are the principles of heating"; those forms are produced in the water, not the fire. Concurrences are neither substances, nor relations, nor real accidents; they add "no entity to things which are causes *in actu*" (*In meta.* 7c8q4§2, 309aB). But "in order that a thing should genuinely change [*mutetur vera mutatione*], even if [the change] is not physical [. . .] it is necessary that it should acquire or lose some entity" (§6, 317bE). So the change from being a cause *in potentia* to being a cause *in*

72. Descartes undoubtedly believed that there were proofs of divine immutability akin to Suárez's (who calls them "quasi" a priori because no attribute of God is, properly speaking, the cause of another). Perhaps the quickest is from God's simplicity: in whatever changes there must, by the arguments of §3.1 and §4.1, be something that remains and something that is new (since otherwise we would have no reason to say there was change rather than annihilation and creation), and the thing that changes will therefore be a composite of the two (Suárez *Disp.* 30§8¶2, *Opera* 26:113).

73. *In meta.* 7c8q4, 3:306a (with reference to Aristotle *Phys.* 5c2text3 for the distinction between causes *in potentia* and causes *in actu*). Cf. Toletus *In Phys.* 8c2q1, *Opera* 4:210ra, 213ra; Coimbra *In Phys.* 8c2q4a2, 2:271.

*actu* is not a genuine change. In particular, when God creates the world, or even if he re-creates it at each moment (a view some commentators ascribe to Descartes), no genuine change occurs in him.[74]

Even though the act of creation requires no change in God, still it could be, and was, objected that the act required a new volition and would therefore again be inconsistent with his immutability.[75] The response is that through *all* time God can will that an action should occur at a *certain* time: "Nor is it surprising that the volition of God to produce the world should always have existed, even though the world has not; for God has had from eternity a volition to create the world such that he did not have the volition to create it from eternity."[76] The logical point is obvious: willing that something should occur always (or in 4004 B.C.) is distinct from willing always (or in 4004 B.C.) that it should occur. But beyond that is a view of divine volition brought out in Fonseca's remark: "God, when he begins to be a cause *in actu,* does not begin to be otherwise in himself [*non aliter se habere coepit in se ipso*], but with respect to [*in ordinem ad*] creatures, formally by various relations of reason that suppose nothing in him, and virtually by the divine essence itself, which by its virtue and eminence contains all the possible modes of being toward creatures" (Fonseca *In meta.* 7c8q4§7, 3:320aC). A bit later Fonseca adds that even though it is logically possible, for some propositions, that God could either have willed that the proposition be true or that it not be true, it does not follow that he can at one time will that it be true and at another time will that it not be true: "even if in God there was eternally a *potentia logica* toward opposite free acts of will and intellect, still it was not toward them successively but only as alternatives" (7c9q5§6, 3:332aE).

---

74. Garber writes: "If God really is immutable, then, it would seem, the world should not change *at all,* and one state should be exactly like the one that preceded it; a genuinely immutable God should create an immutable world, it would seem" (1992:282). Since the successive re-creation of the world in different states—if that is indeed what Descartes meant by 'conservation'—raises no difficulty not already raised by the first act, it should be clear that the Aristotelians anticipated such questions and went to some lengths to answer them. Their answers should put us on guard against assuming that an agent must vary if its effects do.

75. "If [. . .] God began to produce the world [*de novo produxit mundum*]: therefore he began to have some proportion [to the effect, i.e., the world], which he did not have before: therefore he has changed: if not according to place, at least according to *concupiscentia* and will, which he newly has in producing the world" (Toletus *In Phys.* 8c2q2, *Opera* 4:210ra; cf. Coimbra *In Phys.* 8c2q4a2, 2:271). Objections to the creation of the world *de novo* that do not come from Aristotle himself are ascribed to Proclus, Simplicius, Avicenna, and Averroes. A brief outline of radical Aristotelian, Augustinian, and Thomist views is given in Grant 1978:269f.

76. "Nec mirum, quod voluntas divina producendi mundum semper extiterit, nec tamen semper mundus fuerit; sic enim Deus ab æterno voluntatem habuit mundum procreandi, ut non habuerit voluntatem eum procreandi ab æterno" (Coimbra *In Phys.* 8c2q4a4, 2:273; cf. Toletus *In Phys.* 8c2q2, *Opera* 213va; Fonseca *In meta.* 7c8q4§7, 3:319bC gives a very clear exposition of the point). Among the authors cited on the question are Justin Martyr, St. Isidore, and Hugh of St. Victor: it was not at all new.

The language is complicated, but the thought is reasonably straightforward. The varying relations of God to the world that result from mundane change are nothing in God but modes of being—"pure" modes, Fonseca calls them, to distinguish them from figure, intensity, and so forth, which are also called modes. But insofar as they are not nothing, they are, like corporeal things themselves, "eminently" or "virtually" contained in the divine essence.[77] A proposition describes a state of affairs, to which God would stand as cause to effect. That relation corresponds to a pure mode in God. For each proposition that is not logically true or false, God eternally wills either that it be true or that it be false, since it is inconsistent with his immutability that he should will one and then the other. Of those propositions that are willed some are, from the point of view of a temporally ordered world, "executed" at a particular time and place. But that does not count against their having been eternally willed.[78]

2. *Concurrence and the ordained power of God.* It is still a long way from this to the argument of §36. The immutability of God, as I have just presented it, has two components: the ontological claim that the varying dependencies of things on God are pure modes of his essence, and the formal claim that whatever God wills, whether it be an eternal truth or a change occurring in time, he wills eternally. Neither component entails that God's action will result in the conservation of any quantity. The missing link can be found in the phrase 'ordinary concurrence' [*concursum ordinarium*]: Descartes says that God "by his ordinary concurrence alone" conserves as much motion and rest in the whole world as when he created it. I will set aside for the moment the reasons why quantity of motion *especially* is said to be conserved. The problem then is how the additional notion of the ordinary concurrence of God, now conjoined with the immutability of God, is supposed to yield a law according to which the total quantity, whether of matter or of motion, remains constant through time.

The ordinary concurrence of God is an exercise of his *potentia ordinata*, one result of which is the conservation of the amount of motion with which the world was created. In what follows I will briefly attest the relation of

---

77. To use a traditional analogy (Dionysius *Div. nom.* 4§4, *Opera* 697/698; Thomas *ST* 1q104a1, *Opera* [Parma] 1:400): that a plant should be sustained by the sun is true by virtue of the plant's nature, not the sun's; and if the plant is at first sustained and then destroyed by the sun's heat, that need not imply any change in the sun's action, or in the sun. Yet the "eternal" possibility of being a sustainer of plants is contained in the sun's essence, since if it had a different essence it might not be even virtually a sustainer of plants.

78. One common way to put the point was that God's action is "terminated" in a temporally located effect, but its principle in God remains eternal. "Divine causalities, eternal in themselves, we apprehend as temporal, since we conceive them to be terminated in temporal effects" (Fonseca *In meta.* 7c8q4§6, 3:316bA). A person's will can be determined to some action long before the action is performed; in such a case the action will be "terminated" at a different time than the determination (ib. §7, 3:319aF).

*potentia ordinata* to law and then trace a path from concurrence to the conservation of motion. *It is* at that point that the identity of creation and conservation will become relevant.

One needn't look far to find *potentia ordinata* (or *potentia ordinaria*) juxtaposed with law. According to Suárez, the Theologians call

> the power of God 'absolute' [when it is considered] in itself without any other supposition or determination of his will, and apart from any respect toward the nature of things or toward other causes. The ordinary power they explicate in two ways: the first is that the absolute power of God is called 'ordinary' insofar as it is joined with the knowledge and will, by which God has decreed that these effects and not others shall be done, whatever order they belong to—whether natural or supernatural. The other more usual way is to call 'ordinary' the power of God insofar as it operates according to the common laws and causes which he has established universally. (Suárez *Disp.* 33§17¶32, *Opera* 26:216, citing Thomas *ST* 1q25a5, *Opera* [Parma] 1:114)

The intended contrast is not between two kinds of power, but between the same infinite power, considered first only in itself, and then in relation to what God, in accordance with his goodness, has chosen *ordinarily* or *regularly* to do. In the Aristotelians, the notion of law, though sometimes used to explain the "order" God has imposed on his own will, is not emphasized.[79] Words like 'usual' (*usitatus*) and 'customary' (*consuetus*) are as common.

In Aristotelianism the appeal to concurrence had two motives. One was to achieve a balance between the efficacy that, according to the central texts, must on empirical and moral grounds be attributed to second causes (see §7.3) and the equally compelling thesis that the infinite power of God is limited only by logical impossibility. The other motive, more familiar, was to reconcile the liberty of rational agents, and the evils that sometimes result from their free actions, with the perfect goodness of God. That problem belongs rather to metaphysics than to physics, but it is worth remembering that the Aristotelians used the same tool for both.

Concurrence stands to action as conservation to existence. Many of the arguments for concurrence rest on that analogy. Supposing that the necessity of conservation has been proved, Suárez argues that "if the cause depends on God in existence, so too does the effect, because each is a being by participation [in God's being]; as the cause depends [on God] in the instant in which it acts, so the effect does in the instant in which it is brought

---

79. The Coimbrans write that if God acts according to his ordinary power, then "if fire should be properly applied to a suitable matter, God cannot but concur with it in burning, since that is demanded by the common and ordinary law, which produces a concurrence of this sort in created things." But they define the *potentia ordinaria* of God as that which "respects the common and usual course and order ingrained in things" (Coimbra *In Phys.* 2c7q16a1, 1:287).

about, because in that instant too each is a being by participation; every effect, therefore of a second cause depends on God in becoming [*in fieri*], and consequently a second cause does nothing without the concurrence of God" (Suárez *Disp.* 22§1¶9, *Opera* 25:804). Every coming-to-be is analogous to creation. Second causes have the power to generate substance from substance, or to bring about accidental change; nevertheless the new being they effect depends on God for its existence no less than do things brought about immediately by God.

Some philosophers, notably Thomas and his followers, held that concurrence is an action of God distinct from and prior to the action of a second cause. According to the Coimbrans, they believed that "all second causes before they operate receive from God a certain influx and *motus* [. . .], by which they are incited to produce their actions," in the same way that an artisan moves his tools: "though the ax be sharpened and ready to cut the wood, it will certainly not cut unless it is set in motion by the impulse and *motus* of the artisan."[80] The central texts all dissent from this opinion, for two main reasons. The "premotion" of the second cause by God is superfluous and cannot be assigned an object. In rational agents, moreover, if there were such a premotion, it would be irresistible, and no agent would act freely.[81] Instead they hold that God's action and the action of the second cause are identical: "God acts with a creature by an action numerically identical to that performed by the creature, in the same way that a carpenter not only moves the saw [. . .] but also acts with the saw by an action numerically identical [with that of the saw], namely the cutting itself."[82]

God's contribution to that unique action, according to some authors, is the ultimate determination of the action of the second cause:

> The action of second causes is determined *per se* and first by God with respect to singularity. [. . .] This fire, for example, of itself is not determined toward a certain heating, numerically [distinct from others], both because it can elicit many [heatings], as is clear; and because nature does not definitely intend this singular [heating] [. . .] Therefore, in order that this heat should be determinately produced, it must be determined otherwise [than by the fire]. But among second causes there is nothing by which it could be determined, so one must revert to the First Cause [. . .]

---

80. "Namque esto dolabra præacuta sit, & ad expoliendum lignum idonea, haudquaquam tamen id expoliet, nisi artificis impulsu, motúque cieatur" (Coimbra *In Phys.* 2c7q13a1, 1:279). Cf. Thomas *ST* 1q105, 2pt1q109a1, *Opera* (Parma) 1:405, 2:429.

81. See, for example, Abra de Raconis *Phys.* 151; Coimbra *In Phys.* 2c7q13a2, 1:279ᶠ. Suárez refutes the opinion while denying that Thomas himself adhered to it (Suárez *Disp.* 22§2¶7,49, 52, *Opera* 25:811, 823, 824). Fonseca likewise refutes the view while attributing it only to the Thomist Ferrarius (Fonseca *In meta.* 5c2q9§3, 2:142bF).

82. Fonseca *In meta.* 5c2q9§3, 2:143aD; cf. Suárez *Disp.* 22§3¶2, *Opera* 25:826.

for whom it is indeed an ordinary and established law to supplement this defect, so to speak, of nature. (Coimbra *In Phys.* 2c7q15a2, 1:284)[83]

What precisely requires determination is not entirely clear to me. But the doctrine seems to be that when fire heats something, for example, the specific effect—that it is heating, not cooling—is determined by the nature of the fire, but the degree of heating and perhaps its duration are not fully determined either by the nature of the fire, by the place, or by the matter acted on. The concurrence of God, therefore, is required to put the final touches on the actual effect.[84]

One last point will be useful. Divine concurrence is not necessitated by second causes, even when all the conditions needed for their operation are present. God "by his will supplies that concurrence to second causes"; his will, of course, is free, and he can withhold concurrence if he chooses. When the sons of the Hebrews were made to walk through the furnace of Nebuchadnezzar, God withheld his concurrence from the flames, and they emerged unscathed.[85] Nevertheless concurrence is *natural* in the same way in which divine action in generation is natural: on the supposition that God acts according to his ordained power, and within the limits just mentioned, it is entirely *determinate* what God will do in any particular situation. Suárez writes that divine concurrence is "according to nature" [*ad modum naturæ*] for two reasons: "first, because in concurring [God] accommodates himself to the natures of things, and to each supplies a concurrence accommodated to its power [*virtus*]; second, because after he decreed that he would bring about and conserve second causes, by an infallible law he concurs with them in their operation; that law, if it is taken absolutely, without supposing any particular and definite will of God, does not induce necessity but is only like a sort of debt of connaturality [*quasi debitum quoddam connaturalitatis*]" (Suárez *Disp.* 22§4¶3, *Opera* 25:829). To which Suárez adds that the necessity of an effect is "natural" with respect to the second cause, but "only a necessity *ex suppositione* or of immutability" with respect to the first cause. The relation of cause to effect is necessary, in other words, only on the supposition that God has bound himself to concur with the agent so as to produce such and such an effect at such and such a time; the volition to do so is, as we have seen, present in him from eternity (25:830).

83. Both the Coimbrans and Suárez speak of this view as nominalist, although Thomas is also cited (Suárez *Disp.* 22§2¶ 10 & 30, *Opera* 25:812, 818; Thomas *ST* 1q105a1, *Opera* [Parma] 1:403).

84. The connection between the Cartesian use of *determinatio* and the passages from Thomas cited in n.83 was noted by Gabbey (1980:249ᶠ). Gabbey uses it in discerning what *determinatio* might have meant; the role of *determinatio* in discussions of concurrence is not mentioned by him.

85. See Oakley 1984 for information on this standard example and its role in debates about God's absolute power.

To summarize: variation in the world implies no mutability in God. The dependence of things on his power, whether in creation, conservation, or concurrence, is in God a pure mode. The volitions, moreover, with which he acts in those various ways are in him eternally; whether we can know God's particular volitions or not, we can be certain that they do not change. That formal condition does not of itself entail that any property of the world as a whole must remain constant, since if it were to change (as indeed it would, for example, if God were to annihilate any part of matter), that too would have been willed eternally. But God has chosen to act according to a certain order or law. In particular he has chosen, eternally, to concur in the actions of natural agents. On the supposition that he has done so, and that his will is immutable, the actions of natural agents follow by *natural* necessity from their natures. Divine concurrence is, to echo the words of Soto in his theory of grace (§7.1), *owed* to things. It follows that if we are given the nature of the cause and the conditions under which it acts (including the nature of the patient it acts on), the effect will—at least *in specie*, if not in all its singularity—be determined. That determination is characteristic of a certain sort of supernatural intervention in nature, notably in the generation of animals and humans (§7.1).

To return now to the argument of *Principles* 2§36. In the Aristotelian account of concurrence, the word 'law' seems to designate two somewhat different orders: the *mundane order*, also called the "ordinary" or "customary" course of nature; and the *divine order*, or the limits God freely abides by in the exercise of his power. The mundane order depends jointly upon the divine order and on the immutability of the will that binds itself to it. Descartes's intention, however, seems to be to derive from the *formal* condition of immutability, which would seem to yield only the permanence of the mundane order, but not substantive claims about that order itself.

In Aristotelianism, substantive claims about the mundane order concern not natural laws per se, but the effects consequent upon a thing's nature, which is defined, as we have seen, to be the principle of its rest and motion. The ordinary course of nature—that humans give birth to humans, or that fire heats—rests on the natures common to the things we call *human* or *fire*. Divine concurrence is determined, in the sense specified above, by the natures of agents and, to a lesser degree, of patients. Descartes, on the other hand, denies that corporeal things have natures in the Aristotelian sense, though in the *Principles* he contrives a nominal agreement with the Aristotelian definition (*PP2§23* AT 8/1:53). The argument of §36 is intended to yield a basis for natural regularities consistent with the conception of body as *res extensa*.

From that conception it does not follow that any body moves. Experience alone shows us that bodies do move, from which, since there is no principle of motion in bodies themselves, we may infer that they are moved by something other than body, that is, by rational agents. From the immutability of God's volitions, it would seem that the first law, at least, can be inferred. Since bodies have no internal principles of change, all change must come from outside; if no external cause acts on a body, it will not change. Yet for all we know, God could eternally will that a body not acted upon by others should nevertheless grow, or alter in shape, or accelerate. The formal condition of immutability yields no natural law except in conjunction with some further condition.

That condition can be found, I think, if we attend to the phrase *quantum in se est* studied by Cohen (see n.18 above). Even though Descartes denies that bodies themselves contain principles of motion and rest, still he takes over from Aristotelianism the thought that God has given to bodies a "nature" that is sufficient to determine how they will act, but now only on the supposition that they were created with whatever motion they have. That nature consists in extension and its modes. Considering it alone, we find in it nothing that would entail change in its modes: "if some part of matter is square, we easily persuade ourselves that it will perpetually remain square, unless something arrives from elsewhere that changes its figure" (2§37, AT 8/1:62). God, in the exercise of his ordained power, will give to things precisely what is owed to them by nature—neither more nor less. Not to concur in the actions to which they are fitted would be to frustrate his own end in creating them. To do more than what is owed would imply that something was lacking in the original act of creation.[86]

But even if the first law can be vindicated, the argument of §36 and the third law, which is supposed to follow from it, remain troublesome. It is here that the identity of conservation and creation comes into play. Since that identity is essential not only to the argument of §36 and the proof of the third law, but also to the ontology of force, it is worth examining Aristotelian and Cartesian views on the question.

3. *Conservation and the third law.* Conservation can be brought about immediately by God, or by God and created things conjointly. The former applies most obviously to substances, but also to accidents created and sustained apart from substances by God. The latter applies to accidents inhering in a substance, or to matter and form; God's part amounts to

---

86. That God should give to things motions not implied in their nature, or in the original donation of movement to the world, would also deprive physics of what certainty it has. Concerning bodies, all we can be certain of is what is contained in our clear and distinct idea of them, which is to say, our conception of them as determinate *res extensæ*, together with what follows from the supposition, warranted by the experience of moving our own bodies, that bodies do move. If bodies not acted upon by others were to change in state, the cause of that change, since it is alien to our conception of them, would be entirely inscrutable; like the "premotion" of the soul effected by grace, it could be known to us only through revelation.

concurrence in the continued production of an effect, as when light is sustained by the sun.

Concerning the first, the Aristotelians concluded that they differ "not by an intrinsic, but only by an accidental difference."[87] The difference is simply that 'creation' adds to the core notion of a thing's depending on something for its existence the connotation that it did not exist before; while 'conservation' adds the connotation that it did (Suárez *Disp.* 21§2¶2, *Opera* 25:791).

Among the reasons given for the view, the most interesting is this: creation and conservation are one continuous action. Suárez writes:

> Who would believe that the Sun, or a lantern, illuminate the air by a different action when it is first applied, and afterwards? What would there be in the cause, such that the first action should be interrupted or cease, and the other begin? Or again: the first action ought to last only for an instant of our time, because any reason according to which it lasted longer would suffice for it to last forever. The subsequent action therefore either would likewise last only an instant, and so there would be two immediately succeeding instants, or else would last through the whole time that the thing was conserved, and so the same or even a stronger reason can be given to show that the first action lasts, and the superfluous multiplication of actions will be avoided. (Suárez *Disp.* 21§2¶3, *Opera* 25:791)

Conservation need no more be a continual creation than holding a weight need be a continual lifting. If a rock were suddenly to appear on my outstretched palm, there would be no reason to say that the action of beginning to hold it and the subsequent action of continuing to hold it are distinct. The cause is unchanged, and the *terminus* of the action is unchanged. Similarly, in creation and conservation, the *terminus* is existence, and the cause, which is God, is immutable. To hold, as Descartes too holds, that creation and conservation are one and the same does not entail, as some authors have supposed, that conservation is the re-creation of the world at each instant; indeed Suárez's argument gives one reason to doubt the coherence of such a view.[88]

87. Coimbra *In Phys.* 8c2q1a4, 2:259. Suárez concludes that conservation is neither really nor modally distinct from creation. It is distinguished "only by a certain connotation or negation" (*Disp.* 21§2¶2, *Opera* 25:791). There was significant dissent from that view, represented by Henry of Ghent (*Quodlib.* 1q7), Gregory of Rimini (*In Sent.* 2d2q6), and, among the central texts, Abra de Raconis (*Phys.* 141).

88. Since I think Descartes agrees with the Aristotelians on this point, and since many recent interpreters of Descartes, notably Gueroult (1970:272ff), have argued that Descartes believes that time is "discontinuous" and that conservation is the re-creation of the world at each instant (ib. 257, 277), I will briefly state my reasons. An instant, in the Aristotelian conception, is the boundary of an interval of time, just as a point is the boundary of a line segment. It has no existence independent of such intervals (though here one finds disagreement and ambiguity, as in Descartes). The term 'moment', used by Descartes in a number of

The other sort of conservation is a joint action of God and a created thing in sustaining the existence of accidents or incomplete substances. It is, as I have said, a kind of concurrence.[89] Matter depends naturally on form for its existence and conversely (see §5.1); if God were, by his absolute power, to conserve matter without form, that would require an *additional* action to supplement the causal activity of form (see Suárez *Disp.* 21§2¶8, *Opera* 25:793). According to the Coimbrans, "The conservation, by which form conserves matter, and the creation of that matter cannot not be distinguished really among themselves. [. . .] Form is said to conserve matter, insofar as it actualizes matter: but that actualization is nothing other that a certain mode, by which form insinuates itself into matter [. . .] and is the same thing as the form; consequently it is really distinct from matter" (Coimbra *In Phys.* 8c2q1a4, 2:259$^f$). The creation of the matter, on the other hand, is not really distinct from the matter. So the conservation of the matter by its form, and the creation of that matter by God, are—as one might expect—distinct, being modes of distinct things.

More generally, the conservation of a thing by a second cause, and its production, can in some cases be distinct, while in others they cannot. Elemental qualities are produced in matter by the heavens; once produced, however, they are conserved by the substantial forms of the elements, which, like all things, strive to preserve themselves. The most interesting case is *motus*. Suárez, trying to explain why some qualities persist after their causes cease acting (as heat in water) and others do not (as light in air), cites the

passages that have been adduced in favor of an "atomistic" view of time, can usually be read as denoting an interval, perhaps short but not infinitesimal (see Garber 1992:270; two clear cases are *To Mersenne* 11 Mar. 1640 and 2 Feb. 1643, 3:36, 614; Mersenne defines 'moment' to denote what we call a second, cf. *Harm. univ.* 1pr7, 1:12). Even the crucial passage in *Meditation* 3 (7:49; cf. 109) does not require that conservation be interpreted as *creatio de novo:* one should not ignore the fact that Descartes writes *quasi rursus creet*, not *rursus creet*. The point is that because intervals of time are really distinct, each moment in which I exist can be conceived of as the first or last moment (i.e., as an interval before which I did not exist, or after which I will not exist), and that the action of God in sustaining my existence is—just as the Aristotelians argue—not distinct except in reason from the action of creation. Indeed Descartes is there taking up a point made by Suárez also: even an eternally existing thing can be created (see *Disp.* 20§5¶11$^{rr}$, *Opera* 25:782$^{rr}$). Finally, although Gueroult believes that continuous re-creation is needed to explain the difference between the Cartesian and the Aristotelian concepts of motion, I think that the definition of *PP* 2§25 suffices to show that he rejects the view that motion is a *fluxus* or *via*. Since the Aristotelian sources uniformly take conservation to be an action continuous *with* creation, and therefore not really distinct from it, and since they do not, so far as I can see, even take notice of the doctrine of continuous re-creation, it seems to me that in the absence of strong reasons to attribute that novelty to Descartes, one should take him to agree with them, even apart from the psychological arguments advanced by Beyssade (see Garber 1992:268$^{rr}$, 361; see also Arthur 1988 for a vigorous refutation, largely on internal grounds, of the traditional view).

89. Descartes, it is worth noting, in a letter of 1641, uses the term *concursus* to denote the action by which substances too are conserved in being (Descartes *To *** [Hyperaspistes]* Aug. 1641, AT 3:429). The term would be appropriate if one thought of the striving of each thing to persist as a kind of cause of its existence, and of God as concurring with that cause.

following: "In *motus* one can easily give a reason why it always depends on an actual agent, namely because its existence is not so much existence as becoming [*ejus esse non tam est esse quam fieri*]; and because becoming cannot be without acting, so *motus* [cannot be] without a mover [*quia ergo fieri non potest esse sine agere, ideo nec motus sine movente*]" (Suárez *Disp.* 25§3¶14, *Opera* 25:797). Suárez himself, in contrast with Thomas, does not take that to be sufficient to explain every case. But it is worth noting that here again conservation and production are identical.[90]

The doctrine I have outlined can be summarized as follows. The action of creation, with respect to complete substances (and to separated accidents or separated incomplete substances like the soul after death), is, like any action, nothing other than the dependence of an effect—in this case the existence of created things—on its cause. That dependence is the participation in being that God, as the only necessary being, has chosen to give to them. Conservation differs from creation only in connotation; it is otherwise identical with, and indeed continuous with, creation. With respect to accidents and incomplete substances, on the other hand, the two divine actions can differ *in re*. The creaturely actions of production and conservation sometimes differ, sometimes not. The conservation of heat in water warmed by fire was held to be distinct from its production; the conservation of light in air illuminated by the sun was held not to be distinct; the conservation of *motus* in projectiles seems to have been decided differently by different philosophers.

Descartes, therefore, when he embarked on the project of proving his laws of motion by way of the traditional three modes of divine efficient causation—creation, conservation, and concurrence—was faced with ambiguity at just the point where he required an unequivocal connection. But he had, it seems, already a firm conviction that the nature of body consists in its being a *res extensa*. Since, as I have said, nothing in *res extensa* entails that it should move, he was thereby already committed to supposing that the motion we find in the world could have come, in its entirety, only by way of the immediate creative act of God. There is, moreover, no second cause in the world that could sustain motion once God imparted it to things. In response to More, Descartes writes that "I consider *matter left freely to itself and receiving no impulse from elsewhere* to be entirely at rest. It is impelled by God, who conserves as much motion or translation in it as he put there in the beginning" (*To More* 30 Aug. 1649, AT 5:404; the passage in italics is a

90. That resemblance may have inspired the analogy made by Bonaventura and Thomas between conservation and illumination (Thomas *ST* 1q104a1, *Opera* [Parma] 1:400). There was no dearth of "light-cosmogonies" in which the creative power of God was taken, more or less literally, to be an illumination—Grosseteste was one prominent example (Hedwig 1980:136f). Descartes was aware of the analogy (*To \*\*\* [Hyperaspistes]* Aug. 1641, AT 3:429).

quotation from *More to Descartes* 23 Jul. 1649, 5:381). 'At rest', as the word *translatio* shows, has the properly Cartesian sense of 'not separating (at a designated instant) from the vicinity of neighboring bodies' (see §8.1). *Motus,* then, like light in air illuminated by the sun, is the sort of quality that cannot exist except through the immediate continued action of the agent that produced it, which is to say, God.

Motion is thus entirely on a par with *res extensa* itself. Both are *immediately* created by God, and must, through the continuation of the creative act, be immediately conserved by God.[91] In a world in which a single body—the world itself—were given *one* motion, as in certain Scholastic thought-experiments (e.g., Buridan *In Phys.* 3q7, 50va; cf. §2.2), that would be all one needed to say. The first law would suffice. But that is not what God did: "it is obvious that God, in the beginning, when he created the world, not only moved its diverse parts in diverse ways, but at the same time also brought it about that some should impel others and transfer their motions to them: so that now, conserving [the world] by the same action, and with the same laws with which he created it, he conserves motion not always implanted in the same parts of matter but rather passing from some to others accordingly as they collide with one another" (*PP* 2§42, AT 8/1:66; cf. *Monde* 7, 11:43). In this vest-pocket version of Genesis, Descartes affirms first that God created matter and motion together; then that he distributed motion differently to different parts; and finally that in that first instant those parts not only moved but impelled others, and *transferred* their motions to them, so that the laws of motion are instantiated even as the world is created.

There is nothing in matter itself that would compel one to suppose that when a body meets a body it would not just stop, like dough thrown against a wall. But if the parts of *res extensa* cannot interpenetrate, it does follow that certain motions of those parts cannot persist together. Descartes supposes that God set things up so that from the very beginning parts of matter in fact have incompatible motions. Motion is transferred, and the law that governs transfer instantiated, at every moment of the world's existence.

But what should that law be? Consider the analogy drawn in the argument of §36 between the quantity of matter in the world and the quantity of motion: "from the very fact that God moved the parts of matter in diverse ways when he first created them, and now conserves that matter in just the

---

91. See Carteron 1922:275. This begins to answer a question raised by Garber, who asks why quantity of motion, rather than various other modes of extension, is conserved (Garber 1992:282). Unlike figure, or (for somewhat different reasons) number, motion alone must be considered to have been created independently along with body; figure, size, and the number of bodies will all be entirely determined once matter and motion are supposed. See n.95 below.

same way and according to the same reason that he first created it, we should suppose that he also conserves always the same amount of motion" (*PP* 2§36, AT 8/1:61; 9/2:83). I think one should take seriously the use of *part* here and elsewhere (notably in the definition of *motus* at 2§25) to denote what might otherwise be called individual material substances; one should also take seriously the use of the singular *action* to denote creation: Descartes in §42 writes of the "immutability of the operation of God, continually conserving the world by the same action by which he once created it," and a bit later of God "conserving it [i.e., the world] by the same action." Although it is true that elsewhere, notably in the *Meditations,* creation is clearly an action that terminates in individuals, and that he speaks of God as "moving" or "impelling" bodies, still I think it is promising to consider the creation of matter in general, and thus of a certain fixed quantity of matter, as a single action.

The creation of matter and its conservation will also be a single continuous action. Although God could withdraw his conserving action from a part of matter,[92] thereby annihilating it, we have no reason to believe that such a thing occurs. The law governing his creative and conserving action we may easily imagine to be invariant with respect to the temporal termination of that action. God, when he created the world, willed that a certain quantity of matter should exist, and, since conservation is continuous with creation, he wills it at every subsequent moment.

Earlier I argued that since the existence of motion, as opposed to its possibility, is not entailed by the existence of matter, the creation of motion was independent of the creation of matter. Although God could not have created motion without matter (motion, where it exists, is a mode of matter), he could have created matter without motion. I now suppose that, by analogy with matter, the creation of motion can be regarded as a single action, namely, as the creation of "a certain & determinate quantity [. . .] which we easily understand to be able to be always the same in the whole universe of things" (*PP* 2§36, AT 8/1:61).[93] That quantity is a whole of

---

92. See *To More* 5 Feb. 1649, AT 5:272, and the general remarks on conservation in *To \*\*\** [Hyperaspistes] Aug. 1641, AT 3:429.

93. It is pleasant to find this interpretation of Descartes's argument confirmed by Régis, who writes: "One must not put in God as many volitions as there are particular movements; for we know that this multitude of volitions is repugnant to the simplicity of the divine nature: or, if we do put several volitions in God, one must conceive that they are distinguished neither really nor formally, but only by a distinction of reason, founded on the fact that because our mind cannot understand the infinite extension of the volition by which God resolved to move bodies, it divides [that volition] into as many parts as there are particular bodies that God wishes to move; which does not accord with the idea of a perfect being whose extreme simplicity excludes every sort of real and formal composition, and admits only composition in reason, which is consequently the only [composition] one may attribute to God. So, for

which the quantities of motion in parts of matter are parts, just as the quantities of the parts of matter themselves are parts of the quantity of the whole.

God, then, when he created the world, willed not only that a certain quantity of matter should exist, but also that a certain quantity of motion should exist. Since creation is continuous with creation, he wills the same at every subsequent moment. The *whole* quantity is conserved, even though its division into parts changes when particles collide. Though it is not a substance, motion shares with body the character of having quantity and of being divisible into parts. Even apart from the influence of earlier attempts to quantify *motus* or *impetus,* it seems to me that the resemblance of motion to body would have made the hydrostatic model attractive to Descartes, and with it the identification of the quantity of motion in a particular part of matter with the product of volume and speed.

The conservation of the total quantity of motion in the world thus reflects the single continuous action, and the single eternal volition, by which God both creates and sustains it in existence. When two parts of matter whose motions are incompatible meet, God cannot execute that volition simply by conserving the quantity of each part. Instead he concurs in the effect of each upon the other so as to resolve the incompatibility, and, if need be, to diminish the motion of one part and augment that of the other so as to sustain the total quantity he first created. That condition, it should be noted, in no way entails the rules of collision that Descartes actually devised. Beeckman's rules, or Régis's, are equally compatible with the conservation of the total quantity of motion. For that one must turn to the history of the problem, some of which I have outlined above.[94]

4. *Force and divine action.* At the beginning of this subsection I noted that there are essentially three views about the ontology of force in Descartes's

---

example, we will not say that God wills rain and fair weather by two particular volitions, but instead we will think that rain and fair weather, whatever opposition there is between them, are two effects of one and the same volition, by which God wills that rain succeeds fair weather and fair weather succeeds rain" (Régis *Cours* 1pt2c9, 1:331).

94. In n.91 I argued in response to a question raised by Garber that figure and size, unlike motion, are entirely determined once matter and the distribution of the total quantity of motion are given (though questions remain concerning the definition of the parts of matter, and though Descartes is quite unclear about the circumstances under which collision will bring about the breaking off of a piece of one of the two bodies rather than absorption or reflection). Number, also mentioned by Garber, is not a mode of extension but only a *modus cogitandi* (*PP* 1§58, AT 8/1:27). In any case the number of parts of matter will, with the reservations just mentioned, also be determined once matter and motion are given. The only conceptually arbitrary aspect of the argument of §36 is the identification of the quantity of motion in a moving part of matter with the product of its volume and speed. Here the history of the problem, and especially Beeckman's collision rules, provide the best explanation (see the analysis by de Waard in Mersenne *Corr.* 3, appendix I, and Gabbey's remarks in Gabbey 1972:380[ff]). For Beeckman, as de Waard notes, quantity of motion is "a primitive possession of matter" (643), whose use he does not attempt to justify.

physics: that it pertains solely to divine action; that it pertains both to divine action and to bodies; that it is a *façon de parler*, a way of describing how bodies will act according to the laws of motion. The conclusion I reach here is that in the metaphysical setting of Descartes's arguments for the laws of motion, the last two views can be reconciled; the first, on the other hand, does not accord with the dominant view that action inheres in the patient. The historically significant fact is not the predominance of one or the other, but their equivocal co-presence, which can in part be traced to the Aristotelian doctrines Descartes made use of. It was left to a later physics to make one or the other predominant.

I will first show in what sense conservation and concurrence were conceived of as instances of *efficient* causation. There is, I think, a significant difference in how God was thought to act. I will then briefly take up the distinction between the static forces that Descartes appeals to in his discussions of machines and in his definition of what is now called "work," and the *dynamic* force attributed to moving bodies and measured by quantity of motion. The distinction parallels the distinction between the conserving and the concurring action of God. Finally I will argue that the distinction I have appealed to several times already between the senses in which a change could be called 'natural' permits the reconciliation of the three views promised above.

In Aristotelianism, creation, conservation, and concurrence are for the most part treated as instances of efficient causation.[95] The *efficient* cause is defined as "that from which comes the first principle of change [*mutatio*] or rest."[96] Suárez glosses the definition in terms of *actio*, which is the dependence of *mutatio* on whatever gives it being, per se rather than accidentally, so that the efficient cause is the "principle from which the effect flows forth or depends by action."[97] While the formal cause is the *terminus* of change, and the material cause that out of which the form is produced, and in which it inheres, the efficient cause is that which alone gives existence to the

---

95. God was also said by some philosophers to concur with the material, formal, and final causes (Fonseca *In meta.* 5c2q12, 2:170ᶠᵗ). Suárez, treating that view rather harshly, argues that secundum proprietatem "we do not recognize in God any other force of causing outside himself except that of omnipotence, which is an effective power only" (*Disp.* 22§6¶28, *Opera* 25:809). The danger in supposing that God could be a material cause is obvious (see the remarks on David of Dinant in §4.1); the danger in supposing that God could be a formal cause is the temptation to regard him as the *anima mundi*. Although God, as the ultimate end, does exercise final causality, there is no reason to suppose that he concurs in the "metaphorical motion" of other final causes.

96. Aristotle *Phys.* 2c3text31, 194b29. Discussions of the definition and *ratio causandi* of the efficient cause include Coimbra *In Phys.* 1:229 and 2c7q7, 1:256ᶠᵗ; Suárez *Disp.* 17§1, §10.

97. "Perinde enim est dicere efficientem causam esse primum principium unde est actio, ac dicere unde est effectus, media actione, seu esse principium a quo effectus profluit seu pendet per actionem" (Suárez *Disp.* 17§1¶6, *Opera* 25:852).

change itself.[98] Unlike the final cause, whose *motio* is only metaphorical, the action of the efficient cause, which is, as we have seen, just the *motus* itself with respect to its dependence on an agent (§2.2), is real.

The creation of substances is not, strictly speaking, a *mutatio* or *motus*, if those terms presuppose a preexisting subject. Suárez writes: "This dependence [of creature on Creator] does not truly satisfy the definition of *mutatio* [*non habet veram rationem mutationis*]; but it does truly satisfy the definition of a way or becoming of the creature, and so it is called a passive creation; it also truly satisfies the definition of an emanation from God, and insofar as it is from him, it can truly and properly be called an action of God himself, formally a transeunt action, by which the creature is produced" (*Disp.* 20§4¶17, *Opera* 25:774; cf. Thomas *ST* 1p45a3, *Opera* [Parma] 1:185). Against Thomas, who argues that creation, since it is not a *mutatio,* is not an actio but only a relation between God and creature, Suárez holds that *mutatio* can be more broadly used of "any action or effecting, which brings some novelty" (Suárez *Disp.* ¶18), which in creation is actual existence itself. In order for creation to be an *actio* it suffices that there be a "flux emanating from God" (¶21, 775).

In keeping with the identification of conservation and creation, Suárez argues that although creation is in the broad sense a *mutatio* of the thing created, it does not entail an actual succession from nonbeing to being, but only what Suárez calls an "imaginary succession."[99] Since God is the only necessary being, we can, for any other thing, at least conceive of it as not existing, and then existing, even if it has in fact never not existed.

Creation *sive* conservation, then, is a peculiar sort of action. Suárez distinguishes two ways in which things, insofar as they persist, depend on God. The first is "permissive": things "remain in existence just so long as they are permitted to remain by God, that is, so long as they are not deprived of their existence, by him who can so deprive them." The other is "positive": continued existence requires an actual influx of being from God to creatures. Suárez's primary argument for the positive way is that "every positive action necessarily tends to some existence," and so if God required some positive action to destroy things, he could only corrupt them, rather than annihilate them. But since he can annihilate them, it follows that he does so

98. This is not to deny that a form can be an efficient as well as a formal cause, but rather that the form attained in change is not numerically the same as that which initiates it.

99. "Even in eternal creation [i.e., the creation of something that exists at every moment of time] creation can be distinguished in reason from conservation. Insofar as such a creation signifies existence absolutely in eternity, as a simple participation of the existence of the created thing [in the existence of God], it satisfies the definition of 'creation'; but insofar as we conceive in it an imaginary succession, then for any designated instant such a succession satisfies the definition of 'conservation'" (*Disp.* 20§5¶20, *Opera* 25:784). It is in that manner that the "generation" of the eternal Word from the divine essence is to be conceived.

by ending the positive action by which he sustains them in existence (*Disp.* 21§1¶14, *Opera* 25:789; the same argument is found in Descartes's reply to Gassendi, AT 7:370).

Creation *sive* conservation, in short, consists in the voluntary donation of being by God to other things. All things exist in God as divine ideas; but only some are permitted, and borne, as it were, into actuality through his power. That permission and positive action are required at every moment of their existence; so that "if for even a moment the concurrence of God in conserving [them] were absent from created things, all would immediately return to nothing" (Coimbra *In Phys.* 2c7q10a2, 1:267; cf. Descartes *To \*\*\* [Hyperaspistes]* Aug. 1641, AT 3:429.10–11). If there were a "force" associated with it (as Gueroult seems to suggest),[100] it would be quite unlike the moving force ascribed to parts of matter or to God himself (*To More* 30 Aug 1649, AT 5:403$^f$). I will shortly compare it with the static force by which a body is held in place when it is in equilibrium in a balance or suspended from a pulley. Although it may be relevant to understanding the force of rest, it is, so far as I can see, quite irrelevant to the force of motion.

Concurrence, on the other hand, is relevant. We have seen that, according to the central texts, the action of the first cause in concurring with that of a second cause is not additional to the action of the second cause, but rather identical to it. Concurrence does not consist in moving the second cause to act, or in a merely instrumental use of creatures by God to accomplish his ends (God can use creatures, but that is not what happens when they act ordinarily). When God concurs with a second cause, the Aristotelians argue, he does so immediately, both in the sense of not employing intermediaries (like the celestial spheres)[101] and in the sense of being present and exercising his power at the time and place where the second cause acts. As Descartes says in his response to More, God is "everywhere" not because he is an infinite extended thing coextensive with the world, but because "his power is exerted, or could be exerted, in extended things"; he is present *per modum potentiæ*, not *per modum rei extensæ* (*To More* 30 Aug. 1649, AT 5:403; cf. 343).

---

100. In Gueroult's interpretation, the conservation of matter and the total quantity of motion consists in their re-creation at each instant (Gueroult 1970:114$^f$, 117). He therefore believes that a *force créatrice* as well as a *force mouvante* is exercised by God. Since, for the reasons given in n.88, I don't think that in Descartes's view God re-creates the world at each instant, I don't think that there is a *force créatrice*. The real question is whether there is a *force conservatrice* corresponding to the "force of resisting" of a body at rest distinct from the force that corresponds, in the manner described below, to the action of concurrence.

101. This would need to be qualified to distinguish among substances and various kinds of accident (see Suárez *Disp.* 21§3, *Opera* 25:794$^{ff}$). Suárez holds of motion in particular that it is always conserved not only by God but by a secondary cause (¶14, 797). Many other kinds of accident (e.g., quantity) and all substances are conserved immediately and solely by God. God's action in concurrence is always immediate.

Most important, the central texts hold, as I have already noted, that the action of second causes is determined by them only *quoad speciem,* with respect to kind; the divine contribution consists in a determination of the action *quoad singularitatem,* with respect to their singularity:

> The reason and manner of his pre-definition [*præfinitio,* that is, God's eternal, and therefore prior, volition that this secondary cause, acting here and now, should produce this singular effect] can be conceived thus. God, seeing that in the given circumstances he ought to concur with such-and-such a species of act according to the ordinary law, by which he has decreed that a created agent shall not be destitute of his assistance, pre-defines the individual concurrence, with which he wishes to give being [to the effect]. But because God and second causes give being to [*influunt*] the effect of second causes not by different but rather by the same concurrence or action [. . .], the concurrence even of created agents is determined in individual degree, so that this rather than that effect of the same species occurs, among those that could be produced in the same circumstances. (Coimbra *In Phys.* 2c7q15a2, 1:285)

The clotted syntax of the passage, which I have done my best not to relieve, should not obscure the essential point. In every action of a second cause, there is some indetermination with respect to its effect, an indetermination that can only be supplemented by God acting *with* the second cause. An archer, say, aims at the bull's-eye; but it is God who, in keeping with that aim, determines the precise point at which the arrow strikes.

I suggested a moment ago that if there is a force associated with the conserving action of God, it would be better compared with the forces referred to in statics than to the forces of moving bodies. St. Bonaventura, the Coimbrans remark, compared conservation by the first cause with "a weight, that is held in the air by the hand" (Coimbra *In Phys.* 2c7q10a2, 1:267). Descartes seems to have thought of God's action as a concurrence in their self-preservation, as if each thing shared, however imperfectly, with God the attribute of being *causa sui* (*To \*\*\* [Hyperaspistes]* Aug. 1641, AT 3:429). A certain divine action, in other words, is required merely to sustain a thing in existence, even as the sustaining of a weight hung from a nail requires a certain force to be exerted by the nail. What is more, the nail, at least in principle, is capable of exerting whatever force is required to sustain the weight. It resists, in other words, to whatever degree necessary the downward force of the weight.

Descartes, as is well known, made a sharp distinction between the forces defined in statics and the force of motion that he refers to in the third law. In a letter to Mersenne, he reproves those who "make use of speed to explain the force of the lever" and other similar machines: "When you say that a force that can lift a weight from $A$ to $F$ in one moment, can also lift it in one moment from $A$ to $G$ [a point twice as far from $A$ as $F$] if it is doubled, I in no way see the reason for that" (*To Mersenne* 2 Feb. 1643, AT 3:615). When a scale has equal weights on both sides, the addition of only a little weight to one side will make that side drop rapidly, and there is no obvious relation between the additional weight and the speed of descent.

Perhaps for that very reason, if Descartes regarded the conserving power of God as a static force, he would, in identifying it with the force of rest, have introduced into his theory of motion a heterogeneous element. That, it seems to me, is just what he did. As Garber notes and as I argued earlier, the force of rest comprises two elements: the natural inertia, present in every body moving or not, and represented by the conversion factor $M_B/M_C$, and an additional element, present only when a resting body successfully resists the action of a moving body according to Rule 4, and which is indefinitely large. Rule 4, as we have seen, seems to have been introduced to account for reflection from a body at rest (such as the earth), which would otherwise always move off in the same direction as the moving body, joined with it. Just as in statics, the fixed point from which a pulley is hung, or the fulcrum of a lever, may be conceived of as being supplied with an indefinitely large sustaining force, so a body at rest, in the situation of Rule 4, receives from God whatever static force is needed for it to remain in place.

The force of a moving body, on the other hand, when it succeeds in causing another body to change its state or its speed, can be referred to the *concurring* action of God in that causing. It is, evidently, the efficient cause in some sense of the change; but since Descartes ascribes no active power to matter per se, the precise sense remains to be determined. I will return to that question shortly, but first I want to follow out the consequences of taking God to concur in the actions of moving bodies on others.

We have seen that in concurrence, the action of God and that of the second cause are identical. God's action, moreover, is in accordance with his ordained power; the *ordinatio* of his will which defines his ordained power we have seen sometimes referred to as a law. Consider now the proof of the second part of the third law: "it is clear that God [. . .] not only moved its diverse parts [i.e., of matter] in diverse ways, but at the same time also brought it about that some should impel others and transfer their motions to them: so that now, conserving [the world] by the same action, and with the same laws [. . .], he conserves motion not always implanted in the same parts of matter, but passing from some to others accordingly as they collide with one another" (*PP* 2§42, AT 8/1:66). We have the following scheme. When the modes of motion or rest of two bodies are incompatible, there must be in some cases a transfer of motion from one to another. This is not a literal passing of numerically the same motion from one subject to another;

like the Aristotelians, Descartes denies that such a thing can naturally occur.[102] Instead God introduces into the two bodies different degrees of motion than those they previously had. Now the bodies themselves, qua bodies, cannot occupy the same space; to that extent the fact that one or the other changes its motion is determined by the nature of the bodies themselves. But there are any number of ways in which their motions—their successive rupturings from their vicinities—could be made compatible. There are even any number of ways consistent with God's conserving the total quantity of motion in the world. God alone determines the precise amount transferred, which he does according to the third law and the definition of quantity of motion.

Thus far God's action is that of a concurring cause in the manner spelled out by the Aristotelians. But there is a joker in the Cartesian deck. To return to the question postponed a moment ago: in what sense is the moving body that changes the motion or rest of another the efficient cause of its motion? Consider the situation of Rule 5, in which a moving body collides with a body at rest and the two thereupon move off together in the direction of the moving body. If God genuinely concurs with the moving body, rather than simply employing it as an instrument to his ends, that body ought also to be an efficient cause. Now there is an obvious sense in which it yields an occasion for God's redistribution of the total quantity of motion in the world consistent with the conservation of that quantity. Its volume and speed, moreover, suffice to determine how God will do so (so long as the two bodies must move as one after the collision). But is it "that from which comes the *principle* of the beginning of motion"? It doesn't seem that it could be. If we set aside God's concurring action and the laws governing it, we are left with no reason at all why the body at rest should move—why it should, in other words, be separated from the vicinity of its neighbors. The other body's motion is no reason, since it could be conserved in other ways—by a change of direction, for example.

Concurrence is, from the Aristotelian point of view, not quite the right term for what is going on. Yet the moving body resembles an efficient cause in many ways: it "acts" by propinquity, its motion is a sine qua non condition of the effect, that effect is proportionate to it, and so forth. It is a perfect simulation of an efficient cause, lacking only the active power that an efficient cause must have. The closest equivalent in Aristotelian natural philosophy is *human generation*. The human soul is not generated when a child is conceived; only the body is generated. The soul is created by God on the occasion of the fertilization of the matrix by the semen. That divine action is

102. *To More* 30 Aug. 1649, AT 5:404ᶠ. The less punctilious Descartes of *Le Monde* thought differently: "The virtue or power of moving oneself that is found in a body can indeed pass in whole or in part into another, and thus no longer exist in the first" (*Monde* 3, AT 11:11).

utterly natural in the sense of being entirely determined by its natural concomitants (on the supposition, of course, that God acts according to his ordained power). But the father has no more the power to create the soul of his child than *res extensa* has to do anything at all.

The force of motion, then, is in God if it is anywhere. More precisely: the power that brings about the genuine change that is change of motion, or from rest to motion, can only be the power by which God concurs in the "actions" of bodies as they collide with one another, the power by which he alters the portion of the total quantity of motion allotted to each body. That attribution is confirmed by the response to More: "The moving force can be of God himself conserving as much translation in matter, as he put into it from the first moment of creation, or even of created substances, like our mind or whatever other thing he has given the force to move body. And indeed that force, in created substances, is one of their modes, but not in God; but because that cannot be so easily understood by everyone, I did not want to deal with this matter in my writings, lest I should seem to favor the opinion of those who consider God to be the soul of the world, united with matter" (*To More* 30 Aug. 1649, AT 5:404). The "moving force" is not a mode in God for the simple reason that God has no modes.[103] It is the volition with which God created and with which he conserves the total quantity, an eternal volition that is unchanging, as we have seen, even though its effects in the temporally ordered world vary.

But for that very reason it cannot be the *vis ad agendum* referred to in *Principles* 2§43. Nor indeed can the *vis ad resistendum* also mentioned there be the power by which God conserves a body. God's power cannot be said to vary, but these *vires* do.[104] In fact that section of the *Principles* tells us something on the face of it quite different: both the force of acting *and* the force of resisting consist "in this alone, that each thing tends, *quantum in se est,* to remain in the same state in which it exists, according to the first law" (AT 8/1:66). That each thing so tends is, of course, brought about by God conserving it. But that in itself does not tells us why forces should have quantities (which they must if they are to be compared as in §36).

The gloss supplied for 'force' resembles the gloss on *conatus* later on: to say that bodies have a *conatus* to recede from the center they resolve around is just to say that they are situated and moved in such a way "that they would

---

103. The "created substances" referred to in the second sentence are the human or angelic minds to which God has given the force to move body, not the bodies themselves (contrary to Prendergast 1975:96). Descartes's point is that they, being mutable, can be said to have modes, but God cannot.

104. In the striking passage I discussed in §6.4, Toletus does speak of God as concurring more strongly with the stronger of two causes, like an emperor who supports one of his kings more strongly than another (Toletus *In Phys.* 2c9q13, *Opera* 4:76rb). It would be interesting to know how widespread that way of thinking was.

in fact go in that direction, if they were not impeded by another cause" (3§46, AT 8/1:108). In both cases it is tempting to say, as Garber does, that the "forces [. . .] can be regarded simply as ways of talking about how God acts, resulting in the lawlike behavior of bodies; force for proceeding and force of resisting are ways of talking about how, on the impact-contest model, God balances the persistence of the state of one body with that of another" (1992:298). The view is, it seems to me, especially attractive in talking about the force of rest, which (as Gabbey and Garber both point out) reveals itself only when the situation of Rule 4 comes about, and then somehow manages to vary itself with the motion of the *other* body.

But that solution I find not entirely satisfying. Earlier Garber himself has favored what he calls the "divine impulse" view of God's action in creating motion. One selling point of that view is that it makes "better sense of the argument for the conservation principle": "what God directly conserves is not the quantity of motion itself, but its *cause,* the *impulsion* he introduced into the world at its creation." A bit later he adds that "the conservation law," according to the divine impulse view, "is a direct consequence of God's commitment to keep pushing, as it were, and to keep pushing as hard as he did when he set the world into motion" (283).[105] The phrase 'as hard' seems to imply a relation between some quantity associated with the impulse and the quantity of motion in the world: less push, less motion. On the same grounds as before one could argue that this is just a way of saying that God's push produces the same quantity at all times. But then it is difficult to see how one could produce an argument for the conservation principle out of the divine impulse view.

What one must preserve first of all, I think, is the absolute simplicity and continuity of the divine volition. The moving force, Descartes said to More, was of "God himself conserving as much translation in matter, as he put into it from the first moment of creation." It is, in other words, the *execution* of that unchanging volition which can have, from the point of view of our conception, different effects at different times. Considered with respect to particular bodies, the execution of that volition will sometimes yield a change of motion, or from rest to motion. In the Aristotelian world, the

---

105. I'm not sure, in fact, that I understand Garber's view. Garber says that God *directly* conserves impulse. Either that impulse is an active power that God has imparted to matter itself (as the Aristotelians thought) or it is God's own action, that is, the dependence of the modes of *motus* on him as their principle of existence. I doubt, for reasons that will become clear, that it could be an active power imparted to matter. But then what God "conserves" is *his own action* (he "keeps pushing"). The situation is exactly parallel to that regarding the quantity of matter. It would be an unmotivated precision to deny that God conserves matter "directly" because he instead conserves—i.e., continues—the act of creation, and likewise to deny that God conserves motion directly because he conserves the act of creating it.

agent of such a change can be a corporeal substance, or an active quality like weight. One customary name for its action is indeed *impulsus.*

Now God can, from our mundane point of view, be said to act on particular things, so long as one remembers that this does not imply any change in God. (In this respect, the God of the Aristotelians resembles a Leibnizian individual substance, all of whose properties are essential, yet "unfold" in succession). Insofar as those actions result in change of speed or direction or in rest becoming motion, they are analogous to Aristotelian *impulsus.* But that is, like the apparent temporality of God's volitions, a fact not about God but about the *termini* of his actions. Even the creation of motion is nothing other than the fact that the instant of creation was for some bodies an instant in which they were separating from the vicinity of their neighbors. It is only a "shove" in the same improper or broad sense that creation of substance is a *mutatio.*

The Cartesian picture thus involves a strange displacement of predicates, which may account for the difficulty of interpreting it. The quantity assigned to the moving force of God that "concurs" with bodies to change the motions of other bodies does not properly belong to the force except derivatively, by way of its *terminus.* On the other hand, the force that is measured by quantity of motion, and whose effects are specified by it, belongs not to the bodies that determine the quantity, but to God, in whom there is neither quantity nor variation in quantity.

An illuminating parallel can be drawn between the problem of force and the problem, mentioned in the discussion of divine immutability, of the temporality of divine action. The effect of divine action in the world is, as Fonseca puts it, conceived by us to be "terminated temporally" (n.78). Yet the principle of that action, a divine volition, is not temporal at all, being part of the divine essence. The action itself, the causal dependence of the effect on the volition, or on God's power to execute his volitions, though it inheres, like any action, in the patient, is a bridge between the eternal and the temporal. If one asks *when* the action occurs, there are two plausible answers: with respect to the creature in which it inheres, it occurs, say, yesterday morning or tomorrow afternoon; while with respect to its divine cause, it can hardly be said to *occur* at all—instead it *exists* eternally as a relation between God and the effect that is contained eminently within him.[106]

The same may be said of force. With respect to the body with which God concurs in moving or resisting another body, we must conceive God's con-

---

106. The same applies to the spatial location of divine action. When Descartes says that the "impelling force" (presumably of God) "applies itself now to some parts of matter, and now to others" according to the rules of collision (*To More* 30 Aug. 1649, AT 5:405), that can only be with respect to the divine action, not to the principle of that action.

currence at each instant to be terminated in that body's instantaneous size and speed jointly; with respect to God as its cause, concurrence exists eternally, being nothing other than a particular expression of the general volition by which matter and motion were created together. Ontologically that action, as Gueroult and Gabbey have argued, is two-faced: a pure mode of the divine essence, *and* a mode of the objects upon which God acts, a mode implied in the nature of matter itself only on the supposition that God has given to various parts of matter motions that cannot jointly persist if God *also* conserves the quantity of matter itself.[107] The second mode can, all the same, be designated in terms of quantity of motion, which determines how God, while conserving the motion he put in the world, will "singularize" the degree to which each body in a collision is caused to persist in its state. To that extent Garber's nominalism is warranted. Ockham argues that the proposition 'motion is in time' means that 'when something moves it does not acquire or lose everything at once but part after part' (Ockham Q. *in phys.* q20, *Opera philos.* 448). So too the proposition "the body A of size M has six degrees of motion' would mean that 'the body A after a collision will do such and such' depending on the circumstances of the collision.

In short: if by 'force' one means, following the loose usage of Descartes and his contemporaries, God's action, then it is either the *motus* itself that is conserved or changed by that action or the relation of that *motus* to God. But if by 'force' one means a mode whose intensity would be measured by the quantity $Mv$ (in a moving body), then there is no such thing. The *motus* itself has a speed (and a direction). The body has a volume. That, in a way, is all—given that God redistributes motion according to the third law.

The difficulty in understanding Descartes's conception of the *vis ad agendum* is a reflex of a much more general and profound difficulty in understanding divine action, especially those actions that are, as the Aristotelians put it, "owed" to nature. God, though absolutely free in all he does, can be thought of as *determined* by the features of the things on which he acts (specifically, at least, if not singularly). But when the features of the father determine God to create a soul for the newly conceived child, or when the quantity of motion in a moving body determines God to set another body in motion, the father or the moving body are not the efficient cause of the soul or the setting in motion. God alone is the efficient cause. Yet the features that determine the specific effect are in things, not in God, if we set aside the "contract" by which God has bound himself to be determined by those features.

---

107. In a letter to More, Descartes holds that the interpenetration of bodies would require the annihilation of matter (and would therefore not be "interpenetration" at all, but a kind of melding; *To More* 15 Apr. 1649, AT 5:342). Since God conserves matter, he does not resolve incompatibilities of motion by allowing bodies to meld.

Aristotelianism, because it imputed active powers even to inanimate things, did not have to confront the problem in its full generality. Even in human generation, the semen does not altogether lack power to organize the material in the womb into a body suitable to receive a human soul. Less perfect beings can, in some instances, effect their reproduction themselves (though God must, of course, concur with their action). Descartes, on the other hand, defines matter as *res extensa* in part just to exclude active powers. Indeed in his physics, if not his psychology, the very notion of *concurrence* begins to lost its grip. Concurrence is *co-action,* not action *simpliciter.* But if bodies do not act, there is no concurrence; there is only the outright effecting of change in the physical world by God alone.

Descartes in fact sometimes speaks of concurrence and conservation as if they were the same action—as do, occasionally, the Aristotelians. If God is the only genuine agent, distributing a fixed quantity of motion among the parts of matter, then the only difference lies in the things to which motion is given, not in God's action. Gueroult's thesis that force is nothing other than existence is so far justified.[108] More precisely: the *static* force of resisting that a body at rest exhibits in the situation of the fourth and sixth collision rules, a force that is additional to its "natural inertia," does look to be a denomination of the divine action of conservation with respect to its effects. That force is, moreover, the conservation not merely of the mode of rest, but of the union of the resting body with its neighbors, and thus of the existence *tout court* of some larger whole. The force of persevering that a body in motion exhibits, on the other hand, cannot simply be "the power that from within posits each [body] in its duration, and that consequently cannot be distinguished from [its] existence" (Gueroult 1970:117). The existence in question cannot be that of the body; it can only be the existence of the instantaneous modes of *motus:* its speed and direction. The action of conserving them is distinct from that of conserving the body they inhere in (see the critical remarks at Garber 1992:296[f]). Descartes can therefore preserve some sort of distinction between conservation and concurrence. But the distinction has been refashioned to suit a physical world without active powers. That would not be the first time a coincidence of terms belied a revolution in concepts.

108. "En réalité, force, durée, existence, sont une seule et même chose (le *conatus*) sous trois aspects différents [Gueroult cites *Principles* 1§55–57], et les trois notions s'identifient dans l'action instantanée par laquelle la substance corporelle *existe,* dure, c'est-à-dire possède la force qui la pose dans l'existence ou durée [1§62]" (Gueroult 1970:87). The sections of the *Principles* cited, however, argue that duration is distinct only in reason from substance itself; they say nothing about force.

# [9]

# Parts of Matter

Near the end of the *Principles,* Descartes settles accounts with rivals to his physics. The only serious rivals are the Aristotelianism of the Schools and the newly revived atomism of Gassendi and others. Descartes compares all three with the "commonest and most ancient" philosophy of all, to which nearly every philosopher through the centuries has adhered. No Aristotelian, no atomist, and certainly not Descartes, ever doubted that "bodies move, and have various sizes and figures, according to whose diversity their motions also vary, and that by their mutual collision large bodies are divided into smaller bodies, and change in figure" (*PP* 4§200, AT 8/1:323). But that common truth is obscured in Aristotelianism by otiose complications and unfounded distinctions, in atomism by incoherence and the suspicion of heresy. Only in Cartesian physics does it find a clear and correctly argued expression.

It is significant that Descartes chooses the nature of *body* as the topic on which to rest his closing statement for the advantages of Cartesian physics over its rivals. The early seventeenth century was one of the few times in which fundamental questions about the nature of body were genuinely open. Among the rivals of Aristotelianism, there were numerous variants of atomism, and theories in which matter was endowed with some sort of vital power or *spiritus.* There were even theories—proposed by Kepler and Fludd, among others—in which light, as the generating principle of space, was the *forma corporeitatis.*[1] By the end of the seventeenth century the alter-

---

1. On this tradition, see Lindberg 1986:29ff (Kepler), Hedwig 1980:130, 163, 177, 218 (Witelo). On Grosseteste, see McEvoy 1982:151ff, 173ff; on Patrizi, see n.20 below. Descartes would undoubtedly have known of Kepler's view; he may also have read Witelo (or Vitellio), whose works were reprinted in Risner's *Opticæ thesaurus* (1572), a standard collection of optical treatises.

natives had narrowed to one or another version of corpuscularian mechanism; but at the beginning there was, outside the Schools, no predominant view, and even within the Schools adherence to Aristotle did not preclude deep disagreement.

Descartes's treatment of body in *Le Monde* and the *Principles,* reticent as ever about sources and external motivations, yields little sense of the contemporary debate. Nevertheless, when he defines the object of physics to be "a certain matter extended in length, width, and depth" (*PP* 2§1), a great deal is packed into the assertion. The opening sections of the second part of the *Principles* repudiate both the predominant Aristotelian view and the views of those *novatores* who distinguished corporeal from incorporeal space.

That is not to say that Descartes did not draw on his rivals. Like Aristotle, Descartes holds that whatever has the dimensions of body is sufficiently like body to exclude others from its place. Space considered generically as the receptacle of matter is not independent of the singular spaces or "internal places" of the bodies contained in it. The conception, which in classical physics became second nature, of bodies as immersed in an ambient preexisting space, is foreign not only to Aristotle's but to Descartes's treatment of body.

Unlike the Aristotelians, however, Descartes holds not only that the nature of matter is quantity, but that the quantity in question is actual spatial extension, and that it is not really distinct from the quantified substance itself. Extension, in his words, *constitutes* corporeal substance; from that Descartes concludes that the only real qualities in it are the modes of extension. Substantial form—or at least the specific form argued for by the Aristotelians (§3.2)—active powers, and qualities other than the modes of extension are chimerical entities whose basis is a misplaced analogy between body and soul.

With some but not all of the *novatores,* Descartes holds that space, rather than being an accident of bodies as the Aristotelians had thought, is substance. His response to the novel treatment of space in philosophers like Telesio, Patrizi, and Campanella is clarified in the correspondence with Henry More that followed the publication of the *Principles.* There it becomes clear that although he shares with them their view that space is substance or akin to substance—and the difficulties presented by that view for an Aristotelian ontology of substance—he differs in taking spatial extension to be *sufficient* for corporeality. Where the *novatores* attempt to distinguish space from body on the grounds that body alone is tangible and impenetrable, Descartes attempts to derive those properties from extension itself. What came to be seen as the failure of that attempt hastened, I believe, the demise of Cartesianism in the latter part of the seventeenth century.

The discussion of body to follow has three parts.[2] In §9.1 I take up the argument of *Principles* 2§4–9, whose vocabulary and topic show that they are addressed primarily to Aristotelians. Descartes attempts to show that the nature of matter is quantity, by which he understands actual extension in space. The heart of his argument is an explanation of rarefaction and condensation, whose purpose is to exclude from matter any quantity of mass distinct from quantity of extension. That exclusion, together with the rejection of intensive quantities, illustrates that in Cartesian physics it is not enough for a physical property to be *representable* quantitatively. Physical properties must satisfy the more stringent criterion of being *literally* presented, or immediately derived from literally presented properties.

I conclude §9.1 with Descartes's proof that the quantity of corporeal substance is distinct only in reason from the quantified substance itself. That thesis, with its nominalist overtones, is the pivot around which the argument of the second part of the *Principles* turns from Aristotelian *quantitas* to Cartesian *extensio,* and from opposing Aristotelian positions to opposing the *novatores.*

In §9.2 I examine the delicate question of the relation between substance and space. From a survey of representative figures in the late sixteenth and early seventeenth centuries it becomes apparent that the *novatores* found it difficult to encompass substantial space within the Aristotelian ontology of substance; that they sought to distinguish substantial space from body on the grounds that body alone is tangible and impenetrable; and that Descartes had to argue against the Aristotelians that space is indeed substance, while rejecting the *novatores'* invention of a noncorporeal substantial space. Like Patrizi and Gassendi, however, Descartes, having come to believe that space is substance, found it necessary to revise the Aristotelian notion of substance to accommodate his new intuition. The ontology of *Principles* 1§51–56 is an attempt to do so, motivated in part by opposition to other philosophers' suggestions that substantial space is not corporeal, or even

2. Largely untouched in this chapter are two questions whose consideration would be essential to a full treatment of Descartes's philosophy of corporeal substance. Except for some remarks on representation and intensive quantities, I leave epistemological matters aside. Such matters have been amply discussed by others (in, for example, Marion 1981 and Garber 1992:75–93). I also set aside the distinction of body and mind, in part because I think that although securing that distinction was one of Descartes's *motives* in holding that the nature of body is extension, and although he was always ready to diagnose his opponents' errors by referring to their failure adequately to make the distinction, much of his argument on the nature of body is independent of his views about mind. My belief is reinforced by the observation that seventeenth-century critics of Descartes like Cordemoy, Leibniz, and Newton attempted to refute Descartes's opinion on physical grounds. The now traditional interpretation of modern philosophy as the history of theories of knowledge has obscured the extent to which Descartes and his near successors do *not* subordinate ontological and physical questions to epistemological ones. This chapter is in part a corrective to the excessive "Kantification" of seventeenth-century philosophy.

that it is somehow spiritual; here too one finds the best argument for identifying corporeal substance and quantity.

The arguments by which Descartes shows that an individual body is just a determinate region of space say nothing about the criterion by which one such body is distinguished from another. Descartes notoriously attempts to define 'individual body' in terms of motion—or, more precisely, *rupture* (§8.1). Although objections to the definition can be answered, the suspicion behind them is not unfounded. Descartes's conception is perched precariously between two views of extension, the first of which takes it to be a "form of body" singular in each individual, the second a continuous underlying stuff of which individuals are parts.

In §9.3 I consider Descartes's efforts to show that spatial extension is *sufficient* to constitute body. In the *Principles* and the letters, three objections emerge: the independence of impenetrability from extension; the existence of vacuum; and the existence of so-called imaginary spaces. Descartes's answers, especially in his correspondence with More, not only enlarged upon some of the arguments sketched in the *Principles,* but also reveal some of the theological underpinnings of the debate about separate space.

## 9.1. Extensive Quantity and the Nature of Matter

1. *Quantity and extension.* Descartes uses the term *extensio* rather than *quantitas* to denote the essential property of corporeal substance. The choice is significant. In Cartesian physics, Aristotelian *quantitas* is stripped down, leaving only the *extensio* representable by geometric objects. From the Aristotelian standpoint this amounts to a twofold reduction (see §4.2). The distinction between potential and actual extension disappears; and intensive quantities—degrees of heat and the like—are banished. What remains is a concept even more austere in certain respects than that of present-day science.

Already in the *Regulæ* Descartes writes that we should "diligently abstain from the term 'quantity', because some philosophers have been so subtle as to distinguish it also from extension," as well as from the quantified thing itself (*Reg.* 14, AT 10:447).[3] In the *Principles,* he notes again that some philosophers are so "subtle that they distinguish the substance of body from its quantity and quantity itself from extension" (2§5, 8/1:42). Suárez was one such philosopher. Against the nominalists he argued not only that quantity is really distinct from that which has quantity, but that a material

---

3. See Marion's note to AT 10:447 (*Reg.* [Fr.] 269); cf. *Monde* 6, AT 11:35. Marion quotes from the Suárez and Toletus passages discussed below; the Toletus passage can also be found in Gilson 1912:257, no.399.

substance can have quantity without being extended (see §4.2). It is not essential to a body having quantity that it actually be extended, but only that its parts should be "ordered among themselves so that if they are not supernaturally impeded, they must also have extension in place" (*Disp.* 40§4¶14, *Opera* 26:547). Contrary to what Ockham and others thought, to have extension is not merely to have nonidentical parts.[4] It is to occupy a place and to resist the entry of other extended things into that place.

Descartes will have none of this. He recognizes only *actually* extended quantity. Determinate quantity can be potential only in the mathematical sense mentioned earlier (see§8, n.48): in a square there are "potential" triangles that would come to be if the square were divided. That is clearly not what the Aristotelians have in mind. For them an extended thing always *would* occupy a place and keep others out of it, but it may be inhibited from doing so by God. Descartes, on the other hand, takes it to be obvious that even God cannot bring it about that two extended things should occupy the same place. Only when More calls that obvious truth into question does he attempt to *prove* that impenetrability can be derived from extension.

Intensive quantity too is ruled out in the *Regulæ*. Of the varieties of quantity, the most easily and distinctly imagined is "the real extension of a body abstracted from all else save that it is figured" (*Reg.* 14, AT 10:440–441). The imagination or phantasia is, after all, itself a "genuine real body, extended and figured," to which the mind applies itself when it imagines. Intensive quantities, because they cannot be *directly* represented, will not do: "for although one thing can be called more or less white than another, or one sound more or less acute in pitch, and so for the rest, still we cannot exactly determine whether such an excess consists in a duple or triple proportion, etc., except by a certain analogy to the extension of figured bodies" (ib. 441). Though Descartes later abandoned the line of argument presented here, his conclusion became, if anything, stronger: sensible qualities, represented as they are by obscure and confused ideas, do not admit even of being classified into positive qualities and privations (*Med.* 3, AT 7:44).[5] We have in fact no reason to believe that sensible qualities exist in bodies. At best we may conclude only that something, we know not what, is their cause (*PP* 1§70, AT 8/1:34). But insofar as they exist in the mind, they

---

4. On the *partes extra partes* argument in Ockham, see §4.2. The phrase *partes extra partes* occurs in a 1649 letter of Descartes to More: see n.5 below.

5. One indication that the earlier view persists is found in a letter to More in 1649. Descartes writes that the predicate 'extended', though it can be used loosely of qualities like weight or to spiritual substances, applies strictly only to actually extended things: "only that which is imaginable as having parts outside of parts which are of a determinate magnitude and figure I call extended, although other things are by analogy also said to be extended" (*To More* 5 Feb. 1649, AT 5:270). Significant here is the invoking of imagination. Descartes takes imagination to be the "application" of the soul to certain parts of the brain, which are extended things and thus *literal* representations of other extended things.

cannot have quantity: quantity implies divisibility, but the soul is not divisible. The analogy noted in the *Regulæ* is set aside; it was left to others to
begin learning how to measure and reason about intensive quantities.[6]

There are exceptions. Although instantaneous change of place, or rupture, can be regarded as modes of extension, velocity cannot (see §8.1). It
can indeed be represented by line segments—as could any intensive
quantity—but it is itself not such a segment. Quantity of motion, the product of velocity and volume, is likewise not itself an extensive quantity. I have
already considered the question in §8.1, but one further point comes out
here. If average velocity is represented in terms of the proportions of motions, or more precisely of distances traveled in equal times, then it is a ratio
of line segments (of the distances, not of distance to time), and so immediately derived from admissible quantities. The same cannot be said
of heat or weight, since their relation to those quantities is much more
indirect.

Descartes's conception is austere by the standards of both Aristotelian
and present-day physics. Its austerity does not consist merely in a restriction
to what can be quantified or represented geometrically: degrees of weight
or temperature are eminently quantifiable and representable. All Descartes
will admit by way of quantities is line segments *themselves,* and quantities
immediately derived from them. His austere conception is not only maintained in the ontology of material substance; it delimits to some degree the
admissible experimental determinations of quantity. Unlike Mersenne,
Descartes does not gather measurements of weight, temperature, or specific
density, or include them (except qualitatively) among the phenomena. In
the *Dioptrics,* the only work that does offer a numerically precise explanation
of observed quantities, the quantities in question are the angles of light rays.

The neglect of measurement suggests that in Cartesian physics the representation of bodies must be not just mathematical, not just geometrical, but
peculiarly literal. If bodies are nothing other than *res extensæ,* and their
properties modes of extension, the only quantities that occur in them exactly as they are represented to us in sense or imagination are figure, size,
and—derivatively—motion. A literal representation of heat could only be a
representation of the configuration of the particles of fire and of their
motion. Once we are apprised of the true nature of heat, it is idle to interject
an analogical representation of its degrees. To do so is to revert to a repre-

6. Descartes takes no notice, for example, of the various "thermoscopes" or thermometers,
some using air, others water, with which Beeckman, Mersenne, and others measured degrees
of heat (see *Rey à Mersenne* 1 Jan 1632, *Corr.* 3:244, 248, *Cornier à Mersenne* 16 Jan 1626, 1:331n,
Beeckman à Mersenne Jun. 1629, 2:232). Since the accuracy of air thermometers was compromised by variations in barometric pressure, Descartes had some reason to neglect them;
even though Rey had by 1632 invented a water thermometer that avoided the problem, liquid
thermometers came to be widely used only after 1654 (*Corr.* 3:248).

sentation that, though mathematical in form, is subject to the pitfalls of the sensory representation that Cartesian physics seeks to replace.

2. *Quantity and matter.* It was generally agreed that matter cannot naturally exist without quantity. Whether quantity is part of the nature of matter was, on the other hand, disputed. Thomists deny that it is; others are not so sure (§4.2). Everyone at least acknowledges the existence of dissenters, notably Averroes, according to whom prime matter is essentially endowed with indeterminate quantity, *quantitas interminata.* One sixteenth-century proponent of the view was Zabarella. In *De prima rerum materia,* Zabarella, having disposed of the Thomist *pura potentia* and the Scotist *forma corporeitatis,* turns to his own view. "Prime matter," he writes, citing Averroes, "according to itself, and prior to the reception of form, has quantity, and has three dimensions, length, breath, and depth, as an internal and inseparable accident, although [its dimensions] are not circumscribed by any limits [*terminis*], but are indeterminate [*interminatas*], and afterwards receive various limits from diverse forms, accordingly as the various natures of natural bodies require them" (Zabarella *De prima rerum materia* 2c6, *De rebus nat.* 191D; cf. Averroes *In Phys.* 1com63). *Quantitas interminata,* like matter itself, can neither be generated nor corrupted. Form cannot be its cause, because form in itself is not divisible; nor can quality, for similar reasons; nor, finally, can quantity itself be its own cause, and so only matter remains (Zabarella ib. c7, 194A). Zabarella adds that quantity can be "abstracted by the mind from every natural form" and is therefore prior to form in thought as well as in things.

With the important qualification that quantity can be potentially but not actually extended, Suárez holds a similar view. In an argument showing that the rational soul has neither matter nor material cause, he holds that "matter and quantity are inseparable and reciprocal," and that no other property of matter is. Like matter itself, quantity is "most apt for receiving and being acted upon, and of itself is ordered to no particular action" (Suárez *Disp.* 13§1¶15, *Opera* 25:460). Later he cites with approval the same passages from Averroes that Zabarella cites; and, as we have seen (§5.3), he argues not only that quantity precedes form in the composition of substance, but that certain other accidents that dispose matter to receive form inhere in matter alone by way of quantity (*Disp.* 14§3¶10ff). Quantified matter is in Suárez's metaphysics close to being an independent substance in which sensible qualities inhere immediately rather than by way of form.

The argument of *Principles* 2§4–23 begins, as I have noted, by taking on the Aristotelians and then shifts to the *novatores.* In 2§4 Descartes starts with the Aristotelian terms *materia* and *quantitas,* only later switching to his own terms *corpus* and *extensio.* The nature of matter, he holds, is quantity, a quantity that will turn out to be nothing other than spatial extension: "we perceive that the nature of matter, or of body universally regarded [*corporis in universum spectati*], does not consist in its being a hard or weighty or colored thing, or in any other mode affecting the senses: but only in its being a thing extended in length, width, and depth" (AT 8/1:42). The implied interchangeability of *materia* and *corpus* is worth noting. Descartes uses *materia, corpus,* and *substantia* (qualified as *corporea* or *extensa*) to designate the thing whose existence and properties are being considered in the opening sections of *Principles* 2. The shift from one term to another was sanctioned by usage. Goclenius, for example, writes that *corpus* has two acceptations, one in the category of substance, the other in that of quantity. Of the first he writes: "That *corpus* which is Substance, is the Subject of threefold dimensions, length, breadth, and depth; it is nothing other than either corporeal Substance, and then it is a genus, or the Material part of corporated substance."[7] *Corpus* as substance is material substance considered under the aspect of the spatial quantity proper to it. *Corpus* as quantity is simply the three dimensions of corporeal substance. Toletus occasionally uses the term *corpus mathematicum* to denote *corpus* as quantity, contrasting it with *corpus naturale* or *corpus physicum.* The first is quantity terminated by figure, considered apart from matter, sensible accidents, and natural change; the second is the substance that not only has quantity, but all the other accidents as well.[8]

*Corpus,* then, does not in the passage at hand denote an *individual* substance, as it sometimes does elsewhere (see 2§25, where *corpus* alternates with *pars materiæ*). It denotes 'body regarded universally' or what is common to all individual bodies qua bodies, and alone persists through every possible natural change—in short, what the Aristotelian calls *matter.* But it denotes matter under the specific aspect of having the three spatial dimensions. Since for Descartes that aspect constitutes matter, it is not surprising that in his writings *corpus* took over from *materia* the function of denoting the substrate of change.

The circulation in Descartes's texts of the terms *materia, corpus,* and *substantia,* though it did not offend against usage, and was unlikely to have confused his readers, does procure a certain rhetorical advantage. In Aristotelian physics, the natures of corporeal substances are jointly determined by

7. 'Corpus, quod est Substantia, est Subiectum triplicis dimensionis, longitudinis, latitudinis, & profunditatis. Estque nihil aliud, quam vel Substantia corporea, ac tunc Genus est: vel Materialis pars substantiæ corporatæ" (Goclenius *Lexicon,* s.v. "Corpus," 481).
8. 'Body is twofold. One is 'natural', the other 'mathematical'. Mathematical body [*corpus mathematicum*] is quantity itself, having the three dimensions of length, breadth, and depth, insofar as it is considered *per se,* stripped of all other sensible accidents, but only terminated by some figure; [mathematical] body, they say, is of the category of quantity. [ . . . ] Physical body [*physicum corpus*] is the substance itself that has such dimensions, but with sensible qualities, and subject to changes; and this body is in things, and is a substance" (Toletus *In Phys.* 4c2q1, *Opera* 4:106rb–107va).

matter *and* form (see §7.2). Individual and specific differences can be referred to either. Zabarella agreed that the nature of matter is quantity; but he did not deny that the natures of individual substances include sensible qualities and active powers. Descartes, however, means to show not only that the nature of body in general is extension, but that all corporeal accidents can be reduced to the divisibility and mobility of extension. *Corpus physicum* has no property not possessed by *corpus mathematicum*. For that it does not suffice to show that the nature of matter is extension, or even that body and space are one. In §9.2 I will examine whether the arguments of the first part of the *Principles* support the conclusion; here it is enough to notice the gap in the argument of the second part, to which I now return.

Taking hardness to be exemplary of the properties that others have mistakenly imputed to matter, Descartes writes: "Nothing else about it is indicated to our senses than that the parts of hard bodies resist the motion of our hands when they run into them. For if whenever our hands moved in some direction all bodies existing there receded with the same speed as our hands approached them, we would never feel hardness" (AT 8/1:42). Yet mere motion cannot change the nature of matter; so it "does not consist in hardness."

The argument misses its target. Those who argued that bodies are impenetrable certainly did not take impenetrability to consist merely in the fact that bodies cause us to have sensations of hardness. Being impenetrable gives bodies the power to resist our touch, a power they cannot lose merely because God never allows them to manifest it. Just as the bowels of the earth, "though they have never been exposed to the sun, nor has anyone descended to them with lamp or torch," are still visible in themselves (More *To Descartes* 5 Mar. 1649; AT 5:300), so too bodies never touched remain tangible.

Descartes takes what is clear and distinct in sensations of hardness to be the fact of resistance. To infer from that fact a *power* of resistance is a step toward mystery, one that Descartes refuses to take. But then he has no explanation of the fact. Only when More compels him to admit that resistance is a genuine property of bodies, one that could exist even if the sensations characteristically caused by it did not, does Descartes bestir himself to show that it indeed follows from being extended. I will return to this question in §9.3.

3. *There is but one quantity in matter.* In 2§5 Descartes finds "two causes [. . .] why it might be doubted that the true nature of body consists in extension alone" (AT 8/1:42). One is that some people explain rarefaction by supposing that one and the same body can, without having any matter added to or taken away from it, have more or less extension. They infer that matter has not only a quantity of extension but another, independent,

quantity as well—quantity of mass. The other is that some people think that certain regions of space, which they call "empty," contain no bodies. Descartes therefore divides the remainder of his task into two parts. He must show that matter (or "corporeal substance") and quantity are distinct in reason only—and thus that spatial extension is the only quantity in matter; and that extension, internal place, and space differ in reason only. The first is carried out in 2§4–9, the second in 2§10–19 (see §9.2–9.3).

For the Aristotelians, rarefaction and condensation were changes in the intensive quantity associated with rarity and density. The standard case is the change of one element into another—the same quantity of water, when transmuted by heat into air, is said to increase tenfold in volume. Following Aristotelian usage (§4.2), I distinguish *quantity of mass*, which is preserved in such transformations, from *quantity of extension*, which changes (we would now speak of mass and volume). It was controversial whether rarefaction and condensation are distinct from augmentation and diminution, and quantity of mass from quantity of extension. Toletus concludes that in rarefaction and condensation there is always change in quantity of extension (*In Phys.* 4c9q11, *Opera* 4:132vb). Thomists, on the other hand, held that "rarity and density is something pertaining to the parts of a body as ordered among themselves [*in ordine ad se*], and not as ordered in place" (John of St. Thomas *Nat. phil.* 3q7a1, *Cursus* 2:709). One of their arguments is that if quantity of extension were the sole quantity in matter, in rarefaction new matter would have to be generated. But experience tells us that the matter of the generated element is derived entirely from that of the corrupted element. Transmutation conserves quantity of mass while augmenting or diminishing quantity of extension. Hence the two quantities are distinct.

Descartes's response is to show how the appearance of rarefaction or condensation can be saved without admitting any quantity other than quantity of extension, and while maintaining the conservation of matter.[9] His explanation was not new.[10] What we ordinarily call a body is typically a composite, like a sponge soaked in water, of a porous body with other

9. The discussion that follows is indebted to Miles 1983. Atomists explained rarefaction by supposing that atoms moved apart; other philosophers thought that there could be "interstitial vacua" within bodies. In 2§6 the target is Aristotelian explanations; in 2§19 Descartes applies his proof of the impossibility of the vacuum to refute the others.

10. John of St. Thomas ascribes similar views to Scotus, Marsilius of Inghen, Hervæus and others (*Cursus* 2:708; his source is Rubio). Descartes himself had affirmed a similar explanation as early as 1629 (*To Mersenne* 8 Oct. 1629, AT 1:25), his immediate source being Sebastian Basso (*Philos. nat.* 332[f]; cf. Mersenne *Corr.* 2:302, 307[f]). Beeckman briefly entertained the hypothesis of spongelike particles in 1620 before shifting to a full-blown atomist explanation (*Journal* 2:157, 230; cf. Mersenne *Corr.* 2:297). By 1627 he had returned to an explanation according to which subtle matter entering the pores of gross matter caused it to expand (*Journal* 3:127).

bodies—assumed to be fluid—filling the pores. The porous body becomes denser when "the intervals between its parts, as those parts approach one another, diminish or even disappear altogether" (2§6, 8/1:43). Its outward dimension decreases, thereby giving the appearance of a decrease in quantity of mass. But what really happens is a volume-preserving transformation of the body into one having smaller outward dimension. That transformation will, of course, also conserve mass. Anyone, then, who "attends to his thoughts, and will admit only what he clearly perceives," will find that to explain rarefaction and condensation he needs only to imagine "a mutation of figure" and not the increase or decrease of some quantity distinct from extension.

That much would justify Descartes in holding that the Aristotelian account is not forced upon us. But he was not satisfied with so modest a conclusion. Instead he holds that "we perceive that rarefaction can quite easily occur in this way, but not in any other" (*PP* 2§7, AT 8/1:44). It is, in fact, "plainly repugnant that something should be augmented by a new quantity or new extension, unless at the same time new extended substance, that is, new body comes to it." If we attend to our thoughts, we will see that it is not "as agreeable to reason to concoct something which is not intelligible [. . .] as to conclude, from the fact that [bodies] are rarefied, that there are pores in them," pores filled with other bodies that we do not perceive.

The unintelligible something is either the quality of density or a quantity of mass distinct from extension. Physics has done without the quality. But quantity of mass has proved to be essential. Recent authors have harped on Descartes's "obvious difficulties in explaining the relative density of different bodies."[11] He himself asserts what is taken to be the problem: "We easily understand that it cannot be [. . .] that there should be more matter or corporeal substance in a vessel when it is full of lead or gold or some other heavy and hard body than when it contains only air [. . .]: since the quantity of parts of matter does not depend on their heaviness or hardness, but on their extension alone, which in the same vessel is always equal (Descartes *PP* 2§19, AT 8/1:51). In short, "the quantity of matter in equal volumes is always equal" (Clarke 1989:77), and the density of every body will be 1.

But Descartes knew that.[12] Why didn't he see the problem? The answer is that 'mass per unit volume', or density as it is now defined, is not a Cartesian notion. The phenomena that the Aristotelians attempted to explain by reference to rarity and density were of three sorts: the evident change of

---

11. Clarke 1989:76; cf. Funkenstein 1986:74.

12. 'Tous les Cors estant de mesme matiere, deux parties de cette matiere, de mesme grosseur & figure, ne peuuent estre plus pesantes l'une que l'autre" (*To Mersenne* 30 Jul. 1640, AT 3:135).

outward dimension which accompanies changes of phase, as in melting or boiling; the more subtle change that accompanies warming and cooling without change of phase; the fact that equal volumes of different substances have different weights. For the first and second, the *explanandum* is the change of outer dimension that is one effect of certain physical processes. That is what Descartes wants to explain in 2§5–6. For the third, one would have to refer to his theory of weight. The force by which a body is pushed toward the center of the earth is not easily calculated. For bodies of fixed outward dimension, it varies directly, though not linearly, with the amount of matter they contain (see 4§20ff, esp. 24–25; AT 8/1:212ff). What is clear is that the Newtonian equivalence of inertial and gravitational mass is absent from Descartes's physics. With it goes the direct comparison of masses by their weights, and the computation of density by weight and volume.

The explanation of rarefaction and condensation shows that physics can do without the qualities of rarity and density. More significantly, it shows that the only quantity one needs to associate with bodies is extension. Quantity of mass is a perfectly respectable physical dimension. Descartes's refusal to admit it, like his proscription of intensive quantities, must be put down to his insistence that all continuous quantities considered in physics should not merely be representable by determinate extension, but should themselves *be* determinate spatial extensions.

4. *Quantity and substance.* Having disposed of the argument from rarefaction and condensation, Descartes applies the same reasoning to show that substance and quantity differ only in reason: "It cannot happen that even the smallest amount should be removed from this quantity or extension, unless the same amount of substance is taken away; nor conversely, that the slightest bit of substance should be taken away without the same amount of quantity being removed" (*PP* 2§8, AT 8/1:45). This, together with a comparison of quantity with number, which likewise differs only in reason from the things enumerated, is Descartes's entire argument.

It is, unfortunately, irrelevant. In the Aristotelian conception, quantity is an accident of corporeal substance. When a substance is deprived of quantity, as in the Eucharist, that quantity is not *deducted;* it is *removed.* What remains is neither bigger nor smaller than it was. Having *no* quantity, it admits no comparison, any more than souls do.

There are, I think, two reasons why Descartes did not offer better arguments at this point in the *Principles.* The first is that the strongest reasons *for* a real distinction between substance and quantity are drawn from the Eucharist (§4.2). To answer them, Descartes would have had to develop his own explanation of transubstantiation, which would have been both imprudent and out of place in a physics textbook. The second reason is that the identification of substance and quantity is an instance of the general truth, asserted in 1§62, that *no* substance is really distinct from its principal attribute. I will examine that claim in §9.2.

However inadequately argued, the thesis of 2§8 is pivotal in the argument of the second part of the *Principles*. In 2§4 the subject is Aristotelian matter; in 2§11 it would, to judge from the examples used there, have shifted to corporeal substance—in Aristotelian physics the composite of matter and form. But to show that the nature of matter is to be extended (2§4) and to show that quantity and substance are distinct only in reason is not quite to show that the nature of corporeal *substance* is to be extended. One might well agree with the nominalist thesis of 2§8, and even the conclusion of 2§4, while denying that the nature—or rather the natures—of corporeal substances consisted merely in their being extended. The Aristotelian believes that the natures of corporeal substances are their forms more than their matter; but here the possibility of a contribution by something other than matter has been silently elided.

## 9.2. Substance and Space

If there is a characteristically Cartesian thesis on body, it is not that the nature of matter is extension or that substance and quantity are one. It is that space and body differ only in reason. That thesis sets Descartes apart both from the Aristotelians, for whom there is a real distinction between matter and quantity, and from the atomists, some of whom agreed that all accidents of body are modes of extension, but denied that corporeal substance and space differ only in reason. Descartes alone holds that space and corporeal substance are one.

In this section I first survey the scene in which that claim was put forward, a scene of dissension and instability. Aristotelians stuck for the most part with Aristotle's rejection of a "separate" space independent of body. Atomists believed as a matter of course in separate space, but were at some pains to explain what sort of entity it is. Other philosophers argued that space is a substance, but unlike Descartes they held that it is distinct from body. They too found it difficult to locate substantial space within Aristotelian ontology, and were therefore inclined to tamper with the category of substance and to contest the analysis of corporeal substance into matter and form.

I then examine Descartes's central argument for identifying corporeal substance with spatial extension, one version of which is found in *Principles* 2§11. That argument is directed not only against the Aristotelians, but against the *novatores,* many of whom held that there are two kinds of extension, the corporeal extension of bodies and the incorporeal extension of space. Here Descartes could justifiably regard himself as the true inheritor of Aristotle: Aristotle too had argued that whatever has spatial "dimensions" will exhibit many of the properties of body, including the crucial property of impenetrability.

But to save Aristotle on that point was to forsake him at another. The ontology of substance, attribute, and mode proposed in the first part of *Principles* departs significantly from that of the *Categories* and from that of material substance in the *Physics*. Among the motivations for the new ontology was that of explaining how body, conceived as extension, could nevertheless be regarded as a substance. I will examine how Descartes's emphasis on existence in his definition of substance, and the notion of *constitution,* which supplies the ground upon which Descartes rests his exclusion of sensible qualities from body.

I conclude with a look at the individuation of bodies. Extension, though it fulfills some of the metaphysical functions of substantial form, cannot, it would seem, supply a principle by which to distinguish one body from another, as form was said to do for complete substances in Aristotelianism. Descartes's definition of 'individual body' or 'part of matter' has met with severe criticism, and indeed there are profound ambiguities in his treatment of space, which serves both as the *materia* common to all bodies and as the theater within which they move and act on one another. But I think that the definition can be vindicated against some of the criticisms made of it, and that in doing so one may see that, contrary to what is often said, physical space is for Descartes not simply the space of geometry made actual.

1. *Background.* Two arguments against the existence of a separate space in the fourth book of Aristotle's *Physics* provided the starting point for later discussions.[13] Aristotle argues that if there were a place and a body in that place, there would be two bodies in one place, which is impossible. The Coimbrans paraphrase the argument thus: "If there were a place of some sort, it would very much seem to be body, if in fact it had the threefold dimensions, length, width, and depth, by which the nature of body is defined: but it cannot be a body, since when that place, and that which is held in place, were together, it would follow that two bodies existed at once and permeated one another" (Coimbra *In Phys.* 4c1, *explanatio,* 2:6). The presupposition of the argument is that whatever has the three dimensions of bodies is itself body. The argument could be turned aside either by denying the presupposition, as Patrizi, Gassendi, and More do, or by denying that the place of a body is distinct from the body itself, as Descartes does.

The second argument occurs a bit later. Aristotle has brought forth an array of reasons to show that motion in a void would be incommensurable

13. On these arguments, see Grant 1981, c.1–2; on Aristotle, see Sorabji 1988:76ff.

with motion in a resisting medium; the ultimate conclusion, of course, is that there can be no void. He then shifts ground: "But even if we consider what is called 'the void' by itself, it will appear to be truly void"—an empty name (4c8, 216a16ff; Coimbra *In Phys.* 2:55):

> If someone (he says) throws a cube into water, the water will immediately cede the cup to it according to the size of the cube; the same holds for the air, even if it escapes our view. But this, which is conceded by all, cannot occur in empty space, if indeed only bodies cede their place. Yet those who affirm the vacuum suppose that there is an interval extended in three dimensions, from which sensible bodies have been removed; they are [thus] compelled to say that there is a penetration of dimensions, which is no less absurd than if the cube were penetrated by water. Nor can they escape [this consequence] if they answer that the dimensions of the vacuum are not affected with sensible qualities. For the reason why several bodies cannot be in the same place at once is not the admixture of sensible qualities, but their threefold dimension. For if the cube also is stripped of all accidents, it will resist the penetration of water no less than it does now. (ib. *explanatio,* 2:56)

Anything, body or not, sensible or not, that has "dimensions" is subject to the principle that two such things cannot occupy one place simultaneously. The supposed void is either sufficiently like body to obey that principle, or else it is nothing at all. Two keys to Descartes's argument are found here: the thought experiment of stripping away accidents, and the claim that impenetrability depends only on having dimensions. In adapting Aristotle's arguments, Descartes in effect defends the genuine Aristotle not only against the *novatores* but also against the Aristotelians themselves.

Dissent from Aristotle's refusal to admit separate space came not only from anti-Aristotelians, but from within Aristotelianism. Among Aristotelian dissenters, I take Fonseca and Suárez as representatives, among anti-Aristotelians, Patrizi and Gassendi. My aim is a sketch of the intellectual landscape within which Descartes's solution should be located.[14]

In a question on whether *locus* is a genuine quantity, we find Fonseca grappling with Philoponus's reply to the arguments of Aristotle just mentioned. Philoponus held (so Fonseca says) that "there exists a space of three dimensions, fixed and immobile, in which together and, so to speak, penetratively every body whatever is contained, and which is called 'space' and 'plenum' insofar as body is contained in it, and 'vacuum' if every body has

---

14. Grant 1981 is the indispensable starting point for any such history. His discussion makes it clear that virtually every position one could imagine remained a live option in the early seventeenth century.

left it" (Fonseca *In meta.* 5c13q7§1, 2:700E).[15] After rehearsing the arguments of Aristotle, Fonseca presents Philoponus's counterargument: "Philoponus, the defender of the being of space [*entitatis spatii defensor*] answers that three-dimensional quantity is of two sorts: one is material and requires space and cannot exist together with another material quantity; the other [is] immaterial, and is space [itself], which because it is immaterial allows an adequate body [to coexist] with it" (ib. 702C). Space is not a real and genuine quantity, but rather what Fonseca calls *imaginary*. It is imaginary not because it depends on our imagination but because "space, which after its fashion truly is, always was, and will be, is not a genuine but a fictive [*ficta*] quantity" (703C). It is the capacity or aptitude for receiving each body into a space of the right volume, a *non repugnantia ad ea capienda*. That capacity is neither in any body nor in any place; Fonseca calls it a "pure negation."

The negation is this: it is *not* the case that God could not have created quantified matter, or more than he did create, or infinitely much. What makes that true is not an accident or mode of matter. To use Fonseca's earlier example: the truth of the proposition 'a human is not a stone' does not rest on our having the "negative property" of not-being-a-stone, but on our having certain positive properties—being animate, say (see 5c5q1, 2:323[f]). Likewise the "existence" of space, or the capacity to receive body, is merely the reverse of the coin whose obverse is God's power to create quantified matter.[16] Space so understood is not only infinite, but eternal. Its eternity does not contradict the doctrine of creation because the capacity to receive body is not a capacity belonging to any *thing:* it presupposes nothing apart from God, and is therefore no exception to the proposition that all things, save God, are created.[17]

Patrizi, unlike Fonseca, unequivocally accepts the distinction between material and immaterial space. According to his doctrine, which deserves to

15. On Philoponus, see Sorabji 1988, c.2 and pp.200–201; on his *fortuna* in the sixteenth century, see Schmitt 1987.

16. The crucial point, according to Grant, is that Fonseca takes even pure negations to exist. The intended contrast is with philosophers (Grant mentions Gabriel Vasquez) who held that the void, being the mere negation of body, is therefore nothing at all (Grant 1981:157[f]). I am not sure that the difference between Fonseca's view and Vasquez's is as "radical" as Grant takes it to be. Fonseca does use the word 'existere' in his treatment of pure negation. But he explicitly defines their "existence" to be the nonexistence of what would render them false: "This pure negation—that man is not stone—is said to exist independently of the operation of the intellect for the sole reason that there exists no man who is a stone" (*In meta.* 5c5q1§4, 2:323C). His intention is to deny existential import to certain "necessary connections" so that their being necessarily and eternally true does not entail the necessary and eternal existence of anything except God.

17. Goclenius sums up the point thus: 'imaginary space' is "sometimes taken also for that which is not absolutely nothing, as for the capacity of the thing there is given this space [ . . . ] Fonseca holds that the connections of things are not genuine things *per se*, and not inventions, but only identities of things, purely negative [ . . . ] and so it is no wonder that they should neither be created nor uncreated" (Goclenius *Lexicon* s.v. 'Spatium', 1068).

be better known among Descartes scholars,[18] everything is in space, and for that reason space precedes all else, save God. *Locus,* or space insofar as it is occupied by body, is "just as prior to body as body to corporeal qualities. For that without which nothing else can exist, while it can exist without anything else, is necessarily prior."[19] It is prior even to the world itself, "brought forth by the First One before all other things, as if breathed out by the breath of his mouth" (227[61c]).

Space is neither a body nor a property of bodies, but it is not *non-ens.* Like body, it has three dimensions, but unlike body it has no figure, nor does it offer resistance to movement (227–228[61c-62a]). Thus Patrizi answers Aristotle's second argument: there can be mutual penetration of dimensions when at least one of the dimensions does not belong to a body (230[62c]).

Space "is not embraced by any of the categories, and is prior to and outside them all." It is a substance in the sense of *id quod per se substat* [that which subsists by itself], of *quæ aliis substat* [that which underlies others], and of *quæ nulla aliarum rerum eget ad esse* [that which needs no other in order to be] (241[65bc]). Here Patrizi's position is much stronger than Fonseca's: space has a positive, even substantial, existence. But it is not a complete corporeal substance, since it is not composed of matter and form. It is a "mean" between the corporeal and the incorporeal, "an incorporeal body and a corporeal non-body," because it is not sensible nor resistant, and yet unlike spirits it has dimensions (241[65c]).[20]

On the face of it, little of this would be congenial to Descartes. But Descartes too gives up matter and form; he too defines 'substance' as that which needs no other to exist (*PP* 1§51, AT 8/1:24). What separates him from Patrizi is his insistence that *space* and *body* coincide, or rather that every

18. Garber 1992 is one exception (see p.128). The work I draw on is Patrizi's *De spacio physico,* first published in 1587 and later incorporated into the Nova de universis philosophia. Not only was the latter placed on the Index in 1592, but the animus of the Church against Patrizi's Platonism was strong enough that the Holy Office recommended after his death that the chair established at Rome for the teaching of Platonism be suppressed. See Brickman's introduction to De spacio physico, Henry 1979, Grant 1981:199[ff], Copenhaver and Schmitt 1992.

Patrizi is mentioned unfavorably in two of Mersenne's early works, the *Quæstiones celeberrimæ in Genesim* (col.738–742) and the *Vérité des sciences* (1:109), once in Beeckman's *Journal* in 1633 (3:289), and at some length by Gassendi (*Syntagma* 2, *Opera* 1:246). Though his name does not occur in Descartes's correspondence, it is unlikely that Descartes would not have known of him. He does mention the other Italian philosophers with whom Patrizi is commonly associated (see Descartes *To Beeckman* 17 Oct. 1630, AT 1:158, where Telesio, Campanella, and Bruno are included among the *novatores;* on the association of Patrizi with these figures, see Henry 1979:550 and the references cited there).

19. Patrizi *De spacio phys.* 231[62c], 239[64d]; Patrizi is quoting—to his own purposes—Aristotle *Phys.* 4c1, 208b34[ff]; cf. Coimbra *In Phys.* 2:5.

20. In another part of the *Pancosmia,* the *De primævo lumine,* Patrizi argues that space is filled first with light, which like space is between the corporeal and the incorporeal. Cf. Henry 1979:556; Grant 1981:203 and n.142; Mersenne *Corr.* 1:334.

delimited region of space coincides with the three-dimensional place of one or more bodies. Impenetrability and having effects on the senses are, for Descartes, necessary consequences of "having dimensions." That is no trivial difference. But neither that nor Descartes's repugnance toward any suggestion of a "mean" between the corporeal and the incorporeal should obscure the structural similarities between Patrizian space and Cartesian matter.

Gassendi, who knew Patrizi's work well,[21] agrees that space, although it is an existing thing, is not comprehended within the Aristotelian categories:

> it is commonly agreed that Place and Time are corporeal accidents; so that if none of the bodies on which they depend existed, there would be neither Place nor Time. To us, however, because it seems that even if there were no bodies, still constant Place and fleeting Time would remain, Time and Place seem not to depend on bodies, and so not to be corporeal accidents. But neither are they on that account incorporeal accidents [. . .] Hence it happens that Being in its most general acceptation is not adequately divided into Substance and Accident; rather Place and Time should be added as two members to the division. (Gassendi *Syntagma, Physicæ* §2lib1c1, *Opera* 1:182a)

Even if all sublunary matter were annihilated by God, still the space enclosed within the lunar sphere would not cease to be (182b, 184a).[22] The supposition, though it may seem absurd to some, is no more so than the supposition that matter should exist without form: yet the Aristotelians allow the second but not the first. The space that is thus proved to exist apart from body Gassendi calls "incorporeal" (183b), not because it is a thinking thing, but because it exhibits no repugnance in the face of penetration by bodies (see *Syntagma, Logicæ* lib1c7, *Opera* 1:55a).[23]

21. The *Syntagma* includes a précis of Patrizi's *magnum opus* (*Opera* 1:246ab). Although the Syntagma was not published until 1658, much of it was written between 1635 and 1645. A draft of the section on space and time is dated by Bloch to 1635–1636 (Bloch 1971:xxix). Bloch observes that the earlier version also affirms the Epicurean definition of time and space as "accidents of accidents," which is inconsistent with the position sketched above. The inconsistency is resolved completely only in the *Syntagma* itself (Bloch 1971:186ff). On Gassendi's theory of space and time, see Bloch 1971, c.6, and Grant 1981:206ff.

22. A similar argument is made in a letter of Gassendi to Samuel Sorbière in 1644. Sorbière wants to know "what can be said against the Cartesian dogma that *There exists no vacuum.*" Gassendi replies with a brief version of the annihilation argument (*To Sorbière* 30 Apr. 1644, *Opera* 6:187; see AT 4:108–109; Sorbière's letter is in Gassendi *Opera* 6:469, and cf. AT 4:108 and Mersenne *Corr.* 13:110). The annihilation argument was used by Suárez to show that *distances* among bodies do not presuppose the existence of intervening bodies (*Disp.* 30§7¶34, *Opera* 26:106).

23. Applied to souls or to God, Gassendi says, the term 'incorporeal' denotes a "true and full substance, and a true and full nature, with which their faculties and actions are consonant" (*syntagma* 183b). Applied to space and time it denotes merely the negation of body. We thus have a partial coincidence with Fonseca's characterization of space as a pure negation. But Fonseca does not take the more radical step of enlarging the categories to include space and time as distinct ways of being alongside substance and accident.

Like Patrizi, Gassendi holds that space exists and is not an accident of matter (or of spirit). He holds, moreover, that the mark of a *corporeal* thing is that it should be impenetrable and affect the senses. But he denies that space is a substance. The disagreement is not so great as it may seem: where Patrizi enlarges the notion of substance, Gassendi denies that 'substance' and 'accident'—in their Aristotelian acceptation—exhaust the ways that a thing can be said to be. Common to both is dissatisfaction with the categorical conception of substance.

From this survey I draw out four themes to help in understanding Descartes's position:

(i)   Aristotle's claim that whatever has dimensions is body, or at least sufficiently similar to body that it is subject to the constraint of not existing in the same place with body;

(ii)  the general agreement among Aristotle's critics that whatever is corporeal will indeed resist penetration by other corporeal things and will be capable of affecting the senses;

(iii) the general agreement among the critics that space, although it has dimensions, lacks the characteristics in (ii), and thus that (i) is false;

(iv)  the general agreement that space is not an accident of bodies but a substance in its own right, that it is not a composite of matter and form, and that it cannot be subsumed under any of the categories, together with a marked lack of agreement concerning the kind of thing space is.

The bones of contention, then, are whether extension alone suffices to make a thing a body; whether space is a substance; and how to characterize the kind of entity space is without using the Aristotelian apparatus. We have seen that quantity, or actual extension, is, in Descartes's view, distinct only in reason from quantified substance. What remains is to show that, contrary to what the *novatores* thought, bodily extension and spatial extension are one and the same; and to revise the metaphysics of form and matter to accommodate the new entity.

2. *Descartes against the* novatores. *Principles* 2§10–11 attempts to do two things at once: to accommodate Aristotle's thesis that, since interpenetration of dimensions is impossible, there is no space apart from bodies, and to refute the *novatores*' claim that corporeal extension and spatial extension are different. The first is accomplished simply by identifying "space, or internal place" with the "corporeal substance contained in it"; the second by showing that "the extension in length, width, and depth that constitutes space is evidently the same as that which constitutes body."

We have, in fact, already been told that "extension [. . .] constitutes the nature of corporeal substance" (*PP* 1§53, AT 8/1:25), that quantity or bodily extension is the nature of matter, and that it is identical to the substance that has it. What remains to be shown is merely that quantity *is* spatial extension:

> And indeed we would easily recognize that the extension which constitutes the nature of body and the nature of space is the same [. . .], if in attending to the idea we have of a certain body, e.g. a stone, we reject from it all that we recognize is not required for the nature of body: namely, we reject first of all hardness, because it would be lost if the stone were liquefied or divided into exceedingly minute bits of dust, and yet the stone would not cease to be body; we reject also color, since we often see stones so pellucid that there is no color in them; we reject heaviness, since fire, although it is very light, is nonetheless held to be body; and finally we reject cold and heat, and all other qualities, since either they are not considered in the stone, or because if they are changed the stone is not thereby judged to lose the nature of body. Then we would realize that plainly nothing remains in the idea of it, except that it should be something extended in length, breadth, and depth: and the same is contained in the idea of space, not only space filled with bodies, but also that which is called 'vacuum' (2§11, 8/1:46).

Although we are asked here to consider an individual body, the object of Descartes's argument is body in general, the *materia* whose nature he has earlier shown to be quantity. What is common to all bodies must persist through every natural change; what does not persist is not part of the nature of body. But reference to change is only a pedagogical device. What counts is whatever belongs to all bodies, whether they can be changed into one another or not. The result of the experiment, of course, is that only extension—being endowed with the three spatial dimensions—belongs to all bodies. But extension also belongs to all spaces, filled or (apparently) not. So the *novatores* are mistaken: space *is* body. Out go the hesitations of Fonseca, the incorporeals of Gassendi, the promiscuous mingling of matter and spirit in Patrizi.

Many of Descartes's contemporaries rejected his conclusion, holding that the nature of body includes impenetrability, or at least that space without body is possible. Even if extension suffices to constitute *substance*, it does not—so they believed—suffice to constitute *body*. The question of the sufficiency of extension is complicated enough that I will treat it separately in §9.3. In the remainder of this section I consider the metaphysical problems that arise from the identification of body in general with extension.

3. *Revising the concept of substance.* In the standard Aristotelian conception, figure and size are accidents of individual concrete substances, resulting from the determination of matter by substantial form. Quantity too is an

accident, whether of matter alone or of the composite of matter and form. Space, so far as it can be distinguished from quantity, is the abstracted dimensions of individual substances and thus exists only in our conception; so far as it can be said to exist outside the intellect, it is simply quantity under another name.

The *novatores*, believing that space and body are separate, recognized that space could not easily be subsumed under the Aristotelian categories; nor could it be analyzed into matter and form. Patrizi can designate it only by oxymorons. Gassendi rejects the highest division of being into substance and accident. But what, an Aristotelian might ask, could a thing be which is neither substance nor accident, and yet real?

It was in this troubled context that Descartes, rewriting the logic of the textbooks, devised the classification of things into substance, attribute, and mode. Although there are adumbrations of it in earlier works, only in the *Principles* does Descartes systematize his vocabulary. Here I deal mainly with his endeavor to maintain the Aristotelian rejection of a space independent of bodies, while foregoing the Aristotelian analysis of corporeal substance into matter and form. Descartes had, of course, further reasons to introduce the new classification; in particular, it is crucial to proving the real distinction between mind and body. Such questions, which have been thoroughly treated by others, I leave aside.

Descartes begins by noting that what "falls under our perception" we consider either as "a thing or an affection of a thing, or as an eternal truth" (1§48, 8/1:22). Of things, the most general are "*substance, duration, order, number*" and the like. These terms apply "to all kinds of things." In particular, "by substance we can understand nothing other than a thing which so exists as to need no other thing to exist" (1§51, 8/1:24). Strictly speaking, God is the only substance. All others exist by the concurrence of God. But some require *only* that concurrence in order to exist, and for that reason Descartes calls them substances also (1§52, 8/1:25), while the modes of created substances require not only divine concurrence but substance in order to exist.[24]

The existence of a thing—Descartes repeats an Aristotelian common-place—by itself can have no effect on us. Substances are known only through one or another mode. From the existence of a mode, established in perception, we infer, according to the "common notion" that nothing is attributed to nothing, the existence of a substance to which it belongs. The

24. Descartes uses both 'attribute' and 'mode' in the general sense of 'created thing whose existence is not independent of the existence of all other things save God'; but he also occasionally restricts the sense of the terms, so that 'attribute' denotes only those dependent beings that do not vary so long as the substance they depend on exists, while 'mode' denotes dependent beings that can vary while the substance they depend on continues to exist. To avoid ambiguity I use 'attribute' only in the more restricted sense.

inference can be made from any mode whatsoever: we are no less certain, Descartes later observes, that a body exists by virtue of its color than by virtue of its figure (1§69, 8/1:34). But for each substance there is one "principal property," or attribute, which "constitutes its nature and essence, and to which all others are referred." By reference to principal attributes we define the two highest kinds [*summa genera*] of created things: "intellectual or cogitative things, that is, those pertaining to mind or to thinking substance," and "material things, or those that pertain to extended substance, that is, to body" (1§48, 8/1:23).[25]

Principal attributes are at first said to be modes: Descartes even uses the term of art *inesse*, which in Aristotelianism denotes the relation between accident and substance, to describe the relation between attribute and substance. But after defining a "distinction of reason" to be that which obtains "between substance and one of its attributes without which it cannot be understood" (or between two such attributes), Descartes writes:

> Thought and extension can be regarded as constituting the natures of intellectual and corporeal substance; and then they should not be conceived otherwise than as thinking and extended substance itself, that is, mind and body; in that way they are understood most clearly and distinctly. Indeed we more easily understand extended substance or thinking substance than substance alone, omitting the fact that it thinks or is extended, since there is some difficulty in abstracting the notion of substance from the notions of extension or thought, which are diverse from it only in reason. (1§63, 8/1:31)

The indistinctness that Descartes later affirms of substance and quantity (2§8, 8/1:44), here holds between principal attributes and substances generally. As I said in §9.1, that may be one reason why the later argument is perfunctory.

Extension, when we take it to "constitute the nature" of body, is nothing other than extended substance itself. But it can also be taken for a mode of substance when we consider that "one and the same body, retaining its same quantity, can be extended in several different ways"—by taking on various figures or shapes (1§64, 8/1:31). We can conceive *res extensa* without conceiving it to be spherically extended. But we cannot conceive of being

25. There is a third class of things that, although they are certainly modes of thought, cannot be fully understood without reference to body: sensations and passions. The French translation comes close to suggesting that they are properties of a third substance—the "intimate union" of body and mind (9/2:45; cf. Cottingham 1986:127–132). But the word *attribuées* is a poor translation of the Latin *referri* (earlier occurrences of which in the same paragraph are translated by *rapporter*). To refer X to Y is not to regard X as an attribute of Y. It is to denominate or conceive of X by reference to its cause (or, as in "referred pain," what we think is its cause), in any of the Aristotelian senses.

spherically extended without conceiving of the thing that is thus extended. Spherical extension and extended substance are, as Descartes puts it, modally distinct. Extension *simpliciter,* on the other hand, and extended substance are not even modally distinct, but only in reason.

The crux, from an Aristotelian standpoint, is the claim that extension could "constitute the nature" of body. The constituents of an Aristotelian body are matter and form; the accidents of body, including quantity, depend on them, and cannot by themselves constitute substance. But that quantity or extension *does* suffice to give being to a thing is just what Descartes means to say. Created substance is defined as that which needs nothing save the concurrence of God to exist. Nothing, however, *merely* exists: everything exists *somehow.* Among created things, there are two ways or modes of existing: the way of extended things and the way of thinking things. Being extended is an "attribute" in the strict sense, which is to say, an *invariant* mode, because it is, for bodies, nothing other than existence itself. A comparison will help make this clear. Duration applies to all things; since a substance (of whatever kind), "if it ceases to endure, ceases also to be," duration differs only in reason from the substance itself, and for that matter, from its existence (1§62, 8/1:30; *To* *** 1645 or 1646, 4:349). So too, a *body,* if it ceases to "extend," ceases to be.

One immediate consequence is that extension, considered as that which constitutes the nature of certain created substances, is *not* an accident of bodies.[26] Only when we consider that the determinate extension a thing now has could be determined otherwise do we regard extension as a mode (1§64, 8/1:31). Extension as mode is a determination of extension as substance, and cannot exist apart from it; but since determinations of extension can vary while extension itself remains, extension as substance can exist apart from each of its modes. Those modes are, suitably enough, only modally distinct from it (1§61, 8/1:29).

Descartes's terminology, however, varies significantly. Sometimes, as in 1§64, he speaks of extension itself as a mode. More often, he speaks of figure and motion as modes of extension, reserving the term 'extension' to denote extension as substance, as the principal attribute of body.[27] When extension is taken to be the attribute of body, its function is akin to that of Aristotelian form. Together with thought, it serves as the basis for classifying created substances; it is the ground from which flow the properties of body—divisibility, figurability, mobility, and impenetrability. When, on the

26. My thinking on this question has benefited from reading Glouberman 1978, a revised version of which is in Glouberman 1986, c.6.

27. 'Thus the extension of a body can indeed admit in itself various modes: its mode is different if the body is spherical than if it is square; but the extension which is the subject of these modes, regarded in itself, is not a mode of corporeal substance, but rather an attribute which constitutes its essence and nature" (*Notæ in programma,* 8/2:348–349).

other hand, extension is regarded as the *subjectum* of figure—and especially the "configuration," as Malebranche called it, which characterizes natural kinds like water or oil—its function is akin to that of matter, and the characteristic figure plays the role of form.

Extension is at once the *differentia* of body, resembling form in its taxonomic role, and the substrate of change, resembling matter. Odd as that might seem from an Aristotelian standpoint, Descartes's conception does yield decisive answers to the questions that troubled Fonseca and other Aristotelians, without risking the distinction of mind and body or sinking into atomism. Space is indeed substance, though not a composite of matter and form. It is the very same substance we call body; it therefore affects the senses and can be known. Because it is body, there is no "penetration of dimensions"—here Descartes preserves what is true in Aristotle—and no possibility of space devoid of body. Atomism is not merely empirically but necessarily false.

Descartes's conception has the further advantage of yielding, or appearing to yield, an a priori destruction of Aristotelian sensible qualities. Here the notion of constitution comes into play. Extension is said to constitute the nature of body, just as thought constitutes the nature of mind. In the Aristotelian vocabulary, the term 'constitute' is applied most properly to the relation between a substance and its intrinsic causes—matter and form. Form is the principal constituent of substance, because (in the common phrase) it "gives being" to substance.[28] To constitute the nature of a thing, then, is to stand to that nature as form to substance, to be the ground of the properties that follow from having that nature. Extension by itself is not the nature of corporeal substance. That nature also includes being divisible, mobile, capable of figure. But all those properties follow, or so Descartes believes, from extension.

Just as importantly *only* those properties follow from extension. Earlier, in considering the argument of *Principles* 2§11, I asked whether Descartes had shown that color or hardness could not be modes of extension, even if they were not part of the nature of body. What is shown there seems to be at most that body can exist without color; but if color cannot be conceived without body, then it would seem to be a mode of extension. If it isn't, it must differ somehow from the acknowledged modes.

One answer is that the idea of color is obscure and confused when referred to something outside of us (1§68). But that smacks of dogma, and would establish at best that we cannot be certain whether color is a mode of extension. It fails, moreover, to address the phenomenology of color. Colors seem always to be *areas* of color; unextended color is as hard to imagine as

28. Goclenius *Lexicon* s.v. 'Constituo', 456.

unextended figure. Colors seem to "presuppose" extension just as much as figure and motion do (1§53), and the Aristotelians attributed, as a matter of course, not only intensive but extensive quantity to them.

The arguments of 2§4 and 2§11 are often taken to address the question. But 2§4 seems to take as proved that colors exist only as modes of thought, while 2§11 presupposes that extension constitutes the nature of body, the only question being whether corporeal extension is spatial extension. It does not show why color or sound could not be modes of spatially extended substance, even if not part of its nature qua body.

I would like to offer an answer that, although not explicitly put forward by Descartes, can be devised with the means at his disposal. I recall one of the arguments used by the Nominalists to show that quantity and substance are one (§4.2). To have quantity is just to have "parts outside of parts" (*partes extra partes*); but material substance of itself has such parts, so that supposing in addition to the substance itself a quantity distinct from it adds nothing. Descartes, though he does not use the argument, does regard extended substance (or space) as having "distinguishable parts" of "determinate magnitude and figure" (*To More* 5 Feb. 1649, 5:270–271; cf. 305); the absence of such parts suffices to show that a thing is not extended (ib. 270; cf. *Med.* 6, 7:86). But if to be extended is to have parts, the possibility of figure—of boundaries between parts—is implied in the very idea of extension.

The possibility of color is not—not if one supposes that the putative colors in things *resemble* our ideas of color (*PP* 1§66, 8/1:32). There is no evident relation between extension and what an idea of color represents: "it is the same," Descartes writes, "when someone says he sees color in some body or feels pain in some limb, as if he had said he sees or feels in it something quite unknown to him—as if, that is, he had said he does not know what he sees or feels" (1§68, 8/1:33). Not knowing how it is possible for an extended thing to have a mode resembling what is represented in a sensation of color, we have no reason to include color in the nature even of an individual body.

In explaining the nature of corporeal substance Descartes had to steer between two alternatives: Aristotelianism, with its baroque apparatus and its unwarranted distinctions; and the theories of the *novatores*, whose common failing was to abandon the easily and completely understood conception of bodily extension, and to propose instead various confused notions of something neither material nor spiritual—notions, moreover, which could lead their proponents onto dangerous theological ground, as we will see in Descartes's correspondence with More. Descartes's doctrine offers a clear-cut resolution of Aristotelian questions about the relation of matter and quantity, and about the nature of space; at the same time it avoids the spiritualization or divinization of space that one finds in authors like Patrizi. It preserved the basic Aristotelian tenet that the only independently existing things are singular individuals, though not without a profound revision of the ontological and physical analysis of substance. Substantial form in particular is, so to speak, fragmented: its role in defining essence is taken over by extension as substance; its role in providing a ground for the properties of natural kinds is taken over by configuration. Extension as substance also takes over the role of matter; the fusion of the two roles contributes, as we will see, to the difficulties that Cartesian physics has in characterizing individual bodies.

4. *Individual bodies.* The argument of the second part of the *Principles* proceeds through a series of identifications: matter is quantity, quantity is substance, the extension that constitutes body is that which constitutes space. In Aristotelian physics prime matter cannot naturally exist without form; so too, we may suppose, extension in general cannot exist except in the determinate extensions of individuals, in regions of definite size and figure.

But what distinguishes one region from another? The question is important not only in its own right, but because it illuminates the relations of physical and geometrical space in Cartesian physics. Descartes is frequently said to have identified physical and geometrical space; their relation is, I will argue, not that simple. One must be wary, moreover, of attributing to Descartes notions of space—geometric *or* physical—that were not his, however familiar they are to us. Once the relation of physical and geometrical space is understood, the definition begins to make sense. Nevertheless, the lack of genuine unity in bodies as Descartes defines them leaves his physics with inadequate means to explain why bodies sometimes fragment in collisions and sometimes not, or under what circumstances a body will be deformed rather than resisting the intrusion of another body.

In Aristotelian questions on individuation, the first issue was whether a principle of individuation is needed. Taking Aristotle's side against Plato, the Aristotelians agreed that all actual substances are "singular and individual" (Suárez *Disp.* 5§1¶4, *Opera* 25:146). But Ockham and his followers went further. In their view, even species and genera, though they are said of more than one thing, are "by themselves and first of all individual" (Fonseca *In meta.* 5c6q1§1, 2:357C). The nature *Man,* in their view, "is nothing other than this man or that," the only difference being that the common noun 'man' refers vaguely to all men, while the singular term 'Peter', say, refers to one.

Though Nominalism was not without its effects on sixteenth-century Aristotelianism, its central thesis was rejected. "Those natures to which we give the names 'universal' and 'common', are real, and truly exist in things themselves," prior to any operation of the mind (Suárez *Disp.* 6§2¶1, *Opera* 25:206). The question therefore arises: if common natures are "of themselves indifferent" to the distinction between one and many, by what principle are they "individuated" in singular beings? If Peter's and Paul's humanness are one, then evidently something else must account for their being two.

It is not my purpose here to descend into the labyrinthine details of Aristotelian disputes over the principle of individuation. I simply note two rejected alternatives, and then briefly characterize the kind of principles typically adopted.

The first rejected alternative was that some collection of *accidents* could serve as the principle (Fonseca *In meta.* 5c6q1§2–4, 2:359–361). Those who favored it noted that although Peter and Paul may share humanness, there are many other accidents they do not share. It might well happen that no other human shares *all* the accidents of Peter, and so that collection could be the principle by which Peter is distinguished from other humans. To this the Aristotelians objected, among other things, that even a collection of accidents is in principle communicable: although in fact there may be no individual of exactly Peter's height and weight, there certainly could be (ib. §3, 2:359E–F).

The second rejected alternative was the celebrated thesis of Thomas: the principle of individuation among corporeal substances is *materia signata* (or *quanta*), a "designated" portion of quantified matter.[29] Thomas's thesis resembles Descartes's solution to his problem closely enough that it is worth examining the reasons for and against it. The strongest reason for it is that "matter is the principle of multiplication and distinction among individuals of the same species"; since matter is incommunicable, Thomas's solution overcomes the objection to individuation by accidents (Suárez *Disp.* 5§3¶3, *Opera* 25:162). Suárez replies that determinate quantity, like the other dispositions of matter, is consequent upon form. Form, not matter or quantity, is the "principle of distinction" among *complete* substances (¶5, 25:163). A corporeal substance can, moreover, be deprived of quantity altogether; yet it will be no less an individual (¶16, 25:167). Those objections lapse, of course, if like Descartes one believes that quantity itself is the only form in bodies, and that body cannot be deprived even by God of its quantity.

The kind of principle typically adopted is not so much defined as designated. Fonseca writes that the principle of individuation is "a certain posi-

---

29. What *materia signata quantitate* signifies was itself disputed. Suárez considers two interpretations: that *materia signata* is matter actually endowed with determinate quantity; and that it is the capacity of prime matter to receive determinate quantity and other dispositions that enable it to receive form (see §5.3; *Disp.* 5§3¶8, 9, 18; 25:164, 168). The essential point is that bodies are individuated by virtue of having (or of including matter capable of having) determinate quantity—quantity endowed with size and figure.

tive difference," intrinsic and nonderivatively [*primo*] incommunicable; added to the species, it "constitutes the individual" and distinguishes it from others "prior to any discrimination by accidents" (ib. q5§1, 2:3381C). Since the positive difference that constitutes an individual cannot be removed or altered without destroying the individual, it belongs to the essence of the individual. From this, Fonseca concludes in Leibnizian fashion that "no power can bring it about that two individuals should be positively similar in everything that pertains to their essence" (ib. §3, 2:385B). It does not, however, contradict his conclusion, so far as I can see, to suppose that two individuals could share all their accidents: positive differences are not accidents.

Suárez's view is more elusive, in part because, as a subtle variant of the nominalists', it comes near to dismissing the question.[30] The principle by which a thing is individuated is nothing other than "its being [*entitas*], or [. . .] the intrinsic principle by which its being is established" (*Disp.* 6§6¶1, *Opera* 25:180). He rests his conclusion on the proposition, argued in an earlier disputation, that "the ground of unity cannot be distinguished from being itself." To be and to be *one* differ only in reason: 'being one' connotes the negation of numerical division, while 'being' does not. The same is true of 'being an individual', which merely connotes the incommunicability of *entitas*. Socrates' soul is not something that Plato can share; but humanity can exist in both. The difference between Suárez and Ockham is that Suárez does not agree that Socrates' *humanity* is essentially individual, or that all that is common to Socrates' humanity and Plato's is their name.[31]

In particular *matter* is not individuated by quantity or any accident of quantity: "The matter which underlies this form of wood is numerically distinct from that which underlies the form of water or man; it is therefore individual and singular in itself. The ground of such a unity is not the substantial form, nor being ordered to this or that form [. . .], because when any form is varied, there always remains numerically the same matter [. . .] Quantity too cannot be the ground of the individual unity of matter, as the same argument shows, if it is true that matter can lose and acquire various quantities as substantial forms vary" (§6¶2, 25:180f). Indeed, God can remove all quantity from the matter of a thing. Even so—contrary to what Ockham believed (§4.1)—its parts will remain distinct in being (*entitative distinctæ*), and the matter itself will remain distinct in being from other matters (§3¶14, 25:167). Suárez grants that "with respect to location [*situm*] one matter is distinguished from another by quantity." But that is not the essential difference: as an existing thing, "it is truly and really

---

30. On Suárez's position, see the lucid and thorough exposition in Gracia 1994.
31. See Gracia 1994:500.

distinguished by its being [*per suam entitatem*]" (§6¶4, 25:181), and by that alone.

Descartes takes the short way with species and genera. Like all universals, they exist only in our conception, and are only modes of thought (*PP* 1§59, 8/1:27). For him, as for the nominalists, there is no problem of individuation because there is no individuation: "When I say 'Peter is a man', the thought by which I think of Peter indeed differs modally from that by which I think of man, but in Peter himself being man is nothing other than being Peter" (*To* *** 1645 or 1646, 4:350). So too the distinction between extension in general and the extension of *this* body is only a distinction in reason: singular extensions alone are real.

The traditional problem is consigned to the Scholastic shades. But new ones pop up. Extension, as we have seen, can be regarded as constituting the nature of body, and as the substrate of natural change, and thus as substance. It can also be regarded as the determinate extension of this or that body, and thus as a mode. The relation between the two aspects of extension is summed up in Descartes's phrase 'part of matter' (*pars materiæ*).

That phrase has, it would seem, two senses. When Descartes speaks of matter as *partibilis*, or of a simple body having many parts, the parts in question are not actual individual bodies. Yet they are, according to Descartes, really distinct, no less than the sun and the moon. How then are they distinct? Since we are not yet thinking of bodies moving, but only of the parts of one body, the question is one of the *static* individuation of *potential* parts of matter.

Suppose that the static problem has been solved. Potential parts can be distinguished. But some of those parts are *actual*, and correspond to what we call individual bodies, some are not. The potential parts of a vase full of water remain the same whether any part moves or not, so long as the vase does not change. But the actual parts will differ a great deal if the water freezes. This is true even according to common sense, and all the more so for Descartes, for whom, as we have seen, motion is defined as the rupture of a body from its neighbors. A second problem, then, is that of the *dynamic* individuation of the *actual* parts of matter.[32]

Of the two problems, the static problem is simpler. But even it has pitfalls. Descartes, in arguing that the potential parts of an extended substance are really distinct, speaks of each part as being "delimited by us in thought [*à*

---

32. I make two simplifications. First, only *simple* bodies will be considered. Descartes calls a vase full of water *one* body; but that "one" body is made up of many smaller actual bodies; the attempt to include compound bodies also in the definition of 'part' raises further difficulties (see Grosholz 1991:68). Second, I omit *sliding* motions. Descartes's definition cannot straightforwardly be applied to sliding parts. Since it will, if it fails in straightforward applications, no doubt fail too in the rest, I will assume that the simple bodies being defined are rupturing from their neighbors.

*nobis cogitatione definitam]*" (*PP* 1 §60, 8/1:28). But how, one might ask, is this to be done? As Emily Grosholz puts it, "in quiescent *res extensa* there is no physical analogue of a boundary"; there are "no articulating parts which themselves have the integrity of shape" (Grosholz 1991:68). Though she has in mind the "great monolith" with which, in an at least conceivable instant before God gave motion to the world, the creation began, her worry applies equally to the interior of any body. A grain of sand, too, is a monolith.

Or again: Descartes writes as if motion were *given to parts*—as if there somehow were parts in motionless matter already, at least in God's understanding of it. But in his official doctrine, motion and parts come into existence simultaneously, just as when a girl is born, a mother and daughter come into existence together. So matter, or "space in general," has no diverse parts to which diverse motions could be given. Like Hegelian being, it threatens to collapse into nothing, a "surd" as Grosholz calls it (1991:70, 66).

Whatever Ockham may have thought, entitative difference does not of itself yield *partes extra partes*. *Extra* signifies not mere difference in being (as one soul might be said to be "outside" another), but existence in distinct locations. It is characteristic of quantity that difference in being among its parts *entails* difference in location, at least naturally. An inch of extension is such that "another inch of extension cannot be added to it without their making together a quantity of two inches, because if they penetrated one another, they would occupy no more space than one alone occupied when they were separated" (Régis *Cours* 1pti c1, 1:282). Were they to occupy the space of one, were they in exactly the same location, they would *be* one. Parts of extension, in short, are individuated by their locations.[33] The *exteriority* proper to spatial parts must, it would seem, rest on a "topology," a system of places prior in definition to the parts themselves.

Some of those places will be known to us because they coincide with the places actually occupied by individual bodies. The rest must be grasped by the imagination, which will impose on them locally the metric of the *Geometry*. In the *Principles* Descartes lays out a series of distinctions that amount to a construction of the topology of space in general from the "singular" spaces of bodies:

(i) The *singular space* of an individual body differs only in reason from the body itself; it can also be called the *place* of that body.

(ii) *Generic space* is the place of a body at a given time which we imagine to remain even when the body moves—the hollow in the sculptor's mold, for example, where the wax was and the bronze is.

33. So Suárez, in passages discussed in §4.2, concludes that "having parts that occupy extended and divisible space" or "extension of parts in order to place" is the "*ratio* and formal effect" of quantity; or, more precisely, having the *potentia* to occupy space (Suárez *Disp.* 40§4¶15, *Opera* 26:547).

(iii) Finally there is *space in general*. At each instant this coincides with the concatenation of all singular spaces; within it all generic spaces are contained.

The construction is incomplete as it stands, since the set of generic spaces depends on the places of actual bodies; we cannot be certain that those will include every place a body *could* be, every part of space in general. Here again the imagination must be put to work: once space in general has been constructed, we can imagine within it arbitrary concatenations and divisions of the bodies that now constitute it. The generic spaces of those imagined bodies I will call *generic in the broad sense*, or *locations*.

Once a complete set of generic spaces has been constructed within space in general, the problem of static individuation can be addressed. Within the singular space of an actual individual body are contained infinitely many generic spaces in the broad sense. A *potential part* is any such space. Like any space, it is a body; it is distinct from other potential parts by virtue of its place. Insofar as it is a substance in its own right, really distinct from potential parts disjoint from it, it is no less a body than the whole. But insofar as its boundaries are not marked, so to speak, by rupture, it is only a potential individual body; our knowledge of it can come only from the imagination, not from the senses.

Potential parts of matter are individuated by reference to the topology implicit in the idea of extension. Motion enters in only by way of exhibiting to us the singular, and thus the generic, spaces through which we come to conceive space in general. *Actual* parts of matter, on the other hand, are defined by Descartes in terms of motion, or, more precisely, in terms of the fundamental mode of rupture.

Descartes's principle of the dynamic individuation of bodies appears first in *Le Monde;* it is repeated, without significant alteration, in the *Principles.* In the earlier work it looks like this: "Observe in passing that I take here and will always take hereafter as a single part all that which is joined together and which is not in the act of separating [*qui n'est point en action pour se séparer*] [. . .]: thus a grain of sand, a stone, a boulder, and even the whole Earth may hereafter be taken for a single part, insofar as we consider in it only one entirely simple and uniform movement" (*Monde* 3, AT 11:15). In the *Principles,* the definition of 'part' is incorporated into the definition of motus: "We can say that [*motus*] is that *translation of one part of matter, or one body, from the vicinity of those bodies that immediately touch it and which are regarded as being at rest, into the vicinity of others*. Where by *one body*, or *one part of matter*, I understand all that which is transferred at once, even if this part may be itself composed from many parts, which have other motions in them" (*PP* 2§25, AT 8/1:53–54). Descartes devotes little argument to the definition; he does nothing to dispel the appearance of circularity when he juxtaposes the definition of 'part' with that of 'motion'. Small wonder that it has irritated critics from Cordemoy and Leibniz to the present.[34]

What follows is not a vindication of Descartes's principle but an explication that reveals not only certain features of his conceptions of body and space, but also some preconceptions about space that seem to impinge on its interpretation by recent critics.

Space in general is, first of all, *not* Euclidean space as that is now understood. Descartes did not tamper with the Aristotelian conception of space: that innovation should be credited to others, like Gassendi and More. What he transformed was the Aristotelian conception of body, making actualized *corpus mathematicum* the object of physics rather than *corpus physicus*. Doing so did not preclude his retaining in large part Aristotelian conceptions of extension, quantity, and space. Cartesian space, like Aristotelian space, has no existence prior to or independent of the existence of bodies; it is, on the other hand, independent of our thought. We have no way of knowing that the world has the structure we expect it to have by virtue of the ideas implanted in us by God, except through interaction with actual bodies. But though our knowledge of the topology of space is derived from our knowledge of actual bodies, and though the existence of space depends on that of actual bodies, space in general has the structure it has independently of our thoughts and of the forms of the actual bodies it contains.

We have, then, space in general, abstracted from the motions of bodies, and thus from their particular shapes. Space in general contains infinitely many generic spaces in the broad sense, of all shapes and sizes. As Newton put it in an early manuscript: "In all directions, space can be distinguished into parts whose common limits we usually call surfaces; and these surfaces can be distinguished in all directions into parts whose common limits we usually call lines; and again these lines can be distinguished in all directions into parts which we call points. [. . .] And hence there are everywhere all kinds of figures" (Newton *Unpub.* 100/132; trans. Hall and Hall). Like Gassendi, and Henry More in his later works, Newton believes that space is immobile and indivisible. It is the scene within which bodies move, and by reference to which motion is measured; it cannot itself be said to move. With that I think most philosophers now would agree. Descartes does not. Individual bodies, he claims, are parts of extension; the extension of a body and the space it occupies are one and the same substance; and bodies move. It follows that parts of space can move, and that space is divisible.

Those are strange claims. Their strangeness is a clue that any attempt to situate the Cartesian definitions of motion and individual body within the framework of classical physics is likely to result in misunderstanding. For Descartes there is no space apart from the singular spaces of individual

34. One recent example is Funkenstein, who writes, "obviously, Descartes lacks a principle of individuation for single bodies as physical entities" (1986:73).

bodies; space in general is, so to speak, the mass composed of all such spaces when one abstracts from their motion and therefore their boundaries. It is tempting to think of that mass as being like Newtonian absolute space, and of bodies as resulting from its partition. Descartes's description of creation lends itself to that interpretation. But to interpret Descartes thus is to overlook the fact that only the singular spaces really exist.

Call two singular spaces *apart* if the boundary of each includes no part of the boundary of the other; call them *together* otherwise. Unlike regions of Newtonian space, singular spaces can be apart at one time and together at another. Hence one can define the *rupture* of singular spaces: two bodies have ruptured if at a certain time they are apart, having for some immediately prior interval been together. Motion, as we have seen, is successive rupture.

That might suffice were it not that Descartes then defines *one body* as "all that which is transferred at once." One might well think that the definition is intended to supply a means of determining which of the infinitely many *potential* parts of matter—of the whole mass of extended substance that God created—are *actual*. But potential parts, it would seem, are defined in terms of the topology of space in general; parts so defined are either always apart or always together. They cannot move: motion is defined only for singular spaces, and thus only for *actual* parts. Though the definitions of part and of motion taken singly might be acceptable, taken together they seem bedevilled by an equivocation on *part*.

Perhaps the best answer is this. At each moment the whole of space consists of many, perhaps infinitely many, singular spaces. Each of those spaces has potential parts; each part can be distinguished from others according to the static principle of individuation. Like the entire singular space of a body, we must suppose, the parts of that space can be designated and reidentified at subsequent moments—as if they too were actual individual bodies. It makes sense, then, to ask at a subsequent moment whether any parts of the body we began with were together and now are not—any that have, in other words, ruptured. If so, then our body will not have been one, since at least two of its parts will have been "in the act of separating," to quote the definition from *Le Monde*.

In *De ipsa natura* (1698), Leibniz argues that there must be, in addition to motion and the figures that result from motion, intrinsic differences among bodies. The alternative is to be deprived not only of any means of discriminating bodies, but of discerning their motion, if by motion one means the "successive existence of the thing moved in diverse places" (Leibniz *Ph. Schr.* 4:512). Supposing that two parts of matter congruent in shape do not differ, and that motion preserves shape, "it is manifest that from the perpetual substitution of indistinguishables it follows that the states of diverse moments in the corporeal world can in no way be discriminated. For it would be only an *extrinsic denomination* by which one part of matter was distinguished from another, namely from the future, in other words, that it would later be in one place and then another; but of things in the present there is no discrimination" (513). Daniel Garber concludes from this that "motion can be used to individuate bodies *only* if there is some way of reidentifying bits of material substance across time, and this can only happen [. . .] if there is something in body over and above extension" (1992:181). That something, it turns out, is the active power or *vis insita* that Descartes intended to get rid of for good when he took the essence of body to be extension.

Garber is right, I think, in holding that there must be some way to reidentify bits of matter from one moment to another. But that there must be something in body "over and above extension," I am not so sure. Among the Aristotelians, Fonseca appears to have argued for a version of the identity of indiscernibles (or, more precisely, the discernibility of nonidenticals). But Suárez argues that the *entitas* that in his view is the principle of individuation is distinct only in reason from the individual substance itself. Two bodies, alike in all their modes, could still be genuinely *two;* God, at least, could discriminate successive states of the world even in the face of the "perpetual substitution of indistinguishables."

Even so, the definition faces serious empirical difficulties. First of all, Descartes has no way of determining what will happen when bodies collide, once he admits that real bodies sometimes fall apart in collisions. Suppose that a body *B* collides with a body *C* that is twice its size and at rest (see Figure 20). According to the fourth rule of collision, *B* will be reflected by *C* and return whence it came. But it is clear that in the third part of the *Principles* Descartes holds that some collisions result in the fragmenting of one or both bodies. In 2§63 we find a reason: each of the two halves $C_1$ and $C_2$ of the larger body "may be considered to be an individual body," and so, since *B* and $C_1$ are equal in size, the sixth rule will apply. *B* will be reflected, giving up some of its motion to $C_1$, which will separate from $C_2$ and move off to the right. But Descartes gives us no criterion for deciding when we should "consider" the body as two half-bodies, and when we should consider it as a one whole. What will happen in particular instances remains, despite this putative explanation, "indeterminate and cannot be specified" (Grosholz 1991:95).[35]

Similarly, even though Descartes regards some simple bodies as deformable, he has no principled way of determining when a body will bend when

---

35. The example is borrowed from Leibniz (*Animadversiones* ad 2§54–55, *Ph. Schr.* 4:386–387). Leibniz notes that the outcome of a collision between one body and a *part* of another (i.e., a collision where contact is made only at part of the surface of the other, as in Fig. 20) ought to depend, if Descartes were right, only on the size of that part. If $C_1$ remains attached to $C_2$ (if, for example, it were larger than *B*), it does so not by virtue of resisting separation from $C_2$, but *per accidens*, as Leibniz says. It will stay where it is even if $C_2$ does not exist. The whole body *C* is *unum* only *per accidens*, not *per se*.

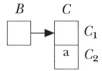

Fig. 20. Collision and division

pushed by another, and when it will break. Squeeze a sponge and its holes get smaller, just as Descartes says in his account of condensation; but try the same experiment with a piece of pumice and the result is a heap of fragments. Descent to the microlevel will not help, since there too we find particles that are supposed to be flexible, while others are supposed to be rigid. At the urging of Mersenne, Descartes does attempt to show why some bodies are resilient. But the explanation presupposes that they can be deformed without breaking; what Descartes explains is at best how through the action of subtle matter an already deformed body is forced back to its original shape.

The two problems are symptoms of a general malaise. Descartes insists, as we have seen, that the only "glue" holding the parts of a body together is their mutual rest. The absence of rupture, he says, is the strongest glue of all; nothing else is needed to explain why bodies hang together. But experience tells us that a die sitting on a table can easily be knocked off, while a protuberant part of the table top of the same size and shape cannot. Descartes is not unaware of the difference. In a description of the making of glass from ash he argues that the particles of ash are separated by particles of the second element, while those of glass are joined along much of their surfaces (see Figure 21; the left-hand figure represents ash, the right-hand figure glass). One could argue perhaps that only processes like melting result in contiguity among particles; setting one body atop another results not in true contact, but in the kind of contact tires make with rain-soaked pavement. Descartes seems to have thought that the force required to sepa-

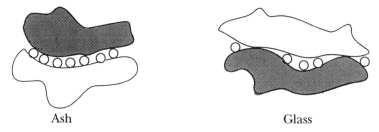

Ash                                    Glass

Fig. 21. Joining of particles (after *PP*4§125, AT 8/1:271)

rate two potential parts of one particle varies directly with the cross section of the particle in the cutting plane.[36]

The rule gives us a sort of ersatz unity—ersatz because it admits degrees. But it does not answer the objections. The point about collisions would stand even if the body $C$ were full of holes, so that its two halves met along only part of the line $a$ in Figure 20. And although it may be easier to deform a porous body than to deform a solid body, porosity alone cannot account for deformation (or fragmentation). Some solid parts of matter, including the particles of the first element, are deformable; and some porous bodies are rigid.

Clarifying the relation between geometrical and physical space, while it brings to light significant differences between Cartesian physical space and the space of classical physics, does not in the end resolve the difficulties that Descarte's principle of individuation brought with it. At best one can hold the conceptual objections of Leibniz at bay; but the empirical insufficiency of the principle remains. Elasticity eludes Cartesian physics, and with it any hope of explaining the outcome of collisions.

## 9.3. Physical Questions: the Sufficiency of Extension

To establish that the nature of body is constituted by extension, one must show that everything in the nature of body follows from extension in the manner suggested above, and that what follows from extension is sufficient to yield substances of the sort we call bodies. To the first point, More objected that impenetrability belongs to the nature of bodies, and yet it does not follow from extension alone. To the second, More and Arnauld in correspondence with Descartes, and Gassendi in later writings, argued that the existence of spaces that neither are nor contain bodies is possible.

1. *Impenetrability.* To show that extension is sufficient not merely to constitute a substance, but to constitute it as *body,* one must show that in bodies as we commonly understand them, there is no property that does not follow from extension. In particular, impenetrability must follow. Descartes seems to have believed that this was obvious.[37]

---

36. 'Nevertheless glass is very fragile, because the surfaces by which its particles touch one another are extremely sparse and small" (*PP* 4§128, 8/1:272). It should be added that fragility in general is also affected by the interweaving of the branches of the particles in certain substances (ib.).

37. The Aristotelians believed that two bodies *could* be in the same place at the same time, on scriptural grounds (the *virginitas intacta* of Mary, Christ's exit from the tomb, his ascension; see Coimbra *In Phys.* 4c5q4, 2:40), on doctrinal grounds (the *corpora gloriosa* of Christ and the saints are said to penetrate other bodies), and on the basis of argument. Since "quantity is not formally contrary to quantity," the exclusion of one by another is not necessary, and can be inhibited by God. Quantity has only the "virtue" of filling space; it is therefore possible for one

He was wrong. In Henry More he found an opponent willing to grant that space is substance, but not that space and body are one. More holds not only that empty space is absolutely possible, but that bodies have at least one property that space does not: they are *impenetrable*. He begins with tangibility, one of the two marks by which Patrizi and Gassendi had distinguished space from body:

> Although matter need not be soft, or hard, or hot, or cold, it is nevertheless most necessary that it should be sensible, or, if you disagree with that, tangible, as Lucretius so well defines it:
>
> > Nothing can touch and be touched, unless it is body.
>
> That notion ought to be all the less abhorrent to you, since your Philosophy most clearly establishes that all sensation [. . .] is touch" (More *To Descartes* 11 Dec. 1648, AT 5:239, quoting De rerum natura 1:304)[38]

Tangibility could be regarded as following upon impenetrability, the other distinguishing mark of body. So the more fundamental objection is that bodies cannot be identical with regions of space because such regions are not impenetrable: "[Tangibility] signifies the mutual contact and the *potentia* of touching, among whatever sort of bodies, animate or not, supposing the immediate juxtaposition of the surfaces of two or more bodies. Which suggests another condition of matter or body, which one could call impenetrability, namely that [a body] can neither penetrate other bodies nor be penetrated by them. From that there is a most obvious difference between divine and corporeal Nature, since [divine nature] can penetrate [corporeal nature], but [corporeal nature] cannot penetrate itself" (AT 5:240). Extension, therefore, being an attribute of both divine and corporeal nature, cannot confer impenetrability on a thing.

In the *Principles,* as in earlier works, Descartes assumes, without proof, that extended substances are impenetrable.[39] That may seem odd. He un-

---

body to be in another's place provided that only one of the two actually fills space (ib. 41). Only if both are actually extended is it impossible for both to be in the same place.

38. To shorten citations, I here give the dates and Adam–Tannery page numbers of the Descartes–More correspondence, all of which is included in AT 5: More *To Descartes* 11 Dec. 16 48, 235–246; Descartes *To More* 5 Feb. 16 49, 267–279; More *To Descartes* 5 Mar. 16 49, 298–317; Descartes *To More* 15 Apr. 16 49, 340–348; More *To Descartes* 23 Jul. 16 49, 376–390; Descartes *To More* Aug. 16 49, 401–405; More "Responsio," Aug. 16 55, 643–647.

The last letter from Descartes to More is a fragment found at Descartes's death; More's "Responsio" is contained in a letter to Clerselier, who had sent him the fragment. The correspondence was included, with scholia by More, in his *Opera* (II.2:227–271). On the exchange, see Gabbey 1980:299n.27 and Garber 1992:145ff.

39. Impenetrability is explicitly ascribed to parts of matter in *Le Monde:* "But let us conceive [matter] as a true body, perfectly solid [ . . . ]; so that each of its parts always occupies a part of this space, so proportioned to its size that it could not fill a larger [space] nor shrink into a smaller, nor allow another, so long as it remains there, to enter [that part of space]" (*Monde* 6, AT 11:33). Cf. also 6 *Resp.* no.9, AT 7:442).

doubtedly knew that some philosophers distinguished corporeal—impenetrable and tangible—extension from incorporeal extension. There are, I think, two reasons why he did not devote more attention to the problem. The first is that the Aristotelians, the primary intended readers of the *Principles,* believed that an *actually* extended thing must exclude other such things from its place. Descartes merely removes the qualification: interpenetration is not just naturally, but absolutely, impossible.

The second reason is that Descartes may have overlooked impenetrability in his efforts to explain hardness and solidity.[40] One indication is that although *durus* occurs both in the argument that the nature of matter is quantity and in the statement of the third law,[41] the hardness of the bodies considered in the laws of motion cannot be the hardness that is rejected from the essence of matter in 2§4; for then the outcome of collisions would depend on how God allowed matter to interact with our senses. Descartes appears not to have noticed the equivocation. It is not surprising, therefore, that in beginning his response to More, he should reiterate that sensible qualities are not in bodies (AT 5:268).

More had emphasized that "tangible" denotes the *power* to affect the sense of touch, and not the qualities we sense (see AT 5:299), and "impenetrable" the *power* bodies have to exclude one another from their respective places. Ultimately Descartes concedes the point. He turns to what looks like a logical quibble. Impenetrability, he says, is not a *defining* feature of matter. A body could be extended but not impenetrable, at least in our conception:

> Tangibility and impenetrability in body is merely, as risibility is in man, proper in the fourth mode, according to the common laws of logic, [and] not a true and essential *differentia,* which I contend consists in extension; and on that account, just as 'man' is not defined by 'risible animal', but by 'rational [animal]', so body (I contend) is not defined by impenetrability, but by extension. This is confirmed by the fact that tangibility and impenetrability have a relation to parts, and presuppose the concept of division or termination; but we can conceive of a continuous body of indetermi-

---

40. On hardness (*durities*), see PP 2§4,11,54, AT 8/1:42,48,70. Impenetrability is a property of every continuous single part of matter; it consists in a body's not allowing others to enter or occupy its internal place. Impenetrability is distinguished from hardness, the power of some bodies to resist the movements of our own and to produce a characteristic sensation; and from solidity (*soliditas*), the power of some bodies to resist division into parts. Hardness and solidity are typically attributed to sensible, "macroscopic," bodies, and explained in terms of the cohesion or connection of their parts; impenetrability is a property of the smallest naturally existing parts of matter and is taken to be primitive or, in Cartesian physics, derived from extension.

41. 'Ita experimur dura quælibet corpora projecta, cùm in aliud durum corpus impingunt', (AT 8/1:65); "si duo tantùm corpora sibi mutuò occurrerent, eaque essent perfectè dura" (ib. 67).

nate or indefinite size, in which nothing except extension is considered.
(AT 5:269)[42]

To consider in an actual finite body only its extension is to abstract from
whatever boundaries the body actually has and from the larger space in
which it is contained. But the idea of impenetrability requires that we conceive at least *two* bodies, and thus not only their extension, but their being
"terminated," as actually divided parts of matter. Though the possibility of
division is clearly contained in the idea of extension, *res extensa* need not be
divided. God could have created one continuous body. Division and termination are in that sense superadded to extension. Impenetrability, too, follows only on the assumption that there are distinct bodies. But bodies in the
plural come about only when various motions are impressed on the parts of
matter. If, leaving motion out, we contemplate only extension itself, we
abstract from impenetrability as well. Extension alone *defines* body.

If, on the other hand, there *are* distinct bodies, they will indeed be impenetrable. The mutual penetration of two parts of space "would imply a contradiction" if no part were destroyed (AT 5:271). More is, quite reasonably,
not persuaded. He cannot imagine how one "part of space" could be transferred to the place of another *without* mutual penetration: "[I cannot] conceive that if they were translated [from one place to another] some parts of
empty space would not absorb others, and coincide with them in their
depths, and penetrate one another" (AT 5:302).[43] Descartes's answer
merely elaborates his point: "if they are absorbed, then the middle part of
space is destroyed & ceases to be; but what ceases to be, does not penetrate
the other."[44]

An individual body has *its* space or "internal place," which is part of the
space in general composed of the spaces of all bodies together. If we try to
imagine one body *B* penetrating another body *C,* we are trying to imagine
*B*'s space overlapping *C*'s. But the putative region of overlap is *one* part of
space in general, *one* location. There can be nothing to indicate that it is, so
to speak, duplicated within itself. God could produce the appearance of

42. On the modes of *propria*, see Coimbra *In Log.* 1:264. An accident proper in the fourth
mode belongs to all members of a species at all times and to them alone. Risibility in humans is
the stock example.

43. More's reply to Descartes's next letter shows what "absorption" might consist in: if one
body invaded another, the second body could immediately *condense* so as to give way to the first.
Condensation, which More distinguishes explicitly from diminution, is decrease in volume
*without loss of substance or quantity* (AT 5:378). In later works More concluded that space, which
he took to be incorporeal and distinct from body, is immobile and indivisible (*Ench. met.* 1c8§9,
*Opera* II.1:167[I]). The problem of explaining how one part of space could penetrate or absorb
another no longer arises.

44. AT 5:342. By 'middle part' Descartes means the part of one or the other of the two
bodies that would have overlapped the other: see Gabbey's note on the passage, AT 5:676, and
Garber 1992:147.

penetration by annihilating some of B's space. But even he cannot bring it about that *one* part of space in general should be the internal place of *two* bodies.

Bernard Williams objects that "it is [. . .] legitimate to think of two geometrical solids, occupying (in one sense) the same space, or parts of them doing so, as when one conceives of two polyhedra constructed on the same base."[45] But suppose that from each polyhedron the part that does *not* belong to the other is truncated. What we have left is two polyhedra whose internal places coincide. But why just two? Why not infinitely many?[46]

The rejoinder may seem to prove too much. Could one not show by similar means that sets cannot intersect? In any case, even if bodies are individuated by location as that term was defined above, locations can overlap. An answer of sorts can be wrested, I think, from the discussion of Descartes's principle of individuation in §9.2. If extension, divided or not, has by nature a "topology" of locations within it, locations that do not exist apart from extension altogether, but which are independent of the particular bodies that compose it, then it is tempting to suppose that "parts" of extension are individuated by their locations. But that opens the way to Williams's objection. We must, despite Descartes's deceptive language, not take bodies to be, or to be solely individuated by, their locations; or at least not without adding the additional condition that the locations of distinct bodies are disjoint.

But why should that be? Location is, in Descartes's terms, generic; bodies are singular. The mode of existence of bodies is to have parts *outside of* parts, where 'outside' means not merely distinct in reason, but really distinct, capable of subsisting separately by the absolute power of God. If we take Williams's polyhedra to be locations, there is no difficulty in supposing that they overlap; the points common to them can indeed be distinguished in reason as belonging first to one, and then to the other. But suppose we take them to be bodies, and consider the larger body that would result from joining them. That body would have at least two real—and not just imagined—parts that were *not* outside one another, and that could not subsist separately.

That this does not again beg the question, I am not sure. The claim is that bodies have just that mode of existence which entails that one body is really distinct from another only if their locations are disjoint. That our conception of location arises from our experience of bodies (both real and imag-

---

45. Williams 1978:229; cf. Funkenstein 1986:74, Shea 1991:262.
46. See Aristotle's argument against separate space: "How, therefore, will the body of a cube and an equal void or place differ? And if two such things [could occupy one place], why should any number of things not also be together?" (*Phys.* 4c8, 216b9[ff]; Coimbra *In Phys.* 4c9tex76, 2:56).

ined) is no objection to defining, once we have the conception, a principle of individuation for them in terms of it. But now the burden rests on the claim about the mode of existence of bodies. It may be that Descartes, like Ockham, took it to be transparently true. Later philosophers disagreed: rather than try to derive impenetrability from extension, they added it to the list of primitive properties of body, and from that concluded that the nature of body does not consist solely in its being extended.

2. *Vacuum.* If the extension that constitutes space is identical to that which constitutes body, there can be no space apart from body, no "separate" space. That is the short way with the vacuum. In the *Principles* Descartes takes a more circuitous route. In 2§10–15 he establishes a vocabulary for place and space, answering along the way an argument occasionally offered on behalf of the existence of void space: that movement would be impossible in a plenum. In 2§16 he argues that there can be no void space, and after telling one of his genealogical tales to explain why some people think there is, he tackles a version of the annihilation argument found in, among others, Gassendi.[47]

The extension that constitutes body is, Descartes tells us, the same as that which constitutes what we call "space," except in the way we conceive it. When it is conceived as the extension of a particular body, it changes whenever and however the body changes. When it is designated as space, or as place, we do not judge it to change when the body that first filled it changes. Instead "it remains one and the same, so long as it remains of the same size and figure, and preserves the same situation [*situm*] among certain bodies outside it," bodies by which we "determine" it (2§10, repeated in 2§12). The traditional example is the space inside a vessel, which whether filled with water, air, or some other substance, we call the *same* space.

Descartes insists, therefore, on the continual reoccupation of the same generic space by successive bodies: "But meanwhile we judge the extension of the place, in which the stone was, to remain and to be the same, although now the place of the stone is occupied by wood or water or air or whatever other body" (AT 8/1:46). And thus that all motion is circular: "From what was noted above, that all places are filled with bodies, and that the same parts of matter are always proportioned exactly to equal places, it follows that no body can be moved except through a circle, so that one body expels another from the place it enters, and that body another, and so on until the last [body] which enters the place left by the first at the very moment the first leaves it" (2§33, AT 8/1:58).[48] The necessary succession of bodies and

---

47. On the history of arguments about the vacuum, see Grant 1981.
48. Descartes's argument holds only for a finite universe constant in volume. Even then the most one can infer is that for each particle $C$ that enters the space previously occupied by the first particle $A$, there will be a continuous path entirely contained within the volumes of

the circularity of motion yield a reply to an argument often urged on behalf of the independence of space. Epicurean in origin, it is found in Gassendi, among others. A body removed from its place "must extrude another body out of its place, or become joint tenant with it and possess one and the same place. Extrude a body out of its possession it cannot, because the Extruded must want a room to be received into; nor can the Extruded dispossess a third, that third expel a fourth, that fourth eject a fifth, &c. since the difficulty sits equally heavy on all." Hence there can be "no beginning of Motion."[49] Descartes has no trouble answering: the "beginning of Motion" and its end are one.

More challenging is the argument from annihilation cited earlier. It will be useful to preface Descartes's treatment of it with a passage from Patrizi.

> The located [*locatum*] body, on entering or leaving a *locus* [i.e., an Aristotelian place, defined as the inward surface of the bodies surrounding the body whose place it is], always takes its volume with it. But the locating [*locans*] body first received within its volume the volume of the located body, and then released that body so that it could take its volume with it. But whither had the volume of the locating body withdrawn while it was receiving the located body? This cannot even be conceived. It is therefore necessary that the locating body, while receiving the located body, depart thence entirely and leave the space, which is fixed there, empty of itself so as to be filled by the entering body. When the latter in turn withdraws completely that space itself, which is immobile, must receive another entering body.[50]

Patrizi is not arguing for the existence of a vacuum, but for the independence and immobility of space. The "volume," or what Descartes would call the internal place, of a body that receives another must be withdrawn so that the entering body can bring *its* volume. So too that body must withdraw its volume when a second enters. The space that the second body enters cannot be the internal place of the first body, since that is gone; its own internal place isn't there yet. So there must be an immobile space, belonging to neither, which admits the volume of the two in succession.

Descartes admits the commonsense way of putting things: the space occupied by the first body "remains" after it leaves. It *is* distinct somehow from *that* body. But not from *every* body, as Patrizi would have it. There is "indeed

---

contiguous displaced particles from *C* to *A*. [—P Spinoza, *ad loc.*]

49. Charleton *Physiologia* 1c4§1, p24; cf. Gassendi *Philosophia Epicuri syntagma* pt2§1c1, *Opera* 3:11; Régis *Cours* 1pt2c11, 1:329, where the connection is made explicit.

50. Patrizi *De spacio phys.* 230 [62b–c]; I have substituted 'located' and 'locating' for Brickman's 'bounded' and 'bounding'. On the role of this passage in Patrizi's argument against Aristotle, see Henry 1979:562.

no connection between the vessel and this or that particular body contained in it," and so we designate the "generic extension [*extensio in genere sumpta*]" within the vase by reference to the figure of the vase. Thus is commonsense accommodated: the generic designation is what an Aristotelian would call a *denomination* of the concatenated singular extensions of the bodies in the vase by reference to their container. What cannot be designated is a space *apart* from substance, "since, as has often been said, there can be no extension of nothing [*nihil nulla potest esse extensio*]" (2§18, AT 8/1:50).

That maxim is an instance of the "common notion" that "there are no attributes of nothing" (1§52, AT 8/1:25). Armed with this notion—which is not so common that it cannot be overlooked[51]—Descartes answers the argument from annihilation: "If it is asked what would happen if God removed every body that was contained in a vessel, and permitted no other to enter the emptied-out place, the answer is that the sides of the vessel would by that act become contiguous. For since nothing lies between the two bodies, necessarily they touch each other; it is manifestly repugnant that they should be distant, or that between them there should be distance, and yet this distance be nothing: because all distance is a mode of extension, and so cannot exist without substance" (2§18, AT 8/1:50). If there is no body between the sides of the vase, there is no *subjectum* in which the distance between them could inhere, and so there is no distance between them, just as I have no hair color if I have no hair.

More calls the argument "slight," Gassendi a "paralogism."[52] More accepts the argument that distance entails an intervening substance; but he thinks that the substance is God.[53] Gassendi, more radical in his departure from the logic of substance and accident, denies that one need attach every accident to a substance; the space measured by the distance between the sides of the vase is "incorporeal." Since More's objection rests on his views about divine extension, I will consider it later. Here I note simply that for all

---

51. 'The difficulty in recognizing the impossibility of the vacuum seems to originate first of all in the fact that we do not sufficiently realize that there can be no properties of nothing" (*To Arnauld* 29 Jul. 1648, AT 5:223).

52. AT 5:241; Gassendi *Syntagma, Pars Physicæ* §1bk2c1, *Opera* 1:184b. Gassendi adds that it is false that "every quantity, and so every dimension is a corporeal Accident," since the void has "incorporeal" dimension and quantity: "insofar as a greater or lesser portion of it can be designated; it admits of being measured, and of all the comparisons that body itself has, insofar as it is quantified."

53. AT 5:302. In the same passage, however, More begins with an argument to show that extension *could* be of a "nonexistent" (thus rejecting the common notion applied by Descartes). Suppose that God annihilated this world and much later created a new one from nothing. "That between-worlds [*intermundium*], or absence of worlds, would have its duration." A nonexistent, therefore, could have duration, "which is a sort of extension." So too a vacuum could be "measured in ells or fathoms'. Nevertheless More concedes the point, since he thinks that distance does have a *subjectum,* namely God.

three, distance is a mode not of the two sides of the vase jointly, but of some one thing.[54]

3. *Divine extension and imaginary space.* The question of God's extension, or more generally of the manner by which he is said to be everywhere present in the world, was bound up in Aristotelianism with that of the existence of a void space, usually called "imaginary," outside the cosmos.[55] The link between the two was the *immensitas* of God, attributed on the basis of scriptural and patristic texts.[56] Immensity, according to Suárez, denotes that mode of existence of the divine essence by which God "can be intimately present to all things and bodies, not only present [i.e., presently existing] but possible, even if they should be increased to infinity in multitude and magnitude."[57] That God is immense would seem to entail only that space be potentially infinite, or that the capacity for receiving matter, as Fonseca called it, should have no limit. But in the view of the Coimbrans, the space in which God exists, though it is not a real or positive being, is actually infinite: "In this imaginary space we assert that God exists *in actu*, not as in some real being but by his immensity, which because the entire world cannot receive it necessarily exists also beyond the heavens in infinite spaces" (*In Phys.* 8c1oq2a4, 2:370; cf. Grant 1981:160ff). Those infinite spaces are not "factitious, or dependent solely on a notion of the mind"; we call them "imaginary" only because "we imagine them in space by a certain proportion to the real and positive dimensions of bodies." But neither are they real. They are not "endowed with genuine three-dimensional quantity," since if they were they could not receive bodies—which was, of course, Aristotle's argument against a space distinct from bodies.

There is, in Aristotelian discussions, a decided reluctance to grant to space any reality distinct from that of the bodies contained in it; and yet also

54. Arnauld suggests that the empty space could be measured, even though, being nothing, it has no properties: "I ask therefore whether, while I considered the wine-jar separately, I could not measure the cavity within and find out the distance in feet from the bottom, and what the diameter of the cylindrical cavity was, and so forth?" (Arnauld *To Descartes* 3 Jun. 1648, AT 5:191). Descartes answers only that it is wrong to refer the cavity of the jar to the sides of the jar, "as if it were not diverse from them" (ib. 194).

55. Mersenne uses the term to denote the "privation of all the bodies that exist in the world," which would occur if God ceased to conserve them, in which case "nothing would remain but the space in which they exist, which is ordinarily called *imaginary*" (Mersenne *Harm. univ.* 1pr4, 1:8). On divine immensity and imaginary space, see Grant 1981, pt.II.

56. See the list given by Suárez (*De div. subst., Opera* 1:48ʳ). Augustine's reply to the Manichaeans that before creation God was "in himself and not in another" is frequently cited; so too a passage from *De civitate Dei* (11c5), in which God's existence in infinite space is compared to his existence throughout eternity (cf. Grant 1981:113 and 328n.51; Coimbra *In Phys.* 8c1oq2a2, 2:366; Suárez *Disp.* 30§7¶28, *Opera* 26:104). Also frequently cited is the well-known saying of Hermes Trismegistus that God is a circle "whose center is everywhere and whose circumference is nowhere" (cf., e.g., Fonseca *In meta.* 5c15q9§4, 2:897A).

57. Suárez *De div. subst.* 2c2§1, *Opera* 1:48; cf. *Disp.* 30§7¶1 and 46, *Opera* 26:95, 109.

a marked desire not to limit God's spatial existence to the finite world of Aristotelian cosmology. The mildest expression of the desire was to hold that God is wherever any body is and could be. But the 'could be' implies a where that is not now the place of any body. That implication puts pressure on the identification of space with the concatenation of bodily places, and thus on the conception of space as an accident of bodies rather than as a container existing independent of them. Little by little, resistance to the separation of space and body was diminishing even in the Schools.

Descartes, needless to say, did not countenance the existence of an extra-cosmic void. In *Le Monde* imaginary spaces are mere machinery to bring onstage his own fabulous world.[58] In the *Principles* he alludes to them in arguing that the world has no limits: "for wherever we suppose [*fingamus*] those limits to exist, indefinitely extended spaces beyond those limits can not only be imagined by us but also be perceived as truly imaginable, that is, real; and so a corporeal substance, indefinitely extended, is contained in them also" (*PP* 2§52, AT 8/1:52). The word 'indefinite' here points to a problem: if space is infinite, and if it is not only substance but body, then an actually infinite thing, the world, will have real and positive being. But the existence of actually infinite *quantity* was thought by many to involve para-dox, and the assertion of an infinite *world* had brought upon more than one philosopher the censure of theologians. The Aristotelians avoided the prob-lem by denying reality to infinite extracosmic space; More by supposing infinite space to be God; Descartes by holding that the world is *indefinite* in size. We can, he agrees, imagine no boundaries to space (and thus to body). But we have no grounds for affirming that any substance other than God is infinite. The idea of God entails infiniteness; the idea of extension does not.[59] Prudence dictates that we say merely that there is no finite space such that a larger space cannot be imagined. That larger space is not void but corporeal. It is not "imaginary" in the Aristotelian sense, nor indeed in the more usual sense.[60]

---

58. 'The philosophers tell us that these spaces are infinite, and we may as well believe them, since it is they themselves who have made them. But in order that this infinity should not hinder or bother us, let us not attempt to go to the end of it; let us enter only so far that we may lose sight of all the creatures God has made five or six thousand years ago" (Descartes *Monde* 6, AT 11:32). On this passage, see Cavaillé 1991:212–218.

59. On the distinction between the indefinite and the infinite, see McGuire 1983:88, 96[f]. Descartes's view is in keeping with the Aristotelian prohibition against actually infinite created things (cf., e.g., Suárez *Disp.* 35§3¶26, *Opera* 26:447, and especially Coimbra *In Phys.* 3c8q1 & 2, 1:374[II]).

60. 'Such a space [beyond the supposed limits of the world] is, according to me, a true body. Nor do I care that by others it is called 'imaginary', and thus that the world is judged to be finite; since I know from which *præjudicia* the error comes" (*To More* 15 Apr. 1649, AT 5:345). The *præjudicium* in question is that according to which a space in which nothing affects the senses is judged to be void (see n.51). Descartes evades the issue raised by More (312), which is that by excluded middle anything that cannot be infinite must be finite, as More exasperatedly

The force behind the doctrine of imaginary space is the view that God is everywhere, as the Coimbrans put it, "by his essence, presence, and power." *Everywhere* includes all imaginable places, in which he is present "according to his real and infinite existence" (Coimbra *In Phys.* 8c10q2a4, 2:370), and imaginary space is not merely boundless but infinite in extent. Only such a space agrees with God's immensity. The keys, then, are divine *presence* and its relation to *space*.

Suárez, subtle as always, distinguishes between "intrinsic" presence, which is not a relation but a mode akin to intrinsic *Ubi* or "whereness" (*Disp.* 51§1¶13ff, see §4.2), and the "bearing" or position of bodies relative to one another:

> when a thing is said to be somewhere, or in some place, two [properties] can be included in that locution: one is the intrinsic and absolute presence [*præsentia*] that the thing has here; the other is a bearing to something else, which touches or is nearby, whether by virtue of quantity of mass, or quantity of active virtue, or the relative presence of its substance or quantity. Both [relations] are necessary in bodies, which are surrounded by a containing body, since [a body] is said to be located among the bodies that circumscribe it, by way of quantitative contact, and has moreover in itself its intrinsic presence, which we explain by its bearing toward the center and poles of the world, and which remains the same even if the circumscribing bodies change or flow, and indeed even if all of them perish by divine power.[61]

God's presence in imaginary space is intrinsic only; he cannot be related to it as to something apart from him, since imaginary space is "nothing" (¶37, 26:107).[62] The effect of Suárez's distinction, then, is to pry apart the connection between presence and extension. God can be present in every actual and possible *Ubi* not only by his effects, but even by his substance, and yet not be an extended thing.

For More the connection cannot be broken. God "seems to be an extended thing; so too the Angels [. . .] And indeed, that God in his way is extended, I judge to be obvious from the fact that he is omnipresent and

---

points out (304).

61. Suárez *Disp.* 30§7¶35, *Opera* 26:106; cf. 51§4¶33, 26:998. I use 'bearing' to translate Suárez's *habitudo*, a term one of whose uses is to denote the relation of a body to what surrounds it (Goclenius *Lexicon* 623).

62. Because of its importance in discussions of the Eucharist, *præsentia* receives a detailed entry in Goclenius's lexicon. He seems not, however, to notice the distinction Suárez makes (*Lexicon* 855), and indeed takes *præsentia* always to be a relation: "est existentia rei cum re, à qua non distat" (857). But he does list the standard distinction appealed to by Descartes between "effective" or "virtual" presence, which consists in having effects at a place, and "substantial" presence, or spatial nearness (856).

occupies intimately the whole machine of the world and all its singular parts."[63] Hence even if all the bodies contained in a vessel were removed, and all others prevented from entering, it is false that there would be *nothing* between the sides of the vessel: "divine extension would be interposed" between them (241). God would be there, just as he is everywhere.

Divine extension, moreover, exists in the same way above the heavens as below: "if we suppose the world to be enclosed within the visible stellar heavens, the center of the divine essence and of his whole presence would be repeated outside the starry heaven in the same way that we clearly conceive it to be repeated and reiterated within" (305). Descartes has earlier granted that the "amplitude" of divine substance is infinite, if not its extension (which is, "properly speaking," null) (275). More's argument is that since the "amplitude" of God is to that of finite spirits as an infinite to a finite sphere, we do have reason to believe that space, the sphere of God's presence, is not merely indefinite but infinite (cf. 304.15–20), and that body, if finite, is not space.

Descartes, in his response, first sets the terms of the debate. Extension in the usual sense is found only in bodies: "By 'extended being' everyone commonly understands something imaginable [. . .] and in that being one can distinguish by imagination various parts of determinate magnitude and figure, of which each is not [the same as] the others in any way" (270). But God cannot be imagined at all, let alone imagined to have "parts with determinate magnitudes and figures." If God is extended, it is not in the usual sense, but merely by virtue of his power, which can be exercised everywhere. Just as fire, "even in a glowing iron, is not ferrous," God's power, even when it is applied to extended things, is not extended—nor is God himself (270; cf. 343, 403).

Descartes later distinguishes "extension in power" (*extensio potentiæ*) from "extension in substance" (*extensio substantiæ*). Extension in power does not belong to the substance that has the power. Only extension in substance does. To prove this, Descartes uses the standard test of separability: "if there were no body, I would understand there to be no space with which the Angel or God were coextensive" (342). Or again: "The extension attributed to incorporeal things is merely extension in power, not in substance; that power, since it is only a mode in the thing to which it is applied, cannot be understood, if the extended thing that coexists with it be destroyed, to be extended" (343). The argument can, by analogy with similar arguments concerning force (§8.3), be construed thus: for God's power, or God him-

---

63. AT 5:238, 305. In More's view that judgment in no way entails that God is or has a body, since body is tangible and impenetrable (240, 301).

self, to be extended, it would have to have really distinct parts, admit of determination in figure and size, and movement in parts. But such modes belong only to the bodies on which God acts, not to his power.

More takes up the last point. Descartes treats divine power as if it were "situated outside of God." But any real mode is always "intimately within the thing of which it is a mode." So divine power and essence, in order to act on a thing, must "by some real mode be united with it" (379). What is united with extended substances must be itself extended. God, the Angels, and the human soul are all *res extensæ*.[64] There is, in short, but one way to *be* at a place, and that is to be extended there. Where for Descartes the distinction is between extension in power, which is extension in name only, and extension in substance, which entails the impenetrability, figurability, and divisibility proper to bodies, for More the distinction is between two kinds of genuine extension—the very proposition against which the argument of *Principles* 2§11 is directed.

Even for More, however, *incorporeal extension* is physically inert, except insofar as it designates a site of operation for spirits and a possible place for bodies; only corporeal extension has all the properties that Descartes assigns to extension simpliciter. Thus put, the difference looks slight enough that More can at last agree that "on God's [. . .] omnipresence there remains no disagreement between us" (AT 5:643).

But perhaps he was merely being conciliatory. Descartes, after all, retains in large part the Aristotelian view. Like Suárez, he distinguishes between *præsentia*, which suffices to answer the question 'Where is it?', and *occupatio*, the invariable concomitant of *præsentia* for bodies. Only *occupatio* entails having parts and thus divisibility, impenetrability, and tangibility. God's presence, even in an extended substance, need not of itself entail that he is extended. For Descartes as for most Aristotelians extension brings with it the other properties of body; to admit that God is extended would be, therefore, to admit that he is or has a body—the long-suppressed heresy of David of Dinant.

More, like the other *novatores,* denies the inference from *extended* to *corporeal.* In doing so, he points toward the absolute space of Newtonian physics. Amos Funkenstein has documented a line of thought leading from certain pantheistic tendencies in the Renaissance to More and Newton. At the end of that line God has a *body*—Newtonian space.[65] One should

---

64. In the *Enchiridium* More offers many pages of argument against the Cartesians or *Nullibistæ* (Nowhere-Men). As in the correspondence with Descartes, the "fulcrum" of the debate between him and the *Nullibistæ* is the contention that every extended thing is material (see More *Ench. met.* 1c27§5–6, *Opera* II.1:309ᶠ).

65. Funkenstein 1986, c.2; see especially pp.96, 116.

note, however, that while More's God has spatial dimensions, space itself is neither body (as Descartes thought) nor an accident of body (as Aristotle thought). So if God can be said to acquire a "body" in the work of More, it is a peculiar sort of body, penetrable and indivisible; its descendant might best be regarded not as matter, but as the ether.

# [10]

## World without Ends

The Cartesian transformation of physics is of the sort that used to excite talk of incommensurability. There can be little doubt that it was profound. Yet when one descends into details, one finds no point at which an Aristotelian would not at least have understood what Descartes was up to, however repugnant his results may have seemed. In this concluding chapter I consider what becomes of finality in the new world of *res extensæ*.

The brief answer is: nothing. The Cartesian world is the world momentarily envisaged by Suárez, in which God simply allows natural agents to act independent of him. In such a world "still the stone would fall down, fire would generate its like," and so forth; there would be no final causality but only "mere natural necessity" (*Disp.* 23§10¶8, *Opera* 25:888). For Suárez the strongest of reasons argue against that world; for Descartes the strongest of reasons favor it. Not just final causes, but the directedness essential to the Aristotelian concept of change, are absent—or rather they are positively excluded. To hold that the nature of corporeal substance is constituted by extension is to deny that corporeal substance could have active powers; to hold that motion is rupture is, as we have seen, to refuse the Aristotelian definition of motion as the *actus* of *potentia*. All this the Aristotelian would have understood well: the truth of Cartesianism, like that of the more familiar atomistic physics of Democritus, Epicurus, and Lucretius, would entail the falsity—the *necessary* falsity, according to Descartes—of what was most basic to Aristotelian physics.

Of Descartes's arguments against the appeal to ends in natural philosophy, I will examine the two that attack that appeal directly.[1] The first is that

---

1. Descartes also argues that it is vain, even impious, to inquire into God's ends, which are

in attributing ends to the actions of natural agents the Aristotelian is treating them as if they had souls. I have already considered to what extent the accusation is just (§6.3). The Aristotelians, far from promiscuously allotting souls even to the rocks that fall, are quite guarded even in attributing to higher animals recognition of their ends. Though finality—being directed toward ends—occurs everywhere in nature, final *causes* need be adduced only in explaining the actions of rational agents. What remains to be considered in Descartes's argument, then, is why, if not through ignorance or malevolence, he came to misrepresent the Aristotelians' position.

The second argument is, from the standpoint of later natural philosophy, the more interesting. Descartes claims he has no need of final causes or of ends in his physics. The directedness of natural change, which so struck the Aristotelians, is illusory; the stability of nature, the cooperation of efficient causes in producing a single result, the fitness of the parts of nature to one another and especially to human needs, are all just the working out of mere natural necessity. The cosmogony of the *Principles* is an implicit polemic against any claim that the present state of the world needs any other account than that provided by bodies moving, dividing, and uniting according to law. From the Aristotelian point of view, that would be disastrous. But proponents of what came to be called "design" could accommodate natural necessity so long as the natural world—however it came about—could be interpreted as fulfilling the intentions of a benevolent creator. If the arrow strikes the bull's-eye, the necessity with which it is impelled is no argument against its having been aimed there; it testifies, rather, to the ingenuity and power of the archer.

But the role of design in Descartes's physics must be restricted still further. The laws of nature themselves are founded not, as Leibniz believed, on God's benevolent choice of the best possible world, but on his immutability. One cannot argue, therefore, that God has chosen those laws so that natural agents, acting in accordance with them, will produce the ends envisioned by him.

The implication, perhaps not fully recognized by Descartes, is that the course of natural events has no end. The most shocking expression of this—to Leibniz, at least—is the hypothesis entertained in the third part of the *Principles:* the world, Descartes says, may be supposed to proceed through all physically possible states. None of those states, one may infer, can be regarded as a state toward which the world tends. But that, in Leibniz's view, would "destroy providence."

---

incomprehensible and beyond our competence to pass judgment on (see *PP* 1§28, AT 8/1:15). That criticism, which was not unknown to the Aristotelians, challenges the utility of teleological reasoning. But a useless project need not be an incoherent project. The two arguments I discuss seem to me more profound.

Here the novelty, or the modernity, of Descartes's natural philosophy reveals itself most fully. The vast enterprise of natural theology that emerged near the end of the seventeenth century, continues into the nineteenth century, and even now reverberates in traditional accounts of the "rise of man," can from this standpoint be regarded as an endeavor to repair the damage done by Descartes, or if not by him then by those who, like Spinoza, were taken to have worked out the hidden implications of his philosophy.

1. *Little souls.* Several times in his writings Descartes accuses the Aristotelians of attributing souls to corporeal substances. Sometimes the basis of the accusation is what he takes to be imaginary distinctions; in *Principles* 2§9, for example, the Aristotelian claim that quantity and substance are really distinct is said to rest on the false attribution of a confused idea of incorporeal substance to bodies.[2] Sometimes the basis is the attribution of ends to the actions of bodies. In the sixth *Replies* Descartes recounts how he had come to think of weight as a "real quality" of bodies, only to realize later that he was thereby mistakenly conceiving of it as soul: "But that the idea of weight was taken partly from the idea I had of mind, is shown principally by the fact that I thought gravity carried bodies toward the center of the earth, as if it contained in itself some knowledge of the center. For that clearly could not occur without thinking, and there can be no thought but in a mind. Nevertheless I attributed certain things to weight that could not be understood in the same way as mind—that it is divisible, measurable, and so forth" (6 *Resp.*, AT 7:442). Whether this is a true confession is unimportant. The significant point is that Descartes believes that to act toward an end (here the attainment of the natural place of heavy bodies) presupposes the capacity to recognize that end, and so also a "little soul," as he says to Mersenne (AT 3:648).

We have seen that the Aristotelians believed that final causes act only on rational agents (§6.3). Not even animals can be said genuinely to recognize the ends toward which, nevertheless, they undoubtedly act. They, and all inanimate things, are acted upon by final causes only by virtue of being the instruments of rational. agents. Where the Aristotelians differ from Descartes is in refusing to restrict *finality* to the actions of agents who recognize the ends toward which they act. To understand the downward motion of a heavy body as a natural change is, as we have seen, to subsume it under a schema in which it can be seen as the step-by-step actualization of the body's *potentia*, whose completion or *actus* consists in arriving at its natural place. The presence in a stone of heaviness links the stone with something else— its natural place. As I argued earlier, the natures of things in the Aristotelian world are bound together through relations of finality: in the nature of the

2. On this argument, see Gilson 1984:168–173 and Garber 1992:97–101.

seed, that of the plant is implied; in the natures of lower animals, those of higher animals are implicated if the lower exist for the sake of the higher.

One reason, I think, that Descartes finds the "little souls" story so persuasive is that he can conceive such relations only by way of the one relation that can connect the natures of distinct things: *intentionality*. His idea of the sun, he writes in the second *Replies*, "is the Sun itself existing in thought— not, indeed, formally, as it exists in the heavens, but objectively, i.e. in the way that things usually exist in thought" (AT 7:102). Having the idea of the sun establishes a relation between me, in whom that idea is a mode, and the thing in the heavens—between two things, in other words, that otherwise are really distinct.

When a stone falls, its relation to the *terminus* of its motion could only be of two sorts. Either the *terminus* is the efficient cause of the motion, which is evidently false, or else the stone is capable of having the *terminus* exist objectively in it—which is to say, is capable of thinking of the *terminus*. But that too is false. So there is no relation between the stone's motion and the *terminus* (now a *terminus* in name only) of its motion, save that it happens to be pushed in the direction of the *terminus* by subtle matter—which cannot, of course, be thinking of the *terminus* either. Despite what the senses seem to show us, we must not think of the actions of inanimate things as directed. Even the dog that leaps toward food is pushed by its animal spirits, which are in turn caused to move by light reflected into its eyes.

2. *Doing without finality*. The replacement of final by efficient causes— ideas where rational agents are concerned, blind pushes where natural agents are concerned, was already occurring among the Aristotelians. Jean Buridan, notably, had argued that in animals the efficient cause suffices; Suárez, as we have seen, has to exercise his subtlety in order to show that ends envisioned by rational agents are not merely efficient causes. Descartes's views here are rather the culmination of a trend than a radical departure.

One further example is worth noting. The Coimbrans instanced the stability of the elemental composition of the world as evidence for what in §6.1 I called "collective" ends. The processes by which the elements are transformed into one another do not yield a bleak and uninhabitable world of fire alone, or water alone. Instead we have a world in which all four elements are present in due proportion and in suitable places, a world in equilibrium.

Descartes's cosmogony yields, so he asserts, both the right proportions of his three elements—enough to form worlds like the earth on whose surface water is found and around which circulates air like the air we breathe. In particular—this is my example—the motions of the particles originally given motion by God, which Descartes supposes for simplicity's sake to have been uniformly sized small cubes, eventually produce the spherical particles of the second element. Second-element particles are stable in the sense that their size and shape is, on the whole, maintained. They are of a size such that collisions with other particles are no more likely to increase than to decrease that size; and because they have no knobs or branches that could easily be broken off, their shape is also unlikely to be altered. They are in a kind of mechanical equilibrium with their surroundings.

To the naive eye, it might seem as if the motions of those and other particles were conspiring, as it were, to keep the particles of the second element spherical. The collective end of those motions, one might say, is to preserve the shape of the second elements. But—Descartes would reply—there is no need to suppose that the particles move on account of the shapes that result from their motion. The preservation of that shape can be explained entirely in terms of the possible actions of efficient causes. One could, of course, say that God has given the parts of matter just the properties that will bring about the preservation of second-element particles. But that claim does no work in physics.

It is when one turns to nature in the large that the thoroughness with which Descartes rejected finality in physics becomes most apparent.

There is, first of all, the hypothesis that so incensed Leibniz. In the third part of the *Principles* Descartes attempts to build the universe from the ground up. Though he casts his account into hypothetical form so as to avoid controverting the book of Genesis, it is, I think, seriously meant. The story begins, as origin myths often do, with chaos—an "entire confusion of all the parts of the universe" (*PP* 3§47). Referring his reader to the unpublished *Le Monde*, he says that he has shown elsewhere how the order that exists at present in the world could have arisen from chaos. In the *Principles,* however, because "it does not agree so well with the sovereign perfection of God to make him the author of confusion rather than order, and because the notion we have of chaos is less distinct," Descartes starts not from chaos but from an initial state in which "all the parts of matter are equal among themselves, both in size and in movement."

That assumption may seem arbitrary. Descartes defends it in the following terms: "It matters very little how I suppose matter to have been disposed at the beginning, since its disposition must afterwards be changed, according to the laws of nature; and one can scarcely imagine any [disposition] from which one could not prove that by these laws it must continually change, until finally it composes a world entirely similar to this one [ . . . ] For since these laws cause matter to take on successively all the forms it is capable of, if one considers all those forms in order, one will be able finally to arrive at [the form] that exists at present in the world" (*PP* 3§47). The universe proceeds, in other words, indifferently through all the states consistent with the quantity of motion initially given to it by God.

Leibniz was horrified. In his polemic against Cartesianism he returns to the passage I have quoted several times. If the world moves successively through all the states it is capable of, then Descartes's God cannot be said to will the good, or to be, as the Aristotelians argued, the ultimate end of the world. Descartes's God is instead nothing other than the "necessity or principle of necessity acting in matter" (Leibniz *Ph. Schr.* 4:299)—in short, the God of Spinoza.

Leibniz was almost right. What the passage from the *Principles* precludes is any appeal to divine goodness to explain either the succession of states of the world or why we find the world in one state rather than another. If God brings about, or permits, every possible state indifferently, then the only object of which we could say that God willed it because it was good is the entire temporal succession. Perhaps it was better for God to have created the world than not; but beyond that there is nothing more to be explained by appealing to his goodness.

There are other passages, notably in the fourth *Meditation,* where the goodness of God seems to have a more substantial role. After explaining that our susceptibility to error arises from the disproportion between our will and our understanding, Descartes notes that God could nevertheless have prevented us from erring if he had enlarged our understanding so as to be sufficient to our needs, or else impressed upon our will a law forbidding us to judge when we do not have clear and distinct ideas. Why, then, didn't God do so? It may be, Descartes answers, that the imperfection of a part of the world is necessary to the greater perfection of the whole.

Though the passage is sometimes taken to be an anticipation of Leibnizian optimism, there are reasons not to do so. The first is that Descartes insists that the good is dependent on God's will:

> If we attend to God's immensity, it is manifest that nothing whatsoever can exist that does not depend on him: not only no subsistent thing, but also no order, no law, nor any reason of truth or goodness; since otherwise [ . . . ] he would not have been entirely indifferent in creating what he created. If a reason of goodness [*ratio boni*] had preceded his pre-ordering [of the world], it would have determined him to do what is best; but the contrary is true, because he determined himself to make the things that now exist, they were for that reason, as it says in *Genesis,* "exceedingly good'—the reason for their goodness, that is, depended on his willing that they be made thus (6 *Resp.,* AT 7:435–436; cf. 7:432 and *To Mersenne* 15 Apr 1630, AT 1:145 and subsequent letters to Mersenne in 1630)

Although one may then say that God, since his will does not change, will maintain the world according to the standard that was created with it, the argument of the fourth *Meditation* has to do with a world—not the actual world—in which human nature is other than it is. To borrow the analogy with royal legislation which Descartes draws on: if justice depends on the decrees of the head of state, then comparison of the laws of one state to those of another is idle, since there is no independent standard. So it is vain for me to complain about my error-prone condition in this world, and to wish that I had a more perfect nature, not because my condition is necessary to the greater perfection of the whole, but because such comparisons have no basis. When Descartes says, therefore, that "we should take no reasons, where natural things are concerned, from the end which God or nature proposed to himself in making them, because we should presume that we are privy to his counsels" (*PP*1§28, AT 8/1:15), the point is not just that God's ways are mysterious, but that the goodness of the world in no way precedes its existence; it therefore explains nothing.

The second reason not to take Descartes to anticipate Leibniz is that the laws of nature are derived not from any "reason of the good" but from the immutability of the divine will. Immutability, like the steadfastness that Descartes recommends in the *morale provisoire* of the *Discourse*, is a formal condition on God's will; it is compatible with any content. Unlike Leibniz's God, who orders the world so that the simplest means will yield the greatest variety of beings, Descartes's God can order the world as he pleases, so long as he conserves that order afterward. Although it is true that for Descartes the simplicity of a hypothesis is a reason to favor it over others that explain the same phenomena, hypotheses chosen according to that criterion can be only morally certain. They are such that, in other words, we know that *we* shall never go wrong in adopting them. But we cannot argue that *God* must operate according to that criterion.

In Aristotelianism, the ordering of the actions of creatures toward their own good, toward ours, and toward that of nature as a whole provides the basis for a natural morality. Suárez, as we have seen, argues from the proposition that woman is ordered to man to the proposition that a husband has dominion over his wife (§6.2). By their nature humans have, in fact, dominion over the entire corporeal world, since they are its most perfect members. The *usus* of the world is ours; it would be a kind of disruption of the natural order for us *not* to exercise our dominion.

In the Cartesian world, there is no such order, and therefore no such law. There is indeed a kind of fitness of the human body to the human soul; that the body is such as to be useful to us is obvious. Perhaps by extension one could say that there is a kind of fitness of the natural world to the needs of beings like us. But one cannot truly say of the human body, or of any other corporeal things, that it has us as its end. Bodies have no ends; what governs their behavior is not individual, collective, or cosmic ends but a set of laws whose ground is the purely formal condition that whatever God wills he wills always and everywhere. The relation of our nature—that we are thinking, *willing* things—to the natural world amounts to nothing more than that we have the power to initiate motion in it, a power no corporeal thing has, and that God, by acting on corporeal things, can act on us. The Aristotelian hierarchy of created things collapses into a single step: from soul to body. The only morality, it would seem, to be gleaned from the natural world so understood consists in the unique admonition: do what you will. An Aristotelian who genuinely conceived such a world would not find it incomprehensible. But he would find it *unheimlich*.

# Bibliography

*Note:* Essays in collective volumes are cross-referenced to the volume in which they appear (e.g., *Filosofia della natura*).

## Primary Sources

Abra de Raconis, C. F. *Tertia pars philosophiæ, seu physica.* Lyons: Irenæi Barlet, 1651. (Orig. pub. 1617, with twelve subsequent editions, mostly in Paris, through 1651. Also in *Manuscripta* list 101 reel 3 pt.6.)

Ægidius Romanus. *Commentaria in octo libros phisicorum Aristotelis.* Venice, 1502. (Facs. repr. Frankfurt: Minerva, 1968.)

Agrippa, Cornelius. *De occulta philosophia.* In his *Opera*, v.1. Lyon: Beringi, s.d. [1600]. (Facs. repr. with intro. by Richard Popkin. Hildesheim: Olms, 1970.)

Alain de Lille (Alanus de Insulis). *Anticlaudianus.* Ed. R. Bossuat. Paris: J. Vrin, 1955. (Written ca. 1150.)

Albertus Magnus. *De cælo.*

Albertus Magnus. *Opera omnia.* Aschendorff: Monasterii Westfalorum, 1951–.

Albertus Magnus. *Physica.* Ed. P. Hossfeld. Aschendorff: Monasterii Westfalorum, 1987. (In *Opera* [Inst.] 4, pt.1.)

Alembert, Jean le Rond d'. *Discours préliminaire.* In Diderot *Encyclopédie* 1:i–xlv.

Aristotle. *De anima.* Trans. and notes by R. D. Hicks. Cambridge: Cambridge University Press, 1907. (Repr. Amsterdam: Hakkert, 1965.)

Aristotle. *De cælo et De mundo.* Trans. J. Tricot. Paris: Vrin, 1949.

Aristotle. *Catégories.* Trans. and notes by J. Tricot. Nouvelle édition. Paris: Vrin, 1977. (Aristotle's *Organon* 1.)

Aristotle. *De generatione et corruptione.* Trans. and notes by C. J. F. Williams. Oxford: Clarendon, 1982.

Aristotle. *De la génération et de la corruption.* Trans. and notes by J. Tricot. Paris: Vrin, [3]1971.

Aristotle. *Organon Graece.* Ed. and comm. by Theodore Waitz. Leipzig: Hahn, 1896.

Aristotle. *Physics.* Trans. and intro. by Philip H. Wicksteed and Francis M. Cornford. Cambridge: Harvard, [2]1957. (Loeb Classical Library.)

Aristotle. *Physics, Books III and IV.* Trans. and notes by Edward Hussey. Oxford: Clarendon, 1983.

Aristotle. *Physique.* Ed. and trans. Henri Carteron. Paris: Les Belles Lettres, 1926.

Arnauld, Antoine, and Pierre Nicole. *La logique ou l'art de penser.* Ed. Pierre Clair and François Girbal. Paris: Presses Universitaires de France, 1965. (Based on the 5th ed., Paris, 1683. Orig. pub. Paris, 1662.)

Arriaga, Rodericus de. *Cursus philosophicus.* Antwerp: Balthasar Moretus, 1632. (Ten subsequent editions, including four in Paris, through 1669. Also in *Manuscripta* list 84 reel 7.)

Augustine. *Confessionum libri XIII.* Ed. Lucas Verheijen. Turnhout: Brepols, 1981. (Corpus Christianorum, Series Latina 27.)

Augustine. *De genesi ad litteram.* Ed. Joseph Zycha. Prague: Academia Litterarum Cæsareæ, 1894. (Corpus scriptorum ecclesiasticorum latinorum 28.)

Augustine. *De trinitate.* Ed. W. J. Mountain. Turnholt: Brepols, 1968. (Corpus Christianorum, Series Latina 50.)

Averroes [Ibn Rushd]. *Aristotelis de physico auditu [. . .] cum Averrois Cordubensis variis in eosdem commentariis.* In *Opera* v.4.

Averroes [Ibn Rushd]. *Aristotelis metaphysicorum [. . .] cum Averrois Cordubensis in eosdem commentariis, et epitome.* In *Opera* v.8. (Includes Theophrastus, *Metaphysicorum liber;* M. A. Zimara, *Solutiones contradictionum.*)

Averroes [Ibn Rushd]. *Aristotelis opera cum Averrois commentariis.* Venice, 1562. (Facs. rep., Frankfurt am Main: Minerva, 1962.)

Avicenna [Ibn Sina]. *Opera in lucem redacta.* Venice, 1508. (Facs. repr. Frankfurt am Main: Minerva, 1961.) (*Sufficientia* is the Latin title given to a collection of eight treatises on natural philosophy known in Arabic as *al-Shifa'.*)

Bacon, Francis. *De augmentis scientiarum.* 1623. In *Works* 1:415–837.

Bacon, Francis. *Works.* Ed. James Spedding, Robert Leslie Ellis, and Douglas Denon Heath. London: Longman, 1858. (Facs. repr. Stuttgart–Bad Cannstatt: Friedrich Frommann, 1963.)

Basso, Sebastian. *Philosophiæ naturalis adversus Aristotelis libri xii.* Geneva, 1621.

Bayle, Pierre. *Dictionnaire historique et critique.* Amsterdam, ⁵1740. (Selected articles rep. in *Œuvres diverses,* Supplement 1, pt.1–2.)

Bayle, Pierre. *Œuvres diverses.* The Hague, 1727. (Facs. repr. with intro. by Élisabeth Labrousse. Hildesheim: Olms, 1982.)

Beeckman, Isaac. *Journal tenu par Isaac Beeckman de 1604 à 1634.* The Hague: M. Nijhoff, 1939.

Boethius. *De duabus naturis.* Paris: Garnier/Migne, 1877 (Patrologia latina, series latina 64.)

Boyle, Robert. *Selected Philosophical Papers of Robert Boyle.* Ed. and intro. by M. A. Stewart. Manchester: Manchester University Press, 1979. (Includes "The origin of forms and qualities" [1666], 1–96; "About the excellency and grounds of the mechanical hypothesis" (1674), 138–154; "A free inquiry into the vulgarly received notion of nature" (1686), 176–191.)

Boyle, Robert. *The Works of the Honourable Robert Boyle.* Ed. Thomas Birch. London, 1744.

Bruno, Giordano. *Opera latine conscripta.* Ed. F. Tocco, H. Vitelli. Florence: Le Monnier, 1879–1891. (Facs. repr. Stuttgart–Bad Cannstadt: Friedrich Frommann, 1962.)

Buridan, Jean. *Acutissimi philosophi reverendi Magistri Johannis buridani subtillissime questiones super octo phisicorum libros Aristotelis . . .* [Paris, 1509]. (Facs. repr. Frankfurt am main: Minerva, 1964, under the title *Kommentär zur Aristotelischen Physik.* Composed ca. 1328.)

Buridan, Jean. *Expositio de anima.* In *Traité de l'âme* 5–163. (Patar dates the work to 1337.)

Buridan, Jean. *In Metaphysicen Aristotelis Quæstiones* . . . Paris, 1518. (Facs. repr. Frankfurt am Main: Minerva, 1964.)

Buridan, Jean. *Le traité de l'âme de Jean Buridan. (De prima lectura.)* Ed. and intro. by Benoît Patar. Louvain–La–Neuve: Éditions de l'institut supérieur de philosophie, 1991.

Calicidius. *Timæus a Calcidio translatus commentarioque instructus.* Ed. J. H. Waszink. London/Leiden:Warburg Institute/E. J. Brill, 1962.

Capreolus, Joannes. *Commentaria in quattuor libros sententiarum* . . . Venice: Scotus, 1483.

Charleton, Walter. *Physiologia Epicuro–Gassendo–Charltoniana: or a fabrick of science natural, upon the hypothesis of atoms.* London: Thomas Newcomb, 1654. (Facs. repr. with intro. by Robert H. Kargon, New York: Johnson Reprint, 1966.)

Chauvin, Stephanus. *Lexicon philosophicum.* Leeuwarden: F. Halma, 1713. (Facs. repr. with intro. by L. Geldsetzer. Düsseldorf: Stern–Verlag Janssen, 1967. The first edition was published in 1692.)

Coimbra [Collegium Conimbricensis]. *Commentarii Collegii Conimbricensis e Societate Iesu: in universam dialecticam Aristotelis Stagiritæ.* Cologne, 1607. (Facs. repr. with intro. by W. Risse, Hildesheim:Olms, 1976.)

Coimbra [Collegium Conimbricensis]. *Commentarii Collegii Conimbricensis [. . .] in octo libros physicorum Aristotelis.* Coimbra, 1594. (Facs. repr. Hildesheim: Olms, 1984. N.B.: in this edition the page numbers 331 to 352 are used twice. I have marked pages from the second series with an asterisk.)

Cordemoy, Gerauld de. *Œuvres philosophiques.* Paris: Presses Universitaires de France, 1968. (The *Discernement du corps et de l'âme* was first published in its entirety in 1666; the text in the *Œuvres* is that of the 4th edition [1704].)

Cusanus, Nicolaus. *Idiota de mente.* In *Werke* v.5. (Written in 1450.)

Cusanus, Nicolaus. *Werke: Neuausgabe des Strassburger Drucks von 1488.* Berlin: Walter de Gruyter, 1967.

David of Dinant. *Quaternulorum fragmenta.* Ed. M. Kurdzialek. Warsaw, 1963. (Studia Mediewistyczne 3).

Descartes, René. *La Dioptrique.* Published with the *Discours,* 1637.

Descartes, René. *Discours de la méthode.* Leyden: Jean Maire, 1637. In AT 6.

Descartes, René. *Meditationes de prima philosophia, in qua Dei existentia et Animæ immortalitas demonstratur.* Paris: Michel Soly, 1641. (The second edition has the subtitle: *In quibus Dei existentia, et animæ humanæ a corpore distinctio, demonstrantur.* Amsterdam: Louis Elzevier, 1642.)

Descartes, René. *Le Monde.* In AT 11. (This work, which is sometimes called the *Traité de la Lumière,* was first published in 1664 in Paris.)

Descartes, René. *Œuvres.* Ed. Ferdinand Alquié. Paris: Garnier, 1963–1973.

Descartes, René. *Œuvres de Descartes.* Ed. Charles Adam and Paul Tannéry. Nouvelle présentation. Paris: Vrin, 1964.

Descartes, René. *Les principes de la philosophie.* Trans. Claude Picot. Paris: Henri le Gros, 1647. In AT 9. (For a discussion of the status of the translation as a Cartesian text, see the "Avertissement" in this volume. Adam concludes that the passages where the French adds substantially to the Latin are by Descartes himself.)

Descartes, René. *Principia philosophiæ*. Amstedam: Elzevier, 1644. In AT 8, pt.1.

Descartes, René. *Règles utiles et claires pour la direction de l'esprit en la recherche de la vérité*. Trans. Jean-Luc Marion, mathematical notes by Pierre Costabel. (Annotated translation, following the "lexicon of Descartes before 1637," of the Crapulli edition.)

Descartes, René. *Regulæ ad directionem ingenii*. Ed. Giovanni Crapulli. The Hague: Martinus Nijhoff, 1966. (Latin text collated from the *editio princeps* [Amsterdam, 1701], the Hanover manuscript used by Leibniz, and the first Dutch translation [Amsterdam, 1684].) See also AT 10. I have used Crapulli's text but the page numbers are according to AT.

Diderot, Denis, ed. *Encyclopédie ou Dictionnaire raisonnée des sciences, des arts et des métiers*. Paris, 1751.

Dionysius Areopagitica. *De coelesti hierarchia*. In *Opera* 119–370.

Dionysius Areopagitica. *De divinis nominibus*. In *Opera* 586–996.

Dionysius Areopagitica [pseudonym]. *Opera omnia quæ exstant*. Latin and Greek paraphrase by Georgius Pachymeres, notes Balthasar Corderius, intro. D. Le Nourry. Paris: J.-P. Migne, 1889. (Patrologia Græca 3.)

Dupleix, Scipion. *La physique, ou science des choses naturelles*. Ed. Roger Ariew. Paris: Fayard, 1990. (Ariew's text is based on the edition of 1640 [Rouen: Louys du Mesnil]. The first edition was published in 1603.)

Du Plessis d'Argentré, Charles. *Collectio judiciorum de novis erroribus* . . . Paris: Andræas Cailleau, 1736. (Facs. repr. Brussels: Culture et civilisation, 1963.)

Durandus, de Sancto Porciano [Durand de St. Pourçain]. *In libros Sententiarum*. Venice, 1571.

Eustachius, à Sancto Paulo. *Summa philosophica quadripartita* . . . Paris, 1620. (1st ed. 1609.)

Fonseca, Petrus. *Commentariorum Petri Fonsecæ Lusitani [. . .] in Libros metaphysicorum Aristotelis* . . . Cologne: Lazarus Zetzner, 1615. (Facs. repr. Hildesheim: Olds, 1964.)

Galilei, Galileo. *Galileo's Early Notebooks*. Trans. and comm. by William A. Wallace. Notre Dame:University of Notre Dame Press, 1977. (Wallace dates the text to 1590.)

Galilei, Galileo. *Le opere di Galileo Galilei*. Ed. Antonio Favaro. Florence: G. Barbèra, 1890–1909.

Gassendi, Pierre. *Exercitationes paradoxica adversus Aristoteleos*. In *Opera* 3. (Book 1: Grenoble, 1624. Book 2 was published posthumously in the *Opera*.)

Gassendi, Pierre. *Opera omnia* . . . Lyon, 1658.

Gassendi, Pierre. *Syntagma philosophicum*. In *Opera* 1–2.

Goclenius, Rudolphus. *Lexicon philosophicum* . . . Frankfurt: Matthias Becker, 1613. (Facs. repr. Hildesheim: Olms, 1964.

Gregory of Rimini [Gregorius Ariminensis]. *Lectura super primum et secundum Sententiarum*. Ed. A. Damasus Trapp and Venicio Marcolino. Berlin: Walter de Gruyter. (Spätmittelalter und Reformation, Texte und Untersuchungen 6–12.)

Hamilton, William. *Lectures on Metaphysics and Logic*. Vol. 1: *Metaphysics*. Boston: Gould and Lincoln, 1859.

Hobbes, Thomas. *De corpore*. In *Opera* 1.

Hobbes, Thomas. *Critique de De Mundo de Thomas White*. Intro., ed., and notes by Jean Jacquot and Harold Whitmore Jones. Paris: J. Vrin, 1973. (The editors date the composition of this work to 1643; see p.44–45.)

Hobbes, Thomas. *Opera philosophica quæ latine scripsit omnia* . . . Ed. W. Molesworth. London: J. Bohn, 1839.

Huygens, Christiaan. *De motu corporum ex percussione*. In *Œuv.* 16:1–168. (The first draft of the *De motu* was written in 1652; the definitive manuscript, completed after 1673, was first published in the *Opuscula postuma* in 1703.)

Huygens, Christiaan. *Œuvres complètes*. The Hague: Martinus Nijhoff, 1888–.

John of St. Thomas. *Cursus philosophicus Thomisticus*. Ed. P. Beato Reiser. Turin: Marietti, 1930–1937. (Orig. pub. Rome, 1637-1638, with numerous subsequent editions.)

John of St. Thomas. *Naturalis philosophiæ* . . . Ed. P. Beato Reiser. Turin: Marietti. Pt.1 and 3: 1933. Pt.4: 1937. (Part II of the *Cursus*. Orig. publ. separately, 1633–1635.)

Kepler, Johannes. *Gesammelte Werke*. Ed. W. von Dyck and Max Caspar. Munich: C. H. Beck, 1937–.

Kepler, Johannes. *Strena seu de Nive sexangula*. Frankfurt, 1611. In *Werke* 4:259–280.

Le Bossu, René. *Parallèle des principes de la physique d'Aristote et de celle de René Descartes*. Paris: Vrin, 1981. (Facs. of 1674 ed.)

Leibniz, Gottfried Wilhelm. *Discours de métaphysique et correspondance avec Arnauld*. Intro. and notes by G. Le Roy. Paris: Vrin, ⁴1984.

Leibniz, Gottfried Wilhelm. *Philosophische Schriften*. Ed. Gerhardt. Berlin, 1875–1889.

Leibniz, Gottfried Wilhelm. *Sämtliche Schriften und Briefe*. Ed. Preussischen Adademie der Wissenschaften. Darmstadt: Otto Reichl, 1923–. (*Reihe* 6 is the *Philosophische Schriften*.)

Locke, John. *An Essay Concerning Human Understanding*. Ed. P. H. Nidditch. Oxford: Oxford University Press, 1975; repr. with corrections, 1979.

Lucretius. *De rerum natura*. Ed. Martin Ferguson Smith, trans. W. H. D. Rouse. Cambridge: Harvard University Press, 1975.

Maignan, Emanuel. *R. P. Emanuelis Maignan Tolosatis ordinis Minimorum . . . Cursus philosophicus . . .* Lyons: Joannes Gregoir, 1673.

Mersenne, Marin. *Correspondance du P. Marin Mersenne, religieux minime*. Ed. Cornelis de Waard et Armand Beaulieu. Paris: PUF and CNRS, 1932–1988.

Mersenne, Marin. *Harmonie universelle contenant la theorie et la pratique de la musique*. Paris: Sebastien Cramoisy, 1636.

Mersenne, Marin. *L'impieté des deistes, athees, et libertins de ce temps, combatuë, & renversee de point en point par raisons tirees de la Philosophie, & de la Theologie*. Paris: Pierre Bilaine, 1624. (Repr. Stuttgart–Bad Cannstatt, 1975. The second part, published under the title *L'impieté des Deistes, et des plus subtils Libertins decouverte . . .* , is not included in the reprint.)

Mersenne, Marin. *Les nouvelles pensées de Galilée*. Ed. and notes by Pierre Costabel and Michel-Pierre Lerner. Preface by Bernard Rochot. Paris: J. Vrin, 1973. (A partial translation and reworking of Galileo's *Discorsi*.)

Micraelius, Johannes. *Lexicon philosophicum terminorum philosophis usitatorum*. Intro. Lutz Geldsetzer. Düsseldorf: Stern–Verlag Janssen, 1966. (Facs. repr. of the second edition, 1662; orig. pub. 1653.) (Instrumenta philosophica, Series Lexica 1.)

More, Henry. *Enchiridion metaphysicum*. In *Opera* II.1:131–334.

More, Henry. *Opera omnia*. London: J. Macock, 1679. (Part I is the *Opera theologica*, Part II the *Opera philosophia.*; facs. repr. Hildesheim: Olms, 1966.)

Newton, Isaac. *Unpublished Scientific Papers of Isaac Newton*. Ed. A. Rupert Hall & Marie Boas Hall. Cambridge: Cambridge University Press, 1962.

Nifo, Augustino [Niphus]. *Expositiones in Aristotelis libros metaphysices*. Venice: Hieronymus Scotus, 1559. (Facs. repr. Frankfurt am Main: Minerva, 1967.) Bound with the following.

Nifo, Augustino [Niphus]. *Metaphysicarum disputationum in Aristotelis Decem & quatuor Libros Metaphysicorum.* Venice: Hieronymus Scotus, 1559. (Facs. repr. Frankfurt am Main: Minerva, 1967.)

Ockham, William of. *Brevis summa libri physicorum.* In *Opera philos.* 6:1–134.

Ockham, William of. *Opera philosophica.* St. Bonaventure, N.Y.: St. Bonaventure University, 1967–.

Ockham, William of. *Quæstiones in libros physicorum Aristoteles.* Ed. Stephen Brown. In *Opera philos.* 6:395–813. (Written 1323–1324; see Brown's *Introduction,* 41*.)

Ockham, William of. *Summulæ philosophiæ naturalis.* Ed. Stephen Brown. In *Opera philos.* 6:135–394. (Written 1319–1321; see Brown's *Introduction,* 29*.)

Oresme, Nicole. *Nicole Oresme and the Marvels of Nature: The* De causis mirabilium. Ed., trans., and comm. by Bert Hansen. Toronto: Pontifical Institute of Mediæval Studies, 1985.

Oresme, Nicole. *Nicole Oresme and the Medieval Geometry of Qualities and Motions* [*Tractatus de configurationibus qualitatum et motuum*]. Ed. Marshall Clagett. Madison: University of Wisconsin, 1968.

Paré, Ambroise. *Des monstres et prodiges.* Ed. and comm. by Jean Céard. Geneva, 1971. (Travaux d'humanisme et renaissance 115.)

Pascal, Blaise. *Œuvres complètes.* Ed., intro. and notes by Jacques Chevalier. Paris: Gallimard, 1954.

Pascal, Blaise. *Pensées.* Ed. Louis Lafuma. Paris: Seuil, 1962. Also in *Œuvres.*

Patrizi, Francesco. "On Physical Space." trans. Benjamin Brickman. *Journal of the History of Ideas* 4(1943):224–245. (Translation of book I [*De spacio physico*] and parts of book II [*De spacio mathematico*] of *Pancosmia,* the fourth section of the *Nova de universis philosophia* (Ferrara, 1591; the edition used by Brickman is Venice, 1593). I have given page numbers in the original in brackets after the page numbers of the translation.

Paulus Venetus. *Summa philosophiæ naturalis* . . . Venice, 1503. (Facs. repr. Hildesheim: Olms, 1974.)

Ramus, Petrus [Pierre de la Ramée]. *Scholarum physicarum libri octo* . . . Frankfurt: A. Wecheli, 1683. (Facs. repr. Frankurt am Main: Minerva, 1967. Orig. pub. 1562.)

Régis, Pierre Sylvain. *Cours entier de philosophie ou système général selon les principes de M. Descartes.* Amsterdam: Huguetan, 1691. (Facs. repr. with intro. by Richard Watson. New York: Johnson Reprints, 1970. Orig. pub. 1690.)

Reid, Thomas. *The Works of Thomas Reid, D.D.* With preface, notes, and supplementary dissertations by Sir William Hamilton. Edinburgh, [7]1872.

Reisch, Gregor. *Margarita philosophica.* Intro. by L. Geldsetzer. Düsseldorf: Stern–Verlag Janssen, 1973. (Facs. repr. of the Basel, 1517 edition, the last in Reisch's lifetime.)

Rohault, Jacques. *System of Physics.* Trans. and notes by Samuel Clarke. London: James Knapton, 1723. (Repr. with intro. by L. Laudan. New York: Johnson Reprints, 1969.)

Rubius, Antonius. *Commentarii in universam Aristotelis dialecticam una cum dubiis et quæestionibus hac tempestate agitari solitis.* Alcalá, 1603. (Many subsequent editions up to 1641; see Risse 1964, 1:399 and Lohr 1974–1982, 1980:703.)

Scaliger, Julius Caesar. *Exotericarum exercitationum liber xv de subtilitate ad Hieronymum Cardanum.* Paris, 1557.

Soncinas, Paulus. *Quæstiones metaphysicales acutissimæ.* Frankfurt: Minerva, 1967. (Orig. pub. 1588.)

Soto, Domingo de. *De natura et gratia*. Paris: Ioannis Foucher, 1549. (Facs. repr. Ridgewood, N.J.: Gregg, 1965.)

Soto, Domingo de. *Super octo libros physicorum Aristotelis quæstiones*. Salamanca, [2]1555.

Suárez, Franciscus. *De angelis*. In *Opera* v.2. (Orig. pub. under the title *Pars secunda Summæ Theologiæ de Deo rerum omnium creatore: In tres præcipuos tractatus distributa, Quorum primus de Angelis hoc volumine continetur.* Lyons: Jacques Cardon, 1620.)

Suárez, Franciscus. *De anima*. In *Opera* v.3. (Orig. pub. under the title *Partis secundæ Summæ theologiæ Tomus alter, complectens tractatum secundum de Opere sex dierum ac tertium de Anima*. Lyons: Jacques Cardon and Pierre Cavellat, 1621.)

Suárez, Franciscus. *De arcano Missæ sacrificii mysterio*. In *Opera* v.20–21.

Suárez, Franciscus. *Disputationes metaphysicæ*. Hildesheim: Olms, 1965. (Rep. of v.25–26 of the Opera.)

Suárez, Franciscus. *De divina substantia, eiusque attributis . . .* In *Opera* v.1.

Suárez, Franciscus. *Opera omnia*. Ed. D. M. André. Paris: L. Vivès, 1856.

Suárez, Franciscus. *De opere sex dierum*. In *Opera* v.3. (Orig. pub. with *De anima*.)

Thomas Aquinas. *In Aristotelis libros de Cælo*. In *Opera* Parma 19.

Thomas Aquinas. *In Aristotelis libros de Generatione et corruption*. In *Opera* Parma 19.

Thomas Aquinas. *In Aristotelis libros de sensu et sensato*. Ed. R. M. Spiazzi. Turin: Marietti, 1949. Also in *Opera* Leonina 45, pt.1.

Thomas Aquinas. *In Aristotelis libros physicorum*. In *Opera* Leonina 2.

Thomas Aquinas. *In Aristotelis librum de Anima commentarium*. Ed. A. M. Pirotta. Turin: Marietti, 1936.

Thomas Aquinas. *In Metaphysicam Aristotelis commentaria*. Ed. M.-R. Cathala. Turin: Marietti, 1926.

Thomas Aquinas. *In Sententiarum libris quatuor*. In *Opera* Leonina 17–20.

Thomas Aquinas. *Opera omnia*. Rome: Commissio Leonina, 1882–.

Thomas Aquinas. *Opera omnia*. Parma: Petrus Fiaccodorus, 1852–1873. Facs. repr. New York: Musurgia, 1949.

Thomas Aquinas. *De potentia Dei*. Turin: Marietti, 1942. (*Quæstiones disputatæ* 1.)

Thomas Aquinas. *Sentencia libri de Anima*. Rome/Paris: Commissio Leonina/Vrin, 1984. In *Opera* Leonina 45, pt.1.

Thomas Aquinas. *Summa contra gentiles*. In *Opera* Parma 5; Leonina 13–15.

Thomas Aquinas. *Summa theologiæ*. In *Opera* Parma 1–3; Leonina v.4–12.

Toletus, Franciscus. *Commentaria unà cum Quæstionibus in libros Aristotelis de Generatione et Corruptione*. In *Opera* v.5.

Toletus, Franciscus. *Commentaria unà cum Quæstionibus in octo libros Aristotelis de Physica auscultatione*. In *Opera* v.4. (First printed 1572, with numerous subsequent editions.)

Toletus, Franciscus. *Commentaria unà cum Quæstionibus in tres libros Aristotelis de Anima*. In *Opera* v.3.

Toletus, Franciscus. *Commentaria [. . .] in universam Aristotelis Logicam*. In *Opera* v.2.

Toletus, Franciscus. *Opera omnia Philosophica*. Cologne: Birckmann, 1615–1616. (Fasc. repr. with intro. by Wilhelm Risse. Hildesheim: Olms, 1985.)

Zabarella, Jacobus. *Commentarii [. . .] in III: Aristot[elis] Libros de Anima*. Frankfurt: L. Zetzner, 1606.

Zabarella, Jacobus. *Opera logica*. Frankfurt, 1608. (Facs. repr. Frankfurt: Minerva, 1966.)

Zabarella, Jacobus. *De rebus naturalibus libri XXX*. Frankfurt: Zetzner, [2]1607. Orig. pub. 1597. (Facs. repr. Frankfurt: Minerva, 1966.)

Secondary Sources

Adams, Marilyn. *William Ockham.* Notre Dame: University of Notre Dame Press, 1987. (Publications in Medieval Studies 26).

Alexander, Peter. *Ideas, Qualities, and Corpuscles: Locke and Boyle on the External World.* Cambridge: Cambridge University Press, 1985.

Andresen, Carl. *Handbuch der Dogmen- und Theologiegeschichte.* Göttingen: Vandenhoeck and Ruprecht, 1982.

Ariew, Roger. "Damned if You Do: Cartesians and Censorship, 1663–1706." *Perspectives on Science* 2 (1994):255–274.

Ariew, Roger. "Descartes and Scholasticism: The Intellectual Background to Descartes' Thought." In Cottingham 1992:58–90.

Armogathe, Jean-Robert. *Theologia cartesiana: L'explication physique de l'Eucharistie chez Descartes et dom Desgabets.* The Hague: Martinus Nijhoff, 1977. (Archives internationales d'histoire des idées 84.)

Arthur, Richard. "Continuous Creation, Continuous Time: A Refutation of the Alleged Discontinuity of Cartesian Time." *Journal of the history of philosophy* 26(1988):349–375.

Auer, Johann. "Der Wandel des Begriffes 'supernaturalis'." In *Filosofia della natura:* 331–349.

Balme, David. "Teleology and Necessity." In Gotthelf and Lennox 1987:275–285.

Barlow, H. B. and J. D. Mollon, eds. *The Senses.* Cambridge: Cambridge University Press, 1982.

Belgioioso, G., et al., eds. *Descartes: Il metodo e i saggi.* Rome: Istituto della Enciclopedia Italiana, 1990.

Bergson, Henri. *Œuvres.* Intro. by Henri Gouhier, notes by André Robinet. Paris: Presses Universitaires de France, 1959. (*L'Évolution créatrice* was first published in 1907.)

Berti, Enrico. *Aristotle on science: The "Posterior Analytics."* Padua: Editrice Antenore, 1981. (Studia aristotelica, 9.)

Birkenmajer, A. "Découverte de fragments manuscrits de David de Dinant." *Revue néoscolastique de philosophie* 35(1933):220–229.

Bloch, Olivier René. *La philosophie de Gassendi: Nominalisme, matérialisme, et métaphysique.* The Hague Martinus Nijhoff, 1971. (Archives internationales d'histoire des idées 38.)

Blum, Paul Richard. "Des Standardkursus der katholischen Schulphilosophie im 17. Jahrhundert." In Keßler et al. 1988:127–148.

Blumenberg, Hans. *Die Legitimität der Neuzeit.* Frankfurt am Main: Suhrkamp, ²1988.

Blumenberg, Hans. "'Nachahmung der Natur': Zur Vorgeschichte der Idee des schöpferischen Menschen." *Studium generale* 10(1957):266–283.

Bonitz, Hermann. *Index aristotelicus.* In *Aristotelis opera,* ed. Academia Regia Borussica, v. 5. Berlin: G. Reimer, 1870.

Bréhier, Émile. *La théorie des incorporels dans l'ancien stoïcisme.* Paris: Vrin, ²1928.

Brockliss, L. W. B. "Aristotle, Descartes, and the New Science: Natural Philosophy at the University of Paris, 1600–1740." *Archives of Science* 38(1981):33–69.

Brockliss, L. W. B. *French Higher Education in the Seventeenth and Eighteenth Centuries: A Cultural History.* Oxford: Clarendon, 1987.

Brockliss, L. W. B. "Philosophical Teaching in France, 1600–1740." *History of Universities* 1(1981):131–168.

Burnyeat, Miles. "Aristotle on Understanding Knowledge." In Berti 1981:97–139.

Carteron, Henri. "L'idée de la force mécanique dans le système de Descartes." *Revue philosophique* 94(1922):243–277, 483–511.

Cartwright, Nancy. *Nature's Capacities and Their Measurement.* Oxford: Oxford University Press, 1989.

Cavaillé, Jean-Pierre. *Descartes: La fable du monde.* Paris: J. Vrin, 1991.

Céard, Jean. *La nature et les prodiges.* Geneva: Librairie Droz, 1977.

Charlton, William. "Aristotle on the Place of Mind in Nature." In Gotthelf and Lennox 1987:408–423.

Chenu, M.-D. *La théologie au douzième siècle.* Paris: J. Vrin, 1957.

Clarke, Desmond. *Descartes' Philosophy of Science.* University Park: Pennsylvania State University Press, 1982.

Clarke, Desmond. *Occult Powers and Hypotheses: Cartesian Natural Philosophy under Louis XIV.* Oxford: Clarendon, 1989.

Clavelin, Maurice. *La philosophie naturelle de Galilée: Essai sur les origines et la formation de la mécanique classique.* Paris: Armand Colin, 1968.

Close, A. J. "Commonplace Theories of Art and Nature in Classical Antiquity and the Renaissance." *Journal of the history of ideas* 30(1969):467–486.

Cohen, H. F. *Quantifying Music: The Science of Music at the First Stage of the Scientific Revolution.* Dordrecht: D. Reidel, 1984.

Cohen, I. B. "'Quantum in se est': Newton's Concept of Inertia in Relation to Descartes and Lucretius." *Notes and Records of the Royal Society* 19(1964):131–155.

Cooper, John M. "Hypothetical Necessity and Natural Teleology." In Gotthelf and Lennox 1987:243–274.

Copenhaver, Brian, and Charles Schmitt. *Renaissance Philosophy.* Oxford: Oxford University Press, 1992. (A History of Western Philosophy 3.)

Costabel, Pierre. "Essai critique sur la mécanique cartésienne." *Archives internationales de l'histoire des sciences* 20(1967):235–252.

Cottingham, John. *Descartes.* Oxford: Basil Blackwell, 1986.

Cottingham, John, ed. *The Cambridge Companion to Descartes.* Cambridge: Cambridge University Press, 1992.

Crewe, A. V. "Atoms and Irony" (Letter to the Editors). *American Scientist* 81(1993):207.

Dainville, François. *L'éducation des jésuites.* Paris: Minuit, 1978.

Darmon, Pierre. *Le mythe de la procréation à l'âge baroque.* Paris: J.-J. Pauvert, 1977.

Daston, Lorraine, and Katherine Park. "Unnatural Conceptions: The Study of Monsters in Sixteenth- and Seventeenth-Century France and England." *Past and Present* 92(1981):20–54.

Dear, Peter. "Jesuit Mathematical Science and the Reconstitution of Experience in the Early Seventeenth Century." *Studies in the History and Philosophy of Science* 18(1987):133–175.

Dear, Peter. *Mersenne and the Learning of the Schools.* Ithaca: Cornell University Press, 1988.

Dear, Peter. "*Totius in verba:* Rhetoric and Authority in the Early Royal Society." *Isis* 76(1985):145–161.

Dénifle, Henri, and A. Chatelain. *Chartularium Universitatis Parisiensis.* Paris, 1889–1891.

Denzinger, Heinrich Joseph, and Adolf Schönmetzer. *Enchiridion symbolorum: Definitionum et declarationum de rebus fidei et morum.* Freiburg im Breisgau: Herder, [36]1976.

Dijksterhuis, E. J. *The Mechanization of the World Picture: Pythagoras to Newton.* Trans. C. Dikshoorn. Foreword by D. J. Struik. Princeton: Princeton University Press, 1961 (repr. 1969). (English translation of *De Mechanisering van het wereldbeeld*. Orig. pub. Amsterdam, 1959.)

Duhem, Pierre. *Études sur Léonard da Vinci*. Paris, 1906–1913.

Duhem, Pierre. *Origines de la statique*. Paris, 1904–1905.

Emerton, Norma E. *The Scientific Reinterpretation of Form*. Ithaca: Cornell University Press, 1984.

Fiedler, Wilfried. *Analogiemodelle bei Aristoteles. Untersuchungen zu den Vergleichen zwischen den einzelnen Wissenschaften und Künsten*. Amsterdam: B. R. Grüner, 1978. (Studien zur antiken Philosophie 9.)

*La filosofia della natura nel Medioevo*. Milan, 1966. (*Atti del Terzo Congresso Internazionale di filosofia medioevale*, Passo della Mendola, 1964.)

Freddoso, Alfred J. "Medieval Aristotelianism and the Case against Secondary Causation in Nature." In Morris 1988:74–118.

Frede, Michael. "Substance in Aristotle's Metaphysics." In Gotthelf 1985:17–26.

Freeland, Cynthia A. "Aristotle on Possibilities and Capacities." *Ancient Philosophy* 6(1986):69–89.

Funkenstein, Amos. *Theology and the Scientific Imagination from the Middle Ages to the Seventeenth Century*. Princeton: Princeton University Press, 1986.

Furth, Montgomery. *Substance, Form and Psyche: An Aristotelean Metaphysics*. Cambridge: Cambridge University Press, 1988.

Gabbey, Alan. "Explanatory Structures and Models in Descartes' Physics." In Belgioioso 1990, 1:273–286.

Gabbey, Alan. "Force and Inertia in the Seventeenth Century: Descartes and Newton." In Gaukroger 1980:230–320. (A revised and expanded version of Gabbey 1971.)

Gabbey, Alan. "Force and Inertia in Seventeenth-Century Dynamics." *Studies in History and Philosophy of Science* 2(1971):1–67.

Gabbey, Alan. [Review of Scott 1970.]. *Studies in History and Philosophy of Science* 3(1972):373–385.

Garber, Daniel. "Descartes, les aristotéliciens et la révolution qui n'eut pas lieu en 1637." In Méchoulan 1988:199–212.

Garber, Daniel. *Descartes' Metaphysical Physics*. Chicago: University of Chicago, 1992.

Garber, Daniel. "Science and Certainty in Descartes." In Hooker 1978:114–151.

Gaukroger, Stephen. "Descartes' Project for a Mathematical physics." In Gaukroger 1980:97–140.

Gaukroger, Stephen, ed. *Descartes: Philosophy, Mathematics, and Physics*. Sussex: Harvester, 1980.

Gilbert, Neal. *Renaissance Concepts of Method*. New York: Columbia University Press, 1960.

Gill, Mary Louise. "Aristotle on Self-Motion." In Judson 1991:243–265.

Gill, Mary Louise. *Aristotle on Substance: The Paradox of Unity*. Princeton: Princeton University Press, 1989.

Gilson, Étienne. *Études sur le rôle de la pensée médiévale dans la formation du système cartésien*. Paris: Vrin, ⁵1984. (Orig. pub. 1930.)

Gilson, Étienne. *Index Scholastico–cartésien*. Paris: Alcan, 1913. (Facs. repr. New York: Burt Franklin, 1964.)

Gilson, Étienne. *Introduction à l'étude de Saint Augustin*. Paris: Vrin, ²1943.

Gilson, Étienne. *La liberté chez Descartes et la théologie*. Paris: Vrin, 1913. (Facs. repr. Paris: Vrin, 1982.)

Glouberman, Mark. "Cartesian Substances as Modal Totalities." *Dialogue* 17(1978):320–343. Repr. in Moyal 1991, 3:363–384.

Glouberman, Mark. *Descartes: The Probable and the Certain.* Würzburg/Amsterdam: Königshausen und Neumann/Rodopi, 1986. (Elementa 41.)

Gotthelf, Allan. *Aristotle on Nature and Living Things.* Pittsburgh/Bristol:Mathesis/ Bristol Classical Press, 1985.

Gotthelf, Allan. "Aristotle's Conception of Final Causality." In Gotthelf and Lennox 1987:204–242. (Reprint, with additional notes and postscript, of a paper first published in 1976.)

Gotthelf, Allan, and James G. Lennox, eds. *Philosophical Issues in Aristotle's Biology.* Cambridge: Cambridge University Press, 1987.

Gouhier, Henri. *La pensée religieuse de Descartes.* Paris: J. Vrin, 1921. (Études de philosophie médiévale 6.)

Grabmann, Martin. *I divieti ecclesiastici di Aristotele sotto Innocenzo III e Gregorio IX.* Rome, 1941. (Miscellanea historiæ pontificiæ 5: I Papi del duecento e l'aristotelismo, fasc. 1.)

Gracia, Jorge. "Francis Suárez." In: Gracia 1994:475–510.

Gracia, Jorge J. E. *Individuation in Scholasticism: The Later Middle Ages and the Counter-Reformation, 1150–1650.* Albany, N.Y.: State University of New York Press, 1994.

Grafton, Anthony. "Teacher, Text, and Pupil in the Renaissance Class-room: A Case Study from a Parisian College." *History of Uuniversities* 1(1981):37–70.

Graham, D. W. "The Paradox of Prime Matter." *Journal of the History of Philosophy* 25(1987):475–490.

Grant, Edward. "Cosmology." In Lindberg 1978:265–302.

Grant, Edward. "Medieval Explanation and Interpretations of the Dictum That 'Nature Abhors a Vacuum'." *Traditio* 29(1973):237–355.

Grant, Edward. *Much Ado about Nothing: Theories of Space and Vacuum from the Middle Ages to the Scientifc Revolution.* Cambridge: Cambridge University Press, 1981.

Grant, Edward. "Ways to Interpret the Terms 'Aristotelian' and 'Aristotelianism' in Medieval and Renaissance Natural Philosophy." *History of Science* 25(1987):335–358.

Gregory, Tullio. *Anima mundi: La filosofia di Guglielmo di Conches e la scuola di Chartres.* Florence: G. C. Sansoni, 1955.

Grosholz, Emily. *Cartesian Method and the Problem of Reduction.* Oxford: Clarendon Press, 1991.

Gueroult, Martial. *Dynamique et métaphysique leibniziennes. Suivi d'une Note sur le principe de la moindre action chez Maupertuis.* Paris: Les Belles Lettres, 1934. (Publications de la Faculté des Lettres de l'Université de Strasbourg 68).

Gueroult, Martial. *Études sur Descartes, Spinoza, Malebranche, et Leibniz.* Hildesheim: Georg Olms, 1970. ("Physique et métaphysique de la force chez Descartes et chez Malebranche" was first published in 1954.)

Gueroult, Martial. "Physique et métaphysique de la force chez Descartes et chez Malebranche." Revue de métaphysique et de morale 59(1954):1–37, 113–134. In Gueroult 1970:85–143.

Hatfield, Gary. "Force (God) in Descartes's Physics." *Studies in History and Philosophy of Science 10(1979):113–140.* Also in Moyal 1991:123–152.

Hedwig, Klaus. *Sphæra lucis: Studien zur Intelligibilität des Seiendenim Kontext der mittelalterlichen Lichtspekulation.* Münster: Aschendorff, 1980. (Beiträge zur Geschichte der Philosophie und Theologie des Mittelalters n.F. 18).

Henry, John. "Francesco Patrzi da Cherso's Concept of Space and Its Later Influence." *Annals of Science* 36(1979):549–575.

Herivel, John. *The Background to Newton's* Principia: *A Study of Newton's Dynamical Researches in the Years 1664–1684.* Oxford: Oxford University Press, 1965.

Hickey, J. S. *Summula philosophiæ scholasticæ in usum adolescentium.* Dublin: M. H. Gill and Son, 1919.

Hissette, Roland. *Enquête sur les 219 articles condamnés à Paris le 7 mars 1277.* Louvain: Publications Universitaires, 1977.

Hoffman, Paul. "The Unity of Descartes's Man." *Philosophical Review* 95(1986):339–370. Also in Moyal 1991:168–192.

Hooker, Michael. *Descartes: Critical and Interpretive Essays.* Baltimore: Johns Hopkins University Press, 1978.

Hussey, Edward. See under Aristotle, *Physics.*

Hussey, Edward. "Aristotle's Mathematical Physics: A Reconstruction." In Judson 1991:213–242.

Hutchison, Keith. "Supernaturalism and the Mechanical Philosophy." *History of Science* 21(1983):297–333.

Ingegno, Alfonso. "The New Philosophy of Nature." In Schmitt and Skinner 1988:236–263.

Jammer, Max. "The Program and Principles of Descartes' Physics." In Belgioioso 1990:303–334.

Judovitz, Dalia. *Subjectivity and Representation in Descartes: The Origins of Modernity.* Cambridge: Cambridge University Press, 1988.

Judson, Lindsay. "Chance and 'Always or for the Most Part' in Aristotle." In Judson 1991:73–100.

Judson, Lindsay, ed. *Aristotle's Physics: A Collection of Essays.* Oxford: Oxford University Press, 1991.

Kalivoda, Robert. "Zur Genesis der natürlichen Naturphilosophie." In *Filosofia della natura* : 397–406.

Keßler, Eckhard, Charles Lohr, and Walter Sparn. *Aristotelismus und Renaissance: In memoriam Charles B. Schmitt.* Wiesbaden: Otto Harrassowitz, 1988. (Wolfenbütteler Forschungen 40.)

Kosman, L. A. "Aristotle's Definition of Motion." *Phronesis* 14(1969):40ff.

Koyré, A. *Études galiléennes.* Paris: Hermann, 1939. (Actualités scientifiques et industrielles 852–854.)

Kraye, J., W. F. Ryan, and C. B. Schmitt, eds. *Pseudo-Aristotle in the Middle Ages: The Theology and Other Texts.* London: Warburg Institute, 1986.

Kretzmann, Norman, Anthony Kenny, and Jan Pinborg, eds. *The Cambridge History of Later Medieval Philosophy.* Cambridge: Cambridge University Press, 1982.

Kuhn, Thomas. "Reflections on My Critics." In Lakatos and Musgrave 1970:213–278.

Kullman, Wolfgang. "Different Concepts of the Final Cause in Aristotle." In Gotthelf 1985:169–175.

Kurdzialek, M. "David von Dinant und die Anfänge der Aristotelischen Naturphilosophie." In *Filosofia della natura,* 407–416.

Kurdzialeck, M., ed. See David of Dinant.

Lakatos, Imre, and Alan Musgrave. *Criticism and the Growth of Knowledge.* Cambridge: Cambridge University Press, 1970.

Lang, Helen S. "Aristotelian Physics: Teleological Procedure in Aristotle, Thomas, and Buridan." Review of Metaphysics 42(1989):569–91.

Leff, Gordon. *The Dissolution of the Medieval Outlook*. New York: NYU Press, 1976.

Lenoble, Robert. *Mersenne ou la naissance du mécanisme*. Paris: Vrin, 1943.

Lerner, Michel-Pierre. *Recherches sur la notion de finalite chez Aristote*. Paris: Presses Universitaires de France, 1969.

Lindberg, David C. "The Genesis of Kepler's Theory of Light: Light Metaphysics from Plotinus to Kepler." *Osiris* ser. 2, 2(1986):5–42.

Lindberg, David C. *John Pecham and the Science of Optics*. Madison: University of Wisconsin Press, 1970.

Lindberg, David C. *Science in the Middle Ages*. Chicago: University of Chicago Press, 1978.

Lloyd, G. E. R. "The Invention of Nature." In Lloyd 1991:417–434. Also in Torrance 1992:1–24.

Lloyd, G. E. R. *Methods and Problems in Greek Science*. Cambridge: Cambridge University Press, 1991.

Lockwood, Michael. *Mind, Brain, and the Quantum: The Compound 'I'*. Oxford/Cambridge: Blackwell, 1989.

Lohr, Charles. "Renaissance Latin Aristotle Commentaries." *Studies in the Renaissance* 21(1974):228–289; *Renaissance Quarterly* 28(1975):689–741; 29(1976):714–745; 30(1977):681–741; 31(1978):532–603; 32(1979):529–580; 33(1980): 623–734; 35(1982):164–256.

Lubac, Henri. *Surnaturel: Études historiques*. Paris: Aubier/Montaigne, 1946.

MacClintock, Stuart. *Perversity and Error: Studies on the "Averroist" John of Jandun*. Bloomington: Indiana University Press, 1956.

Maier, Anneliese. *An der Grenze von Scholastik und Naturwissenschaft*. Rome: Storia e letteratura, 1952. (Maier's *Studien zur Naturphilosophie der spätscholastik* 3; Storia e letteratura 41.)

Maier, Anneliese. *Ausgehendes Mittelalter: Gesammelte Aufsätze zur Geistesgeschichte des 14. Jahrhunderts*. v.2. Rome: Storia e letteratura, 1967. (Storia e letteratura 105.)

Maier, Anneliese. *Metaphysische Hintergründe der spätscholastischen Naturphilosophie*. Rome: Storia e letteratura, 1955. (Maier's *Studien zur Naturphilosophie der spätscholastik* 4; Storia e letteratura 52.)

Maier, Anneliese. *On the Threshold of Exact Science: Seclected Writings of Anneliese Maier on Late Medieval Natural Philosophy*. Ed. and trans. Steven D. Sargent. Philadelphia: University of Pennsylvania Press, 1982.

Maier, Anneliese. *Die Vorläufer Galileis*. Rome: Storia e letteratura, 1949. (Maier's *Studien zur Naturphilosophie der spätscholastik* 1; Storia e letteratura 22.)

Maier, Anneliese. *Zwei Grundprobleme der scholastischen Naturphilosophie*. Rome: Storia e letteratura, ²1951. (Maier's *Studien zur Naturphilosophie der spätscholastik* 2; Storia e letteratura 37.)

Maier, Anneliese. *Zwischen Philosophie und Mechanik*. Rome: Storia e letteratura, 1958.(Maier's *Studien zur Naturphilosophie der spätscholastik* 5; Storia e letteratura 69.)

Marenbon, John. *Later Medieval Philosophy (1150–1350)*. London: Routledge, 1987.

Marion, Jean-Luc. *Sur la théologie blanche de Descartes. Analogie, création des vérités éternelles et fondement*. Paris: Presses Universitaires de France, 1981.

McEvoy, James. *The Philosophy of Robert Grosseteste*. Oxford: Clarendon, 1982.

McGuire, J. E. "Space, Geometrical Objects and Infinity: Newton and Descartes on Extension." In Shea 1983:69–112.

Méchoulan, Henri, ed. *Problématique et réception du Discours de la méthode et des Essais*. Paris: Vrin, 1988.

Mercer, Christia. "The Vitality and Importance of Early Modern Aristotelianism." In Sorrell 1993:33–67.

Michaud-Quantin, Pierre. *Études sur le vocabulaire philosophique du moyen âge.* Rome: Edizioni dell'Ateneo, 1971.

Michel, Paul-Henry. *La cosmologie de Giordano Bruno.* Paris: Hermann, 1962.

Mignucci, Mario. "Ὡσ ἐπὶ τὸ πολὺ et nécessaire dans la conception aristotlicéenne de la science." In Berti 1981:173–203.

Miles, Murray Lewis. "Condensation and Rarefaction in Descartes' Analysis of Matter." *Nature and system* 5(1983):169–180.

Milton, John. "The Origin and Development of the Concept of the 'Laws of Nature'." *Archives européennes de sociologie* 22(1981):173–195.

Moravcsik, Julius. "What Makes Reality Intelligible?" In Judson 1991:31–47.

Morris, Thomas V. *Divine and Human Action: Essays in the Metaphysics of Theism.* Ithaca: Cornell University Press, 1988.

Moyal, Georges J. D. *René Descartes: Critical Assessments.* London: Routledge, 1991.

Murdoch, John E., and Edith D. Sylla. "The Science of Motion." In Lindberg 1978:206–264.

Murray, Alexander. "Nature and Man in the Middle Ages." In Torrance 1992:25–62.

Nadler, Steven, ed. *Causation in Early Modern Philosophy.* University Park: Pennsylvani University State Press, 1993.

Nardi, A. "Descartes 'presque' galiléen: 18 février 1643." *Revue de l'histoire des sciences* 39(1986):3–16.

Nardi, Bruno. *Studi di filosofia medievale.* Rome: Storia e letteratura, 1960. (Storia e letteratura 78.)

Newton-Smith, W. H. *The Structure of Time.* London: Routledge and Kegan Paul, 1980.

Oakley, Francis. *Omnipotence, Covenant, and Order: An Excursion in the History of Ideas from Abelard to Leibniz.* Ithaca: Cornell University Press, 1984.

O'Neill, Eileen. "*Influxus physicus.*" In Nadler 1993:27–55.

Owen, Joseph. *The Doctrine of Being in the Aristotelian Metaphysics.* Toronto: Pontifical Institute of Mediaeval Studies, ³1978. (First ed. 1951.)

Pellegrin, Pierre. "Logical Difference and Biological Difference: the Unity of Aristotle's Thought." Trans. A. Preus. In Gotthelf and Lennox 1987:313–338.

Perez-Ramos, Antonio. *Francis Bacon's Idea of Science and the Maker's Knowledge Tradition.* Oxford: Clarendon, 1988.

Prendergast, Thomas L. "Descartes and the Relativity of Motion." *Modern Schoolman* 50(1972):64–72. Repr. in Moyal 1991, 4:101–109.

Prendergast, Thomas L. "Motion, Action, and Tendency in Descartes' Physics." *Journal of the History of Philosophy* 13(1975):453–462. Repr. in Moyal 1991, 4:89–100.

Redondi, Pietro. *Galileo Heretic.* Trans. Raymond Rosenthal. Princeton: Princeton University Press, 1987.

Rindler, Wolfgang. *Essential Relativity: Special, General, and Cosmological.* Rev. 2 ed. New York: Springer, 1979.

Risse, Wilhelm. *Logik der Neuzeit.* Stuttgart–Bad Cannstatt: Friedrich Frommann, 1964. (Bd.1: 1500–1640; Bd.2: 1640–1780.)

Rochemonteix, Camille. *Un collège de Jésuites aux xviiᵉ et xviiiᵉ siècles: Le Collège Henri IV de la Flèche.* Le Mans: Leguicheux, 1899.

Rossi, Clément. *L'Anti-nature: Éléments pour une philosophie tragique.* Paris: Presses Universitaires de France, 1973.

Russell, J. L. "Action and Reaction before Newton." *British Journal for the History of Science* 9(1976):25–38.

Salmon, Wesley. *Explanations and the Causal Structure of the World.* Princeton: Princeton University Press, 1984.

Sarnowsky, Jürgen. *Die aristotelisch-scholastische Theorie der Bewegung: Studien zum Kommentar Alberts von Sachsen zur Physik des Aristoteles.* Münster: Aschendorff, 1989. (Beiträge zur Geschichte der Philosophie und Theologie des Mittelalters n.f. 32).

Schmitt, Charles. "Experimental Evidence For and Against a Void: The Sixteenth-Century Arguments." *Isis* 58(1967):352–366. Also in Schmitt 1989, c.7.

Schmitt, Charles. *John Case and Aristotelianism in Renaissance England.* Kongston/Montreal: McGill-Queen's University Press, 1983. (McGill-Queen's Studies in the History of Ideas 5.)

Schmitt, Charles. "Philoponus' Commentary on Aristotle's *Physics* in the Sixteenth Century." In Sorabji 1987.

Schmitt, Charles. "Pseudo-Aristotle in the Latin Middle Ages." In Kraye et al. 1986:3–14; Schmitt 1989, c.1.

Schmitt, Charles. *Reappraisals in Renaissance Thought.* London: Variorum Reprints, 1989.

Schmitt, Charles. "The Rise of the Philosophical Textbook." In Schmitt and Skinner 1988:792–893.

Schmitt, Charles. *Studies in Renaissance Philosophy and Science.* London: Variorum Reprints, 1981.

Schmitt, Charles, and Quentin Skinner, eds. *The Cambridge History of Renaissance Philosophy.* Cambridge: Cambridge University Press, 1988.

Schönberger, Rolf. "Eigenrecht und Relativität des Natürlichen bei Johannes Buridanus." In Zimmermann and Speer 1991:216–233.

Schuster, John. *Descartes and the Scientific Revolution, 1618–1634.* Ph.D. Diss., Princeton University, 1977.

Scott, J. F. *The Scientific Work of Rene Descartes (1596–1650).* London: Taylor and Francis, 1952. (Facs. repr. New York: Garland, 1987.)

Scott, Wilson L. *The Conflict between Atomism and Conservation Theory 1644–1860.* New York: Elsevier, 1970.

Shea, William R. *The Magic of Numbers and Motion: The Scientific Career of René Descartes.* Canton, Mass.: Science History Publications, 1991.

Shea, William R. *Nature Mathematized: Historical and Philosophical Case Studies in Classical Modern Natural Philosophy.* Dordrecht: D. Reidel, 1983. (University of Western Ontario Series in Philosophy of Science 20.)

Sirven, J. *Les années d'apprentissage de Descartes (1596–1628).* Albi: Imprimérie Coopérative du Sud-Ouest, 1928.

Smith, A. Mark. "Getting the Big Picture in Perspectivist Optics." *Isis* 72(1981):568–589.

Sorabji, Richard. *Matter, Space, and Motion.* Ithaca: Cornell University Press, 1988.

Sorabji, Richard. *Philoponus and the Rejection of Aristotelian Science.* Ithaca: Cornell University Press, 1987.

Sorrell, Tom, ed. *The Rise of Modern Philosophy: The Tension between the New and Traditional Philosophies from Machiavelli to Leibniz.* Oxford: Clarendon, 1993.

Speer, Andreas. "Kosmisches Prinzip und Maß menschlichen Handelns: *Natura* bei Alanus ab Insulis." In Zimmermann and Speer 1991:107–128.

Steenberghen, Fernand van. *Introduction à l'étude de la philosophie médiévale.* Preface by Georges van Riet. Louvain: Publications Universitaires, 1974.

Steenberghen, Fernand van. *La philosophie au xiii*ͤ *siècle*. Louvain-la-Neuve:Éditions de l'Institut Supérieur de Philosophie, ²1991.

Théry, G. *Autour du décret de 1210. I. David de Dinant*. Paris, 1925. (Bibliothèque thomiste 6.)

Tocanne, Bernard. *L'Idée de nature en France dans la seconde moitié du xvii*ͤ *siècle: Contribution à l'histoire de la pensée classique*. Paris: Klincksieck, 1978. (Bibliothèque française et romane, Sér. C: 67.)

Torrance, John, ed. *The Concept of Nature: The Herbert Spencer Lectures*. Oxford: Clarendon, 1992.

Verbeek, Theo. *Descartes and the Dutch: Early Reactions to Cartesian Philosophy, 1637–1650*. Carbondale: Southern Illinois University Press, 1992.

Wallace, William A. *Causality and Scientific Explanation*. Ann Arbor: University of Michigan Press, 1972.

Wallace, William A. *Prelude to Galileo: Essays on Medieval and Sixteenth-Century Sources of Galileo's Thought*. Dordrecht: D. Reidel, 1981.

Waterlow, Sarah. *Nature, Change, and Agency in Aristotle's Physics*. Oxford: Oxford University Press, 1982.

Weinberg, Julius R. *A Short History of Medieval Philosophy*. Princeton: Princeton University Press, 1964.

Weisheipl, James A. *Nature and Motion in the Middle Ages*. Ed. William E. Carroll. Washington, D.C.: Catholic University of America, 1985. (Studies in Philosophy and the History of Philosophy 11.)

Westfall, Richard S. *Force in Newton's Physics*. New York: American Elsevier, 1971.

Wild, John. "The Cartesian Deformation of the Structure of Change and Its Influence on Modern Thought." *Philosophical Review* 50(1941):36–59. In Moyal 1991, 4:20–42.

Williams, Bernard. *Descartes: The Project of Pure Enquiry*. Harmondsworth, Middlesex: Penguin, 1978.

Williams, Raymond. *Keywords*. New York: Oxford University Press, ²1985. (1st ed. 1976.)

Wolff, Michael. *Geschichte der Impetustheorie: Untersuchungen zum Ursprung der klassischen Mechanik*. Frankfurt am Main: Suhrkamp, 1978.

Wright, Larry. *Teleological Explanations*. Berkeley: University of California Press, 1976.

Zimmermann, Albert, and Andreas Speer. *Mensch und Natur im Mittelalter*. Berlin: Walter de Gruyter, 1991. 21, pt.1 and 2.)

# Index

Abra de Raconis, C. F., 10, 30; distinguishes creation and conservation, 325; instances of ends in nature, 168; motion does not persist, 275–276; senses of *natura,* 214

absorption, 288–290, 300

accidental form, 73

accidents: are caused by substance materially, formally, and efficiently (Suárez), 160; by definition cannot exist separately from their subject (Boyle, Chauvin), 132; do not persist through corruption (Thomists), 144–145; embellishing, 144; empirical arguments for persistence through corruption, 146–148; existence apart from substance, 129–131; *id cuius esse est inesse,* 131; identified with modes in seventeenth-century authors, 133–134; incapable of producing forms, 162; inhere in the union of matter and form, 145; "primary" and "adverbial," 113; real, 113, 127; as signs (*signa*), 66

action: immanent, 41–43; transeunt, 41–42

action and passion, 40–46; not really distinct from *motus,* 44–46; are identical (Descartes), 259

*actus,* 29, 57; complete and incomplete, 93; of existence distinct from that of informing (Fonseca, Suárez), 126; *metaphysicus,* 93; *physicus,* 93; *terminus ad quem* of change, 63

Adams, Marilyn, 82

Ægidius Romanus, 143; on condensation and rarefaction, 107; on matter and darkness, 83

agent, not affected in acting, 44

agent and patient, 43–44, 47–48, 170

Alain de Lille, 144, 216

Albertus Magnus, 94, 206; inchoate forms, 140

Alembert, Jean le Rond d', 168–169

Alexander, Peter, 75

alteration, 61; definition of, 25; irreversible, 73

'always or for the most part', 208

angels, 230, 238

animals, 61; have natural appetites only (Toletus), 202; sagacity and industry of, 181; self-movement of, 230

*anima mundi,* 125, 181

annihilation: argument from, 359; Descartes' response, 384; is the cessation of a positive divine action, i.e. conservation, 332–333

apart, 374

appetite: adventitious, 236; distinguished from propensity, 235; innate and elicited, 236; intellectual, sensitive, and natural (Toletus), 202; natural, 235–236; necessary *quoad exercitiuim* and *quoad speciem,* 236

Ariew, Roger, 1, 12

Aristotelianism, scope of the term, 7

Aristotle: *Categories,* 29, 87, 110, 148, 151, 381; definition of 'nature', 227; inherence, 130–132; matter, 82; matter,